Emerging Threats to Tropical Forests

Emerging Threats to Tropical Forests

Edited by

William F. Laurance and Carlos A. Peres

The University of Chicago Press
Chicago and London

William F. Laurance is a staff scientist at the Smithsonian Tropical Research Institute, and President of the Association for Tropical Biology and Conservation, the world's largest scientific organization dedicated to the study and preservation of tropical ecosystems. He is the author of *Stinging Trees and Wait-a-Whiles: Confessions of a Rainforest Biologist* and coeditor of *Tropical Forest Remnants: Ecology, Management, and Conservation of Fragmented Communities.* Carlos A. Peres is Reader in Tropical Conservation Biology at the Centre for Ecology, Evolution and Conservation of the School of Environmental Sciences, University of East Anglia, United Kingdom. He has received a Bay Foundation Biodiversity Conservation Leadership Award and, in 2000, was elected an Environmentalist Leader for the New Millennium by *Time* magazine and CNN Network.

The University of Chicago Press, Chicago 60637
The University of Chicago Press, Ltd., London
© 2006 by The University of Chicago
All rights reserved. Published 2006
Printed in the United States of America

15 14 13 12 11 10 09 08 07 06   1 2 3 4 5

ISBN-13: 978-0-226-47021-4 (cloth)
ISBN-13: 978-0-226-47022-1 (paper)
ISBN-10: 0-226-47021-0 (cloth)
ISBN-10: 0-226-47022-9 (paper)

Library of Congress Cataloging-in-Publication Data

Emerging threats to tropical forests / edited by William F. Laurance and Carlos A. Peres.
     p. cm.
   Includes bibliographical references and index.
   ISBN 0-226-47021-0 (cloth : alk. paper) — ISBN 0-226-47022-9 (pbk. : alk. paper)
     1. Forests and forestry—Tropics. 2. Forest conservation—Tropics. 3. Forest management—Tropics. I. Laurance, William F. II. Peres, Carlos A., 1963–

   SD247 .E44 2006
   634.90913—dc22   2006006058

♾ The paper used in this publication meets the minimum requirements of the American National Standard for Information Sciences—Permanence of Paper for Printed Library Materials, ANSI Z39.48-1992.

CONTENTS

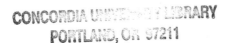

Thomas E. Lovejoy

By any measure, tropical forests continue to be in grave trouble. The pace of tropical deforestation is alarmingly high, varying from roughly 7 to 15 million hectares per year over the past two decades. Yet beyond these disturbing numbers, there is a far more diverse range of threats to forests—from obvious alterations, such as habitat fragmentation and selective logging, to far more insidious and poorly understood threats, including global climate change, regional hydrological alterations, surface fires, exotic pathogens, and growing synergisms among simultaneous, interacting threats.

This book is unquestionably the best and most up-to-date effort to document the chilling panoply of threats to tropical forests. And the dangers are growing. In the Brazilian Amazon, despite various programs to manage and protect forests (including the G-7 Pilot Program for the Brazilian Rainforests), deforestation averaged over 2.3 million hectares per year from 2002 to 2005—among the highest rates ever recorded, equivalent to more than 10 football fields per minute.

Some of the threats to tropical forests are only newly recognized, whereas others are emerging as more complex than previously thought. Clearly the challenge is even bigger, tougher, and more daunting than previously understood.

For example, the notion that large-scale deforestation could undercut the hydrological cycle of the Amazon basin, and probably that of other tropical regions, is not new. The capacity of the rainforest to recycle water through evapotranspiration and thus generate some of its own rainfall has been recognized since the pioneering work of Eneas Salati and others (Salati et al. 1978; Salati and Vose 1984). Major forest loss could initiate a drying trend, with dire consequences for Amazonian forests and parts of Brazil, the Americas, and possibly other regions that receive important moisture from the Amazon.

But only recently recognized is the deceptive way by which local rainfall can increase temporarily as deforestation draws moisture out of the remaining forest. This can create the false illusion that all is well with the hydrological cycle until deforestation and desiccation advance to the point that rainfall declines precipitously. Such complexities highlight an urgent need for models to test the hydrological consequences of potential development projects in the Amazon, and elsewhere in the tropics.

Threats to forests are changing in part because the proximate drivers of forest are also changing. Industrial cattle ranching continues to expand rapidly in Amazonia, which supports nearly sixty million cattle, but is now being rivaled by a huge expansion of soybean farming (Fearnside 2001e) and logging (Asner et al. 2005). Likewise, in Southeast Asia, oil-palm plantations are proliferating at exponential rates.

As habitat conversion and fragmentation proceed apace, the forests become vulnerable to a relatively newly recognized threat: invasive species and pathogens. The invaders are ubiquitous—ranging from fire ants in West Africa, to nearly pantropical plagues of amphibian chytrid-iomycosis and *Phytophthora* infestations of plant communities. Species are disappearing; ecosystems are being profoundly altered. And what we perceive, especially for exotic microbes, may be just the tip of the iceberg.

Such changes foreshadow a newly enlarged sense of negative synergistic effects. For example, major fires from the interaction of El Niño droughts, logging, and slash-and-burn farming first burst on the scene about 25 years ago in Indonesia. Today, the El Niño climatic oscillations are known to be far more extensive geographically, reaching across the Pacific Ocean to South America. Habitat fragmentation, logging, and other land-use practices in Amazonia now interact with El Niño droughts to produce massive fires. In late 1997, such fires were the most extensive in recorded history, triggering changes that severely erode forest biodiversity and ecosystem services (Barlow and Peres 2004).

These negative synergisms are compounded by proliferating infrastructure projects and the spontaneous colonization and habitat conversion that follow (W. F. Laurance et al. 2001b). This threatens a devastating positive feedback among fire, desiccation, and more fire (Nepstad et al. 2001) at an even larger spatial scale. We now know that we must not only manage the Amazon to protect the integrity of its hydrological cycle but also to withstand shocks such as El Niño events and the severe 2005 drought—itself a likely preview of future climatic change.

Increasingly, it appears that the drivers of change in tropical forests are not just local or regional but also global in extent. Rising atmospheric carbon dioxide levels, increasing temperatures, increased storm events, and altered regional patterns of precipitation are just some of the possible consequences of global change. But how will such changes affect tropical forests? The availability of large and long-term data sets, especially for tree communities, makes it possible to take the topic from hand waving to analysis. In recent decades, basic changes in mature tropical forests, such as increasing dynamism, biomass, and floristic composition, are evidently occurring—although it is difficult to identify unequivocally the specific drivers of such changes.

The impacts of climate change will surely become more powerful in the future. At least one climate-change model, from the Hadley Center in the United Kingdom, projects future climate change (at double the pre-industrial level of carbon dioxide, which could easily occur by the end of this century) that would render much of the Amazon unsuitable for rainforest. Throughout the tropics, rising temperatures and the rapid melting of glaciers will drive altitudinal shifts in vegetation and species ranges as well as changes in lotic waters. Montane biotas may be especially vulnerable; in Australian rainforests, endemic montane vertebrates could experience drastic extinctions this century from increasing temperatures.

Collectively, these threats present a fairly grim picture, but recent developments also give reason for optimism and renewed effort. Some of the most important and innovative strategies for conserving tropical forests are highlighted in the latter chapters in this book. Among these is the potential role of trees and forests for managing the global carbon cycle. For example, tree plantations and reforestation sequester atmospheric carbon, provide direct benefits for biodiversity conservation, and can reduce pressures on tropical forests for timber. If properly designed, global carbon trading could also provide major benefits for existing natural forests.

The 2002 World Parks Congress in Durban, South Africa, included announcements of a number of significant new protected areas in tropical regions. Notably among these was the Amazon Region Protected Area (ARPA) program, initiated in Brazil by World Wildlife Fund and the World

Bank. Upon completion, ARPA will bring the total area of Brazilian Amazonia under some form of protection (including indigenous and multiple-use lands) to more than 40%. While this is insufficient in itself to maintain the region's complex hydrological cycle, it is progress beyond the wildest dreams of anyone four decades ago, when Brazil had but one national forest. Tropical forests today face enormous pressures and change, but the situation is not hopeless. While clearly a race to the finish, redoubling our efforts to understand and manage the burgeoning threats to tropical forests could be a key to their long-term survival.

What Are Emerging Threats?

William F. Laurance

A CHANGING WORLD

We live in a world that is changing—fast. When my grandfather was born around 1900 there were only 1.5 billion people on the planet. Today that number has more than quadrupled. The global population is still rocketing upward (fig. 1), largely driven by rapid growth in developing nations, and is expected to reach 7.5 billion by the year 2020 and around 9 billion by 2050 (Anon. 1999b).

In addition, humans are consuming ever more resources and energy. Average consumption, as measured by per capita gross domestic product, rose by an astonishing 460% over the last century (Maddison 1995). Current values are projected to increase by 240% by the year 2050, as ever-larger segments of the world's population adopt the consumptive lifestyles of industrial nations (National Research Council 1999). If this occurs as expected, per capita resource-use will have jumped 11-fold since the year 1900.

So, the world—especially the developing world, where much of the planet's biodiversity resides—is changing at a breakneck pace. Change has become so pervasive, so insidious, that sometimes we don't even recognize it as such. Humans currently appropriate 40% of terrestrial productivity (Vitousek et al. 1986), exploit half of the world's usable freshwater (Tilman et al. 2001), and have destroyed or degraded vast expanses of natural habitat. Global biogeochemical cycles of carbon (Houghton et al.

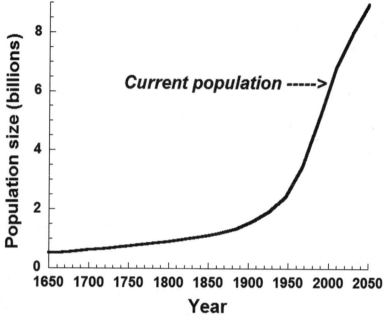

Figure 1. Past and projected growth of the human population, from 1650 to 2050 (using a cubic-spline function to interpolate data from Ehrlich et al. 1995)

2001), nitrogen (Vitousek et al. 1994), and phosphorus (Carpenter et al. 1998) have been fundamentally altered, leading to widespread anthropogenic impacts on climate and natural ecosystems. Exotic species are being introduced to new ecosystems so rapidly that some biologists refer to the present era as the "Homogeocene" (Putz 1998). These myriad changes may be driving a mass-extinction event that could ultimately rival the most catastrophic episodes in Earth's geologic history (E. O. Wilson and Peter 1988; Pimm et al. 1995).

But are things really that bad? Perhaps all these alarmist predictions are largely hype—just a lot of hand waving by ecologists? U.S. President George W. Bush, who only recently, grudgingly acknowledged that global warming is a reality, would have us think so. Bjørn Lomborg, the telegenic Danish statistician who authored *The Skeptical Environmentalist* (Lomborg 2001), would also have us believe this.

Of course, no one can predict the future with any level of certainty. But a clear consensus among environmental forecasters and demographers is that, yes, our circumstances will become far grimmer and more desperate in the foreseeable future. For example, largely as a result of synthetic fertilizers, humans already release as much nitrogen and phosphorus to the environment as all natural sources combined (Vitousek et al. 1994; Ehrlich et al. 1995). But by 2050, current global nitrogen fertilization is projected to rise by 270% while phosphorus use increases by 240%. Agricultural demands for irrigation water are expected to grow by 190% over the same period. Global pesticide production has risen at least 10-fold over the past 40 years and is likely to increase by 270% over current levels by the year 2050 (Tilman et al. 2001).

Land conversion, mostly for agriculture, is also projected to increase markedly over the next 50 years (Tilman et al. 2001). The net areas of pastures and croplands are expected to increase by 540 and 350 million hectares, respectively, despite a predicted loss of 140 million hectares of agricultural land in industrial nations (these lands will be converted to urban or suburban uses, reforested, abandoned, or used in other ways; Tilman et al. 2001). This implies that, during the next 50 years, about one billion hectares of natural ecosystems—an area larger than the United States—will be converted to agricultural land in developing nations (W. F. Laurance 2001). Absolute habitat destruction will be greatest in Latin America and sub-Saharan Africa (Ehrlich et al. 1995; Tilman et al. 2001), with large proportional declines in Asia (Dinerstein and Wikramanayake 1993; W. F. Laurance 1999). These changes could lead to the loss of about a third of all remaining natural tropical and temperate ecosystems and many of the environmental services they provide (Tilman et al. 2001).

The threats to biodiversity from such dramatic changes will be unprecedented, in part because projected population increases will be greatest in species-rich tropical areas with high proportions of endemic taxa. Of 25 identified biodiversity "hotspots" worldwide (Myers et al. 2000), 19 have populations that are growing more rapidly than the Earth's as a whole, and at least 16 have large numbers of people suffering extreme hunger and malnutrition (Cincotta et al. 2000). Already, almost half (45%) of the world's major nature reserves are being heavily used for agriculture (Tilman et al. 2001), and many more are being encircled by encroaching agricultural lands. Such pressures will only rise in the future.

It is important to emphasize, moreover, that natural ecosystems and species are being threatened not only by habitat destruction, but also by the combined effects of many simultaneous environmental changes that often interact additively or synergistically (Myers 1987, 1988a). The collective effects of habitat loss, fragmentation, and biological invasions could pose critical threats for many species (W. F. Laurance and Cochrane 2001). Global climate change will threaten those species that are unable to migrate across hostile landscapes to reach new areas with appropriate climates and habitats (Root et al. 2003; Thomas et al. 2004). Fundamental changes in biogeochemical cycles, water availability, and pesticide release will further damage ecosystem functioning. The world, and especially the tropical world, is changing in myriad and complex ways.

EMERGING THREATS TO TROPICAL FORESTS AND THEIR BIOTA

This book attempts to describe the panoply of threats to tropical forests and their biota, but with an emphasis on "emerging threats." We define these threats in four ways.

The first category of emerging threats are those that have only recently appeared. A prime example is the exotic chytrid fungus that has evidently decimated populations of rainforest frogs dwelling in virtually pristine streams on at least four (and probably more) continents. More than 100 frog species have apparently been driven to global extinction by the fungus, including the unique gastric brooding frogs of Australia and the famous golden toad of Costa Rica, and many others have declined precipitously. The chytrid is but one of many emerging pathogens that are currently degrading tropical ecosystems.

The second category includes threats that are growing rapidly in importance, or are unprecedented in scale. Tropical surface fires are an ideal example of such a threat. Often deceptively unimpressive, surface fires creep slowly across the floor of dense tropical forests, rarely exceeding 30 cm in height. But these cool-burning fires can have disproportionate impacts on rainforests, which are very poorly adapted for fires. Many trees and virtually all vines and forbs are killed, setting in motion an array of ecological changes that can eventually trigger complete forest destruction. Although difficult to detect and monitor, the spatial extent of surface fires is expanding dramatically throughout large areas of the tropics, because of increasing ignition sources and land-use changes that drastically increase the vulnerability of forests to burning.

Our third category includes threats that are poorly understood. The ecological effects of large-scale climatic and atmospheric changes on tropical forests and their biota are a perfect example of this. Many possible impacts of global-change have been postulated for tropical forests—apparent changes in forest dynamics, carbon storage, and species composition; increasing vulnerability of high-elevation endemics; altered rainfall patterns; increasing droughts, storms, and other weather extremes. But the magnitude and direction of long-term changes in global climatic and atmospheric parameters, and their ultimate effects on tropical ecosystems, remain very poorly understood.

The final category is environmental synergisms and interactions, whereby multiple, interacting threats can create exceptional perils for natural ecosystems. One of many examples of this is the devastating one-two punch of habitat fragmentation and chronic overhunting. For targeted species, such as larger mammals and birds and top predators, isolated populations in forest fragments are often driven rapidly to extinction. Moreover, similar processes threaten many species in tropical nature reserves, which are increasingly being severed from other forests and encircled by lands that are hostile to wildlife.

ORGANIZATION OF THIS VOLUME

To my knowledge, this is the first book that attempts to portray the full array of threats to tropical forests, insofar as they are currently understood. The chapters in this book span much of the tropical world, including the Neotropics, African tropics, Southeast Asia, Australia, and Oceania (fig. 2). Some emphasis is placed on lowland rainforests—the most extensive and species-rich of all tropical forest types—but montane forests, seasonal forests, and dry forests and woodlands are also discussed at length.

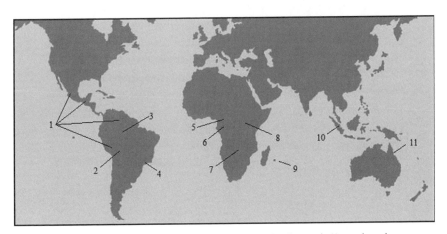

Figure 2. The geographic locations of studies in this volume: 1 = Mexico, Guatemala, Venezuela, and Peru (chap. 22); 2 = the Peruvian Amazon (chap. 13); 3 = the greater Amazon basin (chaps. 1, 4, 5, 6, 12, 14, 19); 4 = Brazilian Atlantic forests (chap. 15); 5 = Republic of Congo (chap. 21); 6 = Gabon (chap. 10); 7 = southern Africa (chap. 3); 8 = Uganda (chap. 7); 9 = Mauritius Island (chap. 11); 10 = Indonesia (chap. 16); 11 = tropical Queensland, Australia (chaps. 2, 20)

The chapters in this book include cutting-edge syntheses by many leading authorities as well as several original, data-rich studies. This book is intended for researchers, students, and conservation practitioners, and no attempt has been made to sweep the considerable uncertainty surrounding the causes and consequences of many emerging threats under the rug. In addition to documenting clear and present dangers to tropical forests, considerable attention is also focused on strategies for combating and mitigating emerging threats.

The book is organized into six parts, each of which is preceded by a general introduction and synthesis of the chapters, written by the editors. Part 1 focuses on emerging threats from climatic and atmospheric change, whereas part 2 examines the synergistic effects of simultaneous, interacting threats. Part 3 evaluates the impacts of exotic pathogens and invaders on tropical forests, while part 4 describes insidious and poorly understood threats. Part 5, the largest in the book, presents a diversity of different strategies to solve and mitigate threats to tropical forests. The final part is a synthesis of key themes and conservation messages in the book.

ACKNOWLEDGMENTS

My co-editor, Carlos Peres, and I would like to acknowledge the many dozens of referees—themselves often leading experts in the field—who critically evaluated our book chapters. Although they must remain anonymous, their efforts in providing rigorous, timely reviews were central to producing a volume of this nature. We also appreciate the hard work of Digna Matias in preparing the list of literature cited.

Finally, we extend our sincerest thanks and appreciation to Christie Henry, Biological Sciences Editor at the University of Chicago Press, and her staff, whose efforts and timely guidance were invaluable. This is my third book with the University of Chicago Press, and I am fully convinced that they are the leading academic publisher in the world.

PART I

Emerging Threats from Climatic and Atmospheric Change

William F. Laurance

INTRODUCTION

A Changing World

Just a few decades ago, one rarely heard concerns about global changes in Earth's climate and atmosphere. Of course, air pollution was widely understood to be a serious problem, but its main impacts were thought to occur in major urban areas, where it takes a grim toll on human health. Acid rain was also a worry—especially in the northeastern United States and parts of Europe—but its effects were regional in scale, not global.

This view began to change in the 1970s with the realization that Earth's protective ozone layer—which shields us from harmful UV-B radiation—was being eroded by man-made compounds like chlorofluorocarbons. These compounds were being produced in large quantities as refrigerants and spray-can propellants and for other industrial uses. The highly reactive chlorofluorocarbons quickly migrated into the stratosphere, where they destroyed parts of a thin ozone layer that had taken millennia to accumulate. The resulting "ozone holes" that formed annually at higher latitudes provoked widespread alarm and provided compelling evidence that air pollutants could have global impacts.

We are fortunate that chemical alternatives to chlorofluorocarbons were available and that we could, with only moderate difficulty, adapt our industries and consumer products to replace them. But the same cannot be said for global carbon emissions. Each year—as a result of industrial activities, automobiles, power generation, cement manufacturing, and tropical deforestation—humankind spews some 8 billion tons of carbon dioxide, methane, and other carbon-based pollutants into the atmosphere.

The net effect, as we all know, has been an alarming rise in air pollutants—particularly carbon dioxide, which has increased by more than a third, from 280 to 380 parts per million (ppm), since the onset of the industrial era. Equally distressing is that these emissions are accelerating, because countries like the United States have failed to rein in their burgeoning emissions and because rapidly developing countries like China, India, and Brazil are increasingly adopting the energy-consumptive lifestyles of industrial nations. From 1800 to 1960, for example, the average annual increase in atmospheric carbon dioxide concentrations was just 0.2 ppm, but this jumped to 1.4 ppm from 1960 to 2000 and has since risen to 2.3 ppm.

Because of its heat-trapping properties in the atmosphere, greenhouse gases like carbon dioxide are increasing global temperatures. But how hot will it get? According to the Intergovernmental Panel on Climate Change, global mean temperatures are expected to rise sharply, from 1.4°C to 5.8°C by the end of this century, with serious implications for the global climate. Although much is uncertain, melting glaciers and polar ice sheets, rising sea levels, large-scale shifts in continental precipitation patterns, and increasing storms and climatic extremes are among the expected consequences. Understanding the effects of such alterations on ecosystems and their biota is among the most active areas of contemporary scientific research.

Are these global changes affecting tropical forests, the world's most biodiverse ecosystems? If so, how? And what are the implications for the future? The chapters in this part grapple with these questions, from differing perspectives. Chapter 1 describes growing evidence that old-growth tropical forests in South America are experiencing concerted long-term changes that may well be driven by global-change phenomena. Chapter 2, from tropical Australia, uses bioclimatic models to predict the potentially severe impacts of future global warming on endemic vertebrates. Chapter 3 describes how feedbacks among varying annual rainfall, grazing by large herbivores, and fire could cause major, potentially irreversible changes in subtropical woodland-steppe ecosystems in southern Africa. The final contribution, chapter 4, assesses how forest destruction in Amazonia—a massive heat engine and carbon stock—might alter the global climate.

Pervasive Changes in South American Forests

In chapter 1, Simon Lewis, Oliver Phillips, Tim Baker, and their colleagues synthesize long-term data from a large network of forest-dynamics plots scattered across undisturbed, old-growth forests in South America. The patterns they find are intriguing. They show that Amazonian forests have become increasingly dynamic in recent decades, with higher rates of tree mortality and turnover. In addition, forest productivity and carbon storage appear to be increasing, evidently because biomass losses from tree mortality are being outpaced by accelerated tree growth. This finding has key implications: the vast Amazon may function as a major "carbon sink"—absorbing perhaps 0.5 to 0.8 billion tons of carbon from the atmosphere each year, thereby helping to slow the onslaught of global warming. Finally, the forests may be undergoing significant compositional shifts; in at least parts of the Amazon, tree communities are changing and lianas—climbing woody vines that often favor disturbed forest—are increasing in size and abundance. Because rainforest plants have complex ecological interrelationships with pollinators, seed dispersers, herbivores, mycorrhizal fungi, and other species, such compositional shifts might trigger important alterations in the Amazonian biota.

Why is the Amazon changing? The most parsimonious explanation, the authors assert, is that global environmental changes have increased forest productivity. They believe that the most likely driver is increased carbon dioxide levels in the atmosphere: plants use carbon dioxide for photosynthesis, and their growth rates and water-use efficiency rise when concentrations of this gas increase. But—as the authors readily admit—their conclusions are controversial. In addition to rising carbon dioxide levels, other large-scale phenomena, such as alterations in regional temperature, rainfall, available solar radiation, or nutrient deposition, may also be driving changes in Amazonian forests. Moreover, it is not inconceivable that local or natural phenomena, including past disturbances or sampling artifacts, could contribute to or even generate some of the observed patterns. On balance, however, the evidence seems increasingly compelling. The Amazon is changing, and because it sustains more than half of the world's remaining tropical rainforest, it is vital to understand both the local and global implications of these alterations.

Threats to Tropical Montane Species

As the ecologist Daniel Janzen noted decades ago, tropical species often show a high degree of specialization for narrow ranges of temperature and elevation. In mountainous areas of the tropics, high-elevation specialists—often locally endemic species—frequently abound. In chapter 2, Steve Williams and David Hilbert use bioclimatic models to project future changes in the geographic distributions of 65 species of endemic upland vertebrates in the rainforests of north Queensland, Australia. These upland specialists invariably have small geographic ranges—and thus are inherently vulnerable to environmental change—given that the rainforests of north Queensland have a relictual distribution, constituting less than 0.2% of Australia's land area. Williams and Hilbert's conclusions are jolting: if mean temperatures should rise by more than 2°C, their models suggest that many of the endemic vertebrates could disappear altogether.

What makes the Williams-Hilbert analysis so compelling is that it is based on sophisticated modeling and probably the world's most detailed data set on the fine-scale habitat associations and physiological ecology of tropical vertebrates. Of course, there are complicating factors in an analysis of this type: some species might adapt to changing temperature regimes or persist in small thermal refugia, and future shifts in rainfall might complicate the effects of rising temperatures on upland rainforests. But the Williams-Hilbert study reveals that many upland specialists in the tropics could be living on a knife edge. As their cool, high-elevation habitats wither and collapse in a hotter world, they may literally "disappear into heaven."

Climate and Land-Use Changes in Africa

In chapter 3, Hank Shugart and his colleagues use state-of-the-art ecosystem models to identify the drivers of vegetation change in subtropical woodlands and savannas of southern Africa. These semi-arid ecosystems are likely to be highly vulnerable to future climatic change because of their extreme sensitivity to moisture regimes. In these systems, interannual variations in rainfall can have a major impact on plant communities. Rainfall variability also interacts with two other key factors—fire and large-herbivore grazing—to determine the distribution of vegetation types. As Shugart et al. argue, feedbacks among rainfall, grazing, and fire could drive important changes in vegetation, with key implications for the conservation of Africa's surviving megafauna.

Unfortunately, predicting the effects of future climate change in semi-arid ecosystems may be extremely difficult. First, at these low-to-intermediate latitudes, global circulation models presently offer only tenuous predictions about the magnitude and direction of future climate change. Second, in these ecosystems even minor shifts in rainfall could drive large changes, potentially causing rapid shifts from woodland to grassland or vice versa. Compounding these complications are often major changes in land-use practices, especially introductions of domestic livestock that can increase grazing intensity and thereby alter fuel loads and fire frequency. Semi-arid ecosystems encompass vast expanses of Africa, Asia, Australia, and North and South America. Studies like that of Shugart et al. are vital to help us understand the impacts on these ecosystems of impending changes in climate and land use.

Local and Global Impacts of Amazon Deforestation

As climatologists are well aware, the Amazon is a vast heat engine, helping to drive global patterns of atmospheric circulation and moving enormous quantities of moisture from tropical to temperate latitudes. How will ever-increasing deforestation affect this great climate engine? This is the focus of chapter 4, by Roni Avissar, Renato da Silva, and David Werth. At local and regional scales, the loss and fragmentation of forests create land covers with very different physical properties. The juxtaposition of cool, moist forests with hot, dry pastures can generate "vegetation breezes," which increase local cloud cover and convective thunderstorms. Thus, paradoxically, forest loss can actually promote local increases in rainfall—but only up to a point. Eventually, regional rainfall is likely to decline—perhaps precipitously—as ever-shrinking forests cause reduced evapotranspiration and smaller inputs of moisture into the atmosphere. The details of these anticipated changes, however, are very poorly understood. How much deforestation can occur before regional rainfall collapses? The answer has profound implications for the survival of Amazonian forests.

An equally important message from Avissar et al. is that changes in Amazonia are likely to have far-flung impacts on remote areas of the world. These so-called climatic teleconnections could significantly alter rainfall elsewhere in the tropics and in other regions as disparate as the American Midwest, the Gulf of Mexico, the Arabian Peninsula, and northwestern India. Although the complex computer-simulation models that generate these predictions are subject to considerable uncertainties, it seems increasingly likely that tropical teleconnections will have important impacts on the global climate. Sadly, by the time these changes become so apparent that their effects are irrefutable, much of the world's tropical forests may already have vanished.

SYNTHESIS

Although the chapters in this part differ markedly in their geographic focus and approach, they share at least three important themes. The first is that anticipated changes in temperature, rainfall, and atmospheric composition in this century could potentially have grave effects on tropical ecosystems and biota. Not all ecosystems are equally vulnerable: Among the most sensitive are high-elevation communities that support biota adapted for cool, moist montane climates. Semi-arid woodlands can also be changed dramatically by shifts in rainfall mediated by related changes in fuel loads, fire, and grazing herbivores. In the humid tropics, large expanses of moderately seasonal forests could experience increased dry-season intensity and possibly more frequent El Niño droughts, promoting potentially catastrophic wildfires (see also W. F. Laurance, chap. 5 in this volume; Barlow and Peres, chap. 12 in this volume). Even in lowland forests that receive abundant rainfall, fundamental alterations in forest dynamics, biomass, and floristic composition may occur as a result of elevated productivity—possibly in response to rising carbon dioxide levels in the atmosphere. Such changes could have important long-term effects on forest biota and ecosystem functioning.

The second theme is that complex interconnections evidently exist among tropical ecosystems at different spatial scales—even among tropical and extratropical ecosystems on different continents. In the Amazon, for example, predicting the effects of deforestation on the regional hydrology first requires an understanding of how land-use changes affect local climate. The effects of the vegetation breeze—whereby clearing and fragmenting forest cover leads to increases in local rainfall—is surprising and counterintuitive; there must be important nonlinearities in the relationship between deforestation and rainfall. Local rainfall may increase with modest levels of forest loss, but eventually it must diminish with increasing deforestation because much of Amazonia's atmospheric moisture arises from forest evapotranspiration. At even larger spatial scales, worsening deforestation could drive changes in albedo (solar-heat reflectance) and evapotranspiration that alter the tropical heat engine. Teleconnections from such changes could impact rainfall regimes in disparate regions of the world.

The final theme is that, in all of these studies, investigators are grappling with daunting uncertainties. The classical, replicated experiments that scientists often favor are of no value here: we have but one Earth to study, and it is being changed in complex ways. There are two main sources of doubt. First, we are far from certain how environmental conditions will change in the future. How hot will it get? How will this affect rainfall patterns? How high will carbon dioxide levels rise? And the smaller the spatial scale, the more difficult it is for simulation models to make even coarse predictions of the future. Second, we are struggling to understand how ecosystems will respond to these changes. It is one thing, for example, to show that tree seedlings in an experimental enclosure grow faster when exposed to elevated carbon dioxide, and quite another to predict how complex rainforest ecosystems—with their bewildering array of ecological interactions—will respond to a series of simultaneous environmental changes. Yet try we must, for the impacts of these changes for the diverse tropical biota are potentially so dramatic.

CHAPTER 1

Impacts of Global Change on the Structure, Dynamics, and Functioning of South American Tropical Forests

Simon L. Lewis, Oliver L. Phillips, and Timothy R. Baker

With contributions by M. Alexiades, S. Almeida, L. Arroyo, S. Brown, J. Chave, J. A. Comiskey, C. I. Czimczik, A. Di Fiore, T. Erwin, N. Higuchi, T. Killeen, C. Kuebler, S. G. Laurance, W. F. Laurance, J. Lloyd, Y. Malhi, A. Monteagudo, H. E. M. Nascimento, D. A. Neill, P. Núñez Vargas, J. Olivier, W. Palacios, S. Patiño, N. C. A. Pitman, C. A. Quesada, M. Saldias, J. N. M. Silva, J. Terborgh, A. Torres Lezama, R. Vásquez Martínez, and B. Vinceti

Ecosystems worldwide are changing as a result of numerous anthropogenic processes. Some important processes, such as deforestation, are physically obvious, whereas others, such as defaunation and surface fires, may be subtler but affect biodiversity in insidious ways (cf. Lewis et al. 2004b; W. F. Laurance, chap. 5 in this volume; Terborgh and Nuñez-Iturri, chap. 13 in this volume). Atmospheric changes, such as increased rates of nitrogen deposition and especially increases in air temperatures and increasing carbon dioxide concentrations, are altering the environment of even the largest and most well-protected areas (e.g., Prentice et al. 2001; Galloway and Cowling 2002; Malhi and Wright 2004). Anthropogenic atmospheric change will certainly become more significant during this century; atmospheric carbon dioxide concentrations are likely to reach levels unprecedented for the past 20 million or even 60 million years (Retallack 2001; Royer et al. 2001). Nitrogen-deposition rates and climates are predicted to move far beyond that of Quaternary envelopes (Prentice et al. 2001; Galloway and Cowling 2002). Moreover, the rate of change in all these basic ecological drivers is likely to be without precedent in the evolutionary span of most species on Earth today (Lewis et al. 2004a). This, then, is the Anthropocene: we are living through truly epoch-making times (Crutzen 2002).

Given the scale of these changes, it is clear that all ecosystems on Earth are already very likely to have been altered by human activities. Recent research suggests that seemingly undisturbed tropical forests that are far from areas of deforestation are indeed undergoing profound shifts in structure, dynamics, productivity, and function. In this chapter, we draw together, review, and synthesize the recent results from a network of long-term monitoring plots across tropical South America that indicate how these forests are changing.

Changes in tropical forest structure, dynamics, productivity, and function are of great societal importance for three reasons: First, tropical forests play an important role in the global carbon cycle and hence the rate of climate change; about 40% of terrestrial carbon stocks lie within tropical forests (Malhi and Grace 2000). Second, because tropical forests house at least half of all Earth's species, changes in these high-biodiversity forests will have a large impact on global biodiversity (Groombridge and Jenkins 2003). Finally, because different plant species vary in their ability to store and process carbon, changes in both climate and biodiversity are potentially linked by feedback mechanisms (e.g., P. M. Cox et al. 2000; Lewis 2005). Overall, small yet consistent changes across the tropical forest biome are likely to have critical impacts on the global carbon cycle, global biodiversity conservation, and the rate of climate change—and hence human welfare.

Recent evidence suggests that remaining South American (and perhaps other tropical) rainforest is currently an important global carbon sink (Malhi and Grace 2000). The evidence is from (1) long-term monitoring plots, which show that forest stands are increasing in aboveground biomass (Phillips et al. 1998, 2002b; Baker et al. 2004a); (2) micrometeorological techniques, which indicate that mature Amazon forests may be a carbon sink (Grace et al. 1995; Araujo et al. 2002; Malhi et al. 2002b), albeit with substantial seasonal and interannual variability; and (3) inverse modeling of atmospheric carbon dioxide concentrations, which shows that tropical ecosystems may contribute a carbon sink of between 1 and 3 gigatons (1 gigaton = 1 billion metric tons) per year (e.g., Rayner and Law 1999; Rodenbeck et al. 2003).

These findings of a substantial tropical carbon sink are consistent with modeling and laboratory studies that imply changes in the productivity of tropical forests in response to increasing carbon dioxide (e.g., Lloyd and Farquhar 1996; Norby et al. 1999; Lewis et al. 2004a). However, the reader should note that all three approaches contain unresolved and currently debated possible areas of error:

1. The possibility exists that some of the biomass increase from forests may reflect forest recovery from episodic disturbance events, which may cause rapid and large biomass losses, whereas recovery is slow and steady;

Simon L. Lewis, Oliver L. Phillips, Timothy R. Baker, et al.

therefore, in this case, if by chance a limited number of monitoring sites fail to include rare, large-disturbance events, they may lead to erroneous extrapolations (Körner 2003).

2. Micrometeorological techniques contain errors in night-time flux measurements, which are not yet adequately addressed (e.g., Baldocchi 2003).

3. The very small number of air-sampling locations and poorly understood atmospheric transport across the tropics leave findings from inverse modeling studies open to debate (Houghton 2003).

In addition, the idea that atmospheric carbon dioxide increases are causing the tropical carbon sink is also controversial (for a recent discussion, see Lewis et al. 2004a). However, because all three lines of evidence suggest the existence of a sink—and efforts to overcome limitations in each line of research have generally confirmed the presence of a sink—it is reasonable to suggest that tropical forests likely provide a substantial buffer against global climate change. Indeed, the results from long-term forest-monitoring plots suggest that intact Amazonian forests have increased in biomass by about 0.3% to 0.5% per year and, hence, if all tropical forests are similarly increasing, they sequester carbon at approximately the same rate that the European Union (in January 2004) emits it by burning fossil fuels (Phillips et al. 1998; Malhi and Grace 2000; Baker et al. 2004a).

Large-scale environmental changes, such as increasing atmospheric carbon dioxide concentrations and rising air temperatures, will alter fundamental ecological processes and in turn will likely effect important changes in tropical biodiversity. In fact, this has already occurred in better-studied temperate areas (e.g., Parmesan and Yohe 2003) and in a well-studied old-growth tropical forest landscape (W. F. Laurance et al. 2004b). The interactive "balance" among tens of thousands of tropical plant species and millions of tropical animal species is certain to shift, even within the largest and best-protected forest ecosystems, which are traditionally thought of as "pristine" wilderness. These areas are vital refugia—where global biodiversity may most easily escape the current extinction crisis—because they are large enough to allow some shifts in the geographic ranges of species in response to global changes and are afforded some protection from industrial development such as logging and agriculture. However, how most tropical forest taxa will respond to rising temperatures and carbon dioxide concentrations, among other global changes, is currently unknown (Thomas et al. 2004).

Biodiversity change has inevitable consequences for climate change because different plant species vary in their ability to store and process carbon. For example, shifts in the proportion of faster-growing light-demanding species may alter the carbon balance of tropical forests. Long-term forest-plot data show that mature humid neotropical forests are a net carbon sink of ca. 0.6 gigatons per year (Phillips et al. 1998; Baker et al. 2004a). However, tree mortality rates have increased by as much as about 3% per year in recent decades, causing an increase in the frequency of tree-fall gaps (Phillips and Gentry 1994; Phillips et al. 2004). A shift in the composition of forests toward gap-favoring, light-demanding species with high growth rates, at the expense of more shade-tolerant species, is plausible (Körner 2004). Such fast-growing species generally have lower wood specific gravity, and hence lower carbon content for a given size (West et al. 1999), than do shade-tolerant trees. An Amazon-wide decrease in mean wood specific gravity of just 0.4% would cancel out the current carbon sink effect that is apparently caused by accelerated plant productivity. Whether such changes are occurring is currently poorly understood, but it is clear that the biodiversity and climate-change issues are closely linked and merit further study.

In this chapter, we present a summary of the latest findings from permanent plots monitored by a large network of Amazon-forest researchers, known as RAINFOR (Red Amazónica de Inventarios Forestales, or Amazon Forest–Inventory Network). The studies associated with RAINFOR have the following goals (Malhi et al. 2002c):

1. Quantify long-term changes in forest biomass and stem turnover.
2. Relate current forest structure, biomass, and dynamics to local climate and soil properties.
3. Attempt to understand the extent to which climate and soils will constrain future changes in forest dynamics and structure.
4. Attempt to understand the relationships among productivity, mortality, and biomass.
5. Use these results to predict how changes in climate may affect the biomass and productivity of the Amazon forest as a whole and to inform basin-scale carbon-balance models.
6. Search for evidence of change in forest composition and biodiversity (tree and liana species) over time.
7. Create a forum for discussion and standardize sampling methods among different research groups.

Here we summarize findings from old-growth forests in terms of (1) structural, (2) dynamic-process, and (3) functional change over the past two decades. Details of the exact plot locations, inventory and monitoring methods, and issues relating to collating and analyzing plot data are largely omitted from this chapter for reasons of space but are discussed in detail elsewhere (Phillips et al. 2002a, 2002b, 2004; Baker et al. 2004a, 2004b; Lewis et al. 2004b; Malhi et al. 2004). In addition, the evolving debates following the discovery that stem turnover had increased across the tropics since the 1950s (Phillips and Gentry 1994; Sheil 1995; Phillips 1995, 1996; Condit 1997; Phillips and Sheil 1997) and that long-term-monitoring plots in Amazonia increased in biomass during the 1980s and 1990s (Phillips et al. 1998; D. A. Clark 2002; Phillips et al. 2002b) are also relevant.

THE PLOT NETWORK

A RAINFOR plot is an area of forest where all trees above 10 cm diameter at breast height (dbh, measured at 1.3 m height or above any buttress or other deformity) are tracked individually over time. All trees are marked with a unique number, measured, mapped, and identified. Periodically (generally, approximately every 5 years) the plot is revisited; all surviving trees are remeasured, dead trees are noted, and trees recruited to 10 cm dbh are uniquely numbered, measured, mapped, and identified. This allows the calculation of (1) the cross-sectional area that trees occupy (termed "basal area"), which can be used with allometric equations to estimate tree biomass (Baker et al. 2004a); (2) tree growth (the sum of all basal-area increments for surviving and newly recruited stems over a census interval); (3) the total number of stems present; (4) stem recruitment (the number of stems added to a plot over time); and (5) mortality (either the number or basal area of stems lost from a plot over time). We present data from 50 to 91 plots, depending on selection criteria for different analyses (most critically, the number of census intervals from a plot and whether only stem-count data or the full tree-by-tree data set is available). The plots span South America (fig. 1.1), including Bolivia, Brazil, Ecuador, French Guiana, Peru, and Venezuela. Most are 1 ha in size and comprise about 600 trees of greater than or equal to 10 cm dbh. The smallest are 0.4 ha and the largest is 9 ha, all large enough to avoid undue influence by the behavior of an individual tree (Chave et al. 2003). Many plots have been monitored for more than a decade, although they range in age

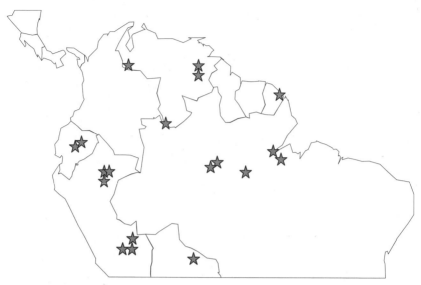

Figure 1.1. Plot locations used in this study. Each star indicates the location of 1–11 plots.

from 2 to 25 years. The earliest plot inventory was taken in 1967, the latest in 2002.

STRUCTURAL CHANGES

Among 59 plots monitored in old-growth Amazon forests where we have access to the full tree-by-tree data, there has been a significant increase in aboveground biomass between the first and last time they were measured. Over approximately the past 20 years, the increase has been 0.61 ± 0.22 tons of carbon per ha per year, or a relative increase of 0.50 ± 0.17% per year (mean ± 95% confidence interval; Baker et al. 2004b). Across all plots, the aboveground biomass change is normally distributed and shifted to the right of zero (fig. 1.2). This estimate is slightly higher than that documented by Phillips et al. (1998), who used a smaller and earlier data set, and similar to that documented by Lewis et al. (2004b), who used a smaller data set, where each plot was monitored over much longer periods of time.

We can crudely estimate the magnitude of the South American carbon sink by multiplying 0.61 tons per ha per year by the estimated area of mature neotropical humid-forest cover (ca. 8,705,100 km²; FAO 1993), which yields a value of about 0.5 gigatons of carbon per year. If we fur-

Simon L. Lewis, Oliver L. Phillips, Timothy R. Baker, et al.

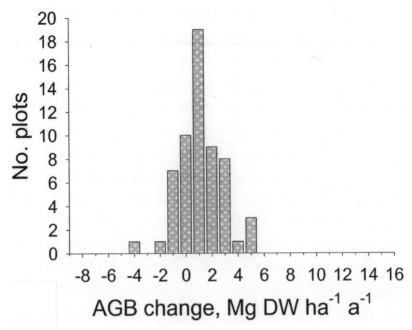

Figure 1.2. Aboveground biomass change (Mg dry weight per ha per year) of trees greater than 10 cm diameter in 59 Amazon plots, based on initial and final stand-biomass estimates calculated using an allometric equation relating individual tree diameter to biomass and incorporating a correction factor to account for variation in wood density among species (from Baker et al. 2004a)

ther assume that the ratio of aboveground to belowground biomass is 3:1 (cf. Phillips et al. 1998), and that belowground biomass is increasing in proportion to aboveground biomass, then the sink increases to 0.7 gigatons of carbon per year. If other biomass components, such as small trees, lianas, and coarse woody debris, are also increasing in biomass, then the sink may be fractionally larger still. However, these estimates depend critically on (1) how representative of South American forests are the 59 plots studied by Baker et al. (2004a) and (2) assumptions about the extent of mature, intact forest remaining in South America. Both areas of uncertainty suggest that we treat these extrapolations with caution.

D. A. Clark (2002) raised two concerns about the original findings of Phillips et al. (1998) that Amazon biomass was increasing, suggesting that (1) some floodplain plots that Phillips et al. considered mature may still be affected by primary succession, and that (2) large buttress trees in some plots may have been measured not above the buttress, as protocols dictate, but around it. However, neither potential error leads to a significant overestimation of aboveground-biomass increases because the carbon sink remains when either all plots on old floodplain substrates or those that may have buttress problems are removed from the analysis (Baker et al. 2004a).

Among 91 plots monitored across South America, there was a significant increase in stem density between the first and last time they were measured. The increase has been 0.84 ± 0.77 stems per ha per year (fig. 1.3; paired t-test, $t = 2.12$, $P = 0.037$), or a $0.15 \pm 0.13\%$ per-year increase (Phillips et al. 2004). Across all plots, stem-change rates are approximately normally distributed and slightly shifted to the right of zero (fig. 1.3). (The number of plots used here is more than that used in the biomass study, largely because complete tree-by-tree data are required to calculate biomass using Baker et al.'s [2004a] methods, whereas stem-change data can often be obtained from published studies.) The same test using 59 plots (from the study by Baker et al. 2004a) shows a similar increase in stem density ($0.16 \pm 0.15\%$ per year), whereas a smaller but much longer-term data set (50 plots from Lewis et al. 2004b) shows a slightly larger increase ($0.18 \pm 0.12\%$ per year).

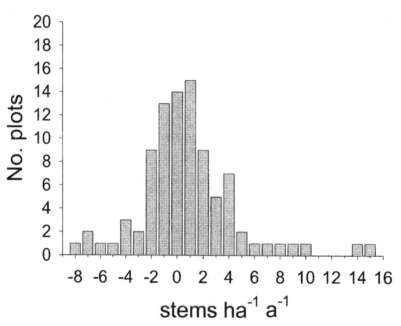

Figure 1.3. Stem-number change per ha per year in 91 plots from across South American tropical forests. Stems were counted during the first and final censuses of each plot (plots are the same as those used by Phillips et al. 2004).

Forest stands are never static. Rather, trees are recruited, grow, and die. By looking only at the stems that are added (stem recruitment) and lost (stem mortality) over time, we can estimate forest dynamics by calculating the rate of stem turnover. Stem-turnover rates between any two censuses can be estimated as the mean of annual mortality and recruitment rates (Phillips and Gentry 1994). However, most surviving trees grow, thus gaining basal area and hence biomass, and no two dying trees are alike in terms of the biomass that they have amassed. Therefore, it is also important to measure the stand-level rate of "biomass growth" (which can be approximated as the rate at which surviving and recruiting trees gain basal area) and "biomass loss" (approximated as the rate at which basal area is lost from the stand via tree death).

Among 50 old-growth plots across tropical South America with at least three censuses (and, therefore, at least two consecutive monitoring periods that can be compared), we find that *all* of these key ecosystem processes—stem recruitment, mortality, and turnover, and biomass growth, loss, and turnover—are increasing significantly (fig. 1.4) when the first and second halves of the monitoring period are compared (Lewis et al. 2004b). Thus, over the past two decades, these forests have become, on average, faster growing and more dynamic. Notably, the increases in the

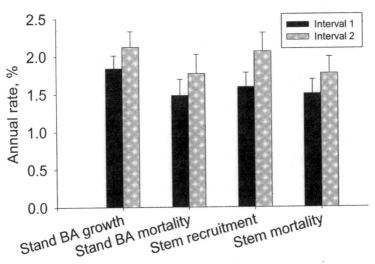

Figure 1.4. Annualized rates of stand-level basal-area growth, basal-area mortality, stem recruitment, and stem mortality from two consecutive census intervals, each giving the mean from 50 plots with 95% confidence intervals. Paired t-tests show that all of the increases are significant. The average midyear of the first and second censuses was 1989 and 1996, respectively. (From Lewis et al. 2004b)

dynamic processes (growth, recruitment, and mortality) are proportionately about an order of magnitude larger than the increases in structural variables (aboveground biomass and stem density; Lewis et al. 2004b).

These and similar results can be demonstrated graphically in a number of ways. In figure 1.5, we plot the across-site mean values for stem recruitment and mortality as a function of calendar year. This shows that the increase has not occurred over the short term (for example, the result of a "spike" around a year with unusual weather), that recruitment rates have on average consistently exceeded mortality rates, and that mortality appears to lag recruitment (Phillips et al. 2004).

Using data for the 50 plots with two consecutive census intervals, we can also separate them into two groups, one faster growing and more dynamic (mostly western Amazonian plots), and one slower growing and less dynamic (mostly eastern and central Amazonian plots). Both groups showed increased stem recruitment, stem mortality, stand basal-area growth, and stand basal-area mortality, with larger absolute increases in rates in the faster-growing and more-dynamic plots than in the slower-

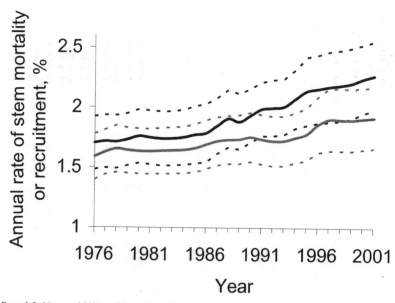

Figure 1.5. Mean and 95% confidence intervals for stem-recruitment and stem-mortality rates against calendar year for plots across South America. Each plot was corrected for differing census-interval lengths. The data set was also corrected for "site-switching": not all plots were monitored across all years, and forests that may have been affected by "majestic-forest bias" were omitted. The same trends hold if these corrections are not applied. Black indicates recruitment, gray indicates mortality, solid lines are means, and dashed lines are 95% confidence intervals. (From Phillips et al. 2004)

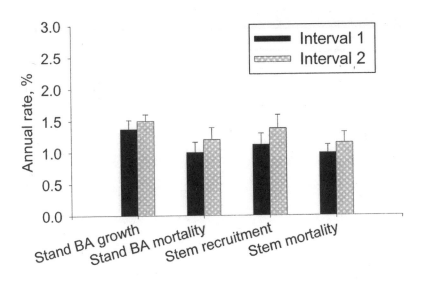

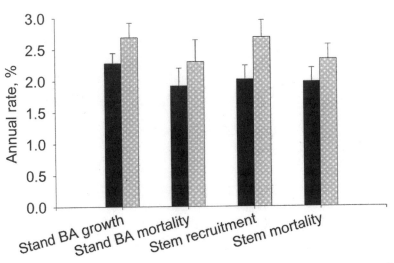

Figure 1.6. Annualized rates of stand-level basal-area growth, basal-area mortality, stem recruitment, and stem mortality over consecutive census intervals for plots grouped into slower-growing less-dynamic (*top*) and faster-growing more-dynamic (*bottom*) forests. Of the slower-dynamics group, 20 of 24 plots are from eastern and central Amazonia, whereas just 2 are from western Amazonia. Of the faster-dynamics group, 24 of 26 plots are from western Amazonia, with just 1 from central Amazonia. The remaining 3 plots are from Venezuela and outside the Amazon drainage basin. Changes have occurred across the South American continent and in both slower- and faster-dynamic forests. (From Lewis et al. 2004b)

growing and less-dynamic plots (fig. 1.6; Lewis et al. 2004b). However, the proportional increases in rates were similar, and statistically indistinguishable, across both forest types (Lewis et al. 2004b). This shows that increases in growth, recruitment, and mortality rates occur proportionately similarly across very different forest types and geographically widespread areas.

FUNCTIONAL CHANGES

Changes in the structure and dynamics of tropical forests are likely to be accompanied by changes in species composition and function. There is, moreover, no a priori reason to expect that large changes in Amazon forests should be restricted to trees. We studied woody climbers (structural parasites on trees, also called lianas), which typically contribute between 10% and 30% of forest leaf productivity but are ignored in many monitoring studies. We found that nonfragmented forests across western Amazonia are experiencing a concerted increase in the density, basal area, and mean size of lianas (fig. 1.7; Phillips et al. 2002a). Over the last two decades of the twentieth century, the density of large lianas relative to trees had increased by approximately 1.7% to 4.6% per year. This was the first direct evidence that intact tropical forests are changing in terms of their composition and function.

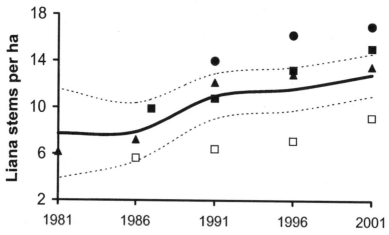

Figure 1.7. Five-year running means (solid line) with 95% confidence intervals (dashed lines) of liana stem density per hectare (>10 cm dbh), with values plotted separately for northern Peru (filled squares), southern Peru (filled triangles), Bolivia (filled circles), and Ecuador (open squares) (from Phillips et al. 2002b)

Simon L. Lewis, Oliver L. Phillips, Timothy R. Baker, et al.

Finally, a recent paper from a cluster of plots in central Amazonia has shown consistent changes in forest composition over the past two decades (W. F. Laurance et al. 2004b). Many faster-growing genera of canopy and emergent stature trees increased in basal area or density, whereas many slower-growing genera of subcanopy or understory trees decreased in density. W. F. Laurance et al. (2004b) provide evidence of pervasive changes in central Amazonian forests: growth, mortality, and recruitment all increased significantly over two decades (basal area also increased, but not significantly so), with faster-growing genera showing much larger absolute and relative increases in growth relative to slower-growing genera. Because this study was confined to a single landscape, further studies are urgently needed to determine whether comparably large shifts in tree communities are occurring throughout the tropics.

WHAT IS DRIVING THESE CHANGES?

What could be causing these continent-wide changes in tree growth, recruitment, mortality, stem density, and biomass? We suggest a parsimonious explanation. The results appear to show a coherent fingerprint of increasing growth (i.e., increasing net primary productivity [NPP]) across tropical South America, probably caused by a long-term increase in resource availability (Lewis et al. 2004a, 2004b). According to this explanation, increasing resource availability increases NPP, which then increases stem-growth rates. This accounts for the increase in stand basal-area growth and stem-recruitment rates, and the fact that these show the "clearest" (statistically most significant) signals in the analyses (Lewis et al. 2004a). Because of increased growth, competition increases for limiting resources such as light, water, and nutrients. Over time, some of the larger, faster-growing trees die, as do some of the "extra" recruits (the accelerated growth percolates through the system). This accounts for the increased losses from the system: stand-level biomass-mortality and stem-mortality rates increase. Thus, the system gains biomass and stems, whereas the losses lag some years behind, causing an increase in aboveground biomass and stems. Overall, this suite of changes may be qualitatively explained by a long-term increase in a limiting resource.

The changes in composition may also be explained by increasing resource availability because the rise in liana density may be either a direct response to rising resource supply rates or a response to greater disturbance caused by higher tree-mortality rates. The changing tree composition in central-Amazonian plots (W. F. Laurance et al. 2004b) is also consistent with increasing resource supply rates; experiments show that faster-growing species are often the most responsive, absolutely, to increases in resource levels (Coomes and Grubb 2000).

What environmental changes could be increasing the growth and productivity of tropical forests? Although there have been widespread changes in the physical, chemical, and biological environment of tropical trees (Lewis et al. 2004a), only increasing atmospheric carbon dioxide concentrations (Prentice et al. 2001), increasing solar radiation inputs (Wielicki et al. 2002), and rising air temperatures (Malhi and Wright 2004) have been documented across South America and may be responsible for increased growth and productivity. However, for none of these three changes do we have good evidence that the driver has actually changed *and* that such a change will accelerate forest growth (Lewis et al. 2004a).

The increase in atmospheric carbon dioxide is our leading candidate because of the undisputed long-term historical increase in cardon dioxide concentrations, the key role of carbon dioxide in photosynthesis, and the demonstrated positive effects of carbon dioxide fertilization on plant-growth rates, including experiments on whole temperate-forest stands (Hamilton et al. 2002; Norby et al. 2002; Lewis et al. 2004a). At present, however, no experiments have assessed the effects of increasing carbon dioxide availability on intact, mature tropical-forest stands, and thus this interpretation is still open to debate (e.g., Chambers and Silver 2004).

Air-temperature increases are also undisputed and could conceivably be causing the changes we document. However, many authors expect that the 0.26°C per decade air-temperature increase over the past two decades (Malhi and Wright 2004) would actually reduce, not increase, forest growth because respiration costs are likely to increase with temperature. However, air temperatures will also increase soil temperatures, which could in turn increase soil mineralization rates and thus nutrient availability (see review by Lewis et al. 2004a). Whether recent rises in air temperature have increased or decreased tropical forest NPP requires further study.

Recent satellite data suggest an increase in incoming solar radiation across the tropics between the mid-1980s and late 1990s as a result of reduced cloudiness (Wielicki et al. 2002). However, because stem turnover has increased steadily across the tropics since the 1950s (Phillips and Gentry 1994; Phillips 1996), increasing solar radiation since the mid-1980s may not have occurred over a long enough period of time to explain the trends in forest-plot data, at least in terms of stem turnover. Furthermore, because the *difference* between stand-level basal-area growth and mortality was similar at the start (1980s) and end (1990s) of the study by Lewis et al. (2004b), the factor causing changes in growth, recruitment, and mortality was probably operating before the onset of the study and, hence, before the observed increase in incoming solar radiation. Nevertheless, the temporary increase in incoming solar radiation, coupled with rising carbon dioxide concentrations, might account for the very large increases in growth (ca. 2% per year) that we observed across South America in the 1980s and 1990s.

Confidently determining which environmental change, or changes, have caused the trends we document across South American tropical forest is very difficult and hence open to debate. However, each environmental change is expected to leave a unique signature, or fingerprint, in forest data because different environmental changes initially impact upon different processes, have different distributions in time and space, and may affect some forests more than others (for example, depending on soil fertility). Future analyses of forest-plot data should allow a further narrowing of potential causes underlying rising productivity across South American tropical forests (Lewis et al. 2004a).

THE FUTURE

For those concerned about future biodiversity losses and global climate change, our analyses suggest both worrying trends and some apparently "good news." South American tropical forests, including the Amazon, the world's largest remaining tract of tropical forest, have shown concerted changes in forest dynamics over the past two decades. Such unexpected and rapid alterations—apparently in response to anthropogenic atmospheric change—raise concerns about other possible surprises that might arise as global changes accelerate over the coming decades (Lewis 2005). Tropical forests are evidently very sensitive to changes in incoming resource levels and may show large structural and dynamic changes in the future as resource levels alter further and temperatures continue to rise (Lewis et al. 2004a). The implications of such rapid changes for the world's most biodiverse region are unknown but could be substantial.

Moreover, old-growth South American forests are evidently helping to slow the rate at which carbon dioxide is accumulating in the atmosphere, thereby acting as a buffer to global climate change—certainly good news for the moment. The concentration of atmospheric carbon dioxide is rising at a rate equivalent to 3.2 gigatons of carbon per year; but without the tropical South American carbon sink, it may increase by 0.5 to 0.8

gigatons of carbon per year more—15% to 25% faster than at present. However, this subsidy from nature could be a relatively short-lived phenomenon. Mature forests may (1) continue to be a carbon sink for decades (Chambers et al. 2001; Cramer et al. 2001), (2) soon become a small carbon source because of changes in functional and species composition (Cramer et al. 2001; Phillips et al. 2002a; Körner 2004; W. F. Laurance et al. 2004b), or (3) become a mega-carbon source, possibly in response to climate change (P. M. Cox et al. 2000; Cramer et al. 2001). Given that a 0.5% annual increase in tropical forest biomass is roughly equivalent to the entire fossil-fuel emissions of the European Union (in January 2004), a modest switch of tropical forests from a carbon sink to a carbon source would have profound implications for global climate, biodiversity, and human welfare.

Finally, it is important to emphasize that climate-based models that project the future carbon balance in Amazonia (and future climate-change scenarios) have made no allowance for changing forest composition. This is likely to lead to erroneous conclusions. For example, lianas contribute little to forest biomass but kill trees and suppress tree growth (Schnitzer and Bongers 2002), and their rapid increase suggests that the tropical carbon sink might shut down sooner than current models suggest. Projections of future carbon fluxes will need to account for the changing composition and dynamics of these supposedly undisturbed forests.

CONCLUSIONS AND IMPLICATIONS

1. Analyses of long-term forest-plot data show that there have been widespread and concerted changes in the structure, dynamics, and functional composition of South American forests in recent decades.

2. The most parsimonious explanation of these changes is that global environmental changes have increased resource supply rates to tropical forests. This would increase net primary productivity, increasing tree growth and recruitment and, in turn, accelerating mortality. Rising atmospheric carbon dioxide concentrations is the environmental change most likely to produce consistently better conditions for growth across South America.

3. Because forest growth exceeds mortality, South American forests have been functioning as a carbon sink. The size of this sink is approximately 0.5 to 0.8 gigatons of carbon per year, depending on assumptions used to scale from plot studies to the continent. This sink is equivalent to

Simon L. Lewis, Oliver L. Phillips, Timothy R. Baker, et al.

between 15% and 25% of the annual increase in atmopsheric carbon dioxide concentrations. Thus, old-growth tropical forests in South America currently buffer the rate of climate change.

4. The future of the South American carbon sink is highly uncertain. The sink is caused by an increase in forest biomass of approximately 0.3% to 0.5% per year. Future climate change or changing species or functional composition, including a documented increase in large liana density, may reduce or reverse the current carbon sink. Even a modest switch from a carbon sink to carbon source for South American forests would have profound implications for global climate, biodiversity, and human welfare.

5. The documented acceleration of tree growth, recruitment, and mortality will affect the interactions of thousands of plant and millions of animal species. The implications of such changes across the world's most biologically rich continent are unknown.

6. Twenty years ago, when most of the plots we study were established, few would have even imagined that the Amazon would show consistent and widespread shifts in structure, dynamics, and function. Protecting remaining forests and expanding on-the-ground monitoring across Amazonia and beyond is essential if we are to better understand the critical interactions between global climate and biodiversity change, two of the key environmental issues of our times.

ACKNOWLEDGMENTS

The results summarized here depended on contributions from numerous field assistants, rural communities, and field-station managers in Brazil, Bolivia, Ecuador, French Guiana, Peru, and Venezuela, and more than 50 grants from funding agencies in Europe and the United States. This essential support is acknowledged in our earlier publications. This work was supported by NERC grant NER/A/S/2000/00532 to Oliver Phillips and Simon Lewis. Simon Lewis is supported by a Royal Society Fellowship. Timothy Baker is a Max-Planck Institut für Biogeochemie Research Fellow. We thank William Laurance and two anonymous referees for comments on the manuscript.

CHAPTER 2

Climate Change as a Threat to the Biodiversity of Tropical Rainforests in Australia

Stephen E. Williams and David W. Hilbert

Forest destruction is thought to be the greatest threat to biodiversity in the tropics, particularly in the Amazon and tropical Asia (W. F. Laurance 1999). Climate change is sometimes discounted as a threat to tropical biotas and has been less studied in the tropics than in temperate, boreal, and arctic ecosystems. However, climate change has already produced significant and measurable impacts on almost all ecosystems around the globe and has altered species distributions, the timing of biological behaviors, assemblage composition, ecological interactions, and community dynamics (L. Hughes 2000; Walther et al. 2002; Parmesan and Yohe 2003; Root et al. 2003, 2005; Pounds et al. 2006). Recent analyses based on bioclimatic-distribution modeling suggested that climate change is potentially a greater threat to global biodiversity, including that in many tropical ecosystems, than is habitat destruction (Thomas et al. 2004).

What are the predicted changes to Earth's climate? Average global temperatures have already risen approximately 0.6°C and are continuing to increase (Houghton et al. 2001). Over the remainder of this century, average temperatures are projected to rise between 1.4°C and 5.8°C in combination with large increases in atmospheric carbon dioxide concentrations and significant changes in rainfall patterns (Houghton et al. 2001). The effects on rainfall patterns are less certain, although it is predicted that rainfall variability and dry-season severity will generally increase (K. J. E. Walsh and Ryan 2000). That is, rainfall will be more variable from month to month with longer dry spells and possibly an increased frequency of disturbance events such as flooding rains and cyclones (Easterling et al. 2000; K. J. E. Walsh and Ryan 2000; Milly et al. 2002; Palmer and Raianen 2002). Additionally, a rise is expected in the average basal altitude of the cloud layer over land (Pounds et al. 1999; Still et al. 1999). Reductions in the amount of moisture captured from clouds by vegetation could exacerbate the effects of longer and more variable dry seasons (Still et al. 1999).

What are the likely impacts of these changes on biodiversity? A changing global climate will likely alter physiological tolerances of species, biotic interactions, and disturbance events and could lead to increased extinction rates and changes in species distributions, assemblage structure, habitat structure, and ecosystem functioning (fig. 2.1; L. Hughes 2000; Lewis et al., chap. 1 in this volume). There is a general perception that climate-change impacts will be more severe in temperate regions than in the tropics, although many believe that montane tropical biotas are extremely vulnerable. Several studies have predicted that the impacts of climate change will largely consist of shifts in latitudinal and altitudinal distributions of species with concomitant complex changes in local assem-

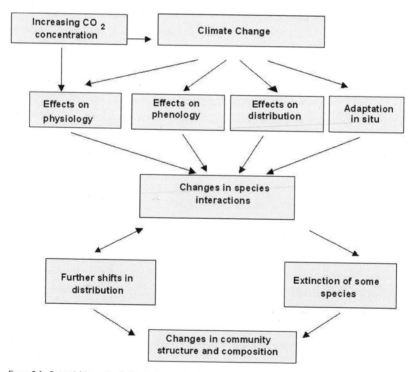

Figure 2.1. Potential impacts of global climate change (modified from L. Hughes 2000)

blage structure (Parmesan 1996; Hill et al. 2002; Peterson et al. 2002; Parmesan and Yohe 2003; Root et al. 2003).

Other impacts include earlier migration and breeding in many bird species (e.g., Visser and Both 2005) and alterations in life history traits, such as clutch size (Visser and Both 2005) and adult body size (F. A. Smith et al. 1998), in a number of vertebrate species. Climatic changes also have the potential to disrupt predator-prey relationships (Durant et al. 2005), increase the susceptibility of species to disease (Burrowes et al. 2004; Pounds et al. 2006), interfere with community dynamics and ecosystem functioning (Sinclair and Byrom 2006), and alter the genetic constitution of local populations (Umina et al. 2005). However, there is also limited evidence suggesting that some species can show evolutionary and behavioral compensation via mechanisms that may mitigate the negative effects of climate change (e.g., Wichmann et al. 2005). The imperative now is to understand, predict, and minimize the ecological and socioeconomic consequences of these impacts.

The necessity of addressing the impacts of climate change on biodiversity is recognized by most governments and is now included in Australian government policy at all levels. It is also a priority for many international conservation organizations. This was particularly evident in the resolutions on climate change in the Durban Accord, which resulted from the International Union for the Conservation of Nature and Nature Resources (IUCN) World Parks Congress in Durban, South Africa, in 2003. Despite such acceptance at the policy level, there has been a general lack of research on impacts of climate change in the tropics and in Australia in particular (Houghton et al. 2001). However, this is changing and there is now increasing interest in research and conservation policy focused on the impacts of climate change in Australia (e.g., Howden et al. 2003). Here we discuss the potential impacts of predicted changes in global climate on one of the world's biodiversity hotspots, the rainforests of the Australian Wet Tropics World Heritage Area.

Climate change is an important long-term threat to biota in the Wet Tropics bioregion. Rainforests in this region were protected as a World Heritage Area in 1988, primarily because of the high biodiversity value of the unique regional biota. Although small in extent (~10,000 km^2), the rainforests of the Wet Tropics sustain high levels of species richness, endemism, and evolutionarily significant taxa. This includes a very large concentration of archaic species, including diverse assemblages of primitive flowering plants, marsupials, birds, reptiles, frogs, insects, and other taxa.

Mountainous regions may be especially vulnerable to global warming. The mountaintops and higher tablelands of the Wet Tropics can be thought of as cool islands in a sea of warmer climates, termed a "mesotherm archipelago" (Nix 1991). These islands are separated from each other by the warmer valleys and form a scattered archipelago of habitat for organisms that are unable to survive and reproduce in warmer climates. Many Wet Tropics–endemic species live only in these cooler uplands.

On a regional scale, the current patterns of vertebrate biodiversity in north Queensland have been largely shaped by Pleistocene contractions in rainforest areas and subsequent expansion episodes (Winter 1988; Williams 1997; Williams and Pearson 1997; Graham et al. 2006). The contraction of rainforests to cool, moist upland refugia probably imposed an extinction filter, resulting in most of the remaining regionally endemic vertebrate species being cool-adapted upland species (Williams and Pearson 1997; Schneider et al. 1998). These factors have predisposed the fauna to being particularly vulnerable to global climate change, for two reasons:

1. Biogeographic history has resulted in an endemic fauna that is mainly adapted to a cool (upland), wet, and relatively aseasonal environment.
2. The impacts of increasing temperatures should be most noticeable across altitudinal gradients, and it is the altitudinal gradient that dominates the biogeography of this region (Nix 1991; Williams et al. 1996).

PREDICTING IMPACTS OF CLIMATE CHANGE

Effects on the Distribution of Rainforest Habitat

Here we discuss predicted changes to the distribution of various structural types of Wet Tropics rainforest with increasing temperature and a changing rainfall regime. In complex tropical forests, very high species richness and our limited understanding of ecological patterns and processes severely limit development of detailed explanatory models that can be applied to climate-change issues. Consequently, empirical-modeling approaches that make use of existing biogeographic and climatic data are necessary. In the Australian Wet Tropics, good maps of forest distributions are available that are based on a well-developed structural-physiognomic classification (L. J. Webb 1959; Tracey and Webb 1975; Tracey 1982). The spatial patterns of climatic, topographic, and soil variables are also reasonably well described. Our approach was to estimate the relative suitability of local environments for a range of recognized forest structural-environmental classes in the Wet Tropics. Using a geographic information system, climate, soil, and topographic information was related to current vegetation patterns. These modeled relationships can then be used to infer the potential distribution of vegetation in different climates of the past or future.

The Forest Artificial Neural Network (FANN; Hilbert and van den Muyzenberg 1999) is a feed-forward artificial neural network (Rumelhart and McClelland 1986) that characterizes the relative suitability of environments for 15 forest classes in the Australian Wet Tropics, from dry open woodlands to a wide range of rainforest types. The modeling approach is similar to discriminant analysis but makes no assumptions about the statistical distributions of variables (D. G. Brown et al. 1998). It also can use categorical independent variables such as soil type, simultaneously model a large number of vegetation classes, and effectively discount outliers (Haung and Lippman 1988). It often outperforms other techniques when faced with complex classification problems.

Under the present climate, there is an overall correspondence of 75% between the classification of forest environments by FANN and the mapped distribution of Wet Tropics forests (at 1 ha resolution over approximately 2 million hectares). This agreement is quite high considering the complexity of the vegetation mosaic, the large number of forest classes, and the fine scale of spatial resolution. Hilbert and van den Muyzenberg (1999) provide a complete assessment of FANN's accuracy and generality, while application techniques are described by Hilbert and Ostendorf (2001). Output from the model also can be used to measure the dissimilarity between the environment at any site and the environment that would be most suitable for the forest type presently at that site (Hilbert and van den Muyzenberg 1999). This provides a measure of the degree of stress caused by climate change during the potentially long lag time before forest structure and physiognomy come into equilibrium with the new climate.

FANN was used to assess the sensitivity of forests in the Australian Wet Tropics to 1°C of warming and five rainfall scenarios ranging from a 10% reduction to a 20% increase in rainfall (Hilbert et al. 2001). Rainforest as an aggregate vegetation class displays almost no sensitivity to a 1°C temperature increase but shows a strong response to rainfall. A decrease in precipitation of 10% would reduce the area suitable to rainforest, whereas increasing rainfall would considerably increase the total area of rainforest. This implies that the total extent of rainforest in the humid tropics is not presently limited by temperature, presumably because the various rainforest classes span a large range of mean annual temperatures.

However, different classes of rainforest are very different in their environmental relationships. The cool-wet rainforests (termed Simple Notophyll and Simple Microphyll Forests and Thickets) that now occur at the highest elevations have by far the largest sensitivity to temperature. With no change in rainfall, the area of these forest types would decrease by half if temperatures increased by one degree (fig. 2.2). This forest class also shows a fairly strong sensitivity to precipitation that is less pronounced

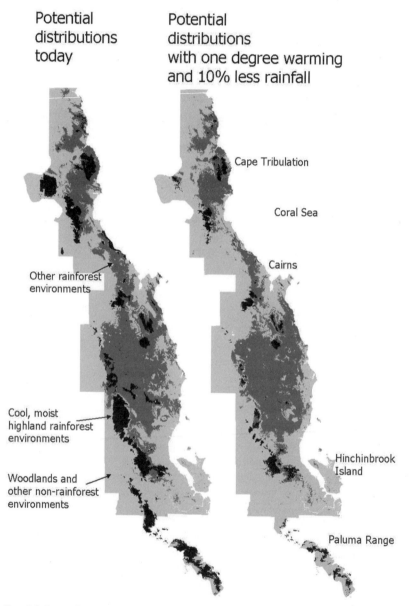

Figure 2.2. Extent of upland rainforest at present and as predicted by neural-net modeling with a 1°C increase in temperature and 10% decrease in rainfall (Hilbert et al. 2001)

when temperature is increased. Thus, for cool-wet rainforests, increased precipitation cannot significantly ameliorate declines due to warming.

A variety of highland rainforest types (Notophyll Vine Forests, Simple Notophyll and Microphyll Forests and Thickets, and Notophyll Semi-evergreen Vine Forests) are strongly affected by even 1°C of warming and a small reduction in rainfall (fig. 2.2). Although rainforest environments overall are not greatly reduced under these conditions, highland rainforests are greatly decreased in extent, become more fragmented, and are eliminated from some parts of the Wet Tropics region. Other studies have demonstrated that a changing atmosphere and climate is already producing significant changes in the physical structure, dynamics, and functioning of rainforests in Amazonia (Phillips and Gentry 1994; Phillips et al. 1998, 2002b; W. F. Laurance et al. 2004b; Lewis et al., chap. 1 in this volume).

Climatic Stress in the Present Forests

If we consider the projected decrease in forest area in response to future climatic change, existing rainforests in Australia's Wet Tropics will experience less-optimal environments and therefore may be stressed to some degree, even though the total potential area of rainforest may depend largely on changes in rainfall patterns (Hilbert et al. 2001). Mesophyll Vine Forests, which predominate in the warmer lowlands, will either increase in area or remain the same, whereas stress in other forest types will depend on changes in precipitation. Complex Notophyll Vine Forests (typical of middle elevations) will be significantly stressed although the future extent of this environment will again depend on rainfall. The highland rainforests (Simple Notophyll and Simple Microphyll Forests and Thickets) will be highly stressed and the extent of their suitable environment would decline significantly, irrespective of rainfall.

Effects on Distribution of Vertebrates

Protecting the biota and ecosystem functions of the Wet Tropics bioregion is only possible if we have some understanding of current patterns of biodiversity as well as factors that maintain ecosystem processes and determine the distributions of species, assemblages, and habitats. Therefore, it is imperative that we gain an understanding of the factors that determine the distribution of species and habitats and the relationships between climate and key ecosystem processes. The distribution and abundance of a species is determined by a number of complex, and often interacting, factors within four general categories (J. H. Brown and Lomolino 1998):

1. biogeographic history (e.g., extinction episodes due to habitat contraction),
2. physiological preferences and tolerances of species and habitats to the abiotic environment (e.g., temperature, rainfall, climatic stability),
3. biotic interactions (e.g., competition, predation), and
4. disturbance (e.g., fire, cyclones).

Bioclimatic modeling of the distributions of all 65 species of regionally endemic, rainforest vertebrates in the Wet Tropics was used to examine the potential ecological impacts of a changing abiotic environment (Williams et al. 2003). These analyses suggest that the impacts of climate change in the region could be catastrophic. The area of core habitat available to individual species is projected to decrease rapidly with increasing temperature (fig. 2.3), drastically increasing extinction rates (fig. 2.4) and reducing overall biodiversity in the region. Increasing temperature is predicted to cause significant

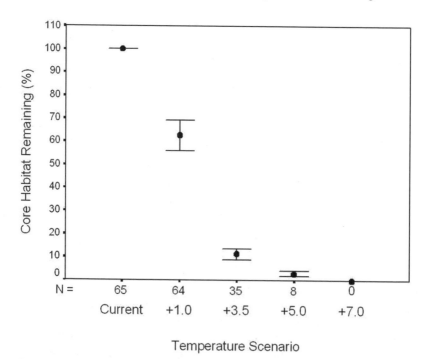

Figure 2.3. Mean core habitat remaining across endemic species with increasing temperature, based on BIOCLIM distribution models of all regionally endemic rainforest species (65 species, approximately 8000 point localities) (Williams et al. 2003)

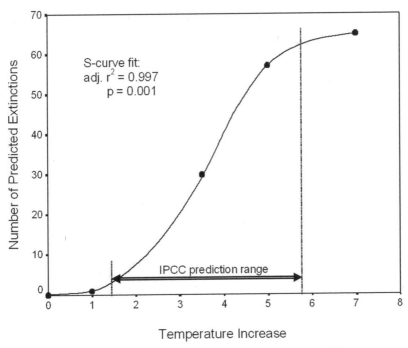

Figure 2.4. Relationship between temperature and number of predicted extinctions (IPCC = Intergovernmental Panel on Climate Change) (Williams et al. 2003)

declines in the size of core environments for all of the endemic vertebrates (Williams et al. 2003). Even a relatively minor increase of 1°C reduces the core environment of each species to about two-thirds of that at present (fig. 2.3). An increase of 3.5°C would cause the complete loss of core habitat for 30 of the 65 species, whereas the remaining 35 species would retain an average of only 11% of their core habitat area. In the Australian Wet Tropics, and in most montane systems worldwide, the dispersal distances involved in moving uphill to cooler habitats are relatively small. However, as montane environments shrink and vanish entirely, the dispersal distances to other suitable habitat will often be too great to allow dispersal.

A study of the potential impact of changing climate on the golden bowerbird (*Prionodura newtoniana*), an endemic, upland species in north Queensland, was undertaken using generalized linear modeling (GLM) with detailed presence/absence data for the species and key climatic variables (mean temperature of the coldest quarter and mean precipitation of the wettest quarter; Hilbert et al. 2004). Predicted habitat for the bird in future scenarios was constrained to fall within the current distribution of rainforest in the region; the accuracy of the model was 67% (Kappa statistic, $n = 30$) for independent validation data.

In the current climate, the total potential habitat for golden bowerbirds is estimated to be 1564 km², scattered across a number of separate patches with distinct subpopulations. This area dramatically declines with increasing temperatures (fig. 2.5). The areas predicted by this GLM model to have suitable habitat were compared with core-habitat areas estimated using another modeling technique, BIOCLIM, for similar warming scenarios (fig. 2.5). Both approaches project an approximately exponential decline in bowerbird habitat with increasing temperatures (which, in part, reflects the rapid decline in total land area with increasing altitude in the Wet Tropics). The BIOCLIM model estimates a considerably larger distribution for golden bowerbirds in today's climate than does the GLM model, partly because BIOCLIM's projected habitat is not restricted to rainforest, unlike that of the GLM model.

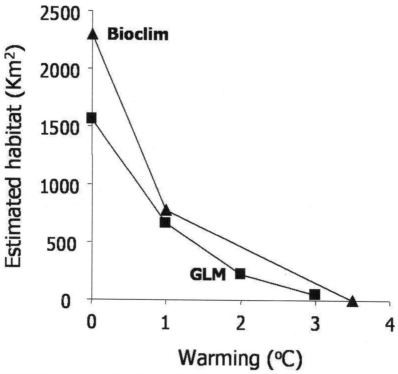

Figure 2.5. Comparison of predicted decline in distribution area with increasing temperature using a bioclimatic model (BIOCLIM; Williams et al. 2003) and a GLM for the golden bowerbird (*Prionodura newtoniana*) (Hilbert et al. 2004)

Nevertheless, the predictions for the two approaches are essentially identical for the two warming scenarios. The impressive similarity between BIOCLIM and the more data-intensive GLM model suggests that the BIOCLIM models are reasonable and, if anything, slightly conservative in their projections.

This example of the golden bowerbird illustrates the general problem for cool-adapted upland species in mountainous tropical regions, and the results are qualitatively typical for many regionally endemic vertebrates in north Queensland. For some species, anthropogenic forcing of climate in this century could completely eliminate their cool, upland rainforest environments.

Relative Species Vulnerability

We can use the relative decline in the core area of each species following 1°C of warming to estimate the relative vulnerability of the endemic vertebrate species to increasing temperature. Identification of the most vulnerable species is important for making informed conservation and management decisions. Twenty-one species exhibited a disproportionately large decrease in their core environment (fig. 2.6). These species (clustered inside the box within fig. 2.6) are all predicted to lose more than 50% of their distribution area with only a 1°C increase in temperature. These 21 species are ranked in order of the expected proportion of their current area that would be lost with a 1°C increase (table 2.1). One montane frog species (*Cophixalus concinnus*) is predicted to have no core environment remaining even with just a 1°C increase. There is no significant taxonomic bias evident among the most vulnerable species, which include 7 species of frogs, 6 reptiles, 5 mammals, and 3 birds (table 2.1). However, two families of ectothermic vertebrates account for more than half of the 21 vulnerable species (5 species of microhylid frogs and 6 species of skinks). All are upland-restricted species with relatively small current ranges (all less than 3000 km^2); however, not all species with small ranges are expected to exhibit such precipitous declines in distribution under warming of 1°C (fig. 2.6).

Predicted Extinctions

If we assume that the complete loss of its core environment will result in the extinction of a species, we can predict the number of likely extinc-

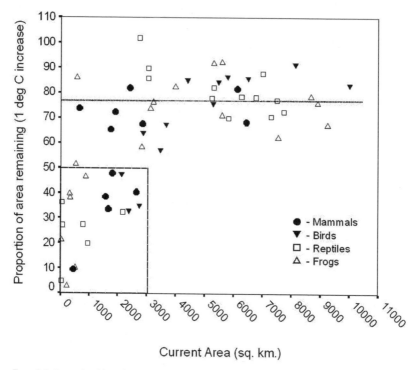

Figure 2.6. Proportional loss of core environment with a 1°C increase in temperature, as a function of the current area of core environment, for all endemic rainforest vertebrates (based on BIOCLIM distribution models). The box indicates the group of species that are most threatened by increasing temperature. The horizontal line indicates the mean loss for those species above the 50% threshold.

tions as a function of increasing temperature (fig. 2.4). Initially, the predicted extinction rate increases slowly as temperature rises, affecting only a few sensitive species. This is followed by rapidly increasing extinctions through the modal region of the curve, with the extinction rate then slowing when only a few hardy species remain. The relationship between increasing temperature and predicted extinctions was best fitted by an S-curve ($R^2 = 0.997$, $P = 0.001$; Williams et al. 2003). This curve suggests that, given the range of likely temperature increases this century suggested by the Intergovernmental Panel on Climate Change (1.4°C–5.8°C), we could potentially see the extinction of between 6% and 96% of the endemic rainforest vertebrates of the Wet Tropics.

It is important to note that extinctions are not expected to increase linearly with rising temperature (fig. 2.4). As a result, the rate of species losses may initially be quite slow and then increase rapidly with further rises in temperature. Although the complete loss of core environment

Stephen E. Williams and David W. Hilbert

Table 2.1 Most vulnerable species, defined as those species that the BIOCLIM modeling predicts will lose greater than 50% of their core environment with a 1°C increase in mean temperatures

	Family	Species	Common
1	Microhylidae	*Cophixalus concinnus*	Thornton Peak nursery frog
2	Microhylidae	*Cophixalus hosmeri*	Pipping nursery frog
3	Scincidae	*Techmarscincus jigurru*	Bartle Frere skink
4	Pseudocheiridae	*Pseudochirulus cinereus*	Daintree River ringtail possum
5	Myobatrachidae	*Mixophyes sp. Nov.*	Northern barred frog
6	Scincidae	*Lampropholis robertsi*	
7	Microhylidae	*Cophixalus neglectus*	Tangerine nursery frog
8	Scincidae	*Eulamprus frerei*	
9	Scincidae	*Saproscincus czechurai*	Czechura's litter skink
10	Scincidae	*Glaphyromorphus mjobergi*	
11	Ptilonorhynchidae	*Prionodura newtoniana*	Golden bowerbird
12	Dasyuridae	*Antechinus godmani*	Atherton antechinus
13	Acanthizidae	*Acanthiza katherina*	Mountain thornbill
14	Scincidae	*Calyptotis thorntonensis*	Thornton Peak skink
15	Microhylidae	*Cophixalus exiguus*	Bloomfield nursery frog
16	Pseudocheiridae	*Hemibelideus lemuroides*	Lemuroid ringtail possum
17	Microhylidae	*Cophixalus monticola*	Mountain top nursery frog
18	Pseudocheiridae	*Pseudochirulus herbertensis*	Herbert River ringtail possum
19	Myobatrachidae	*Taudactylus rheophilus*	Northern tinkerfrog
20	Acanthizidae	*Sericornis keri*	Atherton scrubwren
21	Dasyuridae	*Dasyurus maculatus*	Spotted-tailed quoll

may not cause a species to vanish entirely or immediately, it will certainly make the species extremely vulnerable to subsequent extinction (Lawton 1995), especially if other threatening processes are operating. At worst, the predicted extinction curve will be an underestimate of the extinction rate due to the compounding effects of other anthropogenic impacts and disrupted ecological processes; at best, large numbers of already-rare species could be threatened as their core environments diminish. Even species that retain some core-habitat area may still be "committed to extinction" due to longer-term species–area effects (Thomas et al. 2004).

When assessing extinction risk it is also important to consider the spatial patterns of abundance of individual species, as well as their geographic and elevational ranges. Shoo et al. (2005a, 2005b) incorporated spatial patterns of species abundances into a predictive model and demonstrated that, for three-quarters of the endemic Wet Tropics bird species, as temperature increases their population size is likely to decline more rapidly than their distribution area. Thus, for these species, extinction risk will be more severe than expected from their changing distribution patterns alone. However, not all species fare poorly. Although upland birds and high-elevation endemics such as many microhylid frogs are likely to be threatened by small to moderate increases in temperature, population sizes of some lowland species may increase, at least in the short term, as their area of suitable habitat expands.

Species in the Wet Tropics have survived modest warming episodes during their evolutionary history. For example, temperatures were somewhat warmer than at present during the mid-Holocene (ca. 7000–8000 years ago; Xia et al. 2001). However, the actual increase in mid-Holocene temperature is difficult to determine, and most estimates suggest an increase of only 1°C to 2°C (Nix 1991; Wassen and Claussen 2002). Wassen and Claussen propose that, in the eastern uplands of Australia, the mid-Holocene temperature was only about 1°C higher than at present. Given the likelihood of small thermal refugia in a region like the Wet Tropics with high vegetation diversity and topographic complexity, it seems reasonable to assume that extant species would have survived this small increase in temperature without great difficulty (possibly made even easier by higher rainfall). Another topographic phenomenon that might provide small refugia during minor warming episodes is the *Massenerhebung* (or mountain-mass) effect, whereby upland habitats and species extend their distributions downward to a lower-than-typical altitude (Grubb 1971). The effect generally occurs on small, isolated mountain ranges, particularly those close to the ocean, and is thought to arise when fog and cloud condensation occur at lower altitudes, causing changes to the average air temperatures and soil moisture. Given that the extant species have survived somewhat warmer conditions in the past, and because small thermal refugia are likely to persist in the future, there is reasonable hope that species extinctions will be minimal provided that the temperature increase does not exceed about 2°C. Larger temperate increases, however, are projected to have severe effects on the high-elevation fauna.

Loss of Evolutionary Potential

The complex biogeography and recurring Quaternary contraction and expansion of rainforest in the Wet Tropics has caused many endemic species to have a high degree of phylogeographic structure (Moritz 2002;

Schneider and Williams 2005). Many of the more isolated populations associated with distinct mountain ranges are unique, evolutionarily significant units (Moritz 2002). Therefore, the range contractions predicted from climate change will result in not only species extinctions but also the loss of important genetic diversity and evolutionary potential contained in separate subpopulations. Recent work that examined the potential for adaptation to changing environmental conditions suggested that there may be little or no potential to evolve in habitat-specialist species (Hoffmann et al. 2003). This study also cautioned that genetic diversity was only indicative of evolutionary potential when linked to ecologically relevant traits such as temperature or desiccation tolerance. Genetic diversity of neutral genes was not necessarily a good indicator of adaptive potential.

Direct Effects on Physiology: Temperature

The predicted impacts discussed above rely on the assumption that climatic conditions and the distribution of a particular species or habitat type are significantly linked. This relationship can be driven by a direct physiological mechanism or by an indirect ecological interaction, perhaps because of a direct physiological impact on a prey, competitor, or pathogen species. It has previously been suggested that the limited altitudinal ranges of many species of mammalian folivores are determined by thermal tolerances (Winter 1997; Kanowski et al. 2001, 2003). Direct evidence for such physiological limitations is limited. However, laboratory studies confirm that one arboreal folivore, the Green Ringtail Possum (*Pseudochirops archeri*), is intolerant of high temperatures; its body temperature increases linearly with time when ambient temperature is above 30°C, leading to possibly lethal effects after only 4 to 5 hours (Dr. A. Krockenberger, pers. comm.). Anecdotal reports of unusually high mortality in arboreal folivores following recent periods of record maximum temperatures also suggest that folivores might be temperature limited. Because most models of climate change predict not only rising mean temperatures but also more frequent temperature extremes, it seems likely that the marsupial folivores will be particularly sensitive to predicted climate changes.

Direct Effects on Physiology: High Levels of Carbon Dioxide

Increased carbon dioxide levels have been shown to reduce nutritional value, increase toughness, and increase concentrations of some defense compounds in foliage (Lawler et al. 1997; Kanowski 2001). Such changes could have significant detrimental effects on populations of endemic folivores, including ringtail possums, tree kangaroos, and many insects. Furthermore, predicted upward changes in geographic distribution will shift species from nutrient-rich, basaltic soils and onto increasingly poorer granitic soils at higher elevations (Williams et al. 2003). Rainforests on these poorer soils have already been shown to support lower population densities of arboreal folivores (Kanowski et al. 2001). Thus, climate change could potentially have important effects on the habitat quality and population density of certain species.

Effects of Increasing Climatic Unpredictability

Increasingly unpredictable rainfall may also have a significant negative impact on rainforest fauna (McLaughlin et al. 2002). Variability in rainfall was the most significant factor accounting for regional patterns of bird abundance in the Wet Tropics bioregion, particularly for insectivores and frugivores (Williams 2003). Population density of birds is adversely related to rainfall seasonality, a factor that is predicted to increase under most climate-change scenarios. This correlation may arise because short periods of dry weather limit insect biomass (Frith and Frith 1985) and probably reduce fruit biomass. Increasing dry-season length and greater unpredictability in rainfall would also have significant socioeconomic impacts on human water use in north Queensland.

Effects on Cloud Moisture and Hydrology

"Cloud-stripping" occurs when water condenses onto vegetation from cool, moisture-laden air. In upland forests of the Wet Tropics, cloud-stripping may produce up to 40% of the total water input (Dr. P. Reddell, pers. comm.). A predicted increase in the average basal height of the cloud layer (in concert with rising temperatures) may have serious implications for mountain-top rainforests and their biota. In high-elevation cloud forests of Monteverde, Costa Rica, a lifting cloud bank (Still et al. 1999) reduced critical inputs of mist and water, a phenomenon that was linked with synchronous declines of amphibians and altitudinal shifts in the distribution of birds (Pounds et al. 1999). Evidence of episodic population crashes in terrestrial-breeding rain frogs (*Eleutherodactylus diadstema*) is particularly relevant to the Wet Tropics, where regional patterns of diversity in ecologically similar microhylid species are strongly related to consistent moisture throughout the year (Williams and Hero 2001). A raised orographic cloud base could potentially affect many taxa that require consistently high moisture (e.g., microhylid frogs, litter skinks, soil invertebrates, microbes) and could indirectly affect litter-feeding insectivores (certain birds, skinks, microhylid frogs, dasyurid mammals, bandicoots). It could also alter ecological processes such as nutrient cycling and decomposition.

Fragmentation and Population Connectivity

Habitat fragmentation has previously been shown to increase the extinction proneness of many species within the Wet Tropics bioregion (W. F. Laurance 1991a, 1994). In the context of the analyses presented here, fragmentation has two implications: (1) current patterns of human-caused and natural habitat fragmentation will impede climate-induced shifts in faunal distributions, preventing species from colonizing potentially suitable habitats, and (2) as species' ranges contract farther up mountain ranges, their distributions will become more fragmented, thereby increasing extinction proneness of local populations and exacerbating the impacts of reduced range size. Greater distances between isolated populations and a subsequent decrease in metapopulation connectivity will reduce gene flow, rescue effects, and recolonization, thereby increasing local population extinctions.

Effects on Biotic Interactions

Changes in the abiotic environment, habitat structure, species distributions and abundances, and overall declines in biodiversity will undoubtedly alter existing ecological interactions at all trophic levels. It has been demonstrated that relatively small changes in the phenology of biological interactions can have huge impacts on the species involved (L. Hughes 2000; Parmesan and Yohe 2003). Interactions between disease dynamics and global warming have recently been suggested as a key trigger of extinctions in Neotropical amphibians (Pounds et al. 2006). However, the inherent complexity of such interactions makes the secondary impacts of climate change very difficult to predict, especially in complex tropical ecosystems.

Ecosystem Resilience

Ultimately, the impacts of global climate change will depend on two factors: the magnitude of environmental change and the resilience of the ecosystem in question. (Resilience refers to the ability of a system to withstand or recover from perturbation and is a key management concept in dealing with an unpredictable future; see B. Walker et al. 2002.) The causes of anthropogenic climate change must be addressed at both global and governmental levels by reducing greenhouse gas emissions. Ecological resilience, however, can be addressed locally and immediately. To maximize ecosystem resilience, it is imperative that we maintain key ecological processes and minimize any action that may threaten the ecosystem, such as loss of biodiversity, habitat fragmentation, feral animals, weeds, and diseases. Conservation management can increase the probability of ecosystem survival and recovery by focusing on the maintenance of ecosystem resilience (T. P. Hughes et al. 2003).

Future Research Priorities

Understanding patterns and processes associated with climate-change impacts will only be possible with an integrated, multidisciplinary approach (fig. 2.7). This approach should involve a combination of regional-scale field data, manipulative experiments of environmental variables, laboratory-based physiological experiments, spatial modeling of ecological patterns and processes, and predictive modeling of the impacts of a changing climate on taxa. Key elements should include the following:

- collection of data on species distributions across environmental gradients (particularly altitude) combined with high-resolution spatial modeling;
- long-term monitoring of suitable taxonomic groups to detect early changes and to trigger adaptive management strategies;
- analyses of environmental and habitat determinants of species abundances based on field data and existing databases on vertebrates, invertebrates, and plants;
- molecular analyses to assess patterns of genetic diversity, evolutionary significant units, and phylogeographic patterns for key taxa;
- physiological experiments on key species to assess their tolerances to heat and moisture and to test the efficacy of stress proteins as a monitoring tool;

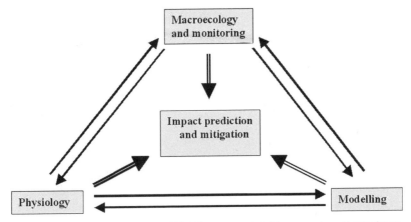

Figure 2.7. Schematic of an integrative, multidisciplinary approach to future research on climatic change and its effects on biotas

- predictive spatial modeling that integrates data on species distributions, habitat associations, and environmental variables;
- field-based and manipulative experiments to improve our understanding of mechanisms underlying the predictive models; and
- feedback loops between the experimental and observational data and modeling efforts to continually refine predictions.

CONCLUSIONS AND IMPLICATIONS

1. Mountain ecosystems around the world, like those in the Australian Wet Tropics bioregion, are often biologically diverse and sustain many restricted endemic species. Despite a high standard of legal protection for the Wet Tropics World Heritage Area, most of its endemic vertebrates could be severely threatened by projected climate changes this century.

2. For Wet Tropics vertebrates, predicted extinction rates are expected to rise nonlinearly with increasing temperature, accelerating rapidly if temperatures rise by more than 2°C.

3. Even if a species shows a greater capacity to adapt to climate change than is suggested by bioclimatic models, it still could be severely threatened by a large reduction of its core environments, which will render it susceptible to other threatening processes.

4. Montane ecosystems in the tropics could be exceptionally vulnerable to projected changes in global temperature, which are expected to

shrink and fragment available habitat for specialized high-elevation species, reduce moisture inputs to forests from cloud-stripping, cause direct physiological stresses for montane species, and alter many ecological and ecosystem processes.

ACKNOWLEDGMENTS

We thank Birds Australia, Queensland Parks and Wildlife Service, Y. Williams, L. Shoo, R. Henriod, J. Winter, and many other individuals for sharing their data, and Steve Turton, Chris Johnson, and William Laurance for comments on the manuscript. This research was supported by the Cooperative Research Centre for Tropical Rainforest Ecology, Australian Research Council, James Cook University, and Commonwealth Scientific and Industrial Research Organisation. Participation by SEW in a National Center for Ecology and Synthesis working group on spatial distribution modeling contributed to the development of this research.

CHAPTER 3

Dynamic Climate and Land-Use Change in the Woodland and Savanna Ecosystems of Subtropical Africa

H. H. Shugart, K. K. Caylor, C. Hély, R. J. Swap, and P. R. Dowty

We are living at a time of exceptional levels of species extinctions and ecosystem change. The actions of technologically advanced, industrial human societies have created novel and alarming perturbations that are outside the realm of conditions experienced in the prior history of most terrestrial ecosystems. Humans transport and introduce new species, modify the patterns of disturbance and baseline conditions, alter the structure and composition of landscapes, and may even contribute to climate change at the regional, if not global, scale (Avissar et al., chap. 4 in this volume). The effect of high-population-density industrial societies on the environment and the biota has been profound. These factors all affect the dynamics of ecosystems and indicate a need to understand ecosystem change as a prerequisite to effective management.

The magnitude of regional effects expected to arise from climatic changes associated with global greenhouse warming in savannas and woodlands is second only to that of arboreal forests (e.g., T. M. Smith et al. 1995). On each continent, the portion of land covered by savannas and woodlands varies with the particular climate but can be as much as 75% of the total land area. Unfortunately, the projected ecosystem effects of climate change—and the degree of climate change itself—are unusually uncertain because savanna and woodland ecosystems are located in regions where even the direction of the projected changes differs among climate models and predictions. These differences arise because of the importance of the moisture regime in these semi-arid regions. The potential for large climatic change attended by high uncertainty in regions that are already substantially influenced by human land uses constitutes an emerging threat of great magnitude. Because woodlands and savannas encompass some of the world's poorest nations, the technological capacity to deal with these changes is often underdeveloped. Certainly, this is the case for Africa.

Africa is unique in that a substantial portion of its large animal fauna still persists from the Pleistocene. The idea that animals can sometimes strongly modify their environments is a important ecological concept that is borne out by differences in grazing that cause sharp differences in the vegetation on different sides of fences (Todd and Hoffman 1999) and in study plots from which animals have been excluded (Kriebitzsch et al. 2000). In this chapter, we present initial model-based investigations of the role of large-animal herbivory as a disturbance mechanism at a spatial scale sufficiently large enough to interact with climate for the southern African landscape. The implications of this study involve potential risks for biodiversity reserves (and the sustainability of cattle grazing) and the potential for atmosphere–surface feedback perpetuating ecosystem change.

Animal interactions with vegetation, mediated by climate, have been reported in several different ecosystems, including boreal forests (Post et al. 1999), tropical grasslands (J. R. Brown and Carter 1998), desert and arid ecosystems (Schlesinger et al. 1990; Oba et al. 2000a, 2000b), and savannas (Dublin et al. 1990). Most of these studies are at a relatively small scale. There also has been considerable speculation about the effects of the massive extinction of large mammals (the "megafauna") that occurred near the end of the Pleistocene about 12,000 years ago, including the effect of these extinctions on the vegetation. For example, recent analyses of changes in the vegetation of Europe (Bradshaw and Mitchell 1999) over the past 13,000 years have associated vegetation patterns with the extinction of large mammalian herbivores. An investigation of the regional effects of animals in African landscapes has direct implications for understanding baseline ecosystem functioning prior to megafauna extinctions in other parts of the world.

STABILITY IN SAVANNA AND WOODLAND ECOSYSTEMS

Savannas and woodlands are second only to tropical forests in their contribution to Earth's terrestrial primary production (Atjay et al. 1987). Savannas are extensive, seemingly stable mixtures of grasses and trees that cover 20% of Earth's surface (and 40% of Africa). They represent an ecological paradox in their shared dominance of plants of radically different life forms. Why is a single life form not "best" for a given environmental condition? Savanna ecosystems vary systematically along gradients of available moisture and soil nutrients (Scholes and Walker 1993). They are profoundly influenced by wildfire, by grazing animals (fig. 3.1), and by humans modifying fire and grazing, as well as by prevailing land uses (B. H. Walker 1983). Savannas are not narrow transition vegetation between conditions that favor grasses and those that favor trees. Rather, they occupy a substantial part of the natural gradients in soil moisture and nutrients.

Walter (1971) hypothesized that, for savannas, a stable mixture of grass and trees resulted from the trees being more deeply rooted than the grasses. This implies that the grasses derive the first use of precipitation and that the trees survive by using water that soaks into the deeper soils. This hypothesis has been explored by digging root profiles, removing trees and grasses (Sala et al. 1989), or simulating two-layer evapotranspiration with models (Eagleson and Segarra 1985). Early in these investigations, the possibility arose, as a consequence of the emerging theoretical understanding of savanna dynamics, that savannas might have more than one stable state (or "phase") for a given environmental condition. It is this multiphase aspect of savanna dynamics in which we are particularly interested (fig. 3.2).

B. H. Walker and Noy-Meir (1982) investigated Walter's (1971) hypothesis on water-resource partitioning between grasses and trees in conjunction with grazing. An important outcome was theoretical grounds to expect trees and grasses that are simultaneously competing for water and under the influence of grazing to have two stable equilibria for the same environmental conditions (B. H. Walker 1983). The feedbacks among nutrients, fire, and grazing should reinforce a biphasic stable-equilibrium system (fig. 3.3) as follows: low nutrients → low herbivore digestibility → higher fuel loads → intense fires → more export of nutrients during fires → low nutrients; versus higher nutrients → higher herbivore digestibility → higher grazing and decomposition rates → lower fuel loads → milder fires → reduced export of nutrients.

Similar feedback can arise between chemical compounds released by plants and an associated alteration of nighttime temperatures. Many compounds emitted by plants (terpenes and phenolic compounds, for example) function as "greenhouse gases," altering local growing conditions for the plants. In drier environments, low inversion layers form at night and trap these compounds close to the surface. This reduces the radiative heat flux to the surrounding air and preserves warmth close to

Figure 3.1. Two modes of recycling nutrients in African ecosystems. In low-nutrient ecosystems, the low-quality forage is not grazed but burns and is dispersed in the atmosphere (*top*). In higher-nutrient ecosystems, animal grazing locally recycles nutrients (*bottom*).

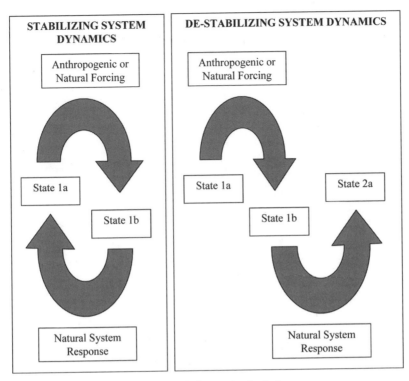

Figure 3.2. Stabilizing and destabilizing dynamics in ecosystem feedback

the surface. Devegetated deserts, however, lack the plant compounds and have plunging nighttime temperatures compared to equivalent vegetated deserts (see Hayden 1998). This feedback implies a potential for vegetation to lengthen the growing season and the reciprocal tendency for an absence of vegetation to shorten the growing season.

These explorations tell us much about the dynamic behavior of a large portion of Earth's surface. Along with such fundamental considerations is an associated practical issue: if savannas have interactions that can produce multiple stable equilibria, then human use and abuse of these systems can potentially shift the ecosystems from one stable-equilibrium condition to another. For example, overgrazing of a savanna can produce a degraded system with feedbacks that resist ecosystem restoration, even when grazing pressure is removed, a finding that has important implications for land-use management.

The notion of multiple stable equilibria is an important concept in savannas; it suggests that "hysteretic dynamics" can arise in ecosystems

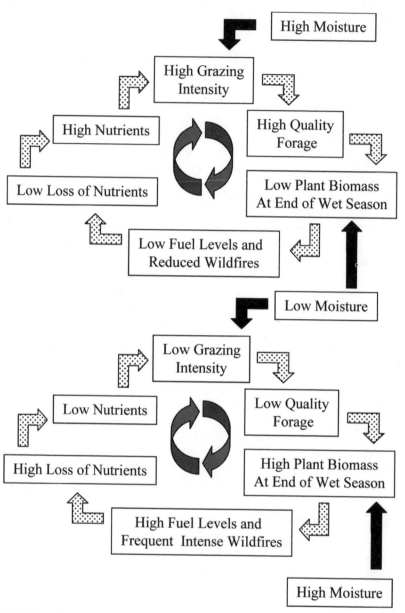

Figure 3.3. Feedbacks among moisture, nutrients, wildfire, and grazing in African savanna and woodland ecosystems

that face altered environmental conditions. Hysteresis is a special type of nonlinear system dynamics in which a system with two equilibrium states under the influence of an external variable shifts abruptly from seeking one equilibrium to seeking the other. For ecosystems, this means that, for a given range of environmental conditions, a particular system could be in one of two different states, depending on its prior history. Once the system is in one state it is difficult to convert it to the other. This possibility of abrupt and nearly irreversible changes produces perplexing management problems. For example, a woodland influenced by overgrazing and a changing climate might suddenly undergo conversion to a savanna. If the conditions were ameliorated, the transition back to woodland would not occur until the climate and grazing regimes were much more favorable than before.

Hysteretic dynamics indicates that ecosystems may change in important ways during a warming (or drying) climate but would not revert to their original condition in a cooling (or wetting) climate. (See Shugart et al. 1980 for a general discussion of the dynamics of hysteretic or multiple-state ecosystems under different directions of climatic change.) Clearly, this has important implications for the management of ecosystems and for the theoretical understanding of ecosystem change.

In southern Africa, one would expect the interaction of wildfire and animal grazing to be a significant force in structuring ecosystems. The interactions among fire, grazing, and nutrient import and export from ecosystems have the potential to promote a stable two-phase landscape (fig. 3.3). Climatic variation, particularly moisture, can have a complicating effect on this hypothesized condition. For example, dry conditions, because they reduce plant growth, may promote fewer wildfires. However, dry conditions also decrease the availability of water, potentially reducing the number of animal grazers and thereby increasing fuel loads and the numbers of fires (fig. 3.3).

Bantu-speaking people have encompassed much of the subtropical zone of southern Africa and their culture is structured around cattle herding. More recent European and African colonizers have similar practices in these regions. Cattle herding is limited by land-holding patterns, by arid locations with inadequate grass, and by the distribution of tsetse flies, which are important disease vectors and pests to cattle. In our initial evaluations of the effects of herbivory, we used a large-scale cattle-grazing database to distribute grazing pressure across southern Africa in conjunction with a regional plant-productivity simulator. In the next section we outline the functioning of the productivity simulator (described in detail by Dowty [1999] and Hély et al. [2003]).

A MODEL OF TREE AND GRASS PRODUCTIVITY ACROSS
SOUTHERN AFRICA

We developed a model of savanna productivity to simulate the seasonal pattern of production of living and dead woody-plant and grass biomass across Africa, south of the equator. We use this model to produce a sub-continental estimate of forage for herbivores. One challenge in simulating productivity across this region is the mixed-life-form, spatially heterogeneous vegetation found in southern African savannas (Huntley and Walker 1982; Bourlière 1983; Scholes and Walker 1993). Because available fuel load is the balance between spatially and temporally explicit processes, such as net primary production (NPP), litterfall, litter decay, and herbivory, a mechanistic modeling approach is necessary.

One class of models, termed production-efficiency models (PEMs), has been useful for predicting NPP at continental to global scales (Prince et al. 1995; Woodward et al. 1995; Ruimy et al. 1996). Because savannas and woodlands are mixtures of woody plants and grasses, usually with different photosynthetic pathways (C3 metabolism for the woody plants and C4 metabolism for grasses), the proportions of the tree/grass mixture can influence NPP. Woody material and grass also burn differently so that the separation of savanna fuel load into tree and grass components also precludes the use of PEMs (or "big-leaf" models) that do not separately resolve productivity of trees and grasses. However, the need to capture sub-continental-scale patterns of fuel load reduces the utility of patch-scale models (e.g., TREEGRASS developed by Simioni et al. [2000]).

We developed a spatially explicit, regional-fuel-load model based on a patch-scale PEM that can be scaled up to the regional level using empirical relationships between patch-scale behavior and multisource remote-sensing data (quantifying spatiotemporal variability of vegetation, radiation, and climatic variables). Model parameters are based on data from existing field sites (Dowty 1999) in the Kalahari region of southern Africa. These sites are distributed across a strong natural rainfall gradient, from the moist tropics in the north to the arid subtropics at the southern extreme. Consequently, the sites exhibit an equally strong gradient in vegetation, which ranges from closed woodland to open shrub savanna. Compared to previous models (where all available observations were used in the calibration and no field data were available to validate them; Potter et al. 1993; Ruimy et al. 1996), the present fuel-load model has been calibrated from measurements recorded in 1996 along the Kalahari transect and verified from independent site measurements recorded during the SAFARI 2000 dry-season field campaign (Swap et al. 2002) and other past campaigns.

Our PEM runs on 24 15-day intervals, from 1 September to 31 August of the following year. The model is driven primarily by the absorption of photosynthetically active radiation and light-use efficiency. These factors determine gross primary production (GPP, as carbon per square meter accumulated over the year, which is converted to biomass using a constant fractional carbon ratio of 0.45; Scholes and Walker 1993). For each time step, the GPP is partitioned into tree and grass components using the leaf-area ratio of trees and grasses. The grass GPP is subsequently reduced to NPP by incorporating respiration costs. Tree NPP, corresponding to nonleaf material, is also calculated. Leaf fall is determined from a leaf-stress ratio that compares the potential evapotranspiration (PET) to the accumulated precipitation over the time step. If PET is greater than precipitation, tree leaves and grass die in proportion to the stress ratio. Leaf fall increases the magnitude of the dead-tree and grass-leaf components of fuel load. Live-green and dead-grass fuel types are also affected by herbivory, which preferentially reduces live grass over dead grass. Fuel load is resolved for each time step as dead-tree litter, dead grass, live grass, and small-diameter twigs. Live-tree leaves are not considered fuel in most fire models. Twig litter is estimated empirically from the percentage of tree cover (Hansen et al. 2000) using a relationship derived from SAFARI-92 data (Shea et al. 1996; Trollope et al. 1996), SAFARI 2000 field data (Alleaume et al. 2006), and additional nonrelated field campaigns (R. J. Scholes, pers. comm.). See appendix 3A for model parameterization for these results.

Our model produces different productivities of grass and woody plants, and accumulates fuel load of dead woody and grassy debris, for evaluating the intensity of wildfires and the materials they release. The model output is in the form of a 2-week-interval dynamic map of the production of live and dead plant material (either grass or woody vegetation) at kilometer-scale resolution for all of Africa south of the equator.

RESULTS AND DISCUSSION

Modeling Effects of Climate, Grazing, and Fire

What environmental factors control plant-production processes over our extensive study area in southern Africa? In sensitivity experiments (Hély et al. 2003), precipitation has the dominant influence on productivity. Higher precipitation levels sustain longer green-grass production. Higher rainfall also leads to greater delays in grass and tree-leaf death, and reduced litter loads, relative to those produced by lower precipitation. Temperature has a relatively slight influence on productivity, where cooler temperature during or toward the end of the rainy season favors green-grass production. A 5°C to 10°C drop in temperature (from 20°C or 25°C to 15°C) may delay tree-leaf death up to 3 months, but it does not influence the total litter load to a great degree. Radiation fluctuations during the year do not affect tree-litter productivity but do slightly influence grass productivity. Lower radiation quantities produce lower grass biomass.

Within a region, differences among sites result from varying tree-cover percentages and exhibit patterns consistent with field observations: higher tree-cover percentages lead to lower total grass load (both live and dead), higher total tree-leaf loads, and delayed onset of grass growth. Differences among regions arise from variation in the satellite-derived Normalized Difference Vegetation Index (NDVI), which measures vegetation "greenness" and the variability in fuel load associated with changes in total biomass and structural heterogeneity.

In arid regions of southern Africa, NPP does not become significant until December, a time that corresponds to a cumulated precipitation of 140 mm. In moister regions, growth starts as early as October, when only 30 to 50 mm of rain has fallen. As shown by the sensitivity analysis, accumulated annual fuel load across the sites is closely linked to the amount of annual rainfall.

Simulated fuel loads at different locations and independent field data both show that fuel-load variability (in terms of composition and amount) is very high across southern Africa (Hély et al. 2003, figs. 9 and 10). Temporally, between the beginning of the fire season (May to July, depending on the termination of rains) and the end of the simulation year in August, the most important change is the death of grass, which increases dead-grass fuels, whereas litter does not change noticeably. Fuel-load field data show the same degree of variability and similar ranges as that of model predictions.

We used our model to simulate the effect of cattle grazing on the vegetation biomass of southern Africa. This was done using data on environmental conditions for both a very wet year (1999–2000 growing season) and a very dry year (1991–1992 growing season). In dry years, large areas that normally carry relatively high cattle densities can have nearly 100% removal of biomass. This implies a considerable potential loss of cattle and an associated decline in human well-being across the region. In the same dry years, the loss of forage would be close to 100%, even if grazing intensities were reduced by half.

According to our model, in a dry year, Africa south of the equator would show increased areas of devegetated land across broad regions. The regions that are not affected would have higher fuel loads for fires in the dry season and increased export of nutrients. The model identifies areas of high short-term risk from climate variation (food shortages) as well as those subject to potential degradation from fires as a result of longer-term climate variations. The implication of these results is that the interaction of grazing (which reduces fuel loads for wildfires and increases local recycling of nutrients, particularly nitrogen) and wildfire is large. The simulated effects of herbivore grazing lead to a substantial reduction in wildfire fuel load across large regions of southern Africa. This underlying ecosystem-level mechanism could maintain multiple stable equilibria across southern African landscapes.

Vegetation and Climate Change

Many readers will be familiar with the models that indicate the potential for an altered climate with extensive forest clearing in the Amazon basin (Avissar et al., chap. 4 in this volume). Just as model experiments in Amazonia have indicated significant feedback between forests and climate, vegetation alteration in other regions can also have potentially disruptive effects on climate.

The conversion of sparsely vegetated, semi-arid shrubland to desert represents a substantial change in the terrestrial surface vis-à-vis the atmosphere. Xue and Shukla (1993) used the SiB model of Sellers (1985, 1987; also Sellers et al. 1986) to investigate the effects of drought on desertification in the sub-Saharan zone, the Sahel. They conducted several model experiments that resulted in reduced rainfall to the Sahel when it was converted to desert. The onset of the rainy season was delayed by 1 month and the zone of maximum rainfall in the Sahel was shifted toward the equator. The net effect was to alter the Sahel toward more desertlike conditions. Schlesinger et al. (1990) noted that internal ecosystem feedback also tended to reinforce the desertification process.

These findings have two implications for southern Africa. First, ecosystems in the region, which might have more than one stable state, respond to climate variation in the current envelope of climatic conditions. If regional or global climatic change were to cause a greater frequency of extreme growing seasons, one would expect potentially nonreversible changes in vegetation over large areas of the continent. Based on models of interactions between the land surface and the atmosphere, the observable changes in vegetation could potentially alter the regional climate. One might expect desertification and overall drying under plausible climate-change scenarios.

Second, our study implies that woodland and savanna ecosystems cannot be easily separated from the other ecosystems that encompass southern Africa. There has been a concerted effort to preserve landscapes and biodiversity in the southern-African region, and some of the world's largest and oldest parks are found there. In many African nations, ecotourism and the associated foreign exchange from tourists have made conservation of their remarkable fauna and flora a priority. However, climatic feedback from altered landscapes and climate change do not recognize park fence-lines. Because of potentially important linkages between vegetation and regional climate, large-scale alteration of southern African woodlands and savannas may affect an array of other ecosystems in the region.

CONCLUSIONS AND IMPLICATIONS

1. According to our model-based results, vegetation productivity can be simulated over the southern-African subcontinent with patterns that resemble observed fuel loads in different locations and different times. Vegetation productivity in southern Africa is strongly dominated by rainfall, and substantial year-to-year rainfall variation is the norm.

2. Animal herbivory can significantly reduce fuel loads and interacts with wildfire to influence patterns of nutrient recycling and transport. Feedback between fire and herbivory has the potential to maintain multiple stable equilibria across the region.

3. A key concept from our work, in terms of understanding threats to forest ecosystems, is the inherent difficulty in separating one landscape system from another. Simulated vegetation changes in southern Africa between wet and dry years indicate considerable and often correlated changes in vegetation-surface features. These effects are concentrated in semi-arid regions that have heavy cattle grazing, but the climatic changes that they imply are regional in their impacts. In this sense, threats to the African grasslands are threats to the forests as well.

APPENDIX 3A

Model Input Variables

Vegetation. The spatial variability of savanna vegetation over the southern-African region is represented by the University of Maryland one-square-kilometer percent tree cover product (Hansen et al. 2000). The temporal variability of vegetation during the 1999–2000 growing season is captured by using 8 km resolution 15-day NDVI product processed by the NASA Global Inventory Monitoring and Modeling Studies (GIMMS) group (Los et al. 1994).

Radiation. Absorbed photosynthetically active radiation is used by the PEM to estimate GPP. Summed 15-day net downward surface shortwave radiation at $2° \times 2.5°$ resolution is taken from version 1 data of the Goddard Earth Observing System atmospheric general circulation model (Schubert et al. 1993). Because data are only available from 1980 to 1993, we used the mean monthly values over the entire time period to be representative of a mean year, and we applied them to the 1999–2000 period.

Meteorological Variables. Monthly mean temperature and monthly cumulative precipitation are needed to compute the light-use efficiency and PET. These monthly meteorological variables were computed from 225 weather stations for the 1999–2000 period using a preformatted subset of climate station data extracted from the National Climatic Data Center (NCDC) Global Surface Summary of Day Data, version 6. This NCDC subset is included in the SAFARI 2000 data CD-ROM (Privette et al. 2001). These monthly data were interpolated over southern Africa (0.5° pixel size resolution) using an inverse distance weighting method.

Grazing Uptake. Herbivory is an important factor that can reduce the potential fuel load of dead and green grass by anywhere from 15% to 80% in unusually productive and nutritious ecosystems (van Wilgen and Scholes 1997; Scholes 1998). To account for the effect of large herbivores (cattle and wildlife) and their spatial distribution, a database from Peter de Leeuw (International Livestock Center for Africa, pers. comm., 1999) was used. The model only takes herbivory for the grass layer into account since data concerning browsing are sparse and the effect of browsing is believed to be negligible (Scholes and Walker 1993). The livestock unit (LSU) was set to a constant 300 kg compared to the 150 kg unit used by Scholes et al. (1996).

Herbivory and Fuel-Load Allocations. In the model, grass (both green and dead) is affected by grazing where herbivory data are available. For a given location, the amount of grass depleted by grazing is assumed to be evenly distributed between months and depends on the number of LSUs, with the conversion of different herbivore types into a single standard LSU. Grazed material is first removed from the green grass. If the predicted grazing needs are unsatisfied, the remainder is removed from the dead-grass compartment. Calculation of the total amount of forage required by herbivores (F_{TOTAL}; g $C/m^2/15$-days) is based on the total energy demand by herbivores (E_{TOTAL}) over a 15-day period (MJ/m²/15-days), defined as (Scholes et al. 1996)

$$E_{TOTAL} = 0.4 \times (M_{LSU})^{0.84} \times N_{LSU} \times 15,$$

where M_{LSU} is the mean mass of an LSU (300×103 g) and N_{LSU} is the number of LSUs per square meter. The total forage required then depends on the grass digestibility, the grass energy content, and the grass carbon fraction according to

$$F_{TOTAL} = \frac{Cfrac_{GRASS} \times E_{TOTAL}}{G_{DIGEST} \times G_{ENERGY}},$$

where the grass carbon fraction ($Cfrac_{GRASS}$) is set to 0.45, the grass digestibility (G_{DIGEST}) to 60%, and the grass energy content (G_{ENERGY}) to 0.018 MJ/g (Scholes et al. 1996). If the green-grass amount is less than the total forage required, herbivores will consume all the green grass and complete their requirement by consuming some dead grass (for which digestibility, energy content, and grass carbon fraction are assumed to be the same as that for the green grass).

ACKNOWLEDGMENTS

This chapter was part of the SAFARI 2000 Southern African Regional Science Initiative. This research was supported by the Southern Africa Validation of NASA Earth Observing System (EOS) program (SAVE-NASA-NAG5-7266) and the Interdisciplinary Science program (IDS-NASA-NAG5-9357). We thank Compton J. Tucker and GIMMS for the NDVI data series, the Data Assimilation Office at the Goddard Space Flight Center's Distributed Active Archive Center for the climatic data, and William Laurance and two anonymous referees for helpful comments on the manuscript.

CHAPTER 4

Impacts of Tropical Deforestation on Regional and Global Hydroclimatology

Roni Avissar, Renato Ramos da Silva, and David Werth

Tropical forests are being cleared, fragmented, and degraded at alarming rates. Forests interact with the atmosphere at various spatial and temporal scales, and the large-scale destruction of forests is expected to have significant impacts on local, regional, and even global climatic phenomena. Here we summarize some of the known and hypothesized effects of tropical deforestation on hydroclimatology, beginning with local-scale interactions and then moving on to regional and global-scale effects.

Local interactions between vegetation cover and the atmosphere can generate "vegetation breezes," which arise as a consequence of physical differences between contrasting cover types (fig. 4.1). Humans clear natural vegetation to create different land-cover types—such as irrigated fields in arid areas, or pastures in deforested areas—that have a different capability than the original vegetation to redistribute radiative energy from sunlight absorbed at the Earth's surface into sensible and latent heat. A land cover capable of "pumping" soil moisture deep from the ground (such as deep-rooted trees) will typically transpire much more than one with shallow roots (such as pasture), which can only use the water stored in the upper soil. When evapotranspiration by plants is limited, the radiative energy absorbed by the land surface is mostly used to heat the atmosphere immediately above it. For this reason, the atmosphere above a well-transpiring forest is relatively cool and moist, whereas that above a dry pasture is relatively warm and dry. This atmospheric pressure gradient resulting from the temperature difference between

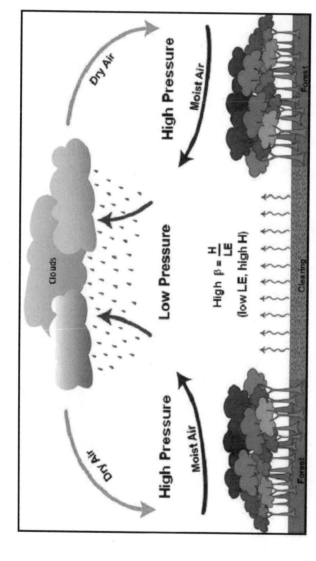

Figure 4.1. Schematic representation of the vegetation breeze (modified from W. F. Laurance, chap. 5 in this volume)

contrasting land-cover types triggers air motion near the land surface, where the air moves from the relatively cool area (which has higher air pressure) toward the relatively warm area (which has lower air pressure).

Using a state-of-the-science atmospheric model, Avissar and Schmidt (1998) found that land-cover patchiness at a scale as small as 3 to 5 km was large enough to trigger such vegetation breezes. In addition, Avissar and Liu (1996) concluded that, in the moist environment typical of tropical regions, vegetation breezes resulting from deforestation could trigger thunderstorms.

Until recently, only anecdotal evidence, mostly provided by the footprints of clouds observed in satellite imagery (e.g., Rabin et al. 1990; Cutrim et al. 1995) and rainfall analyses (Barnston and Schikedanz 1984; Otterman et al. 1990; Moore and Rojstaczer 2001), supported these modeling studies. However, new opportunities to examine vegetation breezes recently became available with the release of data sets collected during regional field campaigns (such as the Atmospheric Radiation Measurement Cloud and Radiation Testbed [Stokes and Schwartz 1994] and the Large-Scale Biosphere–Atmosphere Experiment in Amazonia [LBA; Avissar and Nobre 2002]). Here we focus on tropical forests using recent results from the LBA campaign in Amazonia. The impact of vegetation breezes at midlatitudes is summarized by Weaver and Avissar (2001) and Georgescu et al. (2003).

The effect of above-average temperatures in the eastern and central Pacific Ocean, known as El Niño, has been shown to have a major impact on weather very far away from this region (Shabbar et al. 1997; Trenberth et al. 1998). The anomalously warm ocean surface increases the atmospheric moisture and instability and, as a result, provides conditions that favor increased thunderstorm activity. This propels a large amount of energy to the upper part of the troposphere, where it can propagate around the planet through a mechanism known as "teleconnection" (Namias 1978; Wallace and Gutzler 1981; Glantz et al. 1991). Thunderstorms only occur in a relatively small part of the tropics (mostly in continental areas), and thus a change in their spatial patterns is expected to have global consequences because this affects the movement of energy through the atmosphere.

Sea-surface temperature variations in the tropical eastern Pacific Ocean have three unique properties—large magnitude, long persistence, and spatial coherence—that allow this region to influence the atmosphere effectively (Wu and Newel 1998). Tropical regions are undergoing rapid landscape changes, which are believed to alter the frequency and intensity of thunderstorms (e.g., Avissar et al. 2002; Baidya Roy and Avissar 2002;

Ramos da Silva and Avissar 2006; Ramos da Silva et al., in press). Because land-use change in the tropics has the same three attributes as those observed during El Niño events in the eastern Pacific Ocean, it is expected to cause similar teleconnections (Werth and Avissar 2002; Avissar and Werth 2005).

REGIONAL IMPACTS OF DEFORESTATION

The impact of large-scale deforestation on dry-season regional hydrometeorology in southern Amazonia (Rondônia, Brazil) has been explored by Baidya Roy and Avissar (2002). They simulated a large region encompassing most of Rondônia but focused on a $1° \times 1°$ cell of latitude and longitude (about 100×100 km^2) characterized by a wide band of alternating pasture and forest in its southwestern part along a major highway, Highway BR-364, and local roads (fig. 4.2). This distinctive pattern of deforestation and forest

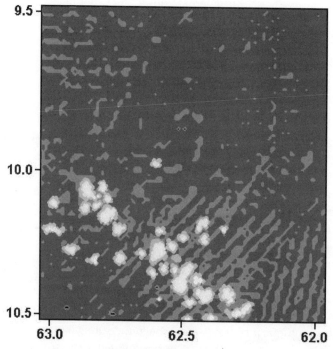

Figure 4.2. Shallow-convection clouds (light gray; derived from the GOES-7 satellite on 17 August 1994) superimposed on a land-cover map of southwestern Rondônia, Brazil (from Calvet et al. 1997). The line of clouds is above cattle pastures (gray) along Highway BR-364 and other local roads. A simulation using the same fishbone pattern of forest clearing (dark gray) produced a similar line of clouds (derived from Baidya Roy and Avissar 2002).

Roni Avissar, Renato Ramos da Silva, and David Werth

fragmentation, which was created by hundreds of government-sponsored colonization projects in the Amazon, is known as the "fishbone pattern" because it resembles the backbone and ribs of a fish.

Vegetation breezes that converge over Highway BR-364 carry moisture that has transpired from the forest, generating shallow-convection clouds. These results are supported by an analysis of images produced by high-resolution satellite imagery (from the Geostationary Earth Observing Satellite 7, or GOES-7). To confirm that these clouds were caused by land-cover heterogeneity arising from local deforestation, Baidya Roy and Avissar (2002) produced three other simulations using exactly the same study area and synoptic meteorological conditions but replaced the observed land cover with (1) a uniform forest, (2) a uniform pasture, and (3) the observed land cover with the topography eliminated (i.e., assuming a flat terrain). Figure 4.3 illustrates the impact of these land-cover types on the intensity and location of vertical wind, which is a good proxy for the intensity of the vegetation breeze. The strong updraft that was generated over Highway BR-364 is clearly eliminated when the fishbone pattern is replaced by either uniform forest or pasture. Topography has only a modest impact on vegetation breezes in this region because the magnitude of the updraft is barely affected by the elimination of topography. Thus, forest conversion in southern Amazonia clearly generates vegetation breezes that can alter local precipitation patterns.

While questions have been raised about the importance of vegetation breezes and their impact on the regional hydroclimate at midlatitudes (see Doran and Zhong 2002, and the response by Weaver and Avissar 2002), their existence during the tropical dry season (winter) is now well accepted (Avissar and Nobre 2002). However, vegetation breezes during the tropical wet season (summer) have not yet been clearly demonstrated. Data collected during the LBA campaign together with state-of-the-science models are being used to address this issue, and the following discussion is based on preliminary results.

A major data-collection effort (the LBA joint-major atmospheric mesoscale campaign in the wet season, and the Tropical Rainfall Measuring Mission validation campaign) was conducted in the 1999 wet season (January–February) in Rondônia. Ramos da Silva and Avissar (2006) and Ramos da Silva et al. (in press) conducted an intensive modeling study in a $1° \times 1°$ domain (fig. 4.4) that was chosen both because of its extensive deforestation and varied terrain and because a weather radar (the S-band dual polarization Doppler radar) was located near its center, thus offering a convenient site for evaluating the performance of simulations in this environment.

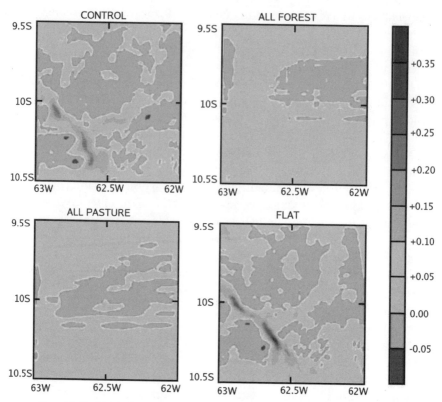

Figure 4.3. *Top left*: Distribution of vertical air motion over the study area in fig. 4.2 (from 17 August 1994 at 3:00 p.m. local time) with observed landscape structure and topography. *Top right*: Model simulation assuming that the entire study area is covered with forest. *Bottom left*: The same simulation assuming that the entire study area is covered with pasture. *Bottom right*: The same simulation with observed landscape structure but no topography. Note the updraft produced by vegetation breezes that converge over Highway BR-364 and other roads in top-left and bottom-right frames. These analyses (derived from Baidya Roy and Avissar 2002) demonstrate that vegetation breezes are created by the land-cover heterogeneity and are affected little by topography.

The simulations duplicate quite well the pattern of clouds and precipitation observed from the GOES-7 satellite and weather radar (fig. 4.5). Some of the rainfall is concentrated near higher-elevation terrain; however, in a simulation of this area with uniform forest cover (fig. 4.4, right panel), most rainfall is concentrated near the eastern boundary of the area, where the terrain is highest (fig. 4.6, right panel). Thus, landscape heterogeneity created by deforestation draws most precipitation to the northern boundary of the area, where convection is strong because of intense deforestation (although a significant pocket of precipitation still occurs near its eastern part [fig. 4.6, left panel]). Interactions between atmospheric flow

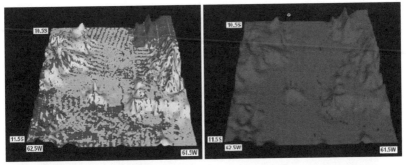

Figure 4.4. Left: Distribution of land cover (pasture in light gray, forest in dark gray) and rendering of topography of the study area in southwestern Rondônia. Right: The same study area before deforestation occurred. (Adapted from Ramos da Silva and Avissar, in press)

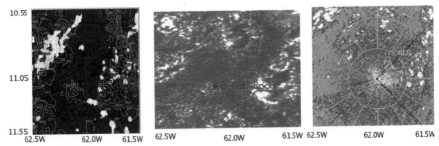

Figure 4.5. Wet-season precipitation in Rondônia (6 February 1999). Left: Simulated rainfall from 3:00 to 4:00 p.m. local time. Middle: Rainfall at 3:00 p.m. local time derived from weather-satellite (GOES) images. Right: Rainfall derived at 3:00 p.m. local time from weather radar (S-POL). (Adapted from Ramos da Silva and Avissar, in press)

forced by topography and strong convective activity over the deforested area cause a shift in the location and intensity of precipitation. It is quite remarkable that the daily mean precipitation averaged over the study area rose by as much as 36% during the wet season as a result of heavy deforestation. In the next section we discuss the significance of this result.

GLOBAL IMPACTS OF DEFORESTATION

Modeling studies performed with general circulation models (GCMs) suggest that the complete deforestation of Amazonia could lead to a significant reduction of the basin's rainfall (e.g., Henderson-Sellers and Gornitz 1984; Dickinson and Henderson-Sellers 1988; Lean and Warrilow 1989;

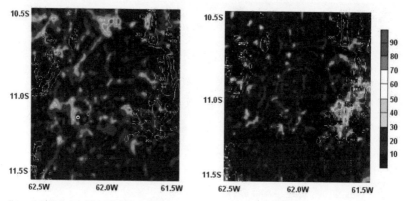

Figure 4.6. Daily rainfall in the wet season in Rondônia (6 February 1999) simulated for the current vegetation cover (*left*) and the original forest (*right*). Precipitation, averaged over the entire study area, rose by 36% in the current degraded vegetation compared to the original forest cover, based on simulations.

Shukla et al. 1990). These studies concluded that global temperature and precipitation were not affected by Amazonian deforestation and, hence, that this process would have only local and regional implications. However, Werth and Avissar (2002; Avissar and Werth 2005) challenged these findings and claimed that large-scale tropical deforestation, via teleconnections that were comparable to those produced during El Niño events, could have a significant impact on rainfall of nontropical regions. In particular, they found that, among continental regions, the U.S. Midwest is most negatively affected by deforestation in Amazonia and Central Africa during spring and summer, when a rainfall decline would severely damage agricultural productivity (fig. 4.7).

Avissar and Werth (2005) used a vegetation map from the 1960s (before heavy deforestation commenced) for their "control" scenario. In three other scenarios, Amazonia, Central Africa, and Southeast Asia were individually deforested and replaced with a mixture of shrubs and grassland. In the fifth case, referred to as the "total deforestation scenario," all three tropical regions were deforested simultaneously. Avissar and Werth assessed the impact of deforestation by calculating, at each grid point of the GCM, the change in monthly mean rainfall (obtained from all years and model runs) relative to the control case.

Figure 4.7 shows the annual cycle and change of monthly mean rainfall at those continental locations significantly affected by tropical deforestation. Although the major impact occurs in and near the deforested regions themselves, a strong impact also propagates along the

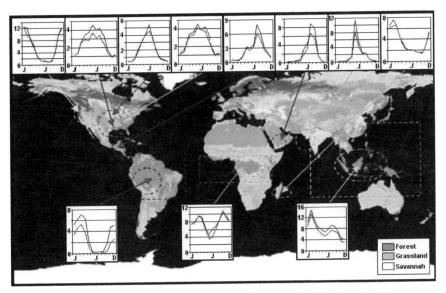

Figure 4.7. Annual cycle of precipitation (mm/day) in continental regions particularly affected by large-scale deforestation of Amazonia (dark gray), Central Africa (gray), and Southeast Asia (light gray). The dark gray curves show simulated mean monthly precipitation before massive tropical deforestation (i.e., the control case), whereas the gray curves show precipitation after deforestation. Ellipses indicate regions where tropical forest (dark gray) was replaced with a mixture of shrubs and grassland. (From Avissar and Werth 2005)

equatorial regions and, to a lesser but still significant extent, to the midlatitudes. Among continental regions, the largest extratropical decrease in precipitation is in the United States, where Amazonian deforestation causes a 10% to 20% decline in spring and summer rainfall. This is especially evident in Texas. In addition, deforestation in Central Africa causes a 5% to 15% decrease of precipitation in the Gulf of Mexico (centered around Louisiana), and deforestation in Southeast Asia reduces winter precipitation in the northwestern United States. However, northwestern India and the Arabian Peninsula receive a considerable increase in summer precipitation as a result of deforestation in Central Africa and Amazonia, respectively.

The total-deforestation scenario reveals that the global impact of deforestation differs from the sum of the three individual regions, which indicates some synergy among the three deforested regions. This is particularly true in eastern Africa and the Arabian Peninsula, which experience enhanced summer precipitation as a result of total tropical deforestation. Also, the reduction of precipitation in the United States appears to be less concentrated than that which occurred when each region was deforested separately.

IMPLICATIONS FOR THE FUTURE

In previous studies we have demonstrated that deforestation of perhaps 20% to 40% of the Amazon forest would lead to enhanced rainfall in the basin (Avissar et al. 2002; Baidya Roy and Avissar 2002; Ramos da Silva and Avissar 2006). This is the likely result of the intensification of thunderstorm activity and water recycling in the basin, which result from increased land-cover heterogeneity in deforested and fragmented land-scapes. On a localized scale, these phenomena might result in a lessening of precipitation over forests, particularly those that are upwind of clearings (see W. F. Laurance, chap. 5 in this volume). If deforestation exceeds the 20% to 40% range, however, a precipitous collapse of basinwide precipitation might be expected (fig. 4.8) as the proportion of land cover dominated by rapidly evapotranspirating forests declines. Although this postulated behavior of regional land–atmosphere interactions remains to be demonstrated, it is a logical linkage between our results—which suggest local increases in precipitation in fragmented regions—and the basinwide GCM studies (e.g., Henderson-Sellers and Gornitz 1984; Dickinson and Henderson-Sellers 1988; Lean and Warrilow 1989; Shukla et al. 1990) that predict major declines in total precipitation with complete Amazonian deforestation. If the relationship between deforestation and total precipitation is nonlinear, as we predict (fig. 4.8), striking increases in the vulnerability of Amazonian forests to droughts and fires would be expected as deforestation approaches some critical threshold

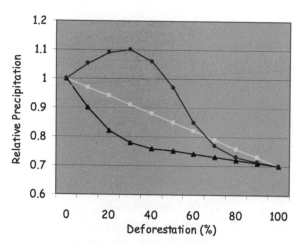

Figure 4.8. Potential impacts of deforestation on relative precipitation in the Amazon basin. The three curves indicate different theoretical models, among many others possible. (From Avissar et al. 2002)

(see W. F. Laurance, chap. 5 in this volume; Barlow and Peres, chap. 13 in this volume). Unfortunately, the nature of this putative threshold is unknown and difficult to predict.

Results from the massive LBA program are currently being used to evaluate and improve our modeling capabilities. As shown in the examples presented here, vegetation breezes and their effects on hydrometeorology affect not only the total area of remaining forest but also the spatial patterns of deforestation and forest fragmentation. Realistic scenarios of forest conversion based on continued development of the Brazilian Amazon, such as those proposed by Nepstad et al. (2001) and W. F. Laurance et al. (2001b; fig. 4.9), should be used in conjunction with carefully tested, state-of-the-science hydrometeorological models to better estimate the potential effects of further deforestation.

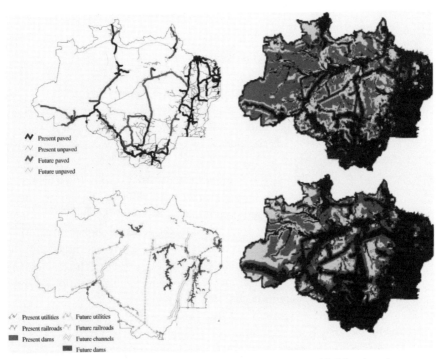

Figure 4.9. Top left: Existing and planned highways in the Brazilian Amazon. *Bottom left*: Other planned infrastructure projects ("utilities" are gas lines and power lines, whereas "channels" are river-channelization projects). *Top right*: Predicted forest degradation by the year 2020 based on an optimistic scenario. *Bottom right*: Predicted forest degradation by 2020 in a nonoptimistic scenario. Black areas are deforested or heavily degraded, including savannas and other nonforested areas. Dark gray areas indicate moderately degraded forest, light gray is lightly degraded forest, and gray is pristine forest. (From W. F. Laurance et al. 2001d)

In addition to such basinwide effects, recent GCM studies (Werth and Avissar 2002; Avissar and Werth 2005) have clearly illustrated the potential for tropical deforestation to affect the global hydroclimate through teleconnections. However, a key element of teleconnections—thunderstorm activity—is neither explicitly simulated nor parameterized in current GCMs, which is an important deficiency of these models. Thus, the major impact of human landscape alteration is not properly represented in these models, and the results obtained thus far likely underestimate the true impacts of tropical deforestation on the global hydroclimate. Despite ever-increasing computing power, simulation of the global hydroclimate at a spatial resolution high enough to represent thunderstorms will not occur in the foreseeable future. Therefore, new parameterizations for Earth-system models that can simulate thunderstorm activity triggered by deforestation-induced landscape heterogeneity are essential for improving our understanding of the consequences of Amazon deforestation on the global hydroclimate.

CONCLUSIONS AND IMPLICATIONS

1. Forest loss and fragmentation create mosaics of land-cover types that have very different physical characteristics. As demonstrated by satellite, weather-radar, and simulation studies, these mosaics generate vegetation breezes, which increase local cloud cover and convective thunderstorm activity.

2. Increased thunderstorm activity can enhance local rainfall and may also have larger-scale effects brought about by altering the movement of energy through the atmosphere. Simulation studies suggest that large-scale tropical deforestation could potentially generate global teleconnections that alter precipitation patterns both in the tropics and in extratropical regions. Significant alterations of rainfall in the tropics, the American Midwest, the Gulf of Mexico, the Arabian Peninsula, and northwestern India are projected in some scenarios.

3. Although Amazonian precipitation is expected to increase at relatively low levels of deforestation, as deforestation increases a sharp decline in precipitation is possible. This decline could result from a nonlinear relationship between relative precipitation and rainfall. The nature of this deforestation threshold is unknown, although recent simulations suggest it could be in the range of roughly 20% to 40% forest loss.

4. These uncertainties highlight the potential risks of large-scale deforestation in Amazonia and other tropical regions. Substantial declines in

precipitation could exacerbate forest vulnerability to droughts and fires, especially in the large areas of Amazonia that already experience relatively strong dry seasons.

ACKNOWLEDGMENTS

This research was supported by the National Aeronautics and Space Administration (under grants NAG5-8213 and NAG5-9359) and by the Gordon and Betty Moore Foundation. We thank William Laurance for detailed editing and two anonymous referees for comments on the manuscript.

PART II

Synergistic Effects of Simultaneous Environmental Changes

Carlos A. Peres and William F. Laurance

INTRODUCTION

As scientists and conservation managers, we are often trained to seek simplicity—to focus on a single environmental change and assess its consequences in painstaking detail. Complexity can be bewildering and incomprehensible; simplicity is equated with insight and elegance. We study the effects of habitat fragmentation, logging, hunting, fires, and climate change as though those environmental insults were divorced from one another—as if the world were neatly divided into quiltlike patches of bright colors, each representing a different kind of environmental change.

Reality is rarely so clear-cut. Few landscapes and species are subjected to just one kind of environmental change; most are reeling from a whole array of anthropogenic alterations, some of which operate in concert. Industrial logging, for example, causes a bevy of important changes to tropical forests: canopy cover is reduced, understory vegetation is disrupted, forest microclimates are altered, disturbance-sensitive species decline, and thin soils are compacted from heavy machinery. Yet these impacts are generally modest compared to the unplanned aftermath of logging—the spontaneous influx of hunters, miners, and slash-and-burn farmers along new logging roads who often destroy forests outright or severely degrade them.

Although highly damaging to forests, the ecological consequences of the logging–forest invasion problem are at least fairly well understood. The effects of other environmental synergisms, such as the interacting effects of global climatic alteration and rampant habitat conversion, are far harder to predict (see part 1). To what degree—and how fast—will climatic and atmospheric conditions change in the future? Will species be trapped as climatic regimes shift latitudinally and elevationally? Will many species fail to track rapidly changing conditions because their geographic ranges have been drastically reduced and fragmented? Could shifts in regional precipitation or increased droughts lead to irreversible declines of tropical forests?

The three chapters in this part focus on combinations of perils that face species and ecological communities in degraded tropical landscapes. These threats result from an amalgam of different anthropogenic disturbances that often exacerbate each other, posing even greater threats to tropical biodiversity. Unfortunately, these synergisms can operate even within nominally protected areas, the last havens for intact ecosystems and the most viable means to preserve representative samples of tropical nature (see part 5).

Habitat Fragmentation, Desiccation, and Wildfires

In chapter 5, Bill Laurance assesses the alarming synergism between habitat fragmentation and fire, synthesizing a diversity of recent studies on recurring wildfires that threaten vast expanses of tropical forest, especially in more-seasonal environments. In fragmented forests, chronically elevated tree mortality near forest edges damages the forest canopy, which leads to increased understory desiccation and an accumulation of fine fuel on the forest floor. In cattle pastures that surround forest fragments in the Amazon, ranchers regularly use fire to control weeds and promote a flush of green grass for cattle, creating an abundance of ignition sources. In drier years, these fires can burn deep into the interior of forest fragments, consuming the dry, accumulated litter on the ground surface.

Evidence from fragmented landscapes of southern and eastern Amazonia suggests that surface fires operate as a large-scale edge effect, penetrating up to several thousand meters into forest interiors. Although slow-burning and unimpressive in appearance, these surface fires are deceptively destructive to rainforest vegetation, which is poorly adapted to survive fire. The occurrence of a single fire event greatly increases the likelihood of subsequent and more intensive fires, because each fire creates more dead fuel than it consumes, creating a positive feedback cycle that frequently causes forest fragments to "implode" over time. In addition, alterations in forest–climate interactions in fragmented landscapes can affect local rainfall and humidity and may also contribute to edge-related fires. Indeed, scientists believe that an unprecedented dynamic of increasingly severe wildfires is being established in many tropical forests—as evidenced, for example, by catastrophic wildfires in Amazonia and Indonesia during the past decade. These fires cause dramatic changes in plant and animal communities and forest ecosystems (see Barlow and Peres, chap. 12 in this volume).

Postfragmentation Disturbances

Fire is not the only peril for species in fragmented landscapes. In chapter 6, Carlos Peres and Fernanda Michalski examine threats to medium- and large-sized vertebrates in fragmented forests, focusing on the degraded frontier of Alta Floresta in the southern Brazilian Amazon. This region is typical of the so-called arc of deforestation in Amazonia, where much forest destruction is concentrated. As a result of Brazilian-government incentives to promote large-scale ranching, in the late 1970s and early 1980s Brazil had one of the world's highest rates of deforestation. Alta Floresta is now a thriving agricultural frontier served by a paved road and boasting eleven head of cattle for every human inhabitant.

Using detailed interviews of local residents, Peres and Michalski surveyed an impressively large number of forest fragments and continuous-forest sites at Alta Floresta. Most forest fragments are surrounded by active cattle pastures and had been disturbed to varying degrees. They found that, although fragment size was an important determinant of vertebrate species richness, much of the remaining variation in species distributions could be explained by past disturbances from surface fires, selective logging, and hunting pressure. For example, persistent hunting in fragments sharply accelerated the local extinctions of large-bodied game species; some disturbance-sensitive vertebrates persisted better in small, undisturbed fragments than in much larger but heavily degraded fragments. These findings clearly reveal that species in fragmented landscapes are battered by interacting suites of threats, depressing the survival of vulnerable species well beyond that predicted by simple species–area relationships.

Human Encroachment in Nature Reserves

How will protected areas in the tropics fare as they are encircled by growing human populations and become ever more isolated from other areas of forest? This is the theme of chapter 7 by William Olupot and Colin Chapman, who describe how Uganda's Bwindi Impenetrable National Park (so named because of its rugged topography) is exploited by villagers densely settled around the park. To assess the effects of human encroachment they conducted a mammoth sampling effort, totaling 287 km of transects across the park.

Uganda is one of the biodiversity jewels of Africa, and Bwindi Park is arguably the most important conservation area in the country because of its exceptional species diversity, which includes the richest faunal community in eastern Africa and a third of the world's dwindling population of mountain gorillas. Forests and wildlife in Uganda have long been subjected to human pressures, but these have greatly intensified over the past half-century because of burgeoning population growth. Expanding at an annual rate of 3.4%, Uganda's population is the fastest growing in the world.

Today, Bwindi Park is assailed by a surrounding human populace that is two to five times denser than the national average in Uganda, exerting a range of different pressures on the park. Olupot and Chapman show that the margins of Bwindi are degraded by elevated tree and sapling mortality from fires and illegal harvesting. As a consequence of such disturbances, exotic plants proliferate near the periphery of the park (see also Benítez-Malvido and Lemus-Albor, chap. 9 in this volume; Baider and Florens, chap. 11 in this volume). Illegal hunting activity, however, is not confined to the park borders but includes occasional forays by poachers right into the heart of the 32,000 ha park. The authors conclude that prior government efforts to foster respect for the park among neighboring communities—while allowing limited resource use, such as harvesting honey from beehives—have largely failed.

Ultimately, the best hope for Bwindi may be the relative inaccessibility of its core, where rugged topography hinders human intrusion to some degree. The same cannot be said for many other protected areas, especially smaller reserves, in Africa and elsewhere in the tropics. Because of the pervasiveness of surrounding threats, the ecologically effective size of many parks for sensitive wildlife and plants may be a mere fraction of their legally gazetted area.

SYNTHESIS

Humans alter ecosystems in many different ways, and for this reason a great diversity of environmental synergisms are possible. The three chapters in this part illustrate some of the most important synergisms that plague fragmented forests or isolated nature reserves. Such synergisms are of special relevance because, at present rates of forest destruction, much of the world's tropical forests will be diminished and fragmented over the coming decades.

Because habitat fragmentation is so ubiquitous worldwide, the study of fragmented ecosystems has grown into a vital subdiscipline of conservation biology. Nonetheless, it is apparent that many fragmentation studies and the conceptual models on which they are based—such as Island Biogeography Theory and Metapopulation Theory—are naïve and oversimplified. This is because they fail to account for a critical reality: in most anthropogenic lands, habitat fragments are not merely reduced and isolated—they are also profoundly affected by further perturbations that can interact additively or synergistically with fragmentation (see W. F. Laurance and Cochrane 2001).

Such synergisms frequently plague ecological communities in forest fragments, as the chapters in this part demonstrate. In Amazonia (chap. 6) and Uganda (chap. 7), intensive harvest pressures and hunting in forest fragments can savage populations of exploited or disturbance-sensitive species and increase local extinction rates, even in the core areas of nature reserves. Moreover, intensive hunting in the modified lands that surround fragments can create a population sink that drains individuals from the fragment, which further increases the likelihood of local extinction.

The synergism between habitat fragmentation and fire is already a severe hazard for tropical forests (chap. 5). The threat is likely to worsen if forest conversion continues apace and if, as some climatic models suggest, global warming provokes more frequent or severe El Niño droughts. Indeed, over the past decade, catastrophic wildfires have arguably become the single most important threat to seasonal tropical forests worldwide. Under certain soil and climatic conditions, such fires can affect even previously undisturbed primary forests (for a neotropical example, see Peres 1999b).

Other environmental synergisms have been only poorly explored. In some areas, pesticides and herbicides, hydrological changes, and livestock grazing degrade fragmented forests. Invasions of exotic species and foreign pathogens are especially likely in fragmented and disturbed habitats and may well increase in importance in the future (see part 3). As discussed earlier, anthropogenic climate change may emerge as an increasingly serious threat to fragmented communities, especially if droughts, storms, and other rare weather events increase in frequency or severity.

Clearly, when evaluating the myriad threats to tropical ecosystems and species, we must explicitly recognize the potential for simultaneous environmental changes to amplify and alter the ecological impacts of one another. We should embrace such complexity rather than ignore it. Some species are being felled by a single stroke, but many more are suffering the death of a thousand cuts.

CHAPTER 5

Fragments and Fire: Alarming Synergisms among Forest Disturbance, Local Climate Change, and Burning in the Amazon

William F. Laurance

In the tropics, as elsewhere, once-intact habitats are being reduced and fragmented at alarming rates. Habitat fragmentation causes myriad changes to biotas and ecosystem functions (W. F. Laurance and Bierregaard 1997; Terborgh et al. 2001), among the most important of which are edge effects—diverse biotic and physical alterations associated with the abrupt, artificial boundaries of habitat remnants (Lovejoy et al. 1986; W. F. Laurance et al. 2002b).

In the past, most edge effects were thought to be confined to the immediate vicinity of forest margins, penetrating from a few meters (e.g., Williams-Linera 1990; Sizer and Tanner 1999) up to a few hundred meters (e.g., W. F. Laurance 1991b; K. S. Brown and Hutchings 1997; W. F. Laurance et al. 1997a, 2000a; Lewis 1998; K. S. Carvalho and Vasconcelos 1999; Didham and Lawton 1999) into forests. Such changes frequently result from physical stresses such as increased desiccation and windshear near forest edges.

An increasing body of evidence, however, suggests that certain edge effects can penetrate much farther into forests (W. F. Laurance 2000). For example, an exotic shrub (*Clidemia hirta*) has invaded at least 2 km into Pasoh Forest Reserve in Peninsular Malaysia, largely as a result of elevated soil damage from hyperabundant wild pigs that thrive by feeding in oil-palm plantations surrounding the reserve (Peters 2001). In Sumatra, large, forest-dependent mammals such as tigers (*Panthera tigris sumatrae*), rhinoceros (*Dicerorhinus sumatrensis*), and elephants (*Elephas maximus*) avoid forest clearings by distances of up to 2 to 3 km (Kinnaird et al. 2003). A large influx of vertebrate seed predators from surrounding degraded lands has led to a virtual cessation of canopy-tree recruitment within at least 10 km of reserve margins in Gunung Palung Park in western Borneo (Curran et al. 1999). In isolated nature reserves, the edge-related mortality of large mammalian carnivores from hunting and human persecution occurs over large spatial scales and can greatly increase their likelihood of local extinction (Woodroffe and Ginsberg 1998; Michalski et al. 2006).

These observations suggest that some edge-related phenomena can penetrate surprisingly large distances into forest interiors. In this chapter, I synthesize studies of another set of interrelated phenomena—edge-related fires, and the ecological and climatic changes in fragmented rain-forests that promote them—and assert that they are becoming among the most destructive of all edge effects in the tropics. These studies reveal a new kind of fire dynamic, one that will almost certainly increase in importance as growing numbers of slash-and-burn farmers, ranchers, and loggers penetrate ever deeper into remote frontier areas of the tropics (Cochrane and Laurance 2002; Cochrane 2003).

My focus is the Amazon, not because edge-related fires are confined to this region, but because much relevant work on fires, fragmentation, and forest climatology has been conducted there. I begin by describing the pattern and pace of Amazonian forest disruption and then relate how interacting environmental changes predispose fragmented forests to fire.

AMAZON FOREST LOSS AND FRAGMENTATION

The Amazon basin encompasses about 7 million km^2 and roughly 60% of the world's remaining tropical rainforest (Whitmore 1997). More than 80% of the basin's forests appear to be intact, insofar as they are classified as unfragmented primary forest on satellite imagery. However, large areas of this forest have been degraded to some degree by selective logging, overhunting, illegal gold mining, and other human activities (W. F. Laurance 1998; Nepstad et al. 1999a, 1999b; Asner et al. 2005).

Absolute rates of forest destruction are higher in the Amazon than in any other tropical region, and they appear to be increasing. In Brazilian Amazonia, which comprises two-thirds of the basin, deforestation rates have accelerated significantly since 1990 (fig. 5.1). Annual deforestation estimates, derived from analyses of satellite imagery by Brazil's National Space Agency (INPE), suggest that mean forest loss from 1990 to 1994 was 1.35 million hectares per year—equivalent to about five football fields per

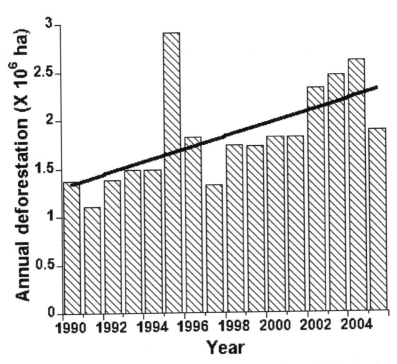

Figure 5.1. Estimated annual deforestation rates for Brazilian Amazonia. The regression line shows the overall trend from 1990 to 2005, during which forest loss accelerated significantly ($F_{1,14}$ = 8.05, R^2 = 36.5%, P = 0.013; linear regression with log-transformed deforestation data).

minute. From 1995 to 2005, the mean rate rose to 2.04 million hectares per year—equal to eight football fields per minute (W. F. Laurance 2003a). Efforts to stabilize the Brazilian currency, which freed up funds for investment and development, were associated with a large peak in deforestation in 1995. More recent increases are evidently being driven by extensive new highway projects, a rapid expansion of cattle ranching and industrial soybean farming, and increasing land speculation (Fearnside 2001e, 2002a; W. F. Laurance et al. 2001a, 2001b, 2002a).

Amazon deforestation is causing widespread fragmentation of forest cover. In the late 1980s, the area of Amazonian forest that was fragmented (i.e., in fragments <100 km^2 in area) or less than 1 km from clearings was over 150% larger than the area that had been deforested (Skole and Tucker 1993). Moreover, prevailing land uses, such as cattle ranching and slash-and-burn farming, tend to create forest fragments with highly irregular shapes and large amounts of forest edge (fig. 5.2). In two fragmented landscapes in eastern Amazonia, for example, every square kilometer of forest

Tailândia   **Paragominas**

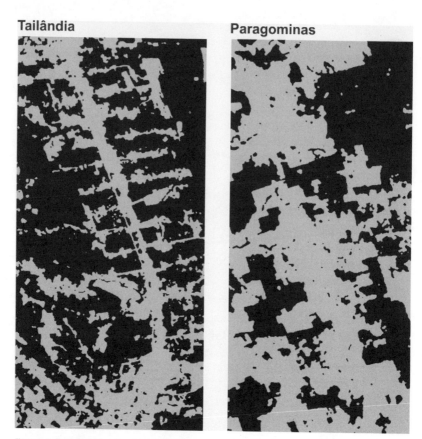

Figure 5.2. Portions of two fragmented landscapes in eastern Amazonia, illustrating the highly irregular shapes of most forest fragments. (Each image shows an area of about 600 km² in area.) Tailândia is a recent government-sponsored settlement project, whereas Paragominas is an older logging and ranching frontier (adapted from Cochrane and Laurance 2002). Each square kilometer of cleared forest created 1.3 to 1.8 km of new forest edge. (Values are for Paragominas and Tailândia, respectively, and were derived by dividing the total amount of forest edge in each landscape by the amount of cleared land.)

that was cleared created 1.3 to 1.8 km of new forest edge (fig. 5.2). Remote-sensing studies suggest that deforestation in Brazilian Amazonia is creating nearly 20,000 km of new forest edge each year (W. Chomentowski, D. Skole, and M. Cochrane, pers. comm.). Thus, vast expanses of Amazonian forest are being fragmented and exposed to edge effects.

EDGE-RELATED CHANGES IN FOREST FUELS

When intact and unaffected by major droughts, tropical rainforests rarely, if ever, burn. This is because rainforests have dense, nearly continuous canopy cover that helps to maintain a shaded, stable, humid microclimate. Under these conditions, potential fuels such as litter and fine woody debris are rapidly broken down by fungal, microbial, and invertebrate decomposers (Luizão and Schubart 1987; Chambers et al. 2000) and tend not to accumulate in the forest understory. Moreover, the fuels that are present are moist and hence difficult to burn.

When forests are fragmented, however, the quantity of dry, flammable fuels rises dramatically, for three reasons. First, forests near edges are subjected to hot, dry winds and increased light from surrounding clearings. On newly created edges, elevated temperatures, reduced humidity, and increased sunlight and vapor pressure deficits penetrate at least 40 to 60 m into fragment interiors (Kapos 1989; Didham and Lawton 1999; Sizer and Tanner 1999). Such changes increase evapotranspiration in the understory, depleting soil moisture and stressing drought-sensitive plants (Kapos 1989; Malcolm 1998). As a result, fuels near forest edges are drier than those in forest interiors.

Second, the amount of flammable fuel increases sharply near edges, as demonstrated by long-term studies at the Biological Dynamics of Forest Fragment Project in central Amazonia. One key cause is that litterfall increases markedly near forest margins (K. S. Carvalho and Vasconcelos 1999; Sizer et al. 2000; Nascimento and Laurance 2004; Vasconcelos and Luizão 2004), especially during the dry season (W. F. Laurance and Williamson 2001), when plants tend to shed leaves to reduce water stress. A second major cause is that rates of tree mortality and damage rise sharply in fragmented forests (W. F. Laurance et al. 1997a, 1998, 2000a; fig. 5.3), both from increased desiccation stress and from elevated wind turbulence near forest margins (D'Angelo et al. 2004; W. F. Laurance 2004a). Higher tree mortality greatly increases the amount of wood debris near fragment edges (Nascimento and Laurance 2004; fig. 5.3) and, by damaging the forest canopy, further exacerbates edge-related desiccation. The net effect is that forest fuels—especially fine, highly flammable fuels like leaves, twigs, and branches—are much more abundant and drier near fragment margins.

Finally, many forest fragments are selectively logged (W. F. Laurance and Cochrane 2001; Peres and Michalski, chap. 6 in this volume). By killing and damaging many trees, logging greatly increases the amount of flammable wood debris in forests, whereas desiccation is increased because of damage to the forest canopy (Holdsworth and Uhl 1997; Uhl and Kauffman 1990). In the Amazon it is likely that most forest fragments have been logged (see Peres and Michalski, chap. 6 in this volume).

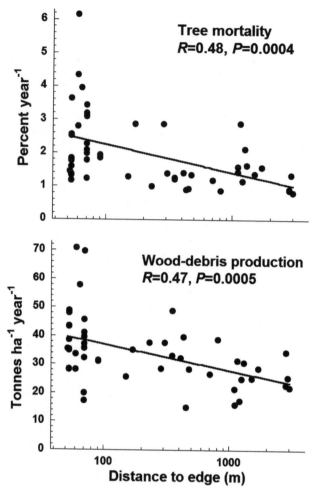

Figure 5.3. Annual rates of tree mortality and production of wood debris as a function of distance from forest edges in central Amazonia (adapted from Nascimento and Laurance 2004)

CLIMATIC CHANGES IN FRAGMENTED LANDSCAPES

Major changes in land cover can have important effects on climate that may increase the likelihood of forest fires. In this section, I focus on local- to regional-scale changes in climate caused by land clearing, which are further discussed by Shugart et al. (chap. 3 in this volume) and Avissar et al. (chap. 4 in this volume). The effects of global-scale climatic alterations on tropical forests (such as those driven by rising atmospheric carbon dioxide concentrations) are treated in other chapters (Lewis et al., chap. 1 in this volume; Williams and Hilbert, chap. 2 in this volume).

Fragmentation and loss of forest cover can alter local and regional climates in several ways. First, habitat fragmentation can help to cause local forest desiccation via a phenomenon known as the vegetation breeze (fig. 5.4). This occurs because fragmentation leads to the juxtaposition of cleared and forested lands, which differ greatly in their thermal characteristics. Air above forests tends to be cooled by evaporative cooling (from evapotranspiration of water vapor), whereas such cooling is much reduced above clearings (this increases the Bowen ratio, which is the ratio of sensible to latent heat). As a result, the air over clearings heats and rises, reducing local air pressure and drawing moist air from surrounding forests into the clearing. As the rising air cools, the moisture it carries condenses into convective clouds that may produce rainfall over the clearing. The air is then recycled—as cool, dry air—back over the forest (Silva Dias and Regnier 1996; Baidya Roy and Avissar 2002). The net effect is that forest clearings promote local atmospheric circulations that may increase rainfall but, paradoxically, draw moist air away from nearby rainforest (fig. 5.4). In regions

The Vegetation Breeze

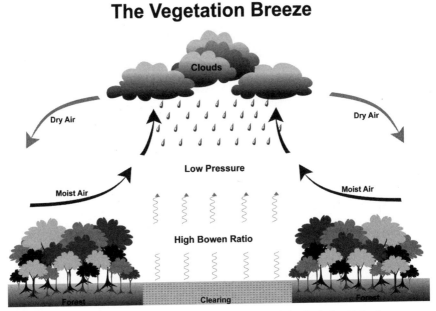

Figure 5.4. The vegetation-breeze phenomenon, which promotes forest desiccation in the vicinity of pastures and clearings

with prevailing winds, some rain generated by the vegetation breeze may fall on downwind forests, not just in clearings, with desiccation being most severe in upwind forests. In the Amazon, vegetation-breeze effects have been observed in clearings as small as a few hundred hectares but appear to peak when clearings are roughly 100 to 150 km in diameter (Avissar and Liu 1996; Avissar et al., chap. 4 in this volume). The vegetation breeze is essentially a large-scale edge effect; satellite observations in Rondônia, Brazil, suggest that the desiccating effects of major clearings can extend up to 20 km into adjoining forests (Silva Dias et al. 2002).

Second, the conversion of forests to pasture or savanna reduces plant evapotranspiration (because grass and shrubs have shallower roots and far less leaf surface area than do forests; Jipp et al. 1998), which could potentially decrease rainfall and cloud cover and increase albedo and surface temperatures. As discussed by Avissar et al. (chap. 4 in this volume), however, the nature of any such declines is still poorly understood. For example, several prominent modeling studies have suggested that Amazonian deforestation could reduce basinwide precipitation by roughly 20% to 30%, but these studies relied on a simplistic assumption of complete, uniform forest clearing (e.g., Nobre et al. 1991; Dickinson and Kennedy 1992; Lean and Rowntree 1993; Gash and Nobre 1997). Model results based on actual (ca. 1988) deforestation patterns in Brazilian Amazonia have been less dramatic: deforested regions are predicted to experience modest (6%–8%) declines in rainfall, moderate (18%–33%) reductions in evapotranspiration, higher surface temperatures, and greater wind speeds (from reduced surface drag) that could affect moisture convergence and circulation (G. K. Walker et al. 1995; Sud et al. 1996). It is even possible that moderate forest loss and fragmentation could *increase* net regional precipitation as a result of the vegetation breeze, although the main effect would be to remove moisture from forests and to redistribute it over adjoining clearings. Nevertheless, if deforestation reaches some critical but unknown threshold, Amazonian rainfall might decline precipitously as the regional hydrological system collapses (Avissar et al. 2002; Baidya Roy and Avissar 2002; Avissar et al., chap. 4 in this volume).

The final two effects of forest loss in changing local climate result from the massive smoke plumes produced by forest and pasture fires. First, smoke hypersaturates the atmosphere with cloud condensation nuclei (microscopic particles in aerosol form) that bind with water molecules in the atmosphere, inhibiting the formation of raindrops (Rosenfeld 1999). Second, by absorbing solar radiation, smoke plumes warm the atmosphere, inhibiting cloud formation (fig. 5.5). As a result of these phenomena, large fires can create rain

Figure 5.5. Thermal-satellite scene of northern Amazonia in early 1998, illustrating the paucity of clouds in the vicinity of heavy smoke plumes (image courtesy of the National Oceanographic and Atmospheric Administration)

shadows that can extend hundreds of kilometers downwind (Freitas et al. 2000). Moreover, because virtually all tropical fires are lit during the dry season, both phenomena reduce rainfall during the critical dry-season months, when plants are already moisture-stressed and are most vulnerable to fire.

INCREASING IGNITION SOURCES

Each year hundreds of thousands of fires are lit intentionally in the tropics—to raze forests for slash-and-burn farming, cattle ranching, and other purposes. Although highly destructive to natural ecosystems (Holdsworth and Uhl 1997; Cochrane 2003), these intentional fires are now being rivaled in terms of their ecological impacts by a subtler menace: accidental forest fires (Nepstad et al. 1999a; Cochrane 2001b, 2003; W. F. Laurance 2003b; Barlow and Peres 2004).

Most fires that penetrate into forest fragments are accidental and originate in adjoining cattle pastures, farming plots, or regrowth forest, any of which can carry a fire after just a few rainless days. Amazonian ranchers burn their pastures at 1- to 2-year intervals to control weeds and to promote a flush of new grass that is favored by cattle (Cochrane et al. 1999).

Prior to the arrival of Amerindians in the late Pleistocene, the only ignition source for Amazonian fires was lightning strikes, and such fires were likely rare because convectional thunderstorms, which generate lightning, also produce abundant rainfall. Many Amerindian groups used fires for swidden farming, but archeological evidence, such as the distribution of potshards and *terra preta* (black earth resulting from repeated burning and mulching), suggests that populations were quite patchily and ephemerally distributed (e.g., Meggers 1994; Woods and Glaser 2001). Nevertheless, during the past two millennia, some significant fire incursions did occur in Amazonia (Sanford et al. 1985; Saldarriaga and West 1986; Piperno and Becker 1996), probably during rare and exceptionally intense El Niño droughts (Meggers 1994).

In the Amazon and throughout the tropics, forest burning has increased drastically today. During a 4-month period in 1997, thermal-satellite images revealed 44,734 separate fires in the Amazon (Nepstad 1998), virtually all of them human-caused. Smoke from forest burning became so severe in regional centers such as Manaus and Boa Vista that airports were temporarily closed and local hospitals reported 40% to 100% increases in the number of patients with respiratory distress (Anon. 1997). Fire incidence is rising both because of the widespread use of fire as an agricultural tool—particularly in cattle ranching and slash-and-burn farming—and because vast new highways and transportation projects are providing increased access to the Amazonian frontier (W. F. Laurance 1998; W. F. Laurance and Fearnside 1999; Nepstad et al. 2001). Efforts to improve frontier governance (Nepstad et al. 2002) and to reduce illegal forest burning (Anon. 1998b, 1999a) have achieved only limited and temporary success (W. F. Laurance et al. 2001a; W. F. Laurance and Fearnside 2002).

Amazonian fires are far from uniformly distributed. Burning is most frequent in the southern, eastern, east-central, and north-central parts of the basin, which have the strongest and most protracted dry seasons (Nepstad et al. 1994, 1999a; Sombroek 2001). These regions, especially the south and east, also support the highest population densities and largest concentrations of roads and infrastructure (W. F. Laurance et al. 2002a). Burning also spikes during periodic El Niño droughts, which typically occur at 3- to 7-year intervals. In the past century 23 El Niño events occurred in the Amazon (Suplee 1999), several of which, such as the 1983 and 1997 droughts, were especially strong and led to catastrophic forest fires (e.g., Barbosa and Fearnside 1999; Nepstad et al. 1999a; Barlow and Peres, chap. 12 in this volume).

The confluence of climatic changes, increased fuel, and abundant ignition sources means that fragmented forests are drastically more vulnerable to fire than are intact forests, especially in more-seasonal areas of the Amazon. Fires that penetrate into rainforests are deceptively unimpressive, creeping along the ground as a thin ribbon of flames burning through the leaf litter (fig. 5.6). Except in areas with unusual fuel structure, the fires may reach only 10 to 20 cm in height (Cochrane et al. 1999) and cover as little as 150 m per day (Cochrane and Schulze 1999). However, these fires are deadly to rainforest trees, vines, lianas, and ground forbs, both because the slow propagation results in long residence times of the flames at the base of encountered stems and because most rainforest trees have thin (<1 cm thick) bark (Uhl and Kauffman 1990; Barlow et al. 2003a). The initial fire in rainforests typically kills 10% to 60% of all trees and most vines and forbs (Barbosa and Fearnside 1999; Cochrane and Schulze 1999; Cochrane et al. 1999; Barlow et al. 2003b; Barlow and Peres, chap. 12 in this volume).

Even more alarming than the initial impacts of a surface fire is the irreversible process of forest degradation that it can set in motion. After the first fire, the forest canopy becomes fragmented and the quantity of dead fuels rises as dead leaves and trees begin to fall. Soon the forest is vastly more prone to subsequent fires because the diminished canopy allows rapid drying and the dying vegetation provides large quantities of combustible fuel (Cochrane et al. 1999).

A second fire is far hotter and more destructive than the initial burn, killing half or more of the remaining trees (Cochrane and Schulze 1999; Cochrane et al. 1999) and overwhelming even large, thick-barked trees (Barlow and Peres 2004b). Moreover, whereas the initial surface fire might require an extensive drought, subsequent conflagrations can occur after just a few weeks of no rain. During the first several fires, more fuels are created than are destroyed, and a positive feedback results in which each fire becomes more likely and intensive (Cochrane et al. 1999; Nepstad et al. 1999a). This process can eradicate rainforest trees and promote extensive grass invasion, converting rainforests into anthropogenic savanna or degraded scrub. In regions with strong dry seasons, this degradation process, once initiated, may be nearly irreversible (Cochrane and Laurance 2002). Even forests that have suffered just a single surface fire exhibit severe structural and floristic changes that have a major influence on faunal communities (W. F. Laurance 2003b; Barlow and Peres, chap. 12 in this volume).

Although several researchers have discussed increased fire incidence near forest edges (Kauffman and Uhl 1990; Cochrane and Schulze 1999;

Figure 5.6. Although slow-moving, surface fires have remarkably destructive effects on rainforests. (Photo by M. A. Cochrane)

Cochrane et al. 1999; Nepstad et al. 1999a; Gascon et al. 2000), the only quantitative studies to date are those of Cochrane (2001b) and Cochrane and Laurance (2002), who used high-resolution satellite imagery to assess the frequency of edge-related fires in two fragmented landscapes in eastern Amazonia. Collectively, the two landscapes, Tailândia and Paragominas (fig. 5.2), contained over 700 forest fragments ranging from 0.1 to over 57,000 ha in area. Areas of forest affected by surface fires were measured using a specialized technique (subpixel linear spectral mixture modeling; Cochrane and Souza 1998) at a spatial scale of 30 m pixels. For each study area, the mean fire frequency was calculated as a function of distance from forest edges based on 12 to 14 years of data.

The incidence of fires was drastically elevated in both landscapes, especially during the severe 1997 El Niño drought. In Paragominas, an area that had experienced intensive logging and cattle ranching for several decades, fire frequency was sharply elevated throughout the entire landscape—to the extent that few if any forest tracts could survive in the long term (Cochrane 2001b; Cochrane and Laurance 2002). In Tailândia, a government-sponsored forest-colonization project that was initiated in the early 1990s, fire frequency more than doubled as far as 2500 m from forest edges and was drastically elevated within about 800 m of edges (fig. 5.7). Under these withering fire regimes, forest margins will collapse inward and forest fragments will progressively "implode" over time (Gascon et al. 2000; Cochrane and Laurance 2002).

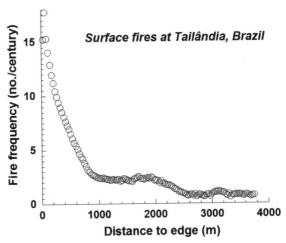

Figure 5.7. The dramatically elevated frequency of surface fires near forest edges in eastern Amazonia (adapted from Cochrane and Laurance 2002). Fire frequency (number per century) was estimated based on 12 to 14 years of satellite observations.

The incorporation of these data into a simple core-area model (W. F. Laurance and Yensen 1991) suggests that even large forest remnants and isolated reserves will be vulnerable to edge-related fires (fig. 5.8). This occurs in part because many forest fragments in Amazonia have highly irregular shapes and thus are particularly vulnerable to edge effects (Cochrane and Laurance 2002). According to the model, even a fragment of 500,000 ha in area would have 25% to 65% of its area vulnerable to edge-related fires, depending on the fragment shape. Smaller fragments are far more susceptible. Based on these results, recent analyses of satellite imagery suggest that at least 45 million hectares of forest in Brazilian Amazonia (more than 13% of the total remaining forest area) is currently vulnerable to edge-related fires (Cochrane 2001a).

FRAGMENTATION, FIRE, AND THE FUTURE OF THE AMAZON

Over the next few decades, forest fires—both intentional burns and escaped fires—are likely to increase in the Amazon, for two main reasons. First, El Niño droughts might be increasing in frequency and intensity

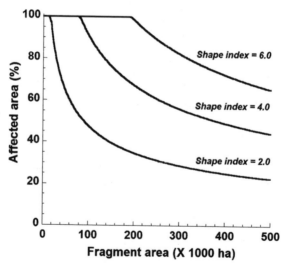

Figure 5.8. A mathematical core-area model illustrating the expected areal impacts of surface fires on Amazonian forest fragments, assuming that surface fires penetrate up to 2.5 km into fragment interiors. Curves are shown for three fragment shapes: somewhat irregular fragments (SI = 2.0), moderately irregular fragments (SI = 4.0), and highly irregular fragments (SI = 6.0). The fragment shape index (SI) is the ratio of the fragment's perimeter length to that of a circle of the same total area; thus, a perfectly circular fragment has an SI of 1.0 and all other shapes have higher values. (Adapted from Cochrane and Laurance 2002)

(Trenberth and Hoar 1996; Dunbar 2000), possibly as a result of increasing global warming (Trenberth and Hoar 1997; Timmermann et al. 1999). El Niño–related droughts affect large areas of the world's tropics, including much of the Neotropics, Southeast Asia, and Australasia (e.g., Meggers 1994; Curran et al. 1999; W. F. Laurance et al. 2001d). During harsh droughts even fully intact, unlogged forests can become vulnerable to surface fires (Barlow and Peres, chap. 12 in this volume).

Second, the Amazon is likely to be fragmented on a large spatial scale as a result of major investments in new highways, railroads, river-channelization projects, and other infrastructure. For example, over 7500 km of highway paving is planned in coming years for the Brazilian Amazon, with highways penetrating deep into the nearly pristine heart of the basin. These projects sharply increase physical accessibility to remote frontier areas and frequently initiate a process of spontaneous colonization, hunting, mining, logging, ranching, and land speculation that is almost impossible for governments to control (W. F. Laurance et al. 2001a; W. F. Laurance and Fearnside 2002). Efforts to project the future condition of Amazonian forests (fig. 5.9) suggest that rates of deforestation and fragmentation will increase dramatically over the next two decades (G. Carvalho et al. 2001; W. F. Laurance et al. 2001b). Past experience indicates that surviving fragments of forest will be vastly more vulnerable to fires, predatory logging, and other exploitative activities. Unless current development trends are altered, the world's largest remaining rainforest is likely to be irreparably diminished and degraded.

CONCLUSIONS AND IMPLICATIONS

1. The incidence of forest fire is rising alarmingly in the tropics as increasing numbers of slash-and-burn farmers, ranchers, and loggers penetrate ever deeper into remote frontier areas. In large areas of the tropics (including much of the Neotropics, Southeast Asia, and Australasia), forest fires may be exacerbated by increasingly frequent El Niño droughts.

2. Because natural fire is a very infrequent event in rainforests, even low-intensity surface fires are highly destructive to trees, vines, lianas, ground forbs, and many faunal species.

3. Evidence from two fragmented landscapes in eastern Amazonia suggests that fire operates as a large-scale edge effect, penetrating considerable distances into forest interiors. In fragmented rainforests, fire may beget more fire because once-burned forests near forest margins appear to be much more vulnerable to subsequent fires, which creates a

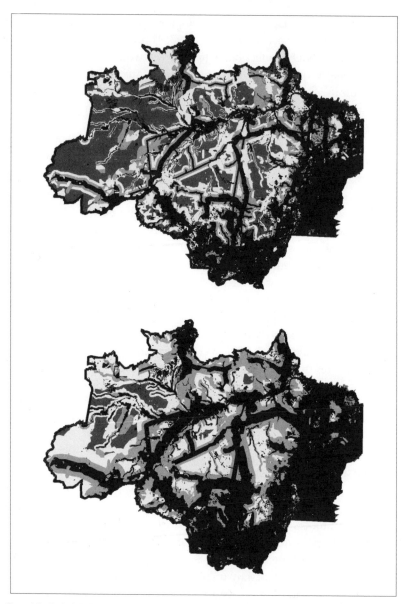

Figure 5.9. Optimistic (*top*) and nonoptimistic (*bottom*) scenarios for forests of Brazilian Amazonia in the year 2020. Black areas are deforested or severely degraded, dark gray areas are moderately degraded, white areas are lightly degraded, and light gray areas are pristine forest. (Adapted from W. F. Laurance et al. 2001b)

positive feedback cycle that can cause a progressive implosion of forest fragments.

4. Edge-related fires are caused by high concentrations of dry, flammable fuel near forest margins and abundant ignition sources in surrounding pastures and degraded lands. Increasing evidence suggests that local climatic changes, caused by the loss and fragmentation of forest cover, also play a significant role.

5. Model simulations suggest that even large forest fragments and isolated nature reserves (>500,000 ha in area) could be vulnerable to fire, especially in more-seasonal areas of the tropics. This occurs both because fragments are often highly irregular in shape and thus inherently vulnerable to edge effects, and because edge-related fires can penetrate considerable distances into fragment interiors.

6. In the future, edge-related fires could increase dramatically in Amazonia as a result of massive new infrastructure projects that will greatly increase access to remote frontier areas for colonists, ranchers, and loggers.

ACKNOWLEDGMENTS

I thank Jay Malcolm, Mark Cochrane, Dan Nepstad, Roni Avissar, Carlos Peres, and Jos Barlow for commenting on drafts of the chapter, and the NASA-LBA program, A. W. Mellon Foundation, and Smithsonian Tropical Research Institute for support. This is publication number 422 in the BDFFP technical series.

CHAPTER 6

Synergistic Effects of Habitat Disturbance and Hunting in Amazonian Forest Fragments

Carlos A. Peres and Fernanda Michalski

Deforestation and forest fragmentation have marched together relentlessly in virtually all humid tropical regions, resulting in both habitat loss and the subdivision of remaining habitat. Echoing classic equilibrium theory of island biogeography (MacArthur and Wilson 1967), species persistence in residual habitat patches has been related primarily to their size and degree of isolation (e.g., Lovejoy et al. 1986; Soulé et al. 1988; Newmark 1996). However, rural people in the tropics almost invariably continue to use forest remnants of varying size, shape, and degree of connectivity for food, firewood, timber, therapeutic plants, and other housing construction materials. These remnants are therefore often exposed to different scales of human-induced habitat disturbance and hunting, which may compound and aggravate the population- and community-level effects of edges and forest patch area on which much of the habitat fragmentation literature has focused.

Patterns of anthropogenic disturbance that puncture the canopy and change the understory light environment of forest patches often come in the form of selective logging (Uhl and Vieira 1989; Fimbel et al. 2001), slash-and-burn agriculture, and single or recurrent surface wildfires that kill a large proportion of the canopy trees (Cochrane et al. 1999; Barlow et al. 2003b). Exploitation of nontimber forest products, such as game vertebrates and a wide range of plant resources, can also accelerate population declines and local extinction rates within forest remnants once the source–sink dynamics of harvested populations break down as a result of the fragmentation process (Peres 2001b). For example, subsistence hunting targeted at large-bodied vertebrates is ubiquitous in many tropical forest remnants, even when they are nominally protected (e.g., Cullen et al. 2000; Brashares et al. 2001; Peres and Lake 2003). Fragmentation ecologists thus require an understanding of how people continue to alter the habitat structure and biological communities of fragmented forest landscapes. Yet most fragmentation studies either ignore the interactions between habitat fragmentation and different scales of human habitat disturbance or treat them as independent variables. As a result, the biotic responses to disturbance within forest remnants—which may be favorable, detrimental, or neutral to the persistence of wildlife populations—remain poorly understood.

In this chapter we first provide a general overview of the interactions among habitat fragmentation, habitat degradation, and hunting in tropical forests. We illustrate how these effects can operate synergistically using a large, original data set from a representative fragmented landscape in southern Amazonia. We then examine levels of vertebrate species persistence and extinction in a statistically robust number of forest patches using species–area relationships (SARs), on which nearly all estimates of local extinction rates due to habitat destruction and fragmentation are based (May et al. 1995). In particular we examine how these relationships are influenced by the effects of human disturbance. We conclude with some general recommendations of how fragmented tropical forest landscapes can be managed to maximize the integrity of their vertebrate assemblages.

HUNTING IN TROPICAL FOREST FRAGMENTS

Most tropical forest fragments are left unprotected and exposed to a wide range of nontimber extractive activities, which include subsistence hunting. This is understandable because the economic value of game vertebrates is second only to that of timber in most tropical forest regions (Peres 2000a). Forest fragmentation could exacerbate the effects of hunting by (1) initially reducing and isolating vertebrate populations averse to the surrounding habitat matrix (Malcolm 1997; Gascon et al. 1999), (2) reducing or precluding recolonization of overharvested areas from neighboring nonhunted areas (Robinson 1996; Peres 2001b), (3) increasing the perimeter-to-area ratio and the amount of core forest habitat accessible to hunters on foot, and (4) further reducing the habitat area suitable for species that are averse to forest edges (W. F. Laurance et al.

2000b; W. F. Laurance, chap. 5 in this volume). These synergistic interactions are potentially ubiquitous in fragmented tropical forest landscapes. In Amazonia, forest fragments in both "old" deforestation frontiers, such as the Zona Bragantina of northeastern Pará, and "new" frontiers along the southern deforestation crescent are typically harvested by game hunters (C. A. Peres, pers. obs.). Spider monkeys, bearded sakis, and brown capuchin monkeys were unable to persist in 100-ha forest patches of central Amazonia following isolation (Rylands and Keuroglian 1988), but these extinctions were facilitated by hunters from neighboring cattle ranches (L. Emmons, pers. comm.). Pressure from hunting can greatly accelerate the local extinction rate of midsized to large-bodied forest vertebrates in fragmented Amazonian forest landscapes where small populations disrupted by loss of connectivity may no longer receive immigrants from unharvested source areas (Peres 2001b). Under the conservative assumption that game hunters severely overharvest a 5 km buffer from the nearest forest edge, an aggregate area of over 1.2 million km^2 of forest in the Brazilian Amazon may have already lost key elements of its large vertebrate fauna due to the combined effects of hunting and fragmentation (estimated from data in Skole and Tucker 1993).

Few neotropical studies have attempted to quantify patterns of large vertebrate persistence or abundance in habitat fragments, but those habitats exposed to persistent hunting pressure often become increasingly impoverished (Oliver and Santos 1991; Chiarello 1999; Cullen et al. 2000; Silva and Tabarelli 2000). On the basis of a comprehensive survey of mid-1800s historical accounts, Dean (1997) attributed many regional extinction events of Brazilian Atlantic forest vertebrates to fragment overhunting. Preferred target species have been extirpated from many small forest patches of Mesoamerican forests in mainland Panama, even where hunting pressure is light (Glanz 1991; Wright et al. 2000). Tapirs, white-lipped peccaries, and brocket deer are rapidly driven to local extinction in hunted Atlantic forest fragments of the State of São Paulo, which average less than 2000 ha in size (Cullen et al. 2000). In the highly fragmented Atlantic forest of southern Bahia, Brazil, some 99% of 418 forest patches surveyed through interviews had lost large ungulates (tapirs and white-lipped peccaries) and midsized to large primates (yellow-breasted capuchins, howler monkeys, and woolly-spider monkeys) within a 37,000 km^2 study region (Pinto and Rylands 1997; L. P. Pinto, pers. comm.). On a more severe scale, the case of the Alagoas curassow (*Mitu mitu*) is an example of a global extinction event in the wild driven by the combined effects of hunting and fragmentation (Sick 1997). Although some relict populations of this endemic species survived in small, inadequately protected fragments of northern Atlantic forest well into the 1950s, one of the last records of this game bird was an animal killed by a hunter in 1984, and the species is now thought to survive only in captivity.

The same trends hold for other tropical forest regions, although data remain scanty. In western Madagascar, where most forest remnants are hunted, the presence of a large, hunter-free core area is the most important determinant of large-bodied lemur persistence (A. P. Smith 1997). Lemur species richness in this region is most severely affected by an index of "cultural impact," defined as the sum of all visible evidence of forest disturbance including hunting (A. P. Smith 1997). Many of the bird and mammal extinctions in the few remaining forest fragments of Singapore—including three species of pheasants, three species of deer, and one species of wild pig—were undoubtedly caused or accelerated by hunting (Corlett and Turner 1997; R. Corlett, pers. comm.). Indeed this negative synergism appears to be ubiquitous in many parts of the humid tropics. Hunting has apparently accelerated extinction rates of bird and mammal populations in fragmented forest landscapes of Sulawesi (M. Kinnaird, pers. comm.), Thailand (Pattanavibool and Dearden 2002), Taiwan (Chun-Yen-Chang 2003), Uganda (C. Chapman, pers. comm.), West Africa (Brashares et al. 2001), Costa Rica (Guariguata et al. 2002), and southern Mexico (A. Cuarón, pers. comm.).

HABITAT DEGRADATION WITHIN FOREST FRAGMENTS

Patterns of tropical deforestation are highly variable in terms of extent, spatial configuration, shrinkage, and attrition and can result in many landscape types. These range from dentritic remnants of primary forest shredded along riparian corridors to forest "fishbones" along branched road systems and those dominated by small, isolated woodlots surrounded by an inhospitable matrix of pastures or monocultures. Whatever the fragmentation process, however, the remaining forest patches usually continue to be exposed to both natural and anthropogenic habitat disturbances that further deteriorate forest-interior conditions. Forest remnants of varying levels of connectivity may be further eroded along edges by elevated rates of tree mortality caused by windthrows (W. F. Laurance et al. 2002b), severe drought and pasture fires (Didham and Lawton 1999; Cochrane and Laurance 2002), slash-and-burn agricultural clearings (Cardille et al. 2002), fine-grained canopy perforation caused by timber extraction (W. F. Laurance et al. 2002b), and pulses of heavy mortality of canopy trees induced by ground fires that often penetrate deep into the fragment (Cochrane and Laurance 2002; Cochrane et al. 2002; Barlow and Peres 2004b). These sources of disturbance both reduce the fragment area of shaded understory underneath a closed canopy and extend or propagate edge effects away from forest margins (Malcolm 2001), further degrading the local habitat quality for edge-intolerant forest vertebrates. In heavily logged fragments, logging gaps, skidder trails, and log yards may be too close to one another, thus creating enough additive edge effects that may permeate into the entire fragment area. If logging gaps are large enough (e.g., <200 m^2), such as those resulting from many mechanized operations, a basal-area reduction (from both removal and collateral damage) as low as 30% to 40% of the stand could erode all nonedge areas regardless of the fragment size or shape (Struhsaker 1997). Amazonian forests typically have only 5% to 20% of their area under conditions of natural treefall gaps at any one time, whereas larger gaps created by selective logging could occupy approximately 50% of the forested area (Uhl and Vieira 1989). This "internal fragmentation" process (Forman 1995; Goosem 1997) can therefore amplify and diffuse the microclimatic conditions typical of peripheral edges, facilitating dispersal of early successional plants and associated gap-dependent animals. Past and ongoing disturbances within a remnant could therefore strongly modulate species responses to forest patch area, particularly for forest-interior taxa (e.g., Dunstan and Fox 1996; M. P. Cox et al. 2003). Ultimately, the history of structural perturbation of forest remnants could affect the slope of SARs, although the time lag of species decay functions remains poorly understood.

Examples of interactions between habitat disturbance and fragmentation are legion and widespread (e.g., Fox and Fox 2000; Laidlaw 2000; Saatchi et al. 2001; Lens et al. 2002), but we shall cite only two cases that are particularly relevant to this study. In the highly fragmented southern Amazonian region of Sinop, Mato Grosso (16,819 km^2), 68% of the remaining forest cover of 12,271 km^2 has been degraded by multiple logging cycles, edge-related fires, or other edge effects (Cochrane et al. 2002). Similar results were obtained in the fragmented region of Paragominas of eastern Pará, where only 6% of an entire Landsat scene (with 66% of forest cover remaining) was unlogged and unburned primary forests (Nepstad et al. 1999b). Although fire disturbance is predominantly an edge effect, it can extend for up to several kilometers into continuous forest, even in previously unlogged areas (Cochrane and Laurance 2002; Peres et al. 2003b). This is aggravated by the fact that the vast majority of neotropical forest remnants are smaller than a few

hectares (Gascon et al. 2000; Peres 2001b) and therefore wholly susceptible to edge-related fire disturbance (Cochrane and Laurance 2002).
Furthermore, new forest edges are formed at an alarming rate, both in
previously fragmented areas and in new frontiers. For example, nearly
20,000 km of new forest edges are created by deforestation every year in
the Brazilian Amazon (M. Cochrane and D. Skole, pers. comm.). Indeed
the long-term persistence of many small rainforest fragments in seasonally dry tropical regions is highly questionable because the average penetration distance of fires from the nearest edge (240 and 700 m in
unlogged and logged forests, respectively; Cochrane et al. 2002) can easily reach the fragment core. These recurrent burns have profound ecological effects on fire-intolerant plant communities and can rapidly
disfigure the forest structure and species composition of forest fragments
with long-term effects (Barlow and Peres, chap. 12 in this volume).

THE ALTA FLORESTA STUDY

Our study area is located in Alta Floresta (09°53′ S, 56°28′ W), a prosper-
ous frontier town in northern Mato Grosso, southern Brazilian Amazonia
(fig. 6.1). This once entirely forested region has been subjected to very
high deforestation rates since the early 1980s and is currently dominated
by open pasturelands, where only 37% of the original forest cover
remains on the left bank of the Teles Pires River.

We conducted detailed interviews on the history of hunting, selective
logging, and wildfires within 144 forest sites, including 129 semi- or
entirely isolated forest patches and 15 "control" sites within continuous,
undisturbed forest (fig. 6.1). All sites were located within a radius of up to
50 km of Alta Floresta and were accessible by river or by paved and
unpaved roads. Forest patches were initially selected using a georeferenced 2001 Landsat Enhanced Thematic Mapper (ETM) image (scene
227/67) on the basis of their size, degree of isolation, and nature of surrounding habitat matrix (Michalski and Peres 2005). When identifying
potential patches, we deliberately targeted a wide gradient of logging
intensity and fire severity, from patches that had never been logged or
burned to those that had been heavily affected by either one or both disturbances. As a key prerequisite, each site was associated with at least one
local informant, usually a long-term resident or landowner, who was
(1) willing to be interviewed, (2) a regular visitor to a previously selected
patch, and (3) thoroughly familiar with the history of disturbance of the
patch and the medium- to large-bodied vertebrate fauna persisting within

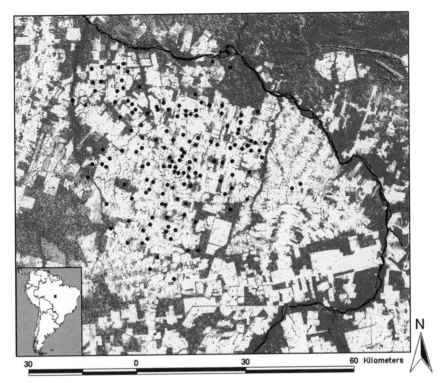

Figure 6.1. Location of the study region in Alta Floresta, Mato Grosso, Brazil, on the west bank of the Teles Pires River, showing the spatial distribution of forest fragments and continuous-forest sites surveyed in this study

that patch. Supplemental information on both the history of human activities in each forest patch and its large vertebrate fauna was often obtained from additional informants if the patch was shared by neighboring landowners. Interviewees had been living next to, or working within, each forest patch for at least 2 years and regularly entered the patch for a range of utilitarian and/or recreational reasons. From June to September 2001 and May to July 2002 we obtained patch occupancy data for a total of 57 species (or functional groups) of midsized to large forest vertebrates, including 44 mammal, 12 bird, and 1 reptile taxa (appendix 6A) that are widely familiar to hunters and other forest users in this region. Occupancy records were specifically related to occurrences within a clearly defined forest patch that could usually be seen from the interviewee's household. In all cases, interviews were aided by color plates in field guides, photographs, and recordings or imitations of vocally conspicuously species. To assess the frequency of type II errors, interviewees were asked to identify which species were present in the patch from a selection of color plates, including 10 species known to be entirely absent from the study region. In more than 1500 trials, interviewees never falsely identified vertebrate species from other neotropical forest regions as present in their forest patches, which gives us confidence in the reliability of their occupancy data. In addition, we used three independent sampling protocols in a smaller subset of 25 forest patches—diurnal line-transect censuses, camera trapping, and fine-sand trackplates—to independently verify the accuracy of species occupancy data obtained from local interviewees (Norris 2002; F. Michalski and C. A. Peres, unpublished data). Our interview approach was again shown to have greater than 98% reliability for common species on the basis of these methods, but performance was better for rare species that may not have been detected by direct short-term census techniques.

In aggregate, species occupancy data reported here for 144 sites are based on 1749 interviewee-years of observational sampling effort. Occupancy records of a given species were defined as patch-level occurrences when interviewees had no doubt about whether the species was locally present at the time of interviews or recent past, whether it was thought to be a full-time resident or an occasional transient within the patch. Local extinction events were conservatively defined as unambiguous records of absences when a species had been reported to have once occurred within a patch but had no longer been sighted, heard, or detected otherwise through tracks, scats, or other signs of their presence for at least 2 years. When compiling conflicting information from more than one interview referring to a given forest patch, we recorded a species as present if it was reported by at least one interviewee even if it had not been reported by another.

LAND-COVER CLASSIFICATION AND ANALYSIS

Following a two-stage unsupervised classification from a Landsat image (ETM 227/067), it was possible to discriminate 10 land-cover classes including closed-canopy forest, open-canopy forest, secondary forest, managed and unmanaged pasture, recent clear-cuts, bare ground, and open water. The image was georeferenced using a guide file from the Brazilian Space Agency (INPE) with an accuracy of 0.29 pixels, each of which had a resolution of 15 m. Locations of interview sites and their respective forest patches were plotted using GPS coordinates obtained in situ. All forest patch and landscape metrics associated with each patch (e.g., total area, core area, percent edge, matrix type, and distance to nearest large patch) were quantified using Fragstats version 3.3 (McGarigal and Marks 1995) coupled with ArcView 3.2. Our landscape metrics for individual forest patches sampled were largely independent from one another because these patches were far apart and rarely shared the same neighborhood (see below).

The intensity and extent of disturbance in each forest patch was ranked using a 5-point scale on the basis of site inspections and information obtained from interviews according to (1) the scale of selective logging activities, (2) the severity and proportional coverage of ground fires, and (3) the degree of hunting pressure. Logging disturbance variables included the method of timber felling and removal, timber species selectivity, amount and extent of the timber offtake, and the number of years since logging practices had been discontinued. Information regarding hunting at each patch included the number of hunters, the length of time (years), and the frequency with which they had been exploiting a patch, hunting techniques, prey selectivity, and a rank order of the five most important target species killed. Subsistence hunters in the Alta Floresta region were both uninhibited by interviewers and largely unaware of legality issues concerning hunting regulations designed to protect sensitive game species as loosely enforced in other parts of Brazil by the Natural Resources Agency (IBAMA). We therefore considered the hunting data to have negligible bias in relation to these restrictions. The land title distribution and resettlement program in Alta Floresta, which dates from the late 1970s, ensured that a wide range of forest patch (and property) sizes and forest disturbance regimes were available for sampling. These included small forest remnants controlled by smallholders and large remnants within large cattle ranches, which were often well protected. A total of 141 landowners were contacted during the study to ensure satisfactory levels of local cooperation and physical access to a robust number of well-distributed study sites.

HABITAT DISTURBANCE WITHIN FOREST FRAGMENTS

Timber extraction in the southern Brazilian Amazon is almost invariably a precursor to large-scale forest conversion into cattle ranching and cash crops and continues to subsidize other economic activities many years after forest patches have been isolated. Of the 129 forest patches surveyed, all but 2 had been affected by some level of manual or mechanized timber felling, often over repeated logging cycles. Fifty percent of the forest fragments were lightly to moderately logged, whereas another 49% had been heavily to very heavily logged, to the point where few or no commercially valuable canopy trees remained (fig. 6.2). Surface-fire disturbance from pasture fires had affected the edges of virtually all patches that were examined. However, nearly half of the patches (49%) had not been affected by extensive ground fires beyond a few meters from forest margins. Those that had been exposed to significant edge-related or core forest fires were evenly split in terms of burn severity; 25% had been subjected to light burns and 26% to severe burns.

Hunting comprises a ubiquitous subsistence or recreational activity for small farmers in Alta Floresta despite the ample availability of domesticated animal protein sourced from small livestock and large numbers of cattle (11 head per person). However, hunting in this region is highly selective toward the five species of forest ungulates (*Mazama gouazoubira, M. americana, Tayassu tajacu, T. pecari,* and *Tapirus terrestris*), the large rodents (e.g., *Agouti paca, Dasyprocta agouti*), the armadillos (e.g., *Dasypus novencinctus*), and the gamebirds (e.g., cracids and tinamids). This is partly because rural settlers are recent immigrants primarily of southern Brazilian (and originally European) origin and have cultural taboos against arboreal taxa such as large primates and sloths, which are widely harvested elsewhere by Amazonian Indians (Jerozolimski and Peres 2003). The vast majority of fragments surveyed (96%) had been subjected to different levels of hunting pressure following isolation (fig. 6.2), and 89 fragments (69%) were still being exploited by hunters at the time of the interviews.

Yet these figures are conservative because throughout the study we consistently sought access to relatively protected forest fragments in different size classes that had been lightly disturbed, but these were often difficult to

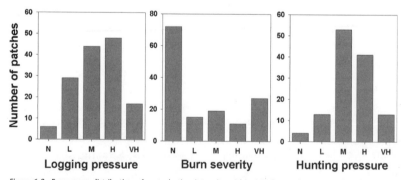

Figure 6.2. Frequency distribution of perturbation intensity within 144 forest sites surveyed in the Alta Floresta region showing the two main forms of anthropogenic habitat disturbance (logging and wildfires) and hunting (N = none; L = light; M = moderate; H = heavy; VH = very heavy)

find. In addition, our questionnaires ignored nontimber extraction of forest resources (e.g., firewood, lianas) other than game vertebrates. This clearly shows that most of the forest fragments we examined, rather than being left alone in the aftermath of fragmentation, are "working forests" fully integrated into different economic activities of rural communities. Fortunately, these sources of disturbance were not strongly intercorrelated and were thus unlikely to be confounded, which allowed a better understanding of the causal mechanisms of species loss. Considering all 129 forest patches surveyed, logging intensity was positively but weakly correlated with burn severity ($r_s = 0.239$, $P = 0.006$) and was uncorrelated with hunting pressure ($r_s = 0.081$, $P = 0.36$), whereas burn severity was not correlated with hunting pressure ($r_s = -0.114$, $P = 0.20$).

SPECIES–AREA RELATIONSHIPS

We can now turn to the effects of fragment area on the number of forest vertebrate species they retained. Fragment area, however, is but one of a number of key patch metrics that could influence the shape and slope of SARs and primarily reflects the effects of lost habitat rather than increased exposure to habitat degradation, overhunting, and changes in floristic composition. We consider our SARs to be relatively immune to systematic sampling artifacts (Rosenzweig 1995; Cam et al. 2002) and primarily a response to patch- and landscape-scale ecological processes, because interviewees did not devote more time and sampling effort to progressively larger fragments. For example, there was no correlation between the amount of time (years) local interviewees had lived next to the fragment on which they reported and either the size of those fragments ($r = -0.113$, $P_{adj} = 1.00$) or the number of vertebrate species they had detected over time ($r = 0.145$, $P_{adj} = 0.62$; $N = 127$ in both cases).

   Forest fragments examined here ranged in size from 0.47 to 13,551 ha (mean = 510 ± 145 ha, $N = 129$) and had been isolated for 2 to 27 years (mean 15.5 ± 5.8 years). These fragments retained from 11 to 55 of the 57 species included in the interview questionnaires (mean = 32.9 ± 10.9, $N = 129$; appendix 6A). In contrast, continuous-forest sites contained between 42 and 57 species (mean = 50.6 ± 5.0, $N = 15$). Habitat area alone explained 62% of the total variation in the number of vertebrate species recorded from 129 forest fragments (fig. 6.3). That relationship becomes marginally stronger when we consider only the core fragment size, calculated here as the aggregate area beyond 100 m from the nearest forest

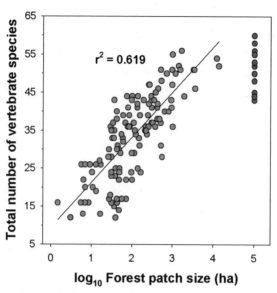

Figure 6.3. Overall relationship between number of vertebrate species and forest area for 129 fragments (light shading) and 15 continuous-forest sites (dark shading, arbitrarily set at 100,000 ha)

edge (R^2_{adj} = 0.651), suggesting that edge habitat was of lower quality for many of the species examined.

There is no evidence to indicate that vertebrate assemblage structure was influenced by large-scale spatial autocorrelation. The straight-line, edge-to-edge distance between any two sampled forest patches varied from 0.8 to 65.7 km (mean = 23.1 ± 11.7 km, N = 10,296), but there was no relationship between the Jaccard coefficient (J') of species similarity between any two patches and the linear distance between them (r = 0.086, N = 10,296 pairwise comparisons). This clearly demonstrates that patches that were farther apart did not necessarily contain more dissimilar assemblages. There is also no evidence to suggest that differences in species composition between sites are due to preexisting differences in forest types and tree species composition. As recently as 1976, the Alta Floresta region was covered entirely by undisturbed Amazonian *terra firme* forest of similar physiognomy (Oliveira Filho 2001), and most of the sampled vertebrate species are wide ranging and require large-scale vegetation mosaics to meet their metabolic needs. An adequate examination of the minimum spatial requirements and life-history correlates of local extinction risk in the sampled vertebrate species is beyond the scope of this chapter (but see Peres 2001b). Nevertheless, several minor-habitat specialists were apparently driven to local extinction in many small fragments due to random or nonrandom truncation of the habitat matrix, and the order with which they were extirpated cannot be attributed to chance events (Michalski and Peres 2005).

The slopes (z) and intercepts (C) of our linear species–area relationships (SARs) were clearly affected by whether we considered all species included in the study, forest specialists only, or game species only (fig. 6.4). These SARs can be represented in the form $\log_{10}S = z \log_{10}A + \log_{10}C$, where S is the number of species in a fragment, A is fragment area, and C is the number of species retained in a fragment of 1 ha. In all cases, the z-values of these relationships fall well within the range 0.13 to 0.33, which describes the linear slope of the vast majority of mainland and island SARs documented to date (Rosenzweig 1995). The positive slope of the forest specialists–area relationship ($z = 0.209$) was steeper than that for all species considered ($z = 0.169$), and steepest still for game species ($z = 0.302$) hunted throughout Amazonian forests (Peres 2000a). The predictive power of these relationships was also marginally improved when the number of considered species was restricted from all species ($R^2_{adj} = 0.560$) to forest specialists ($R^2_{adj} = 0.562$) to game species ($R^2_{adj} = 0.591$). Moreover, both the degree of forest-habitat specificity and whether or not a given species was hunted had significant effects on the relationship between size of fragments and the number of vertebrate species they retained (game/nongame species: ANCOVA, $R^2 = 0.61$, $F_{2285} = 208.8$, $P < 0.001$; forest/nonforest specialists: $R^2 = 0.76$, $F_{2285} = 460.0$, $P < 0.001$). This suggests that (1) forest specialists have smaller effective areas of suitable habitat available to them within a fragment, presumably because they are less edge-tolerant and more matrix-intolerant than other species, and (2) game species have larger spatial requirements or are more frequently driven to local extinction in small fragments by factors that exacerbate demographic stochasticity in small populations, such as hunting pressure.

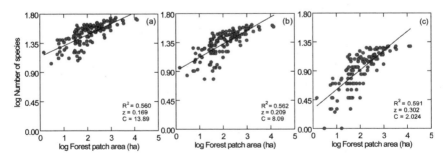

Figure 6.4. Log-log species–area relationships for (a) all 57 vertebrate species included in the study, (b) 35 forest specialists only, and (c) 19 game species only. Slope (z), intercept (C), and R^2 values are shown for each group of species.

The three forms of forest disturbance examined in this study, all of which result directly or indirectly from human activities, had moderate to profound effects on SARs in forest fragments. These can be shown for the entire gradient of selective-logging pressure, burn severity, and hunting pressure experienced by different fragments. We are unable to describe these interactions in detail here, but we illustrate them by focusing on fragment differences in previous levels of game harvest. Levels of hunting pressure profoundly affected SARs, particularly when we considered game species. We distinguished sites that had been exposed to a history of consistently low hunting pressure (including nonhunted and lightly hunted fragments) from those that had been previously subjected to persistently heavy hunting pressure. The slopes of SARs for game species were steeper, and the intercepts much lower, than those for nongame species regardless of the level of hunting pressure (fig. 6.5). However, even nongame species in heavily hunted areas apparently exhibited faster extinction rates ($z = 0.136$) than in lightly hunted areas ($z = 0.090$), probably as a response to other factors correlated with, but independent of, hunting pressure. Finally, the area-dependent local extinction rate of game species in heavily hunted fragments ($z = 0.279$) was much higher than in those that had been lightly hunted ($z = 0.181$). These effects are all the more remarkable because hunters in Alta Floresta are fairly selective and do not kill several harvest-sensitive species, such as large-bodied primates (Peres 1990,

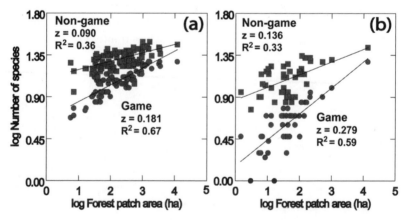

Figure 6.5. Species–area relationships for game (squares) and nongame species (circles) in forest sites that had been subjected to either (a) low or (b) persistently heavy hunting pressure. Slope (z) and R^2 values are shown for each species grouping.

2000a). Yet we distinguished 19 game and 31 nongame species on the basis of regional-scale (rather than local-scale) patterns of game offtake, which is primarily a function of body size (Bodmer et al. 1997; Jerozolimski and Peres 2003).

Stepwise multiple regression models—incorporating 11 forest-patch metrics, 7 landscape metrics, and our measures of within-fragment disturbance—were performed to tease apart the relative importance of different variables on the number of vertebrate species retained in different patches. Variables that were strongly intercorrelated ($r > 0.85$) were excluded from the models. Once other key independent variables had been taken into account, game species richness was markedly depressed by a history of persistent game hunting, although fragment size, forest-habitat quality, distance to Alta Floresta, and presence of water also had significant effects (table 6.1). Likewise, burn severity and a composite index of habitat perturbation derived from the history of both ground fires and logging had significant negative effects on the overall level of species persistence.

DISCUSSION

As far as we are aware, this study is based on the largest number of sites where vertebrate-assemblage composition has been quantified across a wide range of fragment sizes and disturbance regimes in a tropical landscape. We showed that both the area and the regime of human disturbances within habitat fragments in recent times, particularly hunting pressure, are key determinants of local extinction rates in Amazonian forest vertebrates. Logging practices and wildfires, which often co-occur in seasonally dry tropical forests (Cochrane et al. 1999; Nepstad et al. 1999b), are major causes of forest degradation in southern Amazonia. These can rapidly erode remaining areas of forest interior that are characterized by high basal areas of trees and shaded undergrowth underneath a closed canopy. These patterns of perturbation were often severe and widespread and did not mimic the background scale and frequency of natural disturbance events in this region, which has important implications for ecosystem functioning in habitat islands (Pickett and Thompson 1978). Other less-severe forms of patch-scale disturbance induced by human activities in Alta Floresta include small-scale harvesting of firewood, livestock grazing (as shown by a clear browse line in the understory of several fragments used by cattle), and the spread of invasive understory species. However, these neither puncture the forest canopy nor drastically alter microclimatic conditions over the short term.

Table 6.1. Stepwise multiple regression models showing the effects of fragment- and landscape-scale variables and different forms of disturbance on the number of vertebrate species persisting in 129 forest fragments. Variables listed here are those retained by the models (for 52 forest species combined, game species only, and nongame species only) using a backward-elimination procedure at P = 0.15. Variables excluded from the models are not listed.

Effect	Coefficient	SE	Std Coef	Tolerance	t	P
All species (52 taxa)						
$\log_{10}$ fragment area	8.743	0.482	0.850	0.945	18.146	0.000
Habitat quality	3.594	0.562	0.308	0.893	6.392	0.000
Burn severity	0.304	0.344	0.042	0.915	0.885	0.378
Habitat perturbation	−0.288	1.074	−0.024	0.241	−0.268	0.780
Game species (19 taxa)						
$\log_{10}$ fragment area	3.628	0.256	0.775	0.621	14.186	0.000
Habitat quality	1.526	0.252	0.287	0.822	6.050	0.000
Distance to Alta Floresta	0.000	0.000	0.095	0.918	2.110	0.037
Water	2.349	0.882	0.130	0.779	2.665	0.009
Hunting pressure	−0.295	0.149	−0.095	0.804	−1.987	0.049
Nongame species (31 taxa)						
$\log_{10}$ fragment area	4.126	0.348	0.791	0.699	11.864	0.000
Habitat quality	1.681	0.349	0.289	0.861	4.814	0.000
Distance to nearest fragment	−0.008	0.012	−0.042	0.715	−0.641	0.523
Water	−0.081	1.437	−0.004	0.720	−0.056	0.955
Burn severity	0.293	0.229	0.079	0.816	1.278	0.204
Hunting pressure	0.384	0.283	0.084	0.814	1.355	0.178

Several other fragment and landscape variables were quantified and also had important implications for the forest vertebrate fauna in Alta Floresta. These include habitat quality (based on spectral analysis of the forest canopy in each fragment), the proportion of forest edges, the presence or absence of perennial sources of water, and the different measures of forest connectivity including the cost-surface and linear distances to adjacent forest patches. Ultimately, however, the lingering effects of past disturbances and systematic depletion of wildlife can dramatically alter conventional species-richness curves based solely on the effects of fragment size and shape and landscape structure. For example, many relatively undisturbed small forest fragments (e.g., 100–250 ha) retained more species than did larger fragments (e.g., 250–500 ha) with comparable levels of connectivity, if they had not been previously subjected to more-intensive habitat disturbances or hunting (fig. 6.3). Local extinction rates thus appear to be accelerated by habitat degradation, which effectively subtracts from the amount of suitable habitat remaining in a forest fragment. Hunters can also further aggravate declines in game populations that are already reduced in small patches.

We realize that the number of species in a forest fragment is a function of both local extinction and recolonization events, but a more detailed analysis of these processes is beyond the scope of this chapter. We also accept that the levels of species persistence we encountered may reflect preequilibrium conditions because the "relaxation" time since isolation for many fragments may have been insufficient for long-lived taxa (e.g., Ferraz et al. 2003). Hence, many small populations that are presently confined to small fragments may eventually go extinct. Population sizes of several large-bodied, wide-ranging, or rare species in many small fragments are almost certainly not viable in the long run and are estimated to comprise fewer than 50, if not fewer than 10, individuals. Our fragment-occupancy data may therefore severely underestimate local extinction rates over longer intervals. However, the clear responses of species richness to fragment area and several other measures of forest integrity give us confidence that most species respond rather rapidly to both habitat loss and habitat degradation.

Our present analysis also ignores interspecific differences in vulnerability to local extinction that are linked to the life-history traits and ecological plasticity of different species (Purvis et al. 2000). These may include the degree of matrix tolerance (W. F. Laurance 1991a; Dale et al. 1994; Gascon et al. 1999) and sensitivity to current and background disturbance regimes within a patch. Local extinctions documented here were by no means chance events, and the high degree of nestedness of the overall species-by-site matrix indicated that species clearly disappeared from fragments in a nonrandom manner, whereby the most vulnerable species were consistently absent from species-poor fragments (Peres and Michalski, unpublished data). Forest specialists with limited dispersal capabilities presumably have a much lower tolerance of forest fragmentation than highly vagile species, which may perceive the landscape as functionally connected across a greater range of fragmentation severity. The rapid extinction of large-bodied game species likely occurs because they are preferred by hunters and tend to have lower fecundity as well as large spatial requirements (Bodmer et al. 1997; Peres 2001b). Indeed, average extinction rates for game species in forest fragments of 10 to 100 ha were 6 to 10 times faster than those for all other species.

Hunting has been shown to dramatically affect patterns of vertebrate persistence in tropical forest fragments in Africa (Brashares et al. 2001), Southeast Asia (Bennett and Dahaban 1995; Laidlaw 2000), and the Neotropics (Cullen et al. 2000; Peres 2001b). Brashares et al. (2001) showed that extinction rates for 41 large mammal species in six nature reserves in West Africa were 14 to 307 times higher than those predicted by models based on reserve size alone. Externally driven mortality clearly interacts with the size of habitat remnants to determine the viability of insular populations (Woodroffe and Ginsberg 1998). The disturbance–habitat area interaction is aggravated by the fact that, in most tropical landscapes, the amount of remaining forest cover is probably correlated with human population density such that small fragments and reserves are significantly more likely than large ones to be located in densely populated regions (e.g., Harcourt et al. 2001). For this reason, many immediate extinction events in forest fragments may be exacerbated by human disturbances rather than a sole result of effects of habitat area.

The strong trends shown here are unlikely to be restricted to faunal communities and may extend to tree assemblages as well. Logging practices in the humid tropics become increasingly less selective over multiple cutting cycles (Alvarado and Sandberg 2001), and many forest fragments we surveyed had already been logged two or three times. Likewise, differences among tree species in susceptibility to recurring ground fires (Barlow et al. 2003a) can cause variable mortality rates and a strong selective pressure against fire-sensitive species, such as many taxa that bear fleshy fruits (Barlow and Peres 2004b, 2006). This is likely to impoverish the overall tree assemblage and the year-round resource base for frugivorous vertebrates, especially in forest fragments lacking unlogged or unburned refugia. Similar effects on plant communities have also been shown in other forest ecosystems. For example, in 50 fragments of dry open forest studied in southeastern Australia, anthropogenic disturbance coupled with fragmentation had a stronger and more immediate effect in reducing native-plant-species richness and increasing exotic-species richness than did fragment size alone (Ross et al. 2002).

We would therefore add to conventional geometric and landscape principles of reserve design (Diamond 1975) by recommending that forest remnants in severely degraded tropical landscapes should, wherever possible, be spared additional onslaught from intensive habitat disturbances and hunting. In heavily degraded regions like Alta Floresta, the importance of even relatively small forest remnants should not be underestimated in terms of their biodiversity and functional value. Forest patches of any size can act as sources of large-seeded plant species, stepping stones for animal dispersal, buffer strips for riparian corridors, firebreaks to deter pasture fires, and regulators of local and mesoclimate. Nonetheless, we recognize that large, relatively undisturbed forest patches are the biological "crown jewels" of the landscape mosaic (Peres 2005). Their role as refugia and sources of emigrants for many disturbance- and area-sensitive species should thus be recognized by private landowners and by government agencies involved in managing forest frontiers.

CONCLUSIONS AND IMPLICATIONS

1. The vast majority of tropical forest remnants worldwide experience varying anthropogenic disturbances and will continue to be used for a variety of extractive purposes. Ecologists thus require an understanding of how people alter the habitat structure and biological communities of fragmented forest landscapes.

2. The "internal fragmentation" process, which results from a proliferation of large canopy gaps generated by selective logging and fire-induced tree mortality, often propagates edge effects into remaining core areas of fragments, degrading forest habitat for edge- and matrix-intolerant species that are already subdivided into small, isolated populations.

3. Persistent hunting accelerates the local extinction rate of largebodied game species in forest remnants and vastly increases their spatial requirements if they are to be sustainably harvested. Both hunting pressure and fragment size are key determinants of vertebrate species loss in southern Amazonia.

4. Species–area relationships are strongly influenced by postfragmentation human disturbances. The local extinction rate of many areasensitive species is considerably faster in disturbed and/or hunted remnants than in similar-sized or even smaller patches that have been protected or left alone.

5. Lessons from this study can be applied to both the design of future agricultural resettlement schemes in forest frontiers and the management of remaining forest patches in landscapes that have already been heavily degraded.

APPENDIX 6A

Table 6A.1 Species (or functional groups of species) surveyed in the Alta Floresta region

Taxonomic group	Latin name	English name
Ungulates	*Tapirus terrestris*	Lowland tapir
	Tayassu tajacu	Collared peccary
	Tayassu pecari	White-lipped peccary
	Mazama americana	Red brocket deer
	Mazama gouazoubira	Gray brocked deer
Primates	*Ateles marginatus*	White-fronted spider monkey
	Alouatta belzebul	Red-handed howler monkey
	Chiropotes albinasus	White-nosed bearded saki
	Callicebus moloch	Dusky titi monkey
	Cebus apella	Brown capuchin monkey
	Aotus azarai and *A. nigriceps*	Azara's night monkey and black-headed night monkey
	Callithrix emiliae and *Callithrix melanurus*	Snethlage's marmoset and black-tailed marmoset
Rodents	*Hydrochaeris hydrochaeris*	Capybara
	Agouti paca	Paca
	Dasyprocta agouti	Red-rumped agouti
	Sciurus aestuans	Guianan squirrel
	Coendou sp.	Tree porcupine
Lagomorphs	*Sylvilagus brasiliensis*	Brazilian rabbit
Didelphid marsupials	*Didelphis* sp.	Common opossum
Edentates	*Priodontes maximus*	Giant armadillo
	Euphractus sexcinctus	Six-banded armadillo
	Dasypus kappleri	Greater long-nosed armadillo
	Dasypus novemcinctus	Nine-banded armadillo
	Cabassous unicinctus	Southern naked-tailed armadillo
	Bradypus variegatus	Three-toed sloth
	Choloepus didactylus	Two-toed sloth
	Myrmecophaga tridactyla	Giant anteater
	Tamandua tetradactyla	Collared anteater
	Cyclopes didactylus	Silky anteater
Carnivores	*Potos flavus*	Kinkajou
	Eira barbara	Tayra
	Nasua nasua	Coati
	Galictis vittata	Greater grisson
	Procyon cancrivorus	Crab-eating raccoon

Taxonomic group	Latin name	English name
	Cerdocyon thous	Crab-eating fox
	Atelocynus microtis	Small-eared dog
	Speothos venaticus	Bush dog
	Herpailurus yaguarondi	Jaguarundi
	Leopardus wiedii	Margay
	Leopardus pardalis	Ocelot
	Puma concolor	Puma
	Panthera onca	Jaguar
	Pteronura brasiliensis	Giant otter
	Lontra longicaudis	Neotropical river otter
Cracids	*Mitu tuberosa* and *Crax fasciolata*	Razor-billed and bare-faced curassow
	Penelope jacquacu	Spix's guan
	Pipile cujubi	Red-throated piping-guan
Psophiids	*Psophia viridis*	Dark-winged trumpeter
Woodquails	*Odontophorus gujanensis*	Marbled wood-quail
Tinamids	*Tinamus major, T. tao,* and *T. guttatus*	Great tinamou, gray tinamou, and white-throated tinamou
	Crypturellus spp.	Small tinamous (all congeners)
Toucans	*Ramphastos vitellinus* and *R. tucanus*	Yellow-ridged toucan and white-throated toucan
	Pteroglossus spp.	Aracaris (all congeners)
Large parrots	*Ara ararauna*	Blue-and-yellow macaw
	Ara macao and *A. chloroptera*	Scarlet and red-and-green macaw
	Amazona spp.	Amazon parrots (all congeners)
Reptiles (Testudines)	*Geochelone denticulata* and *G. carbonaria*	Yellow-footed and red-footed tortoise

ACKNOWLEDGMENTS

This study was funded by the Natural Environment Research Council of the United Kingdom through a grant (2001/834) to CAP. We wish to thank the Center for Applied Biodiversity Sciences of Conservation International and the John Ball Zoological Society for supplementary funds. FM is funded by a PhD studentship from the Brazilian Ministry of Education (CAPES). Fieldwork in Alta Floresta would not have been possible without the generous cooperation of numerous smallholders, cattle ranch managers, and landowners. We are especially indebted to Vitória da Riva Carvalho. Geraldo Araújo and Iain Lake provided invaluable assistance during the fieldwork and GIS analysis, respectively. We thank Bill Magnusson, Pedro Develey, and Bill Laurance for constructive comments on the manuscript.

CHAPTER 7

Human Encroachment and Vegetation Change in Isolated Forest Reserves: The Case of Bwindi Impenetrable National Park, Uganda

William Olupot and Colin A. Chapman

Human modification of ecosystems is threatening biodiversity on a global scale (Whitmore 1997; W. F. Laurance 1999; Nepstad et al. 1999b; Chapman and Peres 2001). This is illustrated by the fact that an estimated 65.1 million hectares of forest were destroyed between 1990 and 1995 (FAO 1999). Many other areas are affected by forest degradation that involves logging, fire, and hunting; as a result, many conservation efforts have focused on protected areas such as national parks and reserves. However, less than 5% of the world's tropical forests are legally protected from human exploitation, and many of these are subjected to illegal exploitation (Oates 1996; Peres and Lake 2003; Fagan et al., chap. 22 in this volume). Given that a high rate of deforestation is occuring globally, and that the extent of protected areas is limited, it is important to understand how well such protected areas are protecting biodiversity.

Recent analyses have shown that areas of outstanding conservation importance often coincide with dense human settlement or impact (Dobson et al. 1997; Balmford et al. 2001; Harcourt and Parks 2003). For example, Balmford et al. (2001) found that species richness of birds, mammals, snakes, and amphibians was positively correlated with human population density in sub-Saharan Africa. Furthermore, these trends held for widespread, narrowly endemic, and threatened species. Thus, it is likely that many important biodiversity conservation areas are experiencing intense pressures from adjacent dense human settlements. Given current rates of human population growth and patterns of migration, those areas that are not currently experiencing this pressure are likely to do so in the near future. Effectiveness of conservation programs should grow in parallel to the growing threats: protected areas currently surrounded by dense human populations should have programs in place to deal with these intense pressures, whereas protected areas that are currently surrounded by few people should have plans constructed for a changing future. Effective conservation is only possible when accurate, timely information is available regarding the threats that face protected areas. Consequently, conservation biologists should seek novel approaches to determine threats and to assess the degree to which current programs have been successful.

The pressures being experienced by these protected areas will likely be concentrated along their edges. As a result, protected-area management plans should incorporate the rich data available from studies of forest fragments and edge effects (Kapos et al. 1997; W. F. Laurance and Bierregaard 1997; Benítez-Malvido 1998). However, much of the previous work in fragmented habitats has involved fragments protected from human use (Lovejoy et al. 1986; Malcolm 1994; Tutin et al. 1997). In reality, most fragments are not protected; they are on land managed by private residents who depend on them for fuel wood, medicinal products, or game (Chapman et al. 2003; Peres and Michalski, chap. 6 in this volume). In addition, the edges of protected areas in a matrix of high human population density are often not effectively protected and are encroached upon by local people for a variety of resources (Harcourt et al. 2001). The structure and composition of habitat edges will change with varying levels of management effectiveness.

The aim of this chapter is to quantify changes in resource extraction and vegetation structure as a function of distance from the forest edge into Bwindi Impenetrable National Park, Uganda. We consider resource extraction methods such as tree cutting and firewood removal, hunting, burning, grazing, vegetation clearing, and the spread of exotic plants. We also quantify changes in the vegetation structure as a function of distance from the forest edge. It is likely that the patterns we observe will be similar to those observed in other tropical reserves surrounded by dense human populations around the world.

METHODS

Study Site: Bwindi Impenetrable National Park

Data presented here are from a 22-month (May 2001–February 2003 inclusive) study of Bwindi Impenetrable National Park, located in southwestern Uganda (0°53′ –1°08′ S, 29°35′ –29°50′ E). This park is a forest island best known for its uniqueness in bird and butterfly diversity and as home to approximately half of the world's remaining mountain gorillas (*Gorilla gorilla*; approximately 650 individuals). In recognition of these values, it was declared a World Heritage Site in 1994. Topography of the region is extremely rugged and is characterized by numerous steep-sided hills and narrow valleys. Ranging from 1190 to 2607 m (Butynski 1984), the park is broadly classified as a medium- to high-altitude forest. The climate is tropical with two rainfall peaks, from March to May and from September to November. The annual mean temperature range is 7°C–15°C minimum to 20°C–27°C maximum, and annual precipitation ranges from 1130 to 2390 mm (Howard 1991).

Established in an area with a large rural population (200–400 people per km^2; the national average for Uganda is 88 people per km^2), the park boundary is typically an abrupt transition from forest to a matrix of croplands and settlements. The area's status has changed frequently with an increase in protection status and spatial extent (from 207 km^2 in 1932 to 321 km^2 in 1991). The most recent change in 1991 was its transformation from a central forest reserve, which allowed timber extraction, and a game reserve, which allowed the controlled harvest of game, to a national park, which permits very limited extraction.

Prior to its establishment as a national park, the forest was under severe human pressure. Butynski (1984) estimated that between 512 and 1049 people entered the forest daily to illegally remove wood, bamboo, livestock forage, minerals, honey, and meat. Until 1991, timber extraction, gold mining, and hunting were the gravest threats, leading to one of the highest anthropogenically related gap sizes and frequencies known for tropical forests (Babaasa et al. 2001) and the extinction of at least two mammal species from the area: buffalo (*Synerus caffer*) and leopard (*Panthera pardus*; Butynski 1984). Timber extraction, greater than 80% of which was illegal, was widespread throughout the reserve (Butynski 1984), although it was probably more intense along the edge than in the interior (Howard 1991). Evidence of hunting was common throughout the area (Butynski 1984).

Once it achieved national park status, the forest was transformed from an extractive reserve to an officially protected area. Currently, under multiple-use agreements, local communities are permitted limited extraction of weaving and medicinal products in designated areas. The extent to which these and other conservation efforts have reduced illegal activity is not clear. When this study began, most threats were thought to have been reduced. Agricultural encroachment and mining were believed to have been eliminated, but forest burning was considered a continuing threat. In addition, exotic plants were known to occur, but their distribution was unknown.

Field Methods

Fieldwork was conducted with the help of 8 to 10 assistants. Three sampling procedures were used: boundary walks, 100 m edge–interior transects, and 1 km edge–interior transects. Sampling was stratified by the 22 governmental administrative units surrounding the park, known as parishes (fig. 7.1). Boundary walks (146 km) were conducted to obtain information on the distribution of resource extraction within the immediate edge for all parishes and involved sampling an area up to 60 m into the park for signs of human incursions such as tree cutting, firewood extraction, and game hunting. All signs of resource harvest visible within

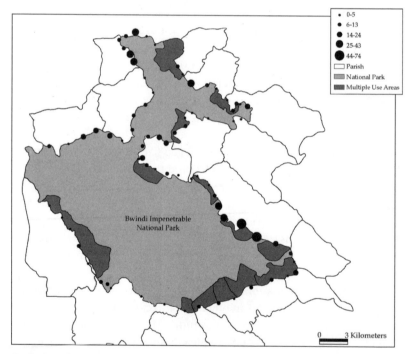

Figure 7.1. Map of Bwindi Impenetrable National Park, Uganda, showing the distribution patterns of forest-resource extraction near the park edge. Size of circles represents the intensity of pole extraction at that point near the edge. Lines demarcating clear spaces indicate parish boundaries. Dark gray areas within the park depict multiple-use zones.

William Olupot and Colin A. Chapman

a 5 m radius of each observed harvest were identified and noted. To do this, we walked a 5 m radius around each harvest sign to establish if there were other harvest locations nearby. When no other signs of harvest were visible, the observer returned to the park boundary line. Instances of resource extraction were detected by walking the boundary and looking for signs, by entering into the forest 5 to 10 m at randomly set distances along the edge, and by following human trails and footpaths from the edge to 20 to 30 m into the park. For resource extraction, only signs less than 2 years old were considered. Aging of felled trees was conducted using a method given by Sheil (1997), which involves scratching the surface of a stump with the tip of a fingernail. A stump was considered to be less than 2 years old if the surface was firm to the scratch. Because extractions are often clumped and more evident near the edge than in the interior, this method yields data more suitable for parish-level comparisons than that obtained from transects.

The team sampled 100 m edge–interior transects while conducting boundary walks. These transects were used to quantify how far different types of resource extractions occur into the park. Transects were 400 m apart along the edge and transect intervals were determined by measuring distance with a hip chain. Along these transects, sampling was done in 5 m radius circular plots, placed at 10 m intervals with the zero mark on the park's boundary. Within each plot, we counted the number of trees (diameter at breast height [dbh] >10 cm), poles (young trees between 5 and 10 cm dbh), and saplings (young trees >2 m high and <5 cm dbh). We also quantified all forms of people–park interactions and counted numbers of dead trees and exotic plants. A total of 375 of such transects were sampled.

Edge-interior transects of 1 km were used to understand how extraction levels and vegetation structure changed from the edge to the interior; we sampled 104 such transects. Eight parallel transects were established along each parish–park boundary more than 5 km long (11 parishes). On two occasions, two adjacent parishes were combined to achieve this minimum length; thus, 13 parish units were sampled. Within each parish, transects were placed equidistantly along, and ran perpendicular to, the park's boundary. Transects were marked at 50 m intervals (corrected for slope). At each 50 m point, we recorded slope (in degrees of inclination), slope location (bottom, lower slope, midslope, upper slope, and ridge top), elevation, and aspect (N, NE, E, SE, S, SW, W, NW). Within 50 × 5 m plots, we counted all forms of human disturbance as well as the number of exotic plants and dead trees. In 50 × 2 m plots, we quantified regeneration patterns by counting the number of poles and saplings. Tree density, size, and diversity were assessed in variable-area plots placed at 100 m intervals starting at the zero mark on the boundary line. Fifteen trees nearest the 100 m mark were identified and dbh measurements were taken. For each plot, sampling was only performed when the nearest tree was less than 5 m from the mark and if the fifteenth tree was no more than 40 m away; otherwise, the plot was recorded as "open." Thus, some plots with very few trees were classified as open.

To explore what determines the intensity of resource extraction, we collated, by parish, data collected through boundary walks and determined population density as estimated during the 1991 population census. Each parish was categorized according to district (table 7.1), whether or not it had a multiple-use program, whether it had no multiple-use program or one based on beekeeping or resource harvesting (the Ugandan Wildlife Authority does not consider beekeeping to be a detrimental form of resource extraction), whether or not settlements in the parish generally tended to be near or far from the edge, and whether the edge ran along some kind of barrier (large river, deep gorge, steep hill).

Using a general linear model (GLM), we employed the backward stepwise regression procedure to determine the best predictors for tree-cutting intensity, pole-cutting, and firewood-harvest intensity. GLM is preferred when the influence of categorical variables or a mixture of con-

Table 7.1 Results of multiple regression analyses to assess factors influencing resource extraction in different parishes at the edge of Bwindi Impenetrable National Park, Uganda. In addition to using data on population density, each parish was categorized according to its district, whether or not it had a multiple-use program (MuP1), whether it had no multiple-use program versus one based on beekeeping or resource harvesting (MuP2), whether settlements in the parish were generally near or far from the edge (proximity), and whether the edge ran along some kind of barrier, such as a large river, deep gorge, or steep hill (accessibility). *$P < 0.05$; **$P < 0.01$; ***$P < 0.001$; ns: not significant.

	Response variables		
Predictors	Trees	Poles	Firewood
Population density	ns	ns	ns
District	ns	ns	ns
MuP1	**	ns	*
MuP2	***	***	***
Proximity	ns	ns	ns
Accessibility	ns	***	***
R^2	0.792	0.522	0.737

tinuous and categorical variables is being examined. In exploring determinants of tree and pole cutting, we also included tree and pole density within 50 m of the edge as determined from 100-m-long transects. Using multiple linear regression, we examined several predictors for edge–interior variation in tree regeneration, density, size, and diversity (as estimated by the Shannon-Wiener index; table 7.2). The predictor variables included distance from the edge, elevation, slope, slope location, and aspect. Statistical analyses were performed using SYSTAT version 10.2. To plot variables measured on different scales on the same axes, each variable was standardized by subtracting the variable's sample mean from each value and then dividing the difference by the sample standard deviation.

Table 7.2 How specific descriptors of the tree community are influenced by five ecological variables in Bwindi Impenetrable National Park, Uganda. This correlational analysis was done over three different distances into the national park from the edge. *$P < 0.05$; **$P < 0.01$; ***$P < 0.001$; ns: not significant.

Response variable	Distance from edge (m)			Trend from edge
	300	600	1000	
Tree density				
Distance from edge	***	ns	*	increasing
Elevation	***	***	***	decreasing
Topographic location	**	***	ns	increasing
Slope	ns	*	**	decreasing
Aspect	*	ns	ns	increasing
N	51	90	140	
Tree species diversity				
Distance from edge	ns	ns	ns	no trend
Elevation	***	**	***	decreasing
Topographic location	ns	*	ns	decreasing
Slope	ns	ns	ns	no trend
Aspect	*	ns	ns	increasing
N	50	88	137	
Tree size				
Distance from edge	ns	ns	ns	decreasing
Elevation	ns	ns	ns	decreasing
Topographic location	ns	ns	ns	increasing
Slope	ns	ns	ns	decreasing
Aspect	ns	ns	ns	increasing
N	50	88	137	

Continued

	Distance from edge (m)			
Response variable	300	600	1000	Trend from edge
Pole regeneration				
Distance from edge	ns	ns	ns	no trend
Elevation	***	***	ns	decreasing
Topographic location	***	***	***	increasing
Slope	ns	**	***	decreasing
Aspect	**	ns	ns	increasing
N	78	156	260	
Sapling regeneration				
Distance from edge	ns	ns	*	decreasing
Elevation	***	***	***	decreasing
Topographic location	***	**	ns	increasing
Slope	ns	ns	*	decreasing
Aspect	***	ns	ns	increasing
N	78	156	260	

RESULTS

Resource Extraction

In 283 km sampled for this study, no fresh mining pits were observed inside the park and only nine trees were recorded as having been pit-sawn for timber. Most cut trees were small (<20 cm dbh), perhaps used for building or felled incidentally during the setting of beehives (total number of cut trees, 313; number of cut poles, 1750). Larger trees ($n = 6$) were cut to extract honey or to make beehives. We established that agricultural encroachment, although minimal (seven new incidences; average size = 1264 m², range = 33–3037 m²), nevertheless still existed, contrary to information available before the study commenced. The agricultural plots were in hidden locations within the park or along the edge. The hidden plots were established to grow marijuana (*Cannabis sativa*), passion fruit (*Passiflora incarnata*), or tree tomato (*Cyphomandra betacea*). Plots along the edge were planted with tea (*Camellia sinensis*) or alfalfa (*Medicago sativa*), but the majority had not been planted with any crop.

Evidence of hunting was found on 31 occasions, all of which involved snares. The 1 km edge–interior transects indicate that, other than hunting, which was still widely distributed in the park (see also McNeilage et al. 1998; Uganda Wildlife Authority, unpublished data), most resource extraction was concentrated near the park's periphery (figs. 7.2 and 7.3). The distribution of resource extraction was clumped (fig. 7.1) and most extraction occurred within 300 m of the park edge (fig. 7.2). Exotic plants, especially cypress (*Cupressus lusitanica*), *Lantana camara*, and tea, were most abundant within 150 m of the edge. The density of exotics within the first 150 m of edges was 2.66 stems and clumps per square meter, whereas deeper into the reserve their density was 0.08 stems and clumps per square meter. Damage from fire was most extensive within 150 m of the forest edge, but there were still areas that were extensively damaged within 500 m of the park's edge. In total, 17% of plots within 150 m of the forest edge had evidence of burning compared to 8% of the plots along the remainder of the transect. Tree/pole mortality was closely correlated with burning ($r = 0.74$, $P < 0.001$).

A regression of tree, pole, and firewood extraction on several predictor variables showed a strong positive relationship between resource extraction and multiple-use programs; extraction was particularly high in areas designated for beekeeping (table 7.1). Population density and the

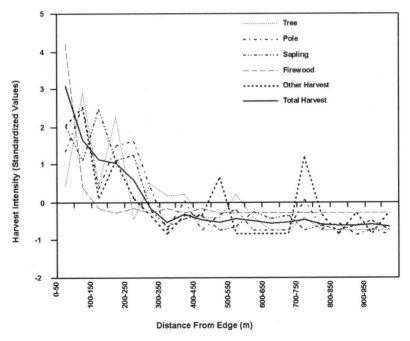

Figure 7.2. Changes in intensity of resource extraction along a gradient from the edge to 1000 m inside Bwindi Impenetrable National Park, Uganda. All values are standardized to facilitate comparisons among different threats.

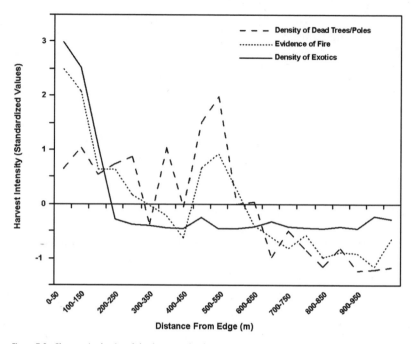

Figure 7.3. Changes in density of dead trees and poles, evidence of fire, and density of exotic plants along a gradient from the edge to 1000 m inside Bwindi Impenetrable National Park, Uganda. All values are standardized to facilitate comparisons among different threats.

other factors considered did not account for much of the variance in intensity of resource extraction (table 7.1).

Global threats of the spread of exotic species are well documented (Williamson 1996). In Bwindi, two small plantations of *Eucalyptus* occurred near the edge, one in the southeast and another in the northern part of the southern sector. Distribution of cypress (*Cupressus lusitanica*) was related to their use as boundary trees. In the northern lower-elevation areas, *Lantana camara* occurred in gaps near the edge, in previously burned sites, and in adjacent old, previously unmanaged tea plantations. Tea and *Acacia mearnsii* plants were predominant in a previously settled area in the northeastern edge in the southern sector. Only *L. camara* and *Camellia sinensis* were known to be reproducing and successfully recruiting without human assistance.

Vegetation Changes as a Function of Distance from the Park Edge

The 100 m edge–interior transects show an abrupt increase in the density of trees, poles, and saplings between the edge and 20 m into the interior,

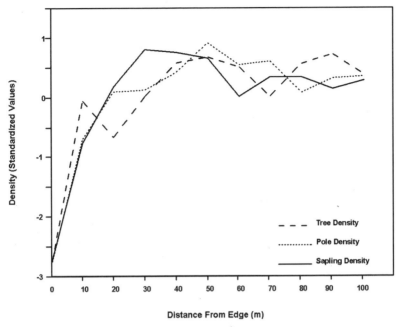

Figure 7.4. Variation in tree, pole, and sapling density along 100 m transects from the edge of Bwindi Impenetrable National Park into the interior of the park. Transects were 400 m apart along the edge.

but there was little change over the remainder of the 100 m transect (fig. 7.4; $n = 375$ transects). Similarly, after about the first 20 m, there were no clear patterns in the density, species diversity, size, or regeneration of trees on the 1 km transects (figs. 7.5 and 7.6; $n = 102$ transects); multiple middistance peaks were apparent for all variables.

Detailed analyses were conducted on data from the 1 km edge–interior transects to describe how environmental factors (e.g., elevation, slope, distance from the edge) were related to descriptors of the tree community. We considered such relationships using three different distance categories (table 7.2). Within 1 km of the edge, only tree density and sapling regeneration were related to distance from the edge and when analyzed only at some of the distance categories. Tree density generally increased toward the interior, whereas sapling density decreased. Within 600 m of the edge, no edge effect was evident. Within the 300 m zone from the edge, tree density increased with distance from the edge, and tree density, tree-species diversity, and pole and sapling regeneration all increased with a change in aspect. Within this zone, north-facing slopes tended to have lower values for each of these variables than slopes facing northeast, which had lower values

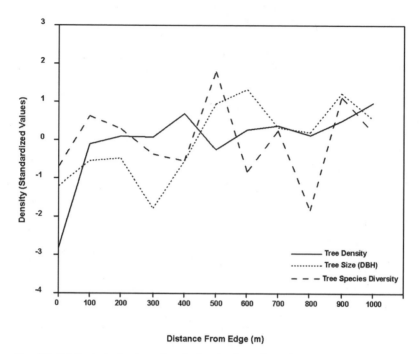

Figure 7.5. Variation in density, size, and species diversity of trees along 1 km transects from the edge of Bwindi Impenetrable National Park into the interior of the park (*N* = 104 transects)

than slopes facing east, and so on. Elevation and slope location (which affect temperature and soil moisture, respectively) had strong influences at any distance from the edge. The influence of slope angle was only important toward the interior.

DISCUSSION

A great deal has been learned about the nature and distribution of edge effects in fragmented habitats (Murcia 1995; W. F. Laurance and Bierregaard 1997). Many species of plants (Helle and Muona 1985; N. R. Webb 1989; Matlack 1994; Camargo and Kapos 1995; Turton and Freiburger 1997; W. F. Laurance 2000) and animals (Kroodsma 1982; K. S. Brown and Hutchings 1997; Schlaepfer and Gavin 2001; Beier et al. 2002) respond to edge proximity. However, in Bwindi Impenetrable Reserve, little was previously known about the effect of edge proximity on species abundance and distribution. Andama (2000) found that side-striped jackals (*Canis adustus*) and African civets (*Viverra civetta*) were

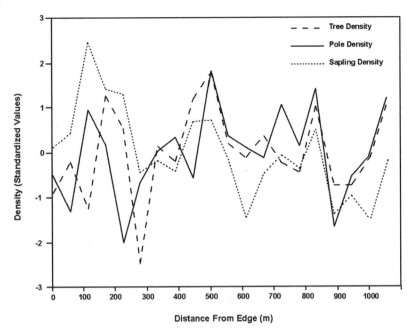

Figure 7.6. Variation in tree, pole, sapling density along 1 km transects from the edge of Bwindi Impenetrable National Park into the interior of the park (*N* = 104 transects)

more frequently located at the edge than were other carnivores, whereas Bataamba (1990) reported that four raptor species were edge dwellers. McNeilage et al. (1998) also found that all monkeys were more common around the edges of the park than in other parts of the forest.

The results of this study suggest that edge effects at Bwindi Reserve are influenced by a combination of abiotic changes and greater levels of resource extraction near the edge. At Bwindi, the increase in tree density with distance from the edge likely reflects the fact that resource extraction was most intense near the park edge and that fires were also frequent within this zone (see also W. F. Laurance, chap. 5 in this volume; Peres and Michalski, chap. 6 in this volume). However, similar trends have been reported near forest edges that did not experience resource extraction (Chen et al. 1992; Malcolm 1994; Camargo and Kapos 1995; Young et al. 1995). Several studies on tree regeneration have documented a decrease in sapling density farther from edges and have suggested that this may represent a regeneration response to disturbance, where many younger individuals are recruited into the locations left vacant by the death of older trees (Malcolm 1994; Camargo and Kapos 1995). The death of older trees may result from changing microclimates and increased wind turbulence associated with forest edges (Malcolm 1994; Camargo and Kapos 1995; W. F. Laurance et al. 1998; W. F. Laurance, chap. 5 in this volume).

Unlike some earlier studies, which found an increase in saplings and small trees near the edges, our results from the 100-m transects show an opposite trend, with saplings and poles both increasing farther from the edges. This probably results from resource extraction that involves harvesting poles, trampling seedlings, and fire encroachment often associated with harvesting honey and likely represents a fundamental difference between forest edges protected from human activities and those that are not. In reserves that are surrounded by dense human populations—especially those that do not have effective procedures to limit forest exploitation—it is likely that all size classes of trees will decline in density near the edges. It is unknown whether saplings and poles will be reduced to a level where they cannot replace adult trees that are harvested by local people or that succumb to edge effects. If they are reduced beyond replacement level, the edge will recede and the habitat area of the park will decline.

At a pantropical level, fire is common around forest edges and is more frequent in cleared areas (Lovejoy et al. 1984; Viana et al. 1997; Nepstad et al. 1999b; W. F. Laurance, chap. 5 in this volume; Peres and Michalski, chap. 6 in this volume). Correspondingly, at Bwindi, fire is likely the most important factor that affects the forest understory (Babaasa et al. 1999; Kasangaki et al. 2001). To stop the high mortality of trees near the edge, burning has to be controlled through implementation of a fire-management plan. Fires in Bwindi are mainly accidental, spreading from neighboring fields or from activities associated with honey harvesting (Babaasa et al. 1999). Some of the measures that need to be developed and implemented to control such fires at Bwindi include educating bee-keepers and developing better methods of handling fire used in honey harvesting, maintaining effective reserve boundaries, educating communities on ways to prevent fire from adjacent fields entering the forest, and, where gaps near the edge predispose the forest to burning, planting trees to facilitate tree regeneration. Because fire is often associated with agriculture (Nepstad et al. 1999b), effective control systems will likely be important in all parks surrounded by high human population density.

In an effort to secure support from communities near protected areas, Uganda adopted a policy of community-based conservation in 1988 (Mugisha 2002). This decision was based on a global trend that advocated a sharing of both the responsibility and the benefits of managing protected areas between government agencies and local communities (McNeely 1989; Brandon 1998). This perspective was a backlash against the purely protectionist approaches that were often previously adopted.

The rationale for establishing community-based conservation projects was based on the assumption that local communities will take on responsibility for resource management when they receive direct benefits from conservation (see also Whitten and Balmford, chap. 17 in this volume).

In the extensive area that adjoins Bwindi Reserve, the presence of multiple-use programs in local parishes had a strong negative impact on tree and pole density and firewood extraction; these effects were especially strong in areas designated for beekeeping. In contrast, human population density accounted for a small proportion of the variance in intensity of local resource extraction, although the presence of nearby settlements was important (table 7.1). This suggests that, although local communities enjoy the benefits of the Bwindi community-based conservation project, they do not necessarily accept responsibility for careful management of the reserve. Hackel (1999) presented a series of case studies from throughout Africa to illustrate that community-based conservation projects do not engender the local communities to comply with conservation laws. In addition, attitudes of residents toward parks in Uganda do not differ between parks that have community-based conservation projects and those that do not (Mugisha 2002). Our findings, in combination with those from Mugisha (2002), suggest that community-based conservation programs in their current form do not meet their objectives in parks surrounded by dense human populations; thus, the programs should be modified. Possible modifications to mitigate threats in Bwindi include better supervision of beekeeping programs, community education, providing alternatives to park resources, deploying effective ranger patrols, and developing ways to improve honey yield.

CONCLUSIONS AND IMPLICATIONS

1. Recent analyses have demonstrated that areas of outstanding conservation importance often coincide with dense human settlements. The edges of forest reserves in such areas will likely experience both negative abiotic effects caused by proximity to the edge and encroachment by local people who extract natural resources.

2. Using 282 km of transect samples, we quantified spatial changes in resource extraction and vegetative structure as a function of distance from the edge into the interior of Bwindi Impenetrable National Park, Uganda. Resource extraction was the most intense within 300 m of the park's edge, as was the density of exotic plant species; however, signs of hunting activity were not limited to the edge of the park.

3. The increase in tree (>10 cm diameter) density with distance from the edge likely reflects the fact that resource extraction was most intense near the park edge and that fires were also frequent within this zone. Unlike previous studies of edge effects in areas where resource extraction does not take place, we additionally found reduced sapling and small-tree density near edges, which was evidently a result of local harvesting and fires. This probably represents a fundamental difference between edges protected from human activities and those that are not and raises the question of whether the Bwindi edge could recede in the future.

4. Regression models showed a strong positive relationship between tree, pole, and firewood extraction and the presence of multiple-use programs, particularly beekeeping. In contrast, population density did not account for much of the variance in intensity of resource extraction. This suggests that, while local communities were receiving the benefits of the community-based conservation project, they were not carefully managing the resources in these parks. For this reason such programs should be modified.

ACKNOWLEDGMENTS

We thank the following for their contributions to this project: Uganda Wildlife Authority for research approval and field support, Uganda National Research Council for research approval, the Wildlife Conservation Society for financial support, and the staff of the Institute of Tropical Forest Conservation for field and general support. Members of the local community around Bwindi Impenetrable National Park and the Uganda security forces are commended for their support in the field. We also thank Lauren Chapman, Michael Wasserman, and the book editors for comments on this manuscript.

PART III

Emerging Pathogens and Invaders

William F. Laurance

INTRODUCTION

Welcome to the Homogeocene

Humankind is altering the Earth in both dramatic and subtle ways. Among the most insidious of all threats is the homogenization of the planet's biota—a direct consequence of the extraordinary mobility of modern civilization and our burgeoning menagerie of domesticated and cultured plants and animals. Introductions of some exotic species are purposeful; many others are accidental. But the net effect is so extraordinary, with such profound implications for the future of global biodiversity, that some biologists propose coining a whole new geological epoch for our present era—the Homogeocene (e.g., see Putz 1998).

Precious few ecosystems on Earth are now free of invading species. Indeed, because invaders have become so ubiquitous, the general trend worldwide is toward increases in local species diversity (despite frequent losses of disturbance-sensitive species) but declines in regional and global diversity. Invading species cause important ecological problems in their own right, and they can also interact synergistically with other threats, such as habitat degradation and climate change, to create even greater pressures on ecosystems. Oceanic islands, which contain many restricted endemic species and are inherently vulnerable to invaders, are of special concern. For these reasons, the field of invasion biology has become one of the most active subdisciplines of applied ecology.

In all but the most pristine landscapes, one can hardly cast their eye about without seeing exotic species. What we actually observe, however, may be merely the tip of the iceberg: for every introduced plant, insect, or vertebrate, there could be scores of introduced microbes and pathogens. We know from our own history that exotic pathogens can have frightening consequences when they encounter immunologically naïve host populations, as demonstrated by the devastating effects of mumps, measles, and influenza on the indigenous New World populace following European colonization, as well as the scourge of syphilis in the Old World, which may have been introduced to Europe by Christopher Columbus and his men (see Wills 1996).

We live in a dangerous era of emerging pathogens. AIDS, SARS, avian influenza, and the Ebola and Marburg viruses are but five notable examples of novel infectious diseases. These pathogens often originate from the tropics, encountering new hosts as humans encroach into the world's last remote frontiers. Such pathogens can jump from one species to another, evolve extraordinarily rapidly, promptly acquire drug resistance, and even borrow genetic material from other species. Humankind's escalating populations and astonishing mobility (one is never more than a quick plane ride from a major city) mean that virulent new pathogens can spread with lethal rapidity.

Yet these examples provide but a glimpse of the real impact of exotic diseases. We are usually aware only of pathogens that attack us or that plague our domestic and cultivated species. In rare instances—probably only the most dramatic cases—we begin to perceive their effects in natural ecosystems. The mass mortality of African ungulates from Rinderpest, the widespread extirpation of Hawaiian birds by avian malaria and pox, and the catastrophic die-offs of Caribbean sea urchins (*Diadema antillarum*) and stream-dwelling rainforest frogs all show that novel diseases can have catastrophic impacts on wild populations. The impacts of exotic pathogens may be one of the great hidden threats to nature—perhaps rivaling the devastating effects of habitat conversion and climate change.

The four chapters in this part provide a compelling overview of the impacts of exotic animals, plants, and pathogens in tropical forests. The first chapter evaluates the impacts of a wide range of emerging infectious diseases in tropical ecosystems. The second focuses on the likelihood that habitat disturbances will predispose tropical plant communities to disease. The third describes critical threats posed to African apes by the combined effects of Ebola-virus disease and hunting. The final chapter, based on a large field study, assesses the impacts of exotic species and extinction of potential seed dispersers on a locally endemic tree species.

Emerging Pathogens in the Tropics

In chapter 8, Andrew Cunningham, Peter Daszak, and Nikkita Patel provide a cutting-edge synopsis of the role of emerging infectious diseases in tropical ecosystems. They first identify the various contemporary factors—such as the dramatic rates at which pathogens are being introduced to new geographic areas and host species—that promote emerging diseases. They also highlight a number of tropical vertebrate, invertebrate, and plant species known to have succumbed to such pathogens. As a case study, they detail the worldwide decline of stream-dwelling frogs, which has been attributed to the global spread of an exotic chytrid fungus. This devastating pathogen—"as lethal to some amphibians as Ebola virus is to humans"—is one of the most important causes of amphibian extinctions today.* The authors conclude by highlighting the rapid proliferation of zoonotic diseases (those vectored to humans by wild animals) in the tropics. Throughout their chapter, they emphasize that major ecological repercussions can occur when key species or entire assemblages of species suddenly vanish from their original ecosystems.

Habitat Disruption and Plant Diseases

In chapter 9, Julieta Benítez-Malvido and Aurora Lemus-Albor assess the role of habitat disturbances in predisposing tropical plant communties to attack from fungal and other pathogens. Although broad in scope, their chapter is informed by their long-term studies of Amazonian forest fragmentation and its impacts on tree seedlings and leaf pathogens.

The authors emphasize that major disturbances such as habitat fragmentation and logging alter many aspects of forest ecology, including habitat structure, plant-species composition, microclimatic conditions, and litterfall. The creation of networks of forest roads also dramatically increases the likelihood of invasion by exotic diseases. These changes in turn affect plant pathogens and the herbivorous insects that vector them and create stresses on native plants that render them vulnerable

* My colleagues and I were among the first to argue that an exotic pathogen was responsible for widespread amphibian declines, although our work was initially attacked as being overly speculative; see, e.g., W. F. Laurance 1995; W. F. Laurance et al. 1996, 1997b.

to disease. In some fragmented landscapes, invading plant species and exotic pathogens—such as the *Phytophthora* fungus that has devastated some forests in Australia—threaten entire plant communities. This chapter provides an excellent summary of the constellation of interacting factors that can influence plant-disease incidence in disturbed tropical forests. Despite our grossly incomplete understanding of plant pathogens in tropical forests, it is apparent that anthropogenic disturbances can sharply alter the incidence and impact of some infectious diseases.

Threats to Apes from Ebola and Hunting

In chapter 10, Peter Walsh provides a riveting view of the recent catastrophic declines of lowland gorillas and chimpanzees in western equatorial Africa. These declines have two main causes—Ebola-virus disease and overhunting—and there is great concern that the last surviving populations of these apes could be further decimated.

As Walsh demonstrates, critical gaps remain in our understanding of Ebola, such as whether the disease is enzootic (locally persistent in some unknown reservoir species, with periodic outbreaks in ape populations) or epizootic (not locally persistent, but moving in epidemic-like waves and possibly involving a range of vertebrate species). Despite such uncertainties, Walsh argues that dramatic measures are needed to combat illegal hunting and to devise strategies for controlling the spread of Ebola. He believes that only tough law-enforcement measures, such as restricting the access of local peoples to critical wildlife areas, will be sufficient to control illegal market hunting. Furthermore, he decries the "paralysis" affecting current efforts to control Ebola, which result in part from scientific uncertainties of its epidemiology. Walsh is outspoken and candid, but the dire circumstances he describes are, literally, a crisis in the making.

Decline of the "Dodo Tree"

In the final chapter in this part, Cláudia Baider and Vincent Florens critically evaluate the case of the "dodo tree," an endemic species on the island of Mauritius in the Indian Ocean, whose populations have declined precipitously. The dodo tree is quite famous—its seeds were believed by some to have been dispersed by the now-extinct dodo, a large, flightless bird that has become a well-known icon for nature conservation. According to this prominent view, the dodo was the obligate seed disperser for the dodo tree, and its extinction led to reproductive failure and a collapse of dodo-tree populations.

Baider and Florens challenge this scientific dogma and propose a contrary explanation. Their conclusions are based on extensive experiments and observations. They assert that, rather than suffering from lost interactions with extinct species, the dodo tree declined from massive habitat loss and fragmentation on Mauritius, greatly amplified by the cumulative impacts of many invasive species. Their experiments reveal that most seeds produced by the few remaining dodo trees are consumed by introduced macaques, rats, pigs, and deer and that any surviving seedlings likely succumb to competition from dense exotic weeds. The authors argue that, given the constellation of exotic species plaguing many islands and some mainland habitats, the collapse of endangered plant populations they observed on Mauritius could occur in many other ecosystems.

SYNTHESIS

The chapters in this part echo important themes and also raise crucial questions. One key theme is that invading plants, animals, and pathogens are seriously degrading many tropical and subtropical forests. In Earth's geological history, instances of significant biotic movement among continents—for instance, the tectonic collision of the Indian Subcontinent and Asian landmass, and the Great American Biotic Interchange between North and South America—were watershed events, reverberating through ecosystems, killing off many species, and permanently altering their evolutionary trajectories. But these events were exceptionally rare—measured on scales of millions to tens of millions of years.

What we are witnessing today is vastly different, in at least two ways. First, the rate of foreign-species invasions into new continents and habitats must literally be millions of times higher than natural background levels. Second, any given ecosystem receives invaders not from a single locality, as was usually the case in the past, but from disparate locations scattered across the globe. The Florida Everglades and its mahogany hammocks, for example, are plagued by a global potpourri of invaders such as South America piranhas and fire ants, Asian swamp eels, Australian paperbark trees, and a diverse but poorly documented array of exotic pathogens. Likewise, the rainforests of north Queensland are beset by neotropical cane toads, a bevy of pantropical weeds like lantana, and proliferating pond-apple trees from the New World. Once established, most invaders are virtually impossible to eradicate, even with the most effective control methods.

The second theme makes it clear that anthropogenic disturbances dramatically increase the vulnerability of tropical forests to invasions. Fragmented forests, for example, are much more vulnerable to invasions from exotic vines and weeds that favor forest edges and disturbed forest understories. Roads penetrating into forests provide avenues for invasions of exotic pathogens like *Phytophthora*, whose spores can be transported by vehicles, and a great diversity of weedy plants and opportunistic animals. Rainforests degraded by surface fires can be colonized by weeds and generalist animals. Such disturbances can increase the invasibility of ecosystems in several ways: by killing off resident plants and animals and thereby freeing up resources for invaders, by stressing the surviving residents and thus increasing their vulnerability to new pathogens or competitors, and by facilitating the spread of invaders and their propagules into formerly intact habitats.

A third theme is that not all tropical forests are equally susceptible to invasions. Those on oceanic islands, for example, are notoriously vulnerable, evidently because they are less species rich and competitive than are continental forests. Deciduous or dry tropical forests appear to be more vulnerable to weed invasions than are evergreen forests because fast-growing weeds can germinate and establish in the understory during the dry season, when the forest canopy is greatly reduced. Forests affected by recurring natural disturbances, such as hurricanes, episodic floods, and droughts, are probably more invasible than are those that are free from such disturbances.

Notwithstanding these themes, much remains poorly understood about biological invasions in tropical forests. One key question is this: How vulnerable are intact, old-growth rainforests to exotic or emerging pathogens? The rapid, pantropical spread of the chytrid fungus, which recent genetic analyses suggest is a single species, has devastated stream-dwelling frog populations on at least two continents and has been detected on several other continents. Most of the affected frog populations were in virtually pristine upland forests, and many species were driven to global extinction. Equally alarming are the recurring outbreaks of Ebola-virus disease across large expanses of intact forest in western equatorial Africa, which has ravaged populations of gorillas, chimps, and possibly other mammals. Modern scientific methods are direly needed to resolve these and other epidemiological mysteries that affect tropical forest wildlife.

CHAPTER 8

Emerging Infectious-Disease Threats to Tropical Forest Ecosystems

Andrew A. Cunningham, Peter Daszak, and Nikkita G. Patel

The agents of infectious diseases are essential components of tropical forests, regulating spatial and temporal host-population dynamics (e.g., Gilbert 2002) and possibly driving biodiversity evolution through host–parasite interactions (Lively and Apanius 1995; Nunn et al. 2004). Changes in these host–parasite relationships can have profound effects on tropical forest biodiversity and structure, resulting in minor shifts or deep-seated changes within the ecosystem. For example, the emergence of infectious diseases has been shown to cause changes in species assemblages and in the distribution of component populations of tropical forest ecosystems. Such changes can alter the community dynamics of forests, leading to spatial and structural shifts (such as the development of canopy gaps), altering nutrient cycling, or affecting host development or fecundity. Any of these changes can lead to biodiversity modification through both direct and indirect effects (see reviews by Augspurger 1988; Barbosa 1991; Burdon 1991; Alexander 1992; Dickman 1992; Herms and Mattson 1992; Gilbert and Hubbell 1996; Gilbert 2002).

Host–parasite relationships have co-evolved over long evolutionary time scales. Perturbations to these relationships, such as the introduction or extirpation of a given host or parasite, are generally infrequent. Since the fifteenth century and up to the present day, however, we have witnessed a dramatic increase in the number of disturbances caused by human activities that act directly and indirectly on both hosts and parasites (MacPhee and Flemming 1999).

In this chapter we discuss both the role of anthropogenic disturbances in promoting emerging diseases and the implications of such rapid changes for tropical forest ecosystems. We evaluate available evidence on the effects of emerging infectious diseases on tropical plants, vertebrates, and invertebrates and then review in detail the intriguing case of widespread, apparently disease-related declines of amphibians. Finally, we describe the increasing emergence of new zoonotic diseases (those that can be transmitted from wild animals to humans) following the inexorable encroachment of humankind into remote tropical regions.

DEFINING EMERGING INFECTIOUS DISEASES

Emerging infectious diseases (EIDs) are defined as infectious diseases that have recently increased in incidence or geographic range, have recently moved into new host populations, have recently been discovered, or are caused by newly evolved pathogens (Lederberg et al. 1992; Morse 1993a; Daszak et al. 2000). In an earlier review (Daszak et al. 2000), we showed that these criteria, which were initially used to define EIDs of humans, could also be used to identify EIDs in wildlife. These criteria also hold for EIDs of plants (Anderson et al. 2004) and we use them here to investigate and evaluate EIDs as a threat to both plant and animal communities that make up the biodiversity of tropical forest ecosystems. An equally important component of tropical forests is the fungus community (see Benítez-Malvido and Lemus-Albor, chap. 9 in this volume), but we do not address this subject here because there is an almost total lack of knowledge of EID threats to tropical forest fungi. EIDs of cultivated plants and domesticated animals can threaten natural biodiversity, such as by overspilling into naïve hosts (Cunningham et al. 2003), but diseases that solely affect cultivated plants or domesticated animals have not been included in this review, which focuses on EIDs of wild organisms and threats to biodiversity.

ECOLOGICAL EFFECTS OF EMERGING INFECTIOUS DISEASES

Infectious diseases have only recently been recognized as a cause of biodiversity loss despite hypothesized links to the extinction of a number of Hawaiian bird species, the thylacine (*Thylacinus cynocephalus*), and the last wild population of the black-footed ferret (*Mustella nigripes*; Cleaveland et al. 1999). Our recent work suggests that extinction by infection is an underreported phenomenon because of major difficulties involved in tracing the cause of historic extinctions in the absence of pathological samples (Daszak and Cunningham 1999). In wildlife and plant populations, a number of EIDs have been linked to regional population declines and even extinctions. Chestnut blight was introduced into North America from Japan in the late nineteenth century and caused the effective extinction of the American chestnut (*Castanea dentata*) across the Appalachian chain (Milgroom et al. 1996). Although the rootstocks of this species survive, the fungal agent *Cryphonectria parasitica* prevents growth to maturity across the American chestnut's range. Myxomatosis caused rabbit (*Oryctolagus cuniculus*) populations to decline rapidly by more than an order of magnitude over a few months after its introduction in the 1950s to control this species in Australia and the United Kingdom (Anderson and May 1986).

Wildlife diseases that significantly alter the population dynamics of multiple hosts at a given site may also affect ecosystem functioning. For example, rinderpest (caused by a morbillivirus) was introduced into the Horn of Africa in the late 1880s and quickly spread south and west across the continent, reaching the cape by the end of the century (Plowright 1982) and causing significant population declines in ranched and feral cattle and in a range of wild African ungulates. The loss of most of the ungulates in these areas significantly altered grazing pressure; remnant stands of mature *Acacia* trees that escaped grazing during this window of reduced ungulate populations can still be seen in the Serengeti (Dobson and May 1986; Dobson and Crawley 1994). Moreover, it is widely suggested that the decline of cattle and other wildlife hosts led to marked reductions of the tsetse fly, which allowed settlers to colonize areas where the high numbers of flies, and particularly the sleeping-sickness parasites they carry, had previously prevented settlement. This colonization then led to anthropogenic modification of the original habitat for agriculture and other uses.

Recently a series of EIDs has been recorded in tropical forests that affects organisms as diverse as honeybees, gorillas, and eucalypt trees. Some examples of high-profile EIDs affecting tropical forest ecosystems are presented in table 8.1.

EMERGING INFECTIOUS DISEASES OF PLANTS

Most known EIDs in tropical forests are diseases of animals, with few being recorded for plants (see Benítez-Malvido and Lemus-Albor, chap. 9 in this volume). Of the latter, most are reported to affect tropical plants cultivated for forestry or other commercial products. In many instances,

Table 8.1 Some examples of high-profile emerging infectious diseases affecting tropical forest ecosystems

Host	Parasite	EID	Direct effects	Driver
Amphibians (multiple species from multiple families)	*Batrachochytrium dendrobatidis* (fungus)	Cutaneous chytridiomycosis	Widespread amphibian population declines; growing number of amphibian species extinctions	Anthropogenic introduction of causative agent
Gorillas and chimpanzees (*G. gorilla* and *Pan troglodytes*)	Ebola virus (filovirus)	Ebola	Possibly causing population declines	Human encroachment
Mountain gorilla (*G. gorilla beringei*)	Measles virus (morbillivirus)	Measles	Mortality	Human encroachment
Mountain gorilla (*G. gorilla beringei*)	*Sarcoptes scabeii* (mite)	Scabies	Debility and mortality	Human encroachment
Chimpanzee (*Pan troglodytes*)	Polio virus (enterovirus)	Polio	Paralysis and mortality	Human encroachment
Honeybees (multiple species)	*Varroa jacobsoni* (mite)	Varroasis	Deformity and mortality	Introduced vector (European honeybee, *Apis mellifera*)

Host	Pathogen	Disease	Impact	Cause
Wide range of endemic Hawaiian birds	*Plasmodium* spp. (protozoan parasites)	Avian malaria	High mortality rates, possible extinctions, current threat to endangered species	Introduced vector (mosquito, *Culex quinquefasciatus*)
Wide range of endemic Hawaiian birds	Avian poxvirus (virus)	Avian pox	High mortality rates, possible extinctions, current threat to endangered species	Introduced vector (mosquito, *Culex quinquefasciatus*)
Torreya taxifolia (Florida torreya)	*Pestalotiopsis microspora* (fungus)	Canker	Necrosis and mortality	Pathogen introduction or environmental change
Casuarina equisetifolia (Australian pine)	*Trichosporium vesiculosum* (fungus)	Blister blight	High mortality rates	Plantation monoculture
Pinus spp.	*Mycosphaerella pini* (fungus)	Red band needle blight	Extensive defoliation	Pathogen introduction
Eucalyptus spp.	*Phytophthora cinnamomi* (fungus)	Root rot	High mortality rates	Pathogen introduction

the disease affects plants introduced into areas beyond their natural geographic range (e.g., *Eucalyptus* spp. in India and Brazil). Why so few EIDs of wild plants have been recorded is unclear, but it is most likely because of inadequate study of wild-plant diseases rather than a true reflection of their occurrence. Research on plant diseases is generally focused on plants of agricultural and direct economic importance; comparatively very little work is carried out on wild plants (Anderson et al. 2004).

As is the case for humans, wildlife, and domestic animals (Lederberg et al. 1992; C. Brown and Bolin 2000; Daszak et al. 2000, 2001; Cleaveland et al. 2001), the most significant driver of EIDs of plants is "pathogen pollution"—the anthropogenic movement of parasites outside their natural geographic or host-species range (Anderson et al. 2004). Pathogen pollution can occur in several ways, but in each case anthropogenic change results in a parasite crossing an evolutionary boundary, such as a geographic or ecological separation (Daszak et al. 2000; Cunningham et al. 2003). For example, needle blight, a disease caused by the fungus *Mycosphaerella pini*, has been introduced from the Northern Hemisphere to tropical pine forests and has caused extensive damage and defoliation (Gibson 1972; Harrington and Wingfield 2000).

Another fungus, *Phytophthora cinnamomi*, native to Southeast Asia, infected eucalypt forests in Australia following its introduction during European settlement. The pathogen is thought to have been first introduced around 1900 with orange trees from Indonesia and has subsequently destroyed large areas of World Heritage forest ecosystems in western Australia (Duncan 1999). The transportation of infected materials has now spread the disease widely in Australia; for example, it was transported from the Dandenong ranges to the Mount Lofty ranges around 1970 via infected berries, and from the mainland to Kangaroo Island in 1993 via infected soil on vehicles (Broembsen and Kruger 1985; Shearer and Tippett 1989; Dickman 1992).

A different group of fungi—termed endophytes—are hidden but important components of biodiversity and are found living within the tissues of virtually all plants in the wild (Hawksworth 1991). In most plant species, endophytic fungi are symbionts, sometimes producing toxins that are protective both to them and to their hosts (e.g., Clay 1988). However, the probable introduction of the endophytic fungus *Pestalotiopsis microspora* to a new host species, the endangered Florida torreya (*Torreya taxifolia*), resulted in the emergence of the fungus as a primary pathogen that has caused the near disappearance of this tree (Lee et al. 1995; Schwartz et al. 1995).

EMERGING INFECTIOUS DISEASES OF VERTEBRATES

EIDs of tropical forest animals are more widely documented than those of plants and in some cases their effects have been recorded over many decades. For example, recent reports of Ebola-virus outbreaks in nonhuman primates have highlighted a growing awareness of the impact of EIDs on tropical forest wildlife. These reports implicate Ebola virus in the rapid decline of lowland gorilla (*Gorilla gorilla*) and common chimpanzee (*Pan troglodytes*) populations in western Africa (P. D. Walsh et al. 2003; Walsh, chap. 10 in this volume). These findings may be key to discovering the wildlife reservoir of this high-profile zoonotic virus (Leroy et al. 2004).

Perhaps one of the best-known instances of wildlife EIDs of tropical forest species is the occurrence of avian malaria and avian poxvirus in Hawaiian birds. Although avian malaria and pox had probably been sporadically present in populations of migratory bird species in Hawaii prior to human arrival, the absence of a disease vector kept endemic birds ecologically and evolutionarily isolated from these pathogens (Pain and Donald 2002). However, following the introduction of the mosquito *Culex quinquefasciatus* in the early nineteenth century, the ecological divide between infected birds visiting the islands on their migration routes and the endemic birds of Hawaii was bridged. Avian malaria and avian pox are strongly implicated in the decline and extinction of a number of endemic Hawaiian birds, especially at lower elevations where mosquitos are most prevalent (Van Riper et al. 1986). Partly as a consequence of the emergence of these infectious diseases, 75% of historically recorded species of Hawaiian birds are now either extinct or threatened with extinction (Pain and Donald 2002).

EMERGING INFECTIOUS DISEASES OF INVERTEBRATES

Invertebrates comprise more than 90% of the animal biomass in tropical rainforests (Ghazoul and Hill 2001). Although rarely studied, when diseases of wild invertebrates have been investigated, EIDs have been found here, too. The introduction of the European honeybee (*Apis mellifera*) to Asia, and then more broadly, resulted in the spread of Asian-bee pathogens to many new areas and species. For example, the Thai sac brood virus spread from an enzootic focus in Thailand (i.e., where the disease is endemic) via *A. mellifera* to South Asian populations of the Asian hive bee *A. cerana*, where the disease caused massive population declines and local extinctions (Saville 2000). The presence of *A. mellifera* outside its natural range has disrupted co-evolved host–parasite relationships in additional ways, such as increasing the exposure of the giant Asian honeybee *A. dorsata* to its ectoparasitic mite, *Tropilaelaps clareae*.

EMERGING INFECTIOUS-DISEASE THREATS TO TROPICAL FOREST ECOSYSTEMS: THE EXAMPLE OF AMPHIBIAN DECLINES

The Discovery of the Threat of Amphibian Chytridiomycosis

One of the most intriguing examples of apparent EID-related impacts on wildlife is the dramatic and apparently widespread decline of rainforest amphibians (fig. 8.1). Amphibian populations have been declining globally for over two decades (Houlahan et al. 2000), and during this time a number of hypotheses have been developed to explain these disappearances. Initially, these hypotheses largely focused on putative anthropogenic factors such as habitat loss, climate change, increases in UV-B irradiation, chemical pollution, or unknown environmental stressors (Carey 1993; Blaustein et al. 1994; Pounds and Crump 1994; Pounds et al. 1999). However, the cause of a series of rapid and precipitous declines of tropical rain forest frogs in Costa Rica, Panama, and northern Australia remained enigmatic despite considerable attention from scientists in the 1990s. These regions were considered to be outside the sphere of human influence: many of the affected areas were legally protected (e.g., the Wet Tropics World Heritage Area in north Queensland, Australia; Monteverde Cloud Forest Reserve, Costa Rica) and, in most, human development and other anthropogenic insults such as pollution or invasive species were minimal (W. F. Laurance et al. 1996; Mahony 1996; Lips 1998, 1999; Williams and Hero 1998). The amphibian declines in these areas were characterized by mass mortalities of a range of species and the often complete loss of some amphibian populations, particularly among frogs that live in or reproduce in streams (e.g., *Atelopus* spp. in Central America; Lips 1998, 1999; Lips et al. 2003a, 2003b). These local extinction events have persisted, and repeated surveys in many areas show no recolonization by previously present species (Lips et al. 2003a, 2003b). A number of high-profile species went extinct: the golden toad (*Bufo periglenes*) of Costa Rica (Pounds and Crump 1994; Pounds et al. 1999) and approximately seven species of Australian frog (fig. 8.1), including the recently discovered northern (*Rheobatrachus vitellinus*) and southern (*Rheobatrachus silus*) gastric brooding frogs from southeastern Queensland and the sharp-snouted day frog (*Taudactylus acutirostris*) from northern Queensland (Mahony 1996).

Figure 8.1. Many stream-dwelling rainforest frogs worldwide, such as these *Litoria genimaculata* from north Queensland (shown mating), have declined precipitously or become extinct in virtually pristine habitats, apparently as a result of a highly pathogenic chytrid fungus. (Photo by Andrew Dennis)

The pattern of these declines and extinctions resembles the impact of a highly virulent pathogen introduced into naïve populations: rapid declines to extinction or near-extinction, simultaneous declines in connected valley systems, the loss of species assemblages at many sites with some ecologically distinctive species being relatively unaffected, and the presence of a patchy wave of die-offs which moved roughly north to south in Central America and roughly south to north in Australia through the 1980s and 1990s (W. F. Laurance et al. 1996; Daszak et al. 1999, 2003).

In 1997, an international team of pathologists, parasitologists, and ecologists discovered a new genus of fungal pathogen in carcasses collected during mass mortality events at Fortuna, Panama, and the Tablelands National Park, Australia (Berger et al. 1998). A total of 45 carcasses were collected from mass mortalities in Australia and 19 from Panama. Of these, 87% (39/45) of the Australian carcasses and 100% (19/19) of the Panamanian carcasses had high-intensity infections indicating that cause of death was due to this new disease. The causative agent has now been isolated, described, and experimentally proved to be a cause of mortality (Berger et al. 1998; Longcore et al. 1999). It is a new species of nonhyphal zoosporic fungus, *Batrachochytrium dendrobatidis*, belonging to the phylum Chytridiomycota. It is lethal to a range of amphibian species in captivity (Pessier et al. 1999), although some, such as the North American bullfrog (*Rana catesbeiana*), appear to have some degree of resistance (Daszak et al. 2004; Hanselmann et al. 2004). The mechanism by which this disease causes death remains unknown but may be related to its preference for keratinized skin cells; as the disease progresses, it results in marked thickening of infected amphibian skin with possible obstruction of transdermal respiration or osmoregulation, or the possible production of a fungal toxin.

The disease cutaneous chytridiomycosis has now been reported from Europe, North America, Central America, South America, New Zealand, and the Caribbean, and the causative agent, *B. dendrobatidis*, has been recorded in African amphibians (Daszak et al. 2003). It has been reported as newly emerging in southwestern Australia (Alpin and Kilpatrick 2000) and has recently been reported from wild and museum specimens of amphibians from Ecuador (Ron and Merino 2000), Venezuela (Bonaccorso et al., in press), and the Caribbean (Magin 2003).

Most emerging wildlife diseases are driven by anthropogenic environmental changes, such as land-use change or introduction of alien species (Daszak et al. 2000, 2001). Climate change, anthropogenic introduction, and other factors have been hypothesized as drivers of chytridiomycosis emergence (Daszak et al. 1999, 2001, 2003; Pounds et al. 1999, 2006). The temporal and spatial pattern of amphibian declines (Daszak et al. 1999) and a lack of genetic variation in a global sample of *B. dendrobatidis* isolates (Morehouse et al. 2003) support the hypothesis that chytridiomycosis has recently been introduced into naïve populations in multiple regions. The presence of *B. dendrobatidis* in both domestic and international trades of amphibians as pets (Mutschmann et al. 2000), outdoor pond animals (Groff et al. 1991), laboratory animals (Reed et al. 2000), zoo animals (Pessier et al. 1999), and food (Mazzoni et al. 2003; Hanselmann et al. 2004) provides a diversity of potential mechanisms for the anthropogenic introduction of this pathogen.

Consequences of Amphibian Chytridiomycosis for Tropical Forest Ecosystems

The population declines and extinctions associated with chytridiomycosis provide a striking example of the impact that EIDs can have on tropical-forest ecosystems. Chytridiomycosis affected a vertebrate class on a global scale via the mass mortality, dramatic population declines, and species extinctions of multiple species assemblages on four continents. In this way, chytridiomycosis parallels some high-profile human EIDs: *Batrachochytrium dendrobatidis* appears to be as virulent to some amphibian species as Ebola virus is to humans; it has a global-scale impact similar to influenza virus (although it does not spread as rapidly); and it seems to be moving on a wave of human activities such as land-use change, international trade, and alien-species introductions in the same way that most human EIDs are driven by anthropogenic factors (Krause 1992, 1994; Morse 1993a). Infectious diseases are not fixed in space but can spread from a point of emergence or introduction into new areas or species, thus reaching otherwise pristine habitats. Chytridiomycosis, for example, traveled through Central America as an epidemic, spreading between contiguous populations or via migrating animals.

Daszak et al. (1999, 2003) showed that the impact of chytridiomycosis varies among species and populations of amphibian hosts: some are completely resistant, others experience limited mortality, and only a minority undergo long-term population declines with local or species extinctions. Ecological or life-history traits may predispose this latter group of amphibians to population declines or extinction risk due to factors such as introduced disease. Comparison of the life-history traits and ecology of Australian frogs undergoing declines with sympatric species that are not declining (Williams and Hero 1998) supports this hypothesis. The declining group contains high-altitude, stream-breeding species—traits that predispose them to infection with a pathogen that prefers cool temperatures and is transmitted by a waterborne zoospore stage (W. F. Laurance et al. 1996; Williams and Hero 1998). The susceptible group also has relatively low fecundity and includes habitat specialists, two characteristics that would prevent rapid population recovery in the face of a persistent, virulent pathogen. Analyses of Central American species demonstrate similar predisposing life-history and ecological traits, which have been used to predict which amphibian species are likely to be affected by chytridiomycosis in the future as it continues to spread through the region (Lips et al. 2003b).

These traits, however, may not be enough to completely explain the pattern of extinctions. Knowledge of the life-history characteristics of *B. dendrobatidis* is also needed to allow an understanding of how the pathogen can bring about extirpation of amphibian populations. Classical models of host–pathogen dynamics predict that pathogens that rapidly deplete host populations in a density-dependent manner will reduce populations to a threshold density below which transmission (and hence pathogen persistence within the host population) is not possible (McCallum and Dobson 1995). This occurs because the remnant host population is now composed of immune individuals or is so sparsely distributed that (direct or indirect) contact rates do not allow sufficient rates of transmission to maintain the viability of the pathogen population. However, pathogens that persist at very low host densities often have traits that have evolved to circumvent this threshold-density effect, such as long-lived infective stages that evade host immunity (e.g., many helminths), sexual or vertical transmission, latency, or vector-borne transmission. For chytridiomycosis to cause complete removal of amphibian populations at a site, *B. dendrobatidis* must also evade the threshold-density effect.

Daszak et al. (1999, 2003) proposed two mechanisms by which the causative agent, *B. dendrobatidis*, may persist at low host densities. First, this pathogen can infect the mouthparts of amphibian larvae and therefore survive after removal of all post-metamorphic individuals from a body of water. Amphibian larvae have keratin only in their beaklike mouthparts, and infection by *B. dendrobatidis* has been recorded in these cells. It is thought that larvae with infected mouthparts are able to develop through metamorphosis (Berger et al. 1998), and the only known clinical impact is reduced growth rates (Parris and Baud 2004). Once larvae metamorphose and form keratinized skin, they become susceptible to chytridiomycosis (Berger et al. 1998). Second, *B. dendrobatidis* probably can survive without amphibians by living as a saprobe (an organism that derives its nutrition from the dead remains of other organisms) either on keratin present in the environment, or on another suitable substrate. This occurs in laboratory culture, where the pathogen can persist on agar that lacks keratin (Longcore et al. 1999). Moreover, Johnson and Speare (2003) have recently shown that *B. dendrobatidis* can survive for up to 8 weeks in sterilized pond water. Thus, a scenario exists in which *B. dendrobatidis* is introduced into a susceptible population of adult and larval amphibians and rapidly removes all the adults; the pathogen then persists in larvae until they metamorphose, at which point they are also killed. After extirpation of its amphibian host, the pathogen may be able to persist as a saprobe and infect migrants into the water body, preventing recolonization of the affected site, although the length of time the organism may persist in the environment in the absence of amphibian hosts is unknown.

Although unknown, the impacts of complete removal of an amphibian fauna from tropical forest regions due to chytridiomycosis are likely to be far reaching. For example, K. Lips (pers. comm.) noted overgrowth of aquatic algae in streams following an epidemic of chytridiomycosis in Fortuna, Panama. Many frog larvae are algal grazers, and it is presumed that the complete, multiyear removal of this component of the aquatic community led to algal overgrowth. The experimental exclusion of tadpoles from a neotropical stream led to a significant increase in sediment accumulation and algal biovolume, altered algal community structure, and consequent changes in the distribution and abundance of other species such as grazing insects (Ranvestel et al. 2004). Few studies have investigated these or other possible impacts, but it is reasonable to hypothesize that predators that rely on an amphibian prey base may also undergo population declines in the regions worst hit by chytridiomycosis (Ranvestel et al. 2004).

The publication of the IUCN Global Amphibian Assessment (a study examining population status, life histories, and ecological traits of all known amphibian species along with information about causes of declines) showed that chytridiomycosis is one of the most significant causes of amphibian-population declines and the most significant cause of contemporary amphibian-species extinction (Stuart et al. 2004). In addition to describing past and current threats for amphibians, it is hoped that analyses of the data gathered by the Global Amphibian Assessment will greatly improve our ability to predict the species that are likely to be adversely affected in the future by *B. dendrobatidis*, perhaps enabling proactive measures to be taken to protect these species.

ZOONOTIC-DISEASE EMERGENCE FROM TROPICAL FORESTS

In addition to direct disease threats to plants and wildlife, human perturbations to tropical forest ecosystems have resulted in the emergence of a number of high-profile zoonotic diseases, such as HIV/AIDS, Hendra virus disease, Nipah virus disease, and Ebola virus disease, all of which have no known cure and a high case-fatality rate. (HIV-1 and HIV-2 are zoonotic pathogens if viewed in the broadest sense, having evolved after jumping host from the chimpanzee and the sooty mangabee, respectively, to human beings [Hahn et al. 2000].) Tropical forests contain a diverse pool of known and potential zoonotic pathogens in wildlife reservoirs (Daszak et al. 2000, 2001) that have potentially important implications for tropical wildlife conservation.

EIDs are a major threat to public health (Binder et al. 1999) and recent analyses suggest that the majority (75%) of human EIDs are zoonotic (Taylor et al. 2001). In particular, a series of lethal agents with no known therapies or vaccines has recently emerged, almost all of which have their origin in wild animals. These include Ebola and Marburg viruses, Nipah virus, Lassa fever virus, *Leptospira* sp., SARS coronavirus, avian influenza, and others (Murphy 1998; Mahy and Brown 2000; Daszak et al. 2001). These pathogens emerged following direct human–wildlife contact or via indirect contact with wildlife through domesticated animal hosts. Such contact is likely to be greatest at the forest frontier, where deforestation and encroachment combine with dense human populations; thus, forest loss and fragmentation may exacerbate the emergence of zoonotic diseases. Active encroachment into areas rich in mammalian biodiversity, and hence with a high species richness of mammalian parasites (Morse 1993b), is a key threat for future emergence of novel diseases, particularly where human–animal contact rates are high or where animals are hunted, transported, or eaten. The combination of rapidly expanding human mobility during the past 50 years, particularly in the form of air travel and encroachment into previously pristine regions, has led to the rapid and often global spread of zoonotic pathogens from their endemic foci (Daszak and Cunningham 2003).

The most significant zoonotic pathogen, HIV/AIDS, emerged because of an intensification of bushmeat collection, processing, and consumption, coupled with more efficient transport of bushmeat to local towns. All of these factors were facilitated by road-building, deforestation, and encroachment in central West African forests (Hahn et al. 2000). In Latin America, Africa, and other regions, encroachment into forest for gold-mining operations led to the resurgence of malaria (Sawyer 1993; J. F. Walsh et al. 1993; Molyneux 2003). The impact of malaria is compounded by the use of mercury in gold extraction, which has immunocompromising toxic effects (Silbergeld et al. 2000). The transient and poorly planned development around mines often provides the perfect habitat for the most effective vectors of human malaria. For example, a recent study (Pearson 2003) combined the identification of mosquitoes in pristine and deforested regions in northeast Peru with satellite analysis of the degree of deforestation at these sites. It showed that, for every 1% increase in deforestation, there is a corresponding 8% increase in the number of *Anopheles darlingi* mosquitoes present (Pearson 2003). This mosquito species prefers open, sunlit pond-breeding sites and is an efficient vector of malaria. Gold-mining operations in the Amazon usually involve temporary clearing of forest in remote areas and settlements that normally persist until the mine is unworkable (usually a period of a few years). Unsustainable bushmeat hunting adds to the problems of altered mosquito habitat by removing the natural hosts of mosquitoes and increasing their dependence on human hosts. Mining-town inhabitants also provide an alternative to the natural hosts of vampire bats, which carry rabies in South America (Caraballo 1996). The net effect of these interactions can be a temporary surge in malaria, yellow fever, and rabies associated with encroachment and mining.

Nipah and Hendra viruses, both novel paramyxoviruses, recently emerged in Malaysia and Australia, respectively; zoonotic emergence occurred via domestic animal hosts (Field et al. 2001). In Malaysia, for example, the expansion and intensification of pig farming adjacent to tropical forest habitat led to the emergence of Nipah virus disease, which is lethal to humans (Chua et al. 2000; Field et al. 2001). Both of these viruses have recently been found to have fruit-bat reservoir hosts, which presents complex ramifications for biodiversity conservation due to public and political demands for the eradication of reservoir hosts rather than taking steps to prevent emergence in the first place. The recent large-scale slaughter of civet cats and other presumed wildlife reservoir hosts of the SARS coronavirus in China exemplifies the threat of emerging zoonotic diseases to biodiversity conservation. Other relevant examples include campaigns to exterminate vampire bats in some areas of Latin America in response to the threat of rabies (Mayen 2003) and the translocation of fruit bats in Australia from areas of high to low human population density in response to the threat of bat lyssavirus.

CONCLUSIONS AND IMPLICATIONS

1. An increasing body of evidence suggests that EIDs pose an important threat to wildlife and plant species in the tropics. The importance of EIDs as a cause of wildlife declines is probably underestimated because of

inadequate knowledge about the identification, distribution, and effects of infectious organisms on wildlife populations.

2. For tropical forest frogs, and probably for other groups, newly emerging pathogens have evidently spread throughout much of the world, which has led to large-scale declines of populations and a number of species extinctions.

3. Because tropical forests comprise a complex interplay of species, disease-related losses or declines of species will often have broad impacts on these ecosystems and food webs.

4. Zoonotic diseases (those transmitted from animals to humans) are increasing in importance in the tropics as a direct result of rapid human encroachment into tropical forest regions.

5. The case studies highlighted here demonstrate a close relationship between anthropogenic environmental change and the emergence of wildlife, plant, and human diseases. It is important that both this complexity and the potential consequences of disease emergence arising from its perturbation are understood.

6. EIDs can have profound implications for both plant and animal species and for human health and welfare. Addressing the processes that lead to the emergence of infectious diseases should be a crucial component in the formation of conservation and natural-resource policies.

CHAPTER 9

Habitat Disturbance and the Proliferation of Plant Diseases

Julieta Benítez-Malvido and Aurora Lemus-Albor

Disease-causing organisms drive many ecological and evolutionary processes in natural ecosystems, but the interactions between plants and their pathogens in the tropics has received very limited attention (Gilbert and Hubbell 1996). Pathogens can play an important role in plant recruitment and species composition through their differential impacts on survival, growth, and fecundity and by reducing the competitive ability of affected individuals (Augspurger 1983b, 1984; Augspurger and Kelly 1984; Burdon 1991, 1993a, 1993b; Dobson and Crawley 1994; Jarozs and Davelos 1995; García-Guzmán et al. 1996; Travers et al. 1998; Gilbert 2002). Furthermore, plant pathogens can help maintain high species diversity in tropical rainforests (Janzen 1970; Connell 1971; Augspurger 1983b; Gilbert 2002; Mills et al. 2006), facilitate successional processes, and enhance the genetic diversity and structure of host populations (Gilbert 2002).

Available information suggests that fungal diseases are common in the understory vegetation as well as in the canopy of neotropical forests, affecting plant species at all stages of their life cycles (Augspurger and Kelly 1984; Gilbert et al. 1994; Gilbert 1995; Lodge 1996; Benítez-Malvido et al. 1999; Rodríguez and Samuels 1999; Spironello 1999; Gamboa and Bayman 2001; García-Guzmán and Dirzo 2001). The most familiar are foliar pathogens that can cause necrosis, blights, chlorosis, and deformation of leaves (García-Guzmán and Dirzo 2001; Gilbert 2002; fig. 9.1). Other fungal pathogens present in the tropics can kill seeds and seedlings (Augspurger 1983b; Dalling et al. 1998), affect the vascular system of

Figure 9.1. A severely fungus-diseased leaf (family Araceae) near a tropical rainforest edge at Chajul, Chiapas, Mexico

plants, cause cankers on stems and trunks and wood decay in roots and trunks, and attack the flowers and fruits of adult plants (Travers et al. 1998; Gilbert 2002). For example, a rust fungus (*Aecidium farameae*) in *Faramea occidentalis* (Family Rubiaceae) trees infects their flowers and thereby reduces fruit set (Travers et al. 1998).

Both abiotic (temperature and moisture) and biotic (host identity and density) factors influence the severity of disease in plant populations (Burdon 1987; Agrios 1997), and the former is considered especially important. Environmental conditions at a given site affect the expression of disease symptoms mainly through the effect of the conditions on the plant host prior to infection (predisposition) and on the host–pathogen association once infection has occurred (Burdon 1987). Variation in a particular physical factor (light, temperature, or moisture) can affect germination, growth, or susceptibility of the host plant; survival, germination, and growth of the pathogen (Agrios 1997); the host–pathogen interaction (Colhoun 1973); and the behavior of disease vectors (García-Guzmán and Dirzo 2001).

Disturbances in tropical rainforest can have major effects on vegetation structure (W. F. Laurance et al. 1998), which in turn affects the intensity of light (Kapos 1989), the soil moisture (Camargo and Kapos 1995), the temperature (Williams-Linera et al. 1998), and the rates of litterfall and litter accumulation in the understory (Sizer et al. 2000). All these factors are known to affect seed germination (Vázquez-Yanes et al. 1990), seedling establishment and performance (Chazdon 1988; Molofsky and Augspurger 1992), herbivory, and the incidence of fungal diseases in plants (Augspurger 1983b, 1984; Augspurger and Kelly 1984; García-Guzmán and Benítez-Malvido 2003). The combination of environmental changes (Kapos 1989; Camargo and Kapos 1995; Williams-Linera et al. 1998) and exotic plant and pathogen species (Anagnostakis 1987; Rizzo et al. 2002) could make the vegetation of disturbed habitats more vulnerable to infectious diseases (Brothers and Spingarn 1992; Gilbert and Hubbell 1996; Harvell et al. 2002). Diseased plants generally appear more vulnerable to unfavorable environmental conditions than do healthy plants (Burdon 1987).

In this chapter, we describe how environmental changes produced by man-made disturbances in the Neotropics may affect the interactions of plants with their fungal pathogens. First, we assess the environmental conditions that favor fungal infection in wild plants of "pristine" rainforest. Second, we compare these environmental conditions with those in disturbed forests and explore the consequences of forest fragmentation, edge creation, deforestation, and secondary succession on levels of pathogen attack. Third, we consider the broader implications of these patterns for tropical rainforest biodiversity. Finally, we suggest where researchers should go from here, and we reiterate the importance of understanding plant diseases in light of tropical forest conservation (Lubchenco et al. 1991; Benítez-Malvido and Lemus-Albor 2005; Gilbert and Hubbell 1996). The examples we employ mostly come from natural tropical systems. Although plant pathogens have critical impacts on tropical agriculture, we focus on their effects on wild plants because host–disease dynamics in natural systems may be very different from those in agricultural systems and because the effects of fungal pathogens on wild-plant populations have been seriously neglected.

ENVIRONMENTAL CONDITIONS AND PLANT DISEASES

Temperature, light, and moisture levels clearly affect host susceptibility and the expression of disease symptoms in some plant species (Burdon 1987); many fungal diseases are favored in moist conditions (Bradley et al. 2003). For seedlings of several tropical tree species, damping-off disease (soil-borne fungal pathogens, often the funguslike Oomycetes *Phytophthora* and

Pythium; Garret 1970) was more severe in soils with blocked drainage or in areas of deep shade than in sunnier, drier areas (Gilbert 2002). Seedling mortality declines with increasing light and associated microclimatic changes in treefall gaps (Augspurger 1983b; Augspurger and Kelly 1984). Seedlings of early successional species, which require treefall gaps in which to establish, generally lack strong resistance to fungal pathogens, whereas late-successional species that tolerate shaded conditions (sensu Swaine and Whitmore 1988) are able to escape pathogen infection (Augspurger and Kelly 1984). Lower resistance of early successional species to pathogen attack might result in delayed forest regeneration after disturbance.

Host density can also affect plant vulnerability. On Barro Colorado Island, Panama, seedlings of *Platypodium elegans* (family Papilionoideae) had a higher probability of mortality by damping-off disease both at high seedling densities and when close to conspecific adults (Augspurger and Kelly 1984). In saplings of *Ocotea whitei* (family Lauraceae), fungal-induced canker disease (by *Phytophthora*) decreased with decreasing density of conspecifics and increasing distance to adults (Gilbert et al. 1994). Anthropogenic impacts such as habitat fragmentation and logging often lead to dramatic changes in plant abundances—some species decline whereas others become hyperabundant (W. F. Laurance and Bierregaard 1997; W. F. Laurance et al. 2006b)—and such changes could clearly affect the susceptibility of affected species to plant diseases.

Finally, changes in litter cover can affect seedling regeneration (Vázquez-Yanes et al. 1990; Molofsky and Augspurger 1992) and levels of damage by insects and foliar pathogens (Benítez-Malvido and Kossmann-Ferraz 1999; García-Guzmán and Benítez-Malvido 2003). Natural and human-induced disturbances (landslides, canopy gaps, animal activity, edge creation, forest fragmentation, etc.) produce large variations in litter cover, which range from forest areas completely devoid of litter to those where litter accumulates extensively (Facelli and Pickett 1991; Benítez-Malvido and Kossmann-Ferraz 1999; Sizer et al. 2000). At Los Tuxtlas, Mexico, seedlings of the tropical tree *Nectandra ambigens* (family Lauraceae) growing in deep litter showed greater damage by herbivores and foliar pathogens (*Colletotrichum* sp. and *Phomopsis* sp.) than did those growing in reduced litter (fig. 9.2). The microenvironment created by the experimental addition of litter may have benefited pathogens by creating a wetter environment, whereas the addition of previously infected *N. ambigens* leaves (García-Guzmán and Benítez-Malvido 2003) may also have exposed seedlings to inocula from pathogens (Gilbert 1995).

Available information suggests that foliar pathogens are common in the understory plant community of tropical rainforests but that leaf area

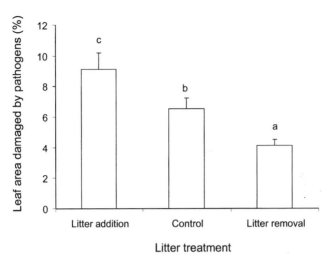

Figure 9.2. Comparison (mean ± SE) of leaf area damaged by pathogens (*Colletotrichum* sp. and *Phomopsis* sp.) in *Nectandra ambigens* (family Lauraceae) tree seedlings after one year of growing in different litter treatments at Los Tuxtlas, Mexico (after García-Guzmán and Benítez-Malvido 2003; Benítez-Malvido and Lemus-Albor 2005). Letters above bars indicate significant differences among treatments.

with apparent disease symptoms is very low, averaging 2% or less of leaf area (Benítez-Malvido et al. 1999; García-Guzmán and Dirzo 2001). Nevertheless, even small reductions in leaf area can have large impacts on seedling survival (D. B. Clark and Clark 1985), and foliar pathogens can cause a large decline in host growth, survivorship, reproduction, and competitive ability (Jarozs and Davelos 1995; Esquivel and Carranza 1996). In Chajul, Mexico, for example, we observed that taller seedlings and those with more leaves had a greater proportion of diseased leaves. Taller, leafy plants probably have a greater surface area for fungal attack and may create a conductive microclimate for disease development, leading to greater disease incidence (G. Gilbert, pers. comm.). Foliar pathogens are also common in rainforest canopies (García-Guzmán 1990; Rodríguez 1994; Gilbert 1995; Lodge 1996; Rodríguez and Samuels 1999; Gamboa and Bayman 2001). At Los Tuxtlas, Mexico, more than 95% of the leaves of *N. ambigens* trees had foliar pathogens (Lemus-Albor 2000).

In tropical plants, the low incidence of disease symptoms on leaves may result from the low density of conspecific adults, which may reduce rates of disease transmission (as has been shown for damping-off and canker diseases; Augspurger 1983b, 1984; Augspurger and Kelly 1984; Gilbert et al. 1994). The abundance of disease vectors such as insects also influences disease transmission (García-Guzmán and Dirzo 2001).

Seasonal differences in pathogen outbreaks are also common, as observed at Los Tuxtlas, where the proportion of plants with foliar pathogens was higher during the dry season (García-Guzmán and Dirzo 2001). Finally, soil nutrients can influence disease incidence, because plants growing on poor soils tend to have high concentrations of defensive compounds such as foliar tannins and phenolics (saponins), which evolved to deter pathogenic microorganisms (Waterman 1983; Nichols-Orians 1991; Agrios 1997; Vasconcelos 1999).

PATHOGENS AND LARGE-SCALE FOREST DISTURBANCE

Forest disturbance alters the physical and biotic environment in which host plants, pathogens, and disease vectors live and interact. Variations in temperature and moisture can affect pathogen development and survival, disease transmission, and host susceptibility (Harvell et al. 2002). Forest edges and fragments, which are becoming dominant features in many tropical landscapes, have hotter (ca. 3°C warmer) and drier microclimates than do forest interiors and nonfragmented forests (Williams-Linera et al. 1998; Sizer and Tanner 1999). Although processes characteristic of forest interior, such as tree regeneration and some biotic interactions, still occur in forest fragments and edges, their dynamics are often influenced by factors that originate outside the fragment margin (Janzen 1983, 1986a; Didham et al. 1996). Forest edges are recognized as the point of entry of external influences such as the invasion of exotic flora and fauna and pathogenic organisms (Janzen 1983; Castello et al. 1995; Gelbard and Belnap 2003; Benítez-Malvido and Lemus-Albor 2005).

In Chajul, seedlings growing close to forest edges had a greater incidence of foliar disease than did those in forest interiors (fig. 9.3), whereas, in an experimentally fragmented landscape of Amazonia, foliar-disease incidence in seedlings likewise increased with decreasing fragment size (fig. 9.4; Benítez-Malvido et al. 1999). Grasslands in eastern Australia showed greater rates of infection by a flower smut fungus (Class Basidiomycetes) at patch edges, which are subjected to greater disturbance, than in core areas (García-Guzmán et al. 1996). In contrast, damping-off disease was negligible in the lighter environments of rainforest edges (Infante 1999). Some pathogens require wounds as a route to successfully infect plants (García-Guzmán and Dirzo 2001). Wounds to plants, such as those caused by increased tree mortality and falling debris in forest fragments (W. F. Laurance et al. 1998; Sizer et al. 2000) and by logging and road construction, can provide infection sites for pathogens; once estab-

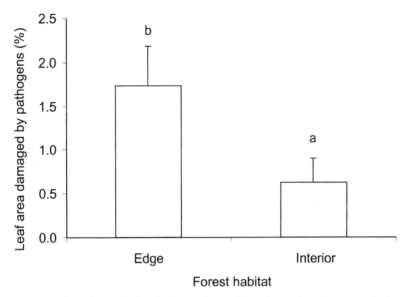

Figure 9.3. Leaf area damaged by fungal pathogens (mean ± SE) on the woody seedling community at forest edges and interiors at Chajul, Mexico. Letters above bars indicate significant differences among treatments.

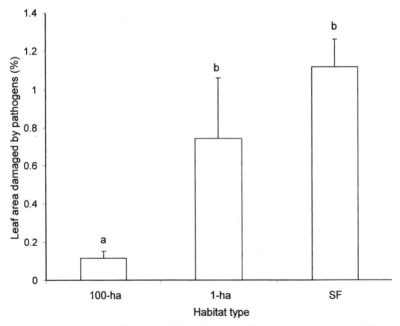

Figure 9.4. Leaf area (mean ± SE) damaged by fungal pathogens (*Chaetomium* sp. and an unidentified species) on seedlings of *Micropholis venulosa* (family Sapotaceae) growing in forest fragments and secondary forest (SF) north of Manaus, Brazil. Letters above bars indicate significant differences among treatments.

lished, these pathogens can spread to neighboring plants (Kellas et al. 1987; Gilbert and Hubbell 1996).

Disturbed or human-dominated lands have microclimates that differ markedly from those in undisturbed forest (Benítez-Malvido et al. 2001; Mesquita et al. 2001) and in turn can affect disease incidence. Although foliar and other pathogens are generally favored by moist, cool, and dark conditions (Bradley et al. 2003), some pathogens such as certain rusts and powdery mildews (G. Gilbert, pers. comm.) are evidently adapted to drier, warmer, and brighter conditions. Moreover, dew formation is generally greater on plants in open environments than in the forest understory (Brewer and Smith 1997), and dew is often crucial for fungal infection of leaves (Bradley et al. 2003). Finally, desiccation and extreme-temperature stresses are greater in disturbed environments, and this can make forest plants more susceptible to disease (Harvell et al. 2002). Cattle pasture at Chajul and in an open secondary forest at Manaus showed greater foliar-disease incidence than did nearby primary forests, where light intensity and temperatures were lower (fig. 9.3). In Puerto Rican forests disturbed by Hurricane Hugo, greater root fungal infection (by *Rosellinia bunodes*) led to elevated mortality of *Psychotria berteriana* (Rubiaceae) seedlings recruited at high densities into treefall gaps. If exposed to full sunlight, individuals of *Theobroma* and *Herrania* spp. (family Sterculiaceae) are often killed by fungal diseases (Lodge and Cantrell 1995).

POTENTIAL CONSEQUENCES OF DISEASE PROLIFERATION

Tropical forests are being rapidly degraded, yet we know very little about the potential consequences of disease proliferation in disturbed habitats. In temperate and other environments, disease epidemics have devastated plant populations and caused drastic changes in community structure and species composition (Kennedy and Weste 1986; Rizzo et al. 2002; Weste et al. 2002; Laidlaw and Wilson 2003; Weste 2003); in some cases, epidemics can completely eliminate susceptible species from infested areas (Weste 2003). Moreover, not only do plant pathogens directly affect their hosts, but they also indirectly affect the associated host fauna and ecological communities. Hence, the extinction of a plant species from a particular area might have cascading effects on other organisms such as specialized phytophagous insects and nesting animals that rely on the host plant for survival (Burdon 1993a; Didham et al. 1996; Harvell et al. 2002). Conversely, a loss of pathogens can affect biodiversity by releasing the host from an important source of population regulation, which often occurs when species are introduced to new geographic areas where their natural pathogens are absent (Burdon 1991; Harvell et al. 2002).

Overall, the bits and pieces of available information strongly suggest that forest disturbances that alter microclimatic and other ecological conditions favor some pathogens, such as certain fungi. Other fungal pathogens, such as damping-off diseases, are evidently not favored by forest disturbance. There are, however, huge holes in our present knowledge of the impacts of fungal pathogens on natural tropical plant populations. In many cases, we do not know the aggressiveness of fungal diseases (whether they are endemic or epidemic), the source of infection (from exotic or native pathogens), or the mode of pathogen dispersal (by direct contact, air or water dispersal, or insect vectors). Key priorities are the identification of key fungi that cause diseases in tropical plants (Gamboa and Bayman 2001) and ecological research on the processes involved in the transmission, emergence, and spread of disease.

Among the greatest concerns is that habitat disturbances and the rapid proliferation of exotic plants (which can carry novel pathogens) in disturbed landscapes may dramatically increase the introduction and proliferation of plant diseases (Anagnostakis 1987; Gilbert and Hubbell 1996; Harvell et al. 2002). Foreign pathogens can be introduced to ecosystems in many ways; for example, bulldozers and other earth-moving equipment can transport damping-off disease, which can persist in infected soil (Weste et al. 2002; Weste 2003). More generally, roads and cars could act as conduits for some pathogens in human-dominated landscapes. In this way, plant pathogens could potentially exacerbate the deleterious effects of habitat disturbance on tropical ecosystems.

CONCLUSIONS AND IMPLICATIONS

1. In the tropics, as elsewhere, habitat fragmentation and other disturbances evidently predispose plant communities to attack from fungi and other pathogens. For example, changes in microenvironmental conditions following disturbance may favor certain pathogens or cause stresses to plants that increase their susceptibility to disease. In addition, successional plant species often increase sharply in abundance following forest disturbance, and such species may be especially vulnerable to pathogen attack, both because they have high population densities and because they are inherently less resistant to certain pathogens than are shade-tolerant plant species.

2. Human-dominated landscapes frequently contain many exotic plant species, which often grow at high population densities in disturbed habitats. These exotic plants may facilitate the introduction of novel pathogens that could have important impacts on native plant species. Roads and vehicles that crisscross human-dominated landscapes can also serve as vectors for some exotic pathogens.

3. Anthropogenic environmental stresses clearly can influence the kinds and intensity of disease development. Long-term monitoring of pathogen damage and its effects on plant fitness is needed to fully assess the synergistic effects of disturbance and disease in rainforest plant communities.

4. We have barely begun to scratch the surface in our efforts to understand the interactions of forest disturbance and native and exotic pathogens in the tropics. Identifying the biological and physical factors that trigger different kinds of plant diseases will improve our ability to predict the full effects of human disturbance on tropical ecosystems.

ACKNOWLEDGMENTS

Fieldwork at Los Tuxtlas, Chajul, and Manaus would have not been possible without the help of A. Cardoso, S. Sinaca, P. Sinaca, R. Lombera, C. Ramos, H. Ferreira, and A. González di Piero. We are very grateful to G. Gilbert and two anonymous reviewers, whose comments greatly improved our manuscript. This work was partly supported by the National Institute for Amazonian Research, the Smithsonian Institution, the Mexican Council for Science and Technology (CONACYT grants 56519 and 36828-V), and the Center for Ecosystem Research, National Autonomous University of Mexico. This is publication number 413 in the Biological Dynamics of Forest Fragments Project technical series.

CHAPTER 10

Ebola and Commercial Hunting: Dim Prospects for African Apes

Peter D. Walsh

The virulent "Zaire" strain of Ebola hemorrhagic fever causes massive internal bleeding, killing about 80% of humans who become symptomatic. Over the last decade, the Zaire strain has repeatedly emerged in human populations in Gabon and the Republic of Congo. What sets events in Gabon and Congo apart from Ebola outbreaks elsewhere in Africa is that they have been linked to the handling of meat from infected gorillas and chimpanzees found dead in the forest (Georges et al. 1999; Leroy et al. 2002, 2004; www.who.int). In fact, literally tens of thousands of western gorillas (*Gorilla gorilla*) and common chimpanzees (*Pan troglodytes*) in these two countries appear to have been killed by Ebola (P. D. Walsh et al. 2003). Substantial proportions of the world populations of these species, which are found only in western equatorial Africa (fig. 10.1), have died. Ebola is now burning through Odzala National Park in northwestern Congo, which has the densest gorilla and chimpanzee populations in the world (Bermejo 1999), and shows no signs of abating. In this region, Ebola is rivaled only by commercial hunting as a threat to the future survival of gorillas and chimpanzees (P. D. Walsh et al. 2003).

In this chapter, I begin by reviewing how the development of a regional economy based on natural-resource export has fundamentally altered both the pattern of ape hunting and the conditions for the transmission of Ebola and other infectious diseases in wildlife populations. I then report on the rapidly deteriorating status of apes in western equatorial Africa. I follow this with a discussion of opposing hypotheses on

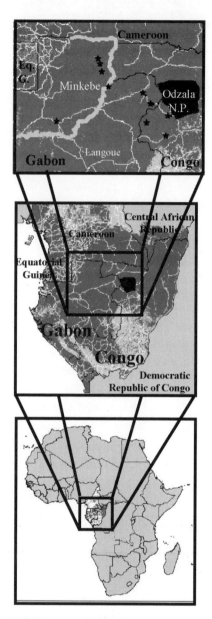

Figure 10.1. Western Equatorial Africa. Primary forest is shaded dark gray and covers an area comparable to that of France or Texas. White lines are major roads. In the top map, human outbreak locations are shown by stars, and the Ogoue-Ivindo-Ayina River system is shown as a thick gray line.

why Ebola has recently emerged, one assuming increased spillover from some (as yet unknown) reservoir species, and the other assuming spatially spreading epizootic diseases (transmitted from animals to humans). The two hypotheses have radically different implications for the prospects of controlling Ebola. I finish with a discussion of priorities for ape conservation in western equatorial Africa, most prominently the need for more aggressive law enforcement to reduce hunting pressure, but I also consider potential strategies for controlling Ebola.

THE HISTORY OF APES IN WESTERN EQUATORIAL AFRICA

The particular blend of threats now facing apes in western equatorial Africa is a product of the unique history of human settlement in the region. Before European arrival, most people in the region lived in small villages distributed fairly homogeneously across the landscape (fig. 10.2). Gorillas and chimpanzees were hunted for food, most heavily in the zone around each village. Thus, subsistence hunting almost certainly fragmented

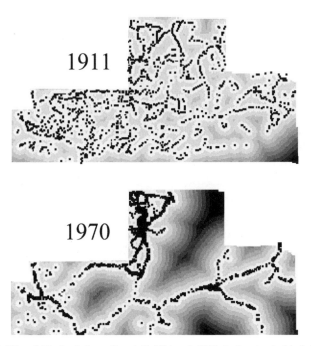

Figure 10.2. Village distribution in the northern half of Gabon. In 1911, the French colonial administration had already started moving villages (dark squares) onto roads. By 1970, "Regroupement" had created large blocks of ape habitat uninhabited by humans. (Adapted from Pourtier 1989)

gorilla and chimpanzee populations and regulated their densities. However, hunting technology was primitive (e.g., crossbows, spears, bows, and pitfall traps), apes were formidable prey, alternate prey was abundant, and human-consumer densities were low.

Ironically, the population fragmentation and density regulation maintained by subsistence hunting may even have protected gorillas and chimpanzees from major epizootics of virulent diseases, such as Ebola, by limiting the rate of contact between infected and susceptible vertebrate hosts. What may have set the stage for Ebola emergence was a profound shift in human-settlement patterns, driven largely by the regional shift from a subsistence economy to a commercial economy based on the export of natural resources. One of the first major export products was slaves. The slave trade depopulated extensive parts of the region's interior up to the mid-nineteenth century (Chamberlin 1978; Gray and Ngolet 1999). During the late nineteenth and early twentieth centuries, large numbers of workers migrated from the interior to work in the timber industry, which was concentrated along major rivers in the coastal zone (Gray and Ngolet 1999). Concentration of human settlements intensified in the 1920s and 1930s when the French colonial administration pursued a policy known as "Regroupement." To increase the profitability of commercial plantation crops such as cacao, villages were forcibly relocated along major roads and rivers (Pourtier 1989). The trend continued during the 1960s, when large numbers of interior residents migrated to coastal areas in search of work in the petroleum industry. In 1987, a transnational railroad was completed in Gabon explicitly to facilitate the export of manganese and timber from the interior. In 1994, the regional currency was devalued and between 1992 and 2000 regional timber exports almost tripled (fig. 10.3, top panel).

This economic transformation was accompanied by intense urbanization. From 1950 through 2000, the rural population of Gabon decreased by half whereas the urban population increased 20-fold (fig. 10.3, bottom panel). About two-thirds of Gabonese now live in the capital city of Libreville. The same trend was repeated across the region, both in major cities and at smaller scales (http://apps.fao.org). People moved from small villages to much larger logging, oil, and mining towns as well as regional transportation hubs.

The combination of forced "Regroupement" and voluntary economic migration created a series of large, lightly settled forest blocks across the region (figs. 10.1 and 10.2). In the interior of these major forest blocks, hunting pressure was relaxed. By the 1980s, populations of apes (e.g., Tutin and Fernandez 1984; Fay and Agnana 1992; Bermejo 1999) and other

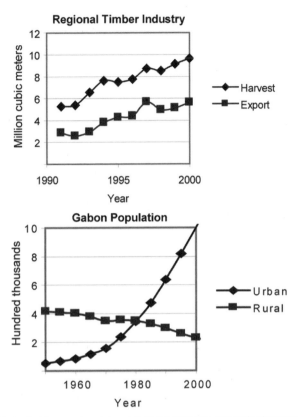

Figure 10.3. Post-colonial economic and demographic transformation. *Top:* Timber harvest and export figures for western equatorial Africa. *Bottom:* Rural and urban human population estimates for Gabon. (Derived from http://apps.fao.org)

wildlife (Barnes et al. 1991; Lahm 1993; Michelmore et al. 1994) in the major forest blocks were impressively dense and continuous. The economic transformation also had a profound effect on the nature of hunting. By the 1990s, what had been mainly a subsistence activity became a major commercial enterprise. Logging roads and vehicles provided access for hunters into the remote forest blocks to exploit the dense wildlife populations that had developed after human outmigration, and allowed hunters to move their product to market towns quickly (D. Wilkie et al. 2000; Tutin 2001; Waltert et al. 2002). To increase efficiency and decrease overhead, commercial hunters reduced prey selectivity, expanding beyond the small animals (e.g., porcupine and blue duiker) preferred by subsistence hunters to large-bodied species such as apes and buffalo (Noss 1998a, 1999; Muchaal and Ngandjui 1999). Demand also increased, not just as consequence of human population growth but also because salaried timber, mining, and oil employees were able to pay for and eat more meat per capita than local villagers could afford (D. Wilkie et al. 2000; Thibault and Blaney 2003).

THE STATUS OF APES IN WESTERN EQUATORIAL AFRICA

For apes, the consequences of these changes in human-settlement patterns have been nothing short of catastrophic (fig. 10.4). Commercial and subsistence hunting has extirpated apes near most major human population centers (P. D. Walsh et al. 2003). The regional timber industry continues to grow (http://apps.fao.org), both as a consequence of depletion of forests in other tropical regions and the gradual shift away from petroleum as oil stocks in Gabon and the Congo dwindle. Commercial logging is now penetrating into the most remote areas, some of which have experienced little or no hunting pressure (W. F. Laurance et al. 2006a). Given that more accessible areas are increasingly being depleted of game animals, these remote areas are virtually certain to come under intense hunting pressure.

The prospects on the Ebola front are less certain, but equally bleak. Ebola has already killed 95% or more of the apes in one of the largest forest blocks created by "Regroupement" (Minkebe; fig. 10.1). It is now burning through Odzala National Park, which has the densest recorded gorilla and chimpanzee populations in the world (Bermejo 1999). If past patterns hold true, Ebola will have killed a large majority of the apes in and around Odzala by the end of 2005.

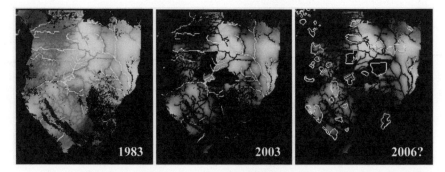

Figure 10.4. Approximate ape distribution in western equatorial Africa. Ape density is proportional to white-shading intensity. Densities were predicted using distance from major urban centers, major roads, and major rivers. Large dark blocks in the center of the 2003 and 2006 maps represent major zones of Ebola impact. The 2006 map assumes that most apes in the Odzala region will be killed by Ebola. Protected areas are indicated by white outlines in the 2006 map.

Lurking in the background is the specter of habitat loss. Although much of the region still remains heavily forested and lightly populated, it seems unlikely to escape the deforestation trends seen elsewhere in Africa and in other tropical forest regions worldwide (Roberts 1998). Not only is the commercial-timber industry expanding, but also human populations continue to grow (http://apps.fao.org). As a result, intact forests will inexorably be cleared for firewood and agriculture. Within perhaps 50 years, the only large blocks of intact forest left will be in protected areas (Q. F. Zhang et al. 2002)

The best available estimate suggests that the combination of Ebola and hunting will cause ape populations in the region to decline by another 80% in about 30 years (P. D. Walsh et al. 2003). However, this estimate is based on very conservative assumptions, and the rate of ape decline is almost certainly higher. If current trends continue, hunting alone seems likely to effectively extirpate apes and other large game from most parts of the region within the next 10 to 20 years, perhaps sooner. If Ebola continues to spread at its past rate, all of the major ape concentrations (i.e., populations with several thousand individuals or more) left in southeastern Cameroon, northeastern Congo, and southwestern Central African Republic will be affected within about a decade. How likely this is depends on the assumptions that one makes about the dynamics of Ebola transmission and the forces promoting Ebola emergence.

HYPOTHESES FOR EBOLA EMERGENCE

The cause of the emergence of Ebola in Gabon and the Congo over the past decade is now a matter of vigorous debate. At issue is whether Ebola is enzootic or epizootic. The enzootic hypothesis (fig. 10.5) assumes that Ebola is locally persistent and of low virulence in some reservoir host species (Georges et al. 1999; Formenty et al. 1999; Gonzalez et al. 2000), and infects primates in entirely independent outbreaks from the reservoir host (Georges-Courbot et al. 1997; Leroy et al. 2004). The epizootic hypothesis (fig. 10.6) assumes that that the virulent varieties of Ebola now killing apes in the Congo do not persist over the long term at a given location but, rather, move through space (Monath 1999; P. D. Walsh et al. 2003) and may involve transmission within and between a variety of different vertebrate hosts (Monath 1999; Reiter et al. 1999; P. D. Walsh et al. 2003). The dichotomy between enzootic and epizootic is somewhat artificial in that if one looks at a small enough spatial scale, most diseases will be seen to be wandering or cycling through space. For our purposes, it

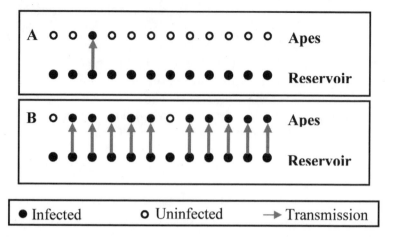

Figure 10.5. The enzootic hypothesis for Ebola emergence. (A) Historically low rates of contact with reservoir species result in few transmissions to apes. (B) Habitat alteration increases the rate of ape contact with the reservoir.

suffices to think of enzootic diseases as those that are persistently present on scales of up to a few thousand square kilometers. Diseases that are only transiently present on such spatial scales will be considered to be epizootic. "Transiently" will be defined to mean 3 to 5 years or less.

The Enzootic Hypothesis

The enzootic hypothesis is implicit in most of the literature on Ebola. The identity of the presumed reservoir host is yet unknown, although a variety of species have been proposed, including bats, rodents, birds, insects, and even plants (reviewed in Monath 1999). The enzootic hypothesis is often accompanied by the corollary assumption that Ebola could only persist enzootically if it was always of low virulence in its reservoir host. Many species have been highlighted (Swanepoel et al. 1996) or dismissed (Georges-Courbot et al. 1997; Formenty et al. 1999; Georges et al. 1999; Morvan et al. 2000) as candidates on the basis of their susceptibility to virulent Ebola infection in the laboratory or the field.

In discussions of the enzootic hypothesis, the increasing rate of Ebola emergence over the last few decades is often explained in terms of environmental change. Two types of environmental conditions are most often cited as influencing Ebola outbreak probability, habitat disturbance by humans (Georges-Courbot et al. 1997; LeGuenno 1997; Georges et al.

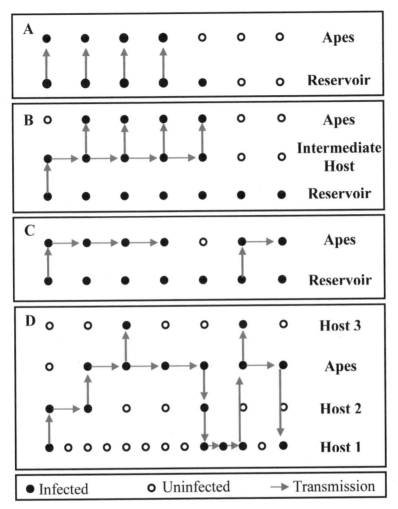

Figure 10.6. Variations on the epizootic hypothesis for Ebola emergence. (A) Spreading waves of infection in the "reservoir," with apes infected directly from the reservoir. (B) The reservoir infects an intermediate host, epizootic waves spread in the intermediate host, and the intermediate host infects apes. (C) Rare infection of apes by the reservoir, with the epizootic spreading through ape-to-ape transmission. (D) Community epidemic. Within-species transmission for Apes, Host 1, and Host 2. Host 3 is a dead end and has no within-species transmission. Apes, Host 1, and Host 2 reciprocally infect each other but spatial spread is dominated by transmission within apes. The smaller ranging scale of Host 1 allows the epidemic to move more slowly through space.

1999; Greiser-Wilke and Haas 1999; Reiter et al. 1999; Morvan et al. 2000; Osterhaus 2000; Tutin 2000; Nakounne et al. 2001) and climatic conditions (Formenty et al. 1999; Tucker et al. 2002; Leroy et al. 2004). Habitat disturbance is assumed to either increase the abundance of the reservoir species or change its behavior so that it comes into more frequent contact with primates. Under the habitat-disturbance hypothesis, the environmental factors precipitating Ebola emergence in the past decade are uptrends in human population size, settlement activity, and commercial-resource extraction rates in Central Africa (Q. F. Zhang et al. 2002; http://apps.fao.org).

The climate hypothesis has two different forms. Some authors, discussing outbreaks through 1996, emphasize that the outbreaks occurred during the rainy season (Formenty et al. 1999; Monath 1999; Tucker et al. 2002). More recent observations suggest that human outbreaks in Gabon and the Congo tend to begin during the two annual dry seasons (Leroy et al. 2004). Formal survey data and anecdotal reports on the number of apes and other mammals found dead in the forest also suggest peaks during the dry season (Leroy et al. 2004), although some caution must be exercised because these data were not corrected for sampling effort. Under the climate hypothesis, increased rates of contact between reservoir and dead-end hosts might be attributed to the reduction in African precipitation observed during the second half of the twentieth century (Mason 1997, 2001; Nicholson 2001; Chase et al. 2003). Of particular note has been the lengthening and intensification of dry seasons (Fauchereau et al. 2003). Intensified dry seasons might augment eruptions in the abundance of a short-lived vertebrate host or arthropod vector (Formenty et al. 1999; Monath 1999). Alternatively, increased reliance of the reservoir and/or dead-end hosts on some common dry-season resource might have caused increased rates of contact rates between the two, independently of their overall abundance (Formenty et al. 1999; Leroy et al. 2004).

Both forms of the enzootic hypothesis have a certain intuitive resonance. They posit Ebola emergence as a sort of retribution for the way in which rapacious commercial activity destroys forest on a local scale (Justice et al. 2001) or disturbs the climate on a regional to global scale (Semazzi and Song 2001). Despite this intuitive appeal, the enzootic hypotheses are based on some tenuous assumptions. The most dubious is that ape deaths in Gabon and the Congo are caused by varieties of Ebola that are spatially persistent, not spreading through space. This is inconsistent with the observation that human and ape mortalities caused by Ebola in Gabon and the Congo show a clear and consistent pattern of spatial spread (see text that follows).

Another questionable assumption is that most ape mortalities *must* be caused by one-way spillover infection from a reservoir species in which Ebola is always nonvirulent. It is true that some viral diseases that cause mass wildlife die-offs have vertebrate reservoirs in which the virus is of low virulence. However, most animals killed in acute outbreaks of these diseases tend not to be infected directly by one-way "spillover" from the reservoir. Rather, the virus occasionally jumps from the reservoir into one or more new host species, in which it is highly virulent, then spreads epizootically within the new host (as we are now seeing with avian influenza). Furthermore, many highly virulent viral diseases with rapid onsets are maintained in enzootic circulation without an alternate vertebrate reservoir in which the virus is always of low virulence. This can occur through long-term reproduction and vertical (parent-offspring) transmission of the virus in an arthropod vector, as with yellow fever in Peruvian monkeys (Bryant et al. 2003). It can also occur when a virus known for causing acute, highly virulent infections in a particular vertebrate species can also cause chronic infection in that species. Examples of this are hemorrhagic disease in rabbits (Forrester et al. 2003), bluetongue disease in ungulates (Stallknecht et al. 1997; Takamatsu et al. 2003), and West Nile virus in birds (Garmendia et al. 2000). Virulent diseases that can persist without a classic vertebrate reservoir host in which infection is normally of low virulence are much more common than is generally appreciated (Kuno 2001a, 2001b).

An additional problem with the hypothesis of one-way spillover from an enzootic reservoir is that high mortality rates in apes necessarily imply high prevalence in the reservoir population. However, to date, infected individuals of the most commonly discussed candidate reservoir species have not been found in any outbreak zone during an acute outbreak. The enzootic hypothesis further implies that Ebola should be prevalent in the reservoir species both before the arrival of mass ape mortalities and long after; and that there should also be a chronic trickle of ape deaths caused by environmental conditions that are similar to, but not as severe as, the episodes driving acute outbreaks. Yet, no infected individuals from candidate reservoirs have ever been found in an outbreak zone before or after an acute outbreak. Furthermore, ape or other large mammal deaths from Ebola in a given local area have never been separated by more than 2 years, suggesting that virulent varieties of Ebola that are killing apes do not persist locally in the longer term.

The habitat disturbance hypothesis is questionable in the context of Gabon and the Congo because most of the human outbreaks and ape die-offs have occurred in areas with very low levels of habitat disturbance, relative to other sectors of these countries or the region as a whole. All outbreaks were initiated in small villages and all but one were in areas with no mechanized logging, more than 100 km from the nearest sizable town.

Another dubious assertion of the enzootic hypothesis is that gene-sequence variability of Ebola virus among human-outbreak sites is too high to have been created during an Ebola epizootic lasting a decade or less, given the observed lack of viral mutations within several human-transmission chains and high sequence conservation over large distances (Leroy et al. 2004). However, epizootic diseases such as West Nile virus can maintain high levels of sequence conservation over large spatial and temporal scales (Charrel et al. 2003), while still diversifying quickly on shorter time scales when invading new areas (Beasley et al. 2003; also see discussions of rabies [Nadin-Davis et al. 1999] and rabbit hemorrhagic disease [Forrester et al. 2003]). Moreover, the current sample of transmission chains (involving about 45 humans) is far too small to draw sweeping conclusions about how observed sequence variability could have been created.

Gene-sequence data for Ebola from the Gabon-Congo border region do, however, have other implications. A strong correlation between genetic distance and geographic distance suggests that Ebola is being spread through some sort of neighborhood contact process. Such a contact process could occur in either an enzootic or an epizootic context. Strong spatial structure is not consistent with a central source of Ebola infection (e.g., a colonial bat roost) that spits out infections in a spatially random pattern, at least not on scales below tens of thousands of square kilometers. Furthermore, Ebola genotypes are structured on a spatial scale that is probably too small (100 km and less) to have been generated in a long-term enzootic if the prime mover of Ebola was a highly mobile species such as a bat or bird. Long-distance dispersal tends to homogenize genetic variation over larger spatial scales, both for hosts (e.g., bats; Ditchfield 2000; Loyd 2003) and the diseases they carry (e.g., birds and West Nile virus; Burt et al. 2002).

The Epizootic Hypothesis

Under the epizootic hypothesis, the gorillas and chimpanzees dying in Gabon and the Congo are not being infected in episodic, entirely independent outbreaks from a reservoir in which Ebola is enzootic. Rather, they are caught up in moving waves of infection. Continuing transmission chains in such epizootic waves might involve only hosts in which Ebola is nonvirulent, with apes and other large mammals infected only as incidental spillover (figs. 10.6A–B). Alternatively, epizootics could be partly or entirely maintained through transmission within or between host species in which Ebola is highly virulent. An arthropod vector might also be involved. At this point, it is not clear which scenario is operating in Gabon and the Congo.

It is also not clear when and where the virulent varieties of Ebola killing apes in Gabon and the Congo originated (Monath 1999). Virus variants lethal to primates could long have been circulating around Gabon and the Congo in epizootics that did not produce conspicuous numbers of deaths in primates or other large mammals. These epizootics might have involved hosts to which Ebola was of either high or low virulence. It is equally possible that the Ebola variants that are killing apes recently immigrated from outside the current outbreak zone in Gabon and the Congo, or recently mutated into forms that are virulent to primates and other large mammals, or even to a "normal" vertebrate reservoir (e.g., a rodent). It is also possible that some change in environmental stressors, host behavior, or host or vector demography has caused chronic Ebola infections that were formally enzootic, to switch to an acute state (Kuno 2001a, 2001b; Forrester et al. 2003), thereby generating epizootic waves (P. J. White et al. 2002).

A common theme shared by these various epizootic scenarios is density dependence. Host or vector density, or both, determine the rate of contact between vertebrate hosts or between vector and vertebrate host and, therefore, the probability that initial infections of a disease can amplify into an epizootic. This is true whether the epizootic involves only spillover infections from low-virulence hosts, transmission between high-virulence hosts, or even a switch from low to high virulence within host species. Regardless of the origins of the virulent strains of Ebola now circulating in Gabon-Congo, increases in the density of hosts or vectors could have increased the probability of epizootic emergence.

The critical role of host density in epizootic dynamics suggests that the history of human-settlement patterns in Gabon and the Congo might have been a critical driver of Ebola emergence. The dense and unfragmented populations of apes and other hunted species created by "Regroupement" would have provided ideal conditions for epizootic amplification and persistence. It is worth noting that a strain of simian immunodeficiency virus (SIV) in chimpanzees made the jump to become human immunodeficiency virus (HIV) in humans in the same region (Gao et al. 1999).

Climate change is also a plausible explanation for the emergence of Ebola epizootics (see Pounds et al. [2006] for a seemingly analogous situation with a virulent frog pathogen). Many of the mechanisms through which climate might influence Ebola transmission rates are density dependent, including changes in host or vector abundance or behavior that locally increase intra- or inter-specific contact rates. Thus, an intensification of dry-season weather could have pushed transmission rates over the threshold necessary to maintain epizootic dynamics. For instance, longer, more intense dry seasons could have raised local Ebola prevalence to a level such that enough chronically infected hosts or vectors survived through the wet season to ensure epizootic reamplification during the following dry season. Once triggered, such a cycle could be self-maintaining.

Because they both involve density-dependent mechanisms, climate trends and long-term changes in human-settlement patterns could have acted complementarily or synergistically. For example, trends in human-settlement patterns could have allowed host densities to rise to a level at which epizootic amplification was possible during the dry but not wet season. This might have occurred without any change in climate, or it could have required a climatic trend.

Spatial Spread

The most compelling evidence in support of the epizootic hypothesis is that ape and human mortalities in Gabon and the Congo are clearly spreading. This is most evident in data from outbreaks occurring in the Gabon-Congo border region from late 2001 to the present. The October 2003 outbreak in Mbandza village, Congo, was 33 km north-northeast of the closest recorded dead ape discovered at Lossi Gorilla sanctuary in February 2003 (implying a rate of spread of 44 km/year), which in turn was 40 km northeast of the human outbreak at Ngoyeboma village in February 2002 (implying a rate of 42.4 km/year). This suggests that the outbreaks observed in the Gabon-Congo border region from 2001 to the present were not independent, episodic spillovers from an enzootic reservoir, but part of an epizootic wave moving to the east and north.

A slightly slower rate of epizootic spread (29 km/year) was apparent between the November 1994 outbreak at the Mekouka gold camp in the Minkebe region of Gabon, and the January 1996 outbreak at the village of Mayibout II, 34 km to the south (Georges-Courbot et al. 1997). In fact, a Baka pygmy village roughly 40 km north of the gold camps was reported to have suffered a disease outbreak with symptoms and mortality rates similar to Ebola in early 1996, although no samples from symptomatic individuals were available to confirm the diagnosis (Lahm 1993). At the same time, ape dieoffs were reported near Minvoul on the northwest corner of the Minkebe forest block (Huijbregts et al. 2003). The human outbreak at Booue, on the southwestern margin of the forest block, occurred about 6 months later (Georges-Courbot et al. 1997).

The almost simultaneous occurrence of outbreaks at the four corners of the enormous (ca. 30,000 km^2) Minkebe forest block suggests that the massive die-off of apes documented throughout most of this block (Huijbregts et al. 2003; P. D. Walsh et al. 2003) was brought on by a single epizootic emanating from the eastern side of the block. That the die-off was spreading, rather than simultaneously occurring everywhere, is evidenced by the fact that the outbreak nearest the eastern edge (the gold camps) occurred more than a year before the others, whereas the outbreak furthest from the center (Booue) occurred later.

The outbreaks that occurred in the Minkebe region have been perceived as being entirely independent of those in the Gabon-Congo border region more than 5 years later. This perception may, in part, be due to the extremely sparse human-settlement pattern in the region separating the two outbreak zones, which lowers the probability of detecting the presence of Ebola. Moreover, there is evidence of a major ape die-off in the intervening region (P. D. Walsh et al. 2003). The Langoue forest block, which showed the highest gorilla nest encounter rates in a 1983 survey of Gabon (Tutin and Fernandez 1984) and remains virtually unhunted, now has very low ape densities (fig. 10.1). The Langoue gorilla population is also strongly skewed toward solitary males, a probable indicator of epizootic passage (see text that follows). Furthermore, if one regresses the distance between the last outbreak in the Minkebe series (Booue 1996) and the Gabon-Congo border outbreaks against the time since the Booue outbreak, the resulting estimate of spread rate is 42.5 km per year. This rate is strikingly similar to the rates of Ebola spread observed within the two outbreak zones. These data are far from conclusive, but they suggest that the Ebola outbreaks observed in Gabon and the Congo over the last decade could all be part of a single epizootic that originated at the center of the Minkebe block around 1992. This view is supported by the fact that the human outbreaks in 2001 and 2002, and the ape carcasses found at the Lossi sanctuary in early 2003 were distributed along north-south axes perpendicular to a line emanating from the putative origin near Booue. This arrangement is suggestive of an advancing wave front.

Rivers as Barriers

Another observation consistent with the epizootic hypothesis involves the distribution of surviving apes with respect to rivers and swamps. In several areas with evidence of Ebola impact, apes occur in moderate to high density on one side of a navigable river or large swamp but exhibit very low densities on the other (Halford and Auzel 2003; Huijbregts et al. 2003; Mabaza 2003; P. D. Walsh et al. 2003; Schaffner Cappello 2004). These observations imply a barrier effect similar to that observed in rabies (D. L. Smith et al. 2002) and other diseases transmitted primarily between terrestrial animals that rarely cross rivers (e.g., primates or rodents). Transmission might be primarily within apes, within both apes and other terrestrial animals, or only within other terrestrial animals, with apes being spillover hosts. A highly mobile, airborne reservoir species, such as a bat or bird, could easily traverse rivers and swamps. This suggests that an airborne species is not the source of most ape infections, although it is not inconsistent with an airborne species moving the virus long distances or occasionally initiating outbreaks that are then amplified in terrestrial hosts.

APE CONSERVATION PRIORITIES

Ape demography is a good place to start in considering ape conservation priorities. Both chimpanzees and gorillas have exceptionally slow reproductive rates. Female chimpanzees do not bear their first offspring until reaching an average age of about 14 and, thereafter, give birth only every 5 to 6 years (Kaplan et al. 2000). Consequently, a long time is necessary for ape populations to recover after they have crashed. Leslie-matrix projections based on extremely optimistic assumptions (i.e., mortality rates from captive animals and no density dependence) suggest an annual chimpanzee-population growth rate of about 1.63% (P. D. Walsh et al. 2003, supplementary online information). This implies that chimp populations that were reduced by 95% by Ebola (a typical reduction) could require almost a century to recover to only a quarter of their pre-Ebola densities (fig. 10.7). According to the model, full recovery could take roughly 180 years.

The above calculation optimistically assumes that there would be no hunting during the recovery period. This assumption is highly unrealistic, particularly because apes in post-Ebola areas are scattered at very low densities over very large areas and, therefore, are very difficult to protect efficiently. Sustained low densities of apes may also lead to negative effects often associated with small population size, including demographic stochasticity and inbreeding depression, which interact with environmental variability to threaten population survival (Burgman et al. 1993). Disruptions of social structure would also be expected to hinder population recovery. The latter effects seem particular likely in light of the peculiarities of gorilla and chimpanzee social systems. Chimpanzees

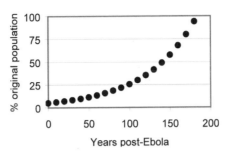

Figure 10.7. Chimp population-recovery rates shown as the time necessary for chimpanzee populations to reach a given percentage of their original size. Projections are based on a Leslie matrix model parameterized using captive survival rates and no density dependence.

live in large communities that literally wage war on each other (Stanford 1998). Mortalities caused by Ebola and hunting, and range shifts forced by logging, have likely upset the balance of power between neighboring communities (L. J. T. White and Tutin 2001). Restoring that balance may require a protracted interval of internecine strife in which survival and, possibly, reproductive rates might be diminished.

Gorillas live in multi-female harems defended by one or, rarely, two adult males (silverbacks). Ebola has apparently had a disproportionate impact on these harems leaving a large surplus of solitary silverbacks. This not only means that post-Ebola gorilla populations have a lower reproductive potential than under normal sex ratios, but also that male reproductive competition within post-Ebola populations is likely to be intense. This has serious implications for reproductive rates, because silverbacks taking over a harem generally engage in infanticide, killing unweaned infants and juveniles (Fossey 1984; Doran and McNeilage 1998). Given the long lifespans and low reproductive rates of gorillas (Steklis and Gerald-Steklis 2001), it might be years or even decades before sex ratios are restored and this hypothesized cycle of social disruption and infanticide dissipates.

This discussion of ape demography is intended to convey two concepts. First, attempting to "regrow" ape populations in areas that have been heavily impacted by either hunting or Ebola is not a promising strategy. Ape populations are simply too slow to recover, particularly for developing governments anxious to make money on ape-based tourism. Second, the issues of ape conservation and sustainable hunting should be entirely decoupled because virtually any hunting of apes is unsustainable. For instance, chimpanzees have such low demographic rates that they produce only a tiny sustainable yield of meat (<1 kg/km^2/year; P. D.

Walsh et al. 2003). Furthermore, hunting of other wildlife species in ape habitat is likely to threaten apes because nonselective hunting methods such as snares maim and kill apes (Quiatt et al. 2002), as well as other protected species (Noss 1998b). Moreover, it is operationally far more tractable to prohibit hunting access in a given area than it is to intensively monitor exactly which species are being killed.

Both slow recovery rates and the unsustainability of hunting suggest that conservation efforts for apes should be focused on protected areas that still have large ape populations where no hunting is permitted. There are still perhaps 10 or 15 parks or reserves in western equatorial Africa that sustain more than one thousand apes. The emphasis should be on protecting apes in these areas. The question is, What is the best way to protect these areas, both from hunting and from Ebola?

Controlling Ape Hunting

Hunting poses a critical threat to ape survival, despite the fact that some evidence suggests that Ebola epizootics might be less severe in areas that are heavily hunted. Mortality from hunting is especially likely to have severe effects on ape populations during their slow recovery from Ebola outbreaks. Pressures from intense hunting, catastrophic disease-related mortality, habitat destruction, and increased physical accessibility to forests from industrial logging are likely to operate synergistically, posing dire threats to dwindling ape populations.

In the case of hunting, the key to protecting apes lies in the fact that most hunting of apes is not a subsistence activity. Most consumers of ape meat are not starving villagers; they are salaried employees in logging camps, towns, and cities. In the larger cities, bushmeat is actually more expensive than other sources of protein (fish or meat from domestic animals). Urban consumers view ape meat and other types of bushmeat not as a vital protein source but as a sentimental link to their rural heritage: something akin to the Thanksgiving turkey or Christmas goose (P. D. Walsh et al. 2003). A public education campaign to deglamorize consumption of ape meat should definitely be a piece of the ape-conservation puzzle. However, given the deep cultural roots of bushmeat consumption, this campaign will be a protracted one and seems unlikely to slow the killing of apes before their populations are further devastated.

A second important point about ape hunting is that it increasingly takes place in remote areas with little or no traditional land tenure, including parks and other formally protected areas. Furthermore, many people who hunt apes are not long-term residents of traditional villages but transients who often travel long distances to work out of newly created logging towns. Although alternative-use programs targeted at people in traditional villages may help to slow ape declines, they cannot stop them. In fact, although some local villagers profit from the bushmeat trade in the short term, they will be victims in the long term. Intense commercial hunting depletes village hunting zones of small game that might have been sustainably harvested if hunting was confined to a subsistence level. By exterminating the larger fauna, commercial hunting also deprives local villagers of the opportunity to profit from sustainable-use programs, such as wildlife tourism.

These considerations suggest that the key to checking the precipitous decline of apes in western equatorial Africa is an increased emphasis on law enforcement, a view shared by most ape researchers working in the region (www.westerngorilla.org). Law enforcement has been critical to the survival of mountain gorillas in East Africa (Steklis and Gerald-Steklis 2001). Intensive law enforcement has also shown promise in western equatorial Africa in the few cases where it has been tried in earnest (e.g., Elkan et al. 2001; chap. 21 in this volume). Unfortunately, most of the countries in the region now have minimal wildlife departments and make only rudimentary efforts at wildlife-law enforcement. For example, in 2002 Gabon's wildlife department had only 17 civil-service-level employees for the entire country and an operating budget of about $60,000, exclusive of civil-service salaries (L. White, pers. comm.). Foreign governments and nongovernmental organizations supplement this substantially via programs such as the U.S. government's commitment of up to $53 million for the Congo Basin Forest Partnership (http://www.state.gov/g/oes/rls/rm/2003/25414.htm). However, this program involves six countries and extends over 4 years, with an average annual investment of only $2.2 million per country. This is a good start but it is less than one-thousandth of the amount spent annually on similar wildlife conservation efforts in the United States. In 2003, the United States budgeted $1.6 billion for the operation of its National Park system (http://data2.itc.nps.gov/budget2/tables.htm), $1.4 billion for the operation of the National Forest system (http://www.fs. fed.us/budget_2004/), and $1.3 billion for the Fish and Wildlife Service (http://budget.fws.gov/). Yellowstone National Park alone had an annual budget of $27 million (http://www.nps.gov/yell/pphtml/facts.html), several times more than that spent each year on wildlife conservation in all of western equatorial Africa. Moreover, much of the money budgeted by the U.S. government is targeted specifically at law enforcement, with $78 million for the U.S. Park Police and $83 million for law enforcement by the Forest Service. Other federal, state, and local agencies also intensively regulate hunting and enforce wildlife law.

Clearly, it is high time for a major investment in the capacity of African ape-range countries to adequately enforce existing laws against ape hunting (which is illegal in all range countries). However, there is no time to wait for national capacity to be fully developed. Ape populations are declining so rapidly that if aggressive enforcement action is not taken quickly, it will be too late to save the vast majority of apes. Outside institutions, including national governments from developed countries, multilateral organizations, nongovernmental organizations, and private donors, need to partner with range-country governments to ensure that wildlife laws are enforced. Because of the political sensitivity of law enforcement, many of these outside actors have in the past been skittish about actively investing in law enforcement. But now they need to become intimately involved at every level, from funding and administration to training and implementation. A further need is to draw individuals and institutions with specific expertise in law enforcement into the ape-conservation community.

Controlling Ebola

To date, the enzootic hypothesis has dominated thinking on the potential for ameliorating the impact of Ebola on apes (http://www.ecofac.org/Ebola/EbolaBrazzaville.htm). If Ebola outbreaks are simply popping up randomly, then there is little hope of success; the cost of intervening everywhere in the region would be prohibitive. However, if we are actually dealing with spreading waves of infection, as is increasingly apparent, then there is hope because interventions could be strategically targeted along the advancing wave front.

Several forms of intervention are possible. The most obvious is vaccination. No Ebola vaccines are yet ready for use, but several independent lines of vaccine research are now advancing rapidly. Ebola vaccines have been successfully tested in mice (Warfield et al. 2003) and monkeys (Sullivan et al. 2003). Two major issues need to be resolved. The first is vaccine delivery. Apes live at low densities and are afraid of people. Therefore, intramuscular vaccination with a dart gun would probably not be feasible for more than a small number of apes. Oral vaccination, which has been highly successful in large-scale rabies control programs (Brochier et al. 1991, 1995; Stohr and Meslin 1996), is probably the most realistic option for vaccinating large numbers of apes or other host species. The second issue is the potential for unintended side effects. Vaccines based on attenuated or recombinant viruses have the potential to cause damaging or fatal infections in target or nontarget animals (Brochier et al. 1991).

Given these constraints, the most promising option may be a vaccine using virus-like particles (VLPs). A VLP vaccine is attractive because it contains only Ebola coat proteins, not a fully functioning virus. Consequently, it is not infective. A VLP vaccine has also been successfully administered nasally to mice (Warfield et al. 2003), suggesting that mucous-membrane contact is adequate for protection. This gives hope that an oral vaccine could be developed using VLPs.

A second form of intervention would be to amplify the barrier effect that rivers apparently have on Ebola passage. Apes and other mammals traverse rivers by climbing across fallen and overhanging trees. Thus, clearing rivers of these trees might decrease the probability of Ebola passage. A variation on this theme is to create psychological barriers to large mammal movement by lighting fires, making noise, etc.

A third option is the translocation of apes from in front of the advancing Ebola wave to areas that have previously been depopulated by either Ebola or hunting. This would not only save the translocated apes, but the induced reduction in ape density might block epizootic spread or, at least, decrease rates of Ebola mortality in remaining animals. Other options are control of insect vectors (if they exist) and culling or birth control in vertebrate hosts, all common disease-control measures.

Of the potential intervention options, river clearing seems most promising in the short term whereas vaccination may be the best medium- to long-term solution. Translocation is probably too expensive to be used in any but a very targeted manner, and carries the risk of ape mortality during handling; nonetheless, the risk of death from Ebola (50%–98%) is much higher for apes living just ahead of the Ebola front than is the risk of mortality during translocation (perhaps 3%–5%). Culling, birth control, and vector control are not feasible at present because we do not know which hosts are critical to the transmission chain or even if there is an arthropod vector. Birth control would be most effective for rapidly reproducing, eruptive species. Culling would not be ethically acceptable without a high level of certainty that it would be effective. However, if future studies strongly indicate that many animals can be saved by culling a few, this option should be seriously considered, particularly if the animals to be culled are not members of highly endangered species such as gorillas and chimpanzees.

The best way to implement these options would probably be in an integrated strategy focused just ahead of the advancing Ebola wave-front and exploiting natural and manmade features that channel Ebola transmission. For instance, vaccination, translocation, culling, or birth control might be concentrated along cleared rivers and in corridors separating the headwaters of rivers. A targeted vaccination program restricted to Belgium's international borders was used to block the introduction of rabies from France after a more widespread vaccination program eradicated fox rabies in Belgium (Brochier et al. 1995). The cost of carefully targeted intervention along the advancing Ebola front would be lower than the cost of intervention across the entire range of gorillas and chimpanzees in the region. Even if a targeted strategy did not totally eradicate the virulent strains of Ebola now circulating in the Congo, it might protect pockets of habitat containing hundreds or even thousands of apes. Given the precipitous decline of apes in Africa, protecting pockets of this size seems well worth the investment. Particular emphasis should be given to Ebola control inside existing protected areas, where future hunting-control efforts can be targeted at protecting surviving apes.

CONCLUSIONS AND IMPLICATIONS

1. In recent years, populations of western lowland gorillas and common chimpanzees have declined precipitously in much of western equatorial Africa, mainly because of Ebola-virus disease and overhunting.

2. Many aspects of Ebola epidemiology, such as possible reservoir species and vectors, or whether the disease is enzootic or epizootic in nature, are still unknown.

3. If aggressive action is not taken to control commercial hunting and Ebola, then the current declines will continue or even accelerate. Over the next decade, populations of gorillas and chimps in western equatorial Africa could collapse to small pockets that have low long-term viability.

4. It is time to get serious about law enforcement with respect to illegal hunting. This is a major challenge that will require fundamental changes in public attitudes in African countries, and restricting the access of local peoples to wildlife resources. It is abundantly clear that if we do proceed aggressively our closest relatives will continue their precipitous decline toward extinction.

5. Whether Ebola can be controlled is unclear. The most obvious Ebola-control strategies all have drawbacks and limitations, and we currently lack the data necessary to make optimal decisions about allocation of resources to Ebola control. But we should not let this continue to paralyze us. Mounting evidence suggests that Ebola transmission in wildlife

populations is amenable to at least partial control. Basic research on transmission dynamics should continue, but should be complemented by applied work on the cost, feasibility, and effectiveness of potential Ebola-control measures.

6. The consequence of further inaction—the loss of a large proportion of the world's remaining gorillas and chimpanzees—is unacceptable.

CHAPTER 11

Current Decline of the "Dodo Tree": A Case of Broken-Down Interactions with Extinct Species or the Result of New Interactions with Alien Invaders?

Cláudia Baider and F. B. Vincent Florens

Tropical forests worldwide share a number of negative anthropogenic impacts, including deforestation for agriculture or settlement and accompanying habitat fragmentation, over-hunting, over-harvesting, and introduction of alien species (Vitousek et al. 1987; Loope and Mueller-Dombois 1989; World Conservation Monitoring Centre 1992). The forests of tropical oceanic islands are well known for their unique flora and fauna (Brockie et al. 1988). They are more susceptible to human impacts than mainland habitats of comparable latitude and ecology; and their species are more vulnerable to the deleterious effects of invasive aliens (Loope and Mueller-Dombois 1989; Simberloff 1995; Enserik 1999), in part because of their unsaturated habitats and nonequilibrial communities (Barret 1998).

Biological invasion is today considered the second most important threat to biodiversity worldwide (Vitousek et al. 1997), preceded only by habitat destruction. Species invasions are causing a striking homogenization of the world's biota (Lodge 1993; Wilcove et al. 1998), to the extent that some biologists have referred to the present geological era as the "Homogeocene" (Putz 1998). This daunting problem is expected to worsen as species introductions are on the increase, given the remarkable mobility of people and expanding international trade (Ewel et al. 1999).

Often, the only hope of reversing the extinction and homogenization trends, especially for tropical oceanic island biota, is through intense and costly conservation management interventions (Cabin et al. 2002). However, we still have only a limited knowledge of how and how quickly the alien-driven degradation of ecosystems is occurring, and their precise impacts on native communities and populations (D'Antonio et al. 2001). In the absence of a good understanding of these effects and mechanisms, conservation management runs the risk of being poorly focused and thereby of failing to achieve its desired objectives.

As an example of this challenge, we identify the factors causing the current decline of the "dodo tree," more properly termed the Tambalacoque tree (*Sideroxylon grandiflorum*, Family Sapotaceae, formerly *Calvaria major*), a species endemic to the island of Mauritius in the Indian Ocean. This tree, like virtually all other native plants on Mauritius, has suffered massive reductions in numbers and range from past deforestation that destroyed 95% of Mauritian forests (Safford 1997). However, even though the felling of Tambalacoques has ceased, the species continues to decline because its natural regeneration is evidently extremely low (Witmer and Cheke 1991). This has prompted several authors to propose explanations for its declining populations. Some have suggested that past faunal extinctions, which have been extensive on Mauritius, are responsible. Temple (1977) suggested that an obligate seed-dispersal mutualism once existed between the tree and the now-extinct dodo (*Raphus cucullatus*), a very large, flightless relative of pigeons and doves formerly found only on Mauritius. Iverson (1987) and Cheke (1987) also evoked broken-down ecological interactions, but with extinct endemic tortoises (*Cylindraspis* spp.); whereas Wyse-Jackson et al. (1988) suggested a more indirect role, whereby vertebrate frugivores helped to clean the endocarp of its fleshy pulp and thereby reduced fungal infestations that could otherwise destroy the seed. Some of these and other authors also blamed alien plants and animals (Vaughan and Wiehe 1941; Owadally 1979; Friedmann 1981; Cheke et al. 1984; Quammen 1996). At present, however, the controversy over the Tambalacoque's decline is unresolved, with different authors holding strikingly contradictory views (e.g., Wenny 2001; Lundberg and Moberg 2003).

Under these uncertain circumstances, it is extremely difficult to address successfully the conservation needs of the Tambalacoque. A critical first step is to test, to the extent possible, the various hypotheses forwarded by previous authors. Our conclusions, we hope, will have broader relevance for the survival of other direly threatened species on tropical islands.

THE SPECIES AND STUDY SITES

Mauritius is a small tropical island (1856 km^2 in area) 700 km east of Madagascar (20°20′ S, 57°30′ W; fig. 11.1A) and is part of the southwestern

Indian Ocean biodiversity hotspot (Myers et al. 2000). Forty-five percent of its angiosperm species, most of which are threatened, are endemic to the island (Strahm 1993). The once-pristine forests of Mauritius have vanished, with many species now extinct, particularly the larger vertebrates, which must have had a significant biotic influence on the various vegetation communities of the island (Cheke 1987). The island's native vegetation ranged from tropical savanna on many coastal lowlands to mossy forests on the highest peaks with azonal edaphic climaxes like marshes and heath in the uplands (Vaughan and Wiehe 1937). Presently, a mere 5% of the island's

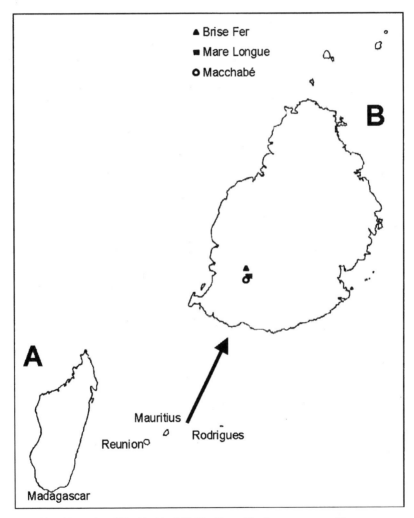

Figure 11.1. (A) Location of Mauritius Island; (B) Mauritius and the three areas surveyed for Tambalacoque trees

native forests remain scattered in several fragments, mainly in the highlands (Safford 1997). These remnants are permeated with hordes of invasive animal and plant species whose interactions with native species are causing their gradual in situ decay (Strahm 1993), such that native plants now dominate only a third of these remnants (Page and D'Argent 1997). However, the mechanisms of this degradation are poorly understood (Safford 1997).

The Tambalacoque is a rare canopy or emergent tree of upland montane forests, which receive 2500 to 5000 mm of rainfall annually. It usually grows into a straight bole ending with a compact crown, with a maximum height of about 20 m; the largest individuals can reach 100 cm diameter at breast height (dbh). Its stout stature and well-developed buttresses confer good resistance to cyclones, which are common on Mauritius. The tree reaches maturity when about 10 cm dbh. Its flowering intensity appears to be enhanced by cyclones, although this is based on the observation of only one tree (Friedmann 1981).

The fruit of the Tambalacoque requires up to 18 months to reach maturity, when it turns from an odorless green color to a mottled yellowish-rusty color with an apple smell. The fruit is spherical and up to 5 cm in diameter, and contains one seed enclosed in a thick, woody endocarp (Friedmann 1981). Germination is said to take 3 to 6 months (Cheke et al. 1984), but the recorded germination rate is very low, from 2.5% (Cheke et al. 1984) up to perhaps 30% (although this latter value is based on merely 10 seeds; Temple 1977).

Tambalacoques are heteroblastic, that is, leaves from adult trees are distinctly different in form from those of juvenile plants. Thus, leaves of saplings can be three times longer than adult leaves, which themselves are usually 7 to 11 cm long. It is a slow-growing species, based on available data. Three 25-year-old plants, grown in a botanical garden within the species' natural range, had dbhs of 4.7 to 6.4 cm, representing an average girth increment of 2.22 ± 0.34 mm per year. For the only adult forest tree with a previous dbh measurement (initial dbh = 50.9 cm; Vaughan and Wiehe 1941), estimation of girth increment over 49 years was just 0.65 mm per year (Strahm 1993).

We conducted field surveys in three sites, at Macchabé, Mare Longue, and Brise Fer (fig. 11.1B), within the Black River Gorges National Park in southwestern Mauritius. This is where most remnant native-upland forest survives today. Each of these sites has one small Conservation Management Area (CMA): Macchabé (0.4 ha), Mare Longue (2.4 ha), and Brise Fer (19.3 ha). CMAs are weeded of alien-invasive plants and fenced to try keep out feral pigs (*Sus scrofa*) and Java deer (*Cervus timorensis*). Other alien mammals, such as long-tailed macaques (*Macaca fascicularis*) and rats (*Rattus rattus* and *R. norvegicus*), are not stopped by the fence. Data from the only known remaining Tambalacoque surviving outside the forest, in a private garden in the town of Curepipe (hereafter, the urban tree), were included in our study.

PRESENT POPULATION STRUCTURE OF TAMBALACOQUES

Tambalacoques have long been known to regenerate poorly (Thompson 1880; Koenig 1914; Vaughan and Wiehe 1941). Indeed, until the present study, seedlings had never been recorded from the forest (Witmer and Cheke 1991). In 1973, only 13 large senescent trees, believed to be more than three centuries old, were thought to survive (Temple 1977). This, however, was not correct (Owadally 1979). Recently, the total population has been estimated at fewer than 500 individuals (Page and D'Argent 1997). However, none of these authors determined the tree's population structure, although all recognized a marked rarity of trees in smaller size classes (<10 cm dbh).

To determine the population structure, we surveyed both weeded and nonweeded areas of our three study sites (fig. 11.1B), in teams of 2 to 4 persons, from October 2001 to April 2003. All Tambalacoques found were tagged and their dbh measured.

Including the urban tree, we measured 296 individuals reaching 1.3 m in height, and located 26 seedlings smaller than 20 cm in height, the first ever recorded in the forest. No seedlings or saplings between 20 cm high and 0.9 cm dbh were found.

All seedlings were in the weeded area of Brise Fer, from the seed crop of 2001, with some seeds germinating around 1.5 years later. They came from three parent trees, germinating in the canopy shadow or nearby (mean distance from bole = 7.0 m; range 1–26 m). The seedlings could confidently be attributed to their mother trees owing to the unique and recognizable shape of endocarps coming from each tree.

The size-class distribution showed a deficit of trees below 20 cm dbh (fig. 11.2), comprising 27.9% ($N = 83$) of the population. If the Tambalacoque population followed the theoretical inverted-J (or type III curve) of Pearl (1925), then nonreproductive trees would range from 84.9% to 99.7% of the population. These two values were extrapolated from the log-log equation of frequency-by-size class of the actual population, based on trees with dbh ≥20 and ≥50 cm, respectively.

The actual population of Tambalacoques is likely to be much larger than the 300 or so individuals we located because we intensively surveyed only about a third of the species' habitat (~300 ha). Assuming our study sites were representative, we believe that the total population is likely to be close to 1000 individuals. (To arrive at this figure, we first estimated the extent of

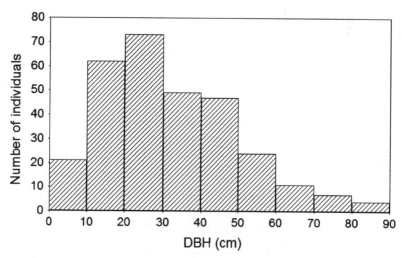

Figure 11.2. Size-class distribution of all Tambalacoque trees surveyed

tree habitat that we did not survey, using the presence/absence data of Page and D'Argent [1997] together with other current observations. This yielded seven habitat areas covering about 625 ha. Assuming an average density of 1 tree per ha, which was typical of the areas we surveyed, we estimated a population of 625 additional trees, giving a total of about 925 trees.)

Our survey demonstrates that the overall Tambalacoque population is somewhat less imperiled than previously thought, when the smallest individual was thought to be 10 cm in girth (Friedmann 1981). However, the population still has very few young trees, creating a real challenge for the long-term survival of the species. Why are younger trees so rare? The best-known explanation is the alleged broken-down mutualism with an extinct species, the dodo. But several other causes are also possible and these will be discussed in light of our new findings.

THREATS TO THE SURVIVAL OF TAMBALACOQUES

Seed Dispersal

Obligate Mutualism with Dodos or the Need for Endocarp Abrasion for Germination

In the 1970s, seed-dispersal studies became very popular following the publication of the seminal Janzen-Connell hypothesis (Janzen 1970; Connell

1971) and renewed interest in mutualistic interactions (McKey 1975). Into this intellectual milieu, Temple (1977) revived an earlier suggestion of Vaughan and Wiehe (1941) that "germination and distribution [of Tambalacoque seeds] were probably assisted by their passage through the alimentary canal of the dodo."

Temple hypothesized that the Tambalacoque evolved a thick endocarp to protect its seed from being digested by the dodo. Such an endocarp would imprison the seed, preventing germination unless abrasion by the stones of a dodo's gizzard weakened it enough to allow it to be split open by the germinating seed. To test his idea, Temple force-fed a turkey—used as a surrogate for the dodo—17 Tambalacoque seeds. Seven were digested and, of the 10 retrieved and sown, 3 germinated. This result, coupled mainly with the belief that successful germination had never before been observed, sealed Temple's idea that an obligate mutualism had existed between the bird and the tree (Temple 1977). This work earned the Tambalacoque the name "dodo tree." Although the idea attracted considerable criticism (Owadally 1979; Cheke et al. 1984; Witmer and Cheke 1991), it still persists today in ecology textbooks and scientific literature (Ricklefs and Miller 1999; Willson and Traveset 2000; Traveset and Verdú 2002), albeit sometimes with a word of caution (P. A. Cox et al. 1991; Mabberley 1997; IUCN 2002). Although they exposed important discrepancies in Temple's experiment, the anti-mutualism proponents had only limited data to support their own views.

We investigated Temple's hypothesis by setting up an experiment to determine the influence of mechanical scarification on seed germination. We collected 224 Tambalacoque seeds from two forest trees. Half of these seeds were scarified by grating on the side from which the radicle sprouts (see Hill 1941), removing 1.36 ± 0.38 mm of pith. The remaining seeds were kept intact as controls. Scarified and intact seeds were sown separately in groups of 7 per pot and arranged in a Latin square-like grid ($N_t = 224$ seeds in 32 pots). Scarification did not enhance seed germination ($\chi^2 = 2.057$, df = 1, $P = 0.15$; Chi-square test) nor did it alter the time taken for germination ($t = -1.844$; $P = 0.072$; t-test). Remarkably, 18 months after initiating the experiment, more non-scarified piths had been split by their germinating seeds (28.6%) than scarified ones (20.5%), although this difference was not statistically significant. These results provide no support for the notion that mechanical scarification improves Tambalacoque germination, leaving unanswered the question of why the tree's germination rate is so low.

The most plausible alternative to Temple's hypothesis is that fungal pathogens kill Tambalacoque seeds whose pulp has not been cleaned by a vertebrate disperser (Friedmann 1981; Wyse-Jackson et al. 1988; Witmer and Cheke 1991; I. M. Turner 2001). After falling to the ground, the pulp of noneaten, ripe fruits would presumably become infested with fungus, which would consume the pulp then move into the endocarp and destroy the seed. The dodo might have been such a beneficial endocarp cleaner, but Iverson (1987) and Cheke (1987) suggested that now-extinct giant tortoises could also have played that role.

To test this idea, we evaluated the effects of pericarp removal on germination of ripe Tambalacoque fruits. Using 32 fruits, half were stripped of their pericarps and the remainder left intact as controls. The seeds and fruits were sown singly per pot in a Latin square design of 16 pots. This experiment was performed using ripe fruits first from the urban tree and a second time using fruits from a forest tree. In neither experiment did the frequency of germinating seeds differ between the two treatments (χ^2 = 0.41, df = 1, P = 0.52 for the forest tree, and χ^2 = 0.29, df = 1, P = 0.59 for the urban tree). Hence, pulp removal had no effect on germination rate, providing no support for this hypothesis.

Low Germination Rate

Yet another possible explanation for poor regeneration of the Tambalacoque was that it produced very few fertile seeds (Koenig 1914), possibly because of pollination problems. Indeed, observed germination rates were very low, ranging from just 2.5% to 30% (Temple 1977; Cheke et al. 1984).

We tested this hypothesis using 427 fresh endocarps collected at the base of various forest trees. From a random subsample of 57 endocarps, we estimated that 10.5% contained live seeds (seeds were classified either as alive if hard and fresh, and dead if a foul-smelling soft mass or liquid was present). The remaining 370 seeds were sown, of which 19.5% started germinating. These two percentages do not differ statistically (χ^2 = 2.64, df = 1, P = 0.103). This result suggests that virtually all living seeds do germinate when sown and that the thick hard endocarp cannot imprison the seed inside. Hence, the Tambalacoque can have a high germination rate if the sown seeds are actually alive. Indeed, in one of the sowing trials, where we used fresh endocarps from the urban tree, a germination rate of 100% was obtained (N = 11). Why then are so many freshly collected seeds already dead?

Interactions with Alien Invaders

The continued decline of Tambalacoque populations has also been linked with the negative impacts of invasive animals. The main incriminated effects are the depredation of unripe fruits by monkeys (Owadally 1979; Friedmann 1981; Cheke et al. 1984; Quammen 1996), seed depredation by rats (Quammen 1996), and destruction of seedlings by deer and feral pigs (Cheke et al. 1984). Extensive invasions of weeds (Owadally 1979) might also suppress Tambalacoque regeneration by competing with seedlings and trees (Cheke et al. 1984). Some of these specific interactions were based on general field observations, others were merely inferred, and none has previously been quantified. We therefore tried to quantify the impacts of different alien species on the reproductive biology of the Tambalacoque.

Reproductive Biology and Alien Species

For all known Tambalacoques, we noted presence or absence of fruits on the tree during the fruiting season and counted all endocarps found in the canopy shadow. From February to May 2002, all trees were revisited to estimate flowering frequency and intensity. This survey was repeated 1 year later.

Searching at the foot of mother trees, we found 7459 fallen fruits, most of which (81.2%) could be classified as dead as they floated in water. Not a single fruit had teeth marks from rats, showing that rats do not gnaw the endocarps. Once the endocarp splits during germination, rats can access and eat the seed, but we observed a mere 0.7% ($N = 300$) of seed mortality in this way under nursery conditions, despite the unnatural crowding of germinating seeds and their full accessibility to rats. The impact of rats in the decline of Tambalacoques thus appears to have been overestimated. Indeed, among endocarps that began germinating in the forest, only about 4% of split endocarps ($N = 26$) had dead seedlings, indicating the maximum impact of rats on these seedlings.

Tambalacoques can bear fruit when around 10 cm dbh, a size we consider as mature in the following analyses. Most of the mature trees we surveyed did not bear fruits (72.7%, $N = 201$). Among fruiting trees ($N = 75$), fruit production was low (<50 seeds in 72% of individuals), with a few having more than 500 fruits (8%). In fact, these estimates of fruit production probably represent the trees' cumulative reproduction for several years because endocarps can resist decay on the forest floor for a few fruiting seasons.

One factor that greatly increases fruit productivity is the control of alien weeds. In weeded areas (CMAs), 40.7% of Tambalacoque trees had fruits, versus only 13.3% in weed-invaded areas (table 11.1). Among fruit-producing trees, those growing in weeded areas had 9 times higher fruit production ($Z_{140,135}$ = -5.39, P < 0.0001) compared with those in weed-infested zones (44.2 vs. 4.8 fruits per tree, on average). In fact, if one outlier from the nonweeded area is removed, the average fruit production there drops to just 1.2 fruits per tree.

Tambalacoque fruits take 12 to 18 months to develop (Friedmann 1981), with little overlap between flowering and fruiting. Trees had a massive flowering event in 2002, possibly triggered by a strong cyclone (maximum gusts of 207 km/hour, 2 to 4 months earlier. More than half (55.5%) of all trees flowered (N = 200), with a higher frequency of flowering in weeded (60.6%) than nonweeded (47.4%) areas (table 11.1).

After pollination, many developing fruits abort at different stages of development because flowers are so tightly packed on branches that only a small proportion of fruits have sufficient space to develop to maturity. The main factor limiting fruit production appears to be the space available for the developing fruits on branches, rather than the rate of pollination of flowers.

The main cause of reduced production of viable seeds is depredation by alien long-tailed macaques. Simply by moving among trees, these monkeys destroy some flowering branches; however, only about 4% of trees flowering at any time are affected and, among these, only about 2% of the flowers on affected trees are destroyed. The monkey's main impact is the depredation of unripe fruits. Virtually all forest trees that were found to bear fruits showed signs of having been visited by monkeys (95.7%, N = 23) during monthly monitoring (from June 2002 to May 2003). The unripe fruits would usually be found on the ground with distinctive teeth marks visible in the pulp. Macaques begin depredating

Table 11.1 Number and percentage of fruiting and flowering Tambalacoque trees (dbh ≥10 cm) in weeded and nonweeded forests

Type of forest	Fruit		Flower	
	Total trees	Trees with fruits N (%)	Total trees	Trees with flowers N (%)
Weeded	140	57 (40.7)	122	74 (60.6)
Nonweeded	135	18 (13.3)	78	37 (47.4)
Total	275	75 (27.3)	200	111 (55.5)

immature fruits when they are as small as 2 cm long, and a visiting troop can sometimes destroy nearly all the fruits on a given tree. In most cases, these fruits are depredated before their seeds have had time to mature. We found that the endocarps of freshly depredated, unripe fruits always contained a developing embryo, indicating that seed inviability was not caused by pollination failure.

To estimate the extent of monkey-caused seed mortality we compared the percentage of viable seeds collected in the forest, where monkeys abound, with that of the urban tree, which was the only known tree completely inaccessible to monkeys. Seeds were planted in nurseries and were classified as being viable upon visible onset of germination (i.e., splitting of the endocarp). Discounting endocarps that had split in situ (i.e., at the foot of the mother tree), on a percentage basis more than 10 times as many seeds germinated from the urban tree (25.7%, $N = 538$) than from forest trees (2.4%, $N = 6806$). Similarly, discounting seeds that germinated in nurseries, the percentage of endocarps that had already split in situ prior to our sowing experiments was about 38 times greater with the urban tree (17.4%, $N = 484$) than with forest trees (0.46%, $N = 6671$).

Human activity also seemed to have an indirect effect on fruit production. In the weeded forest, nearly half (43.8%, $N = 16$) of the fruit-bearing trees located within 20 m of well-used tracks, paths, or gates produced some viable seeds, whereas only 2.4% ($n = 41$) of fruit-bearing trees found deeper in the forest produced viable seeds. Presumably, fruits developing in areas frequented by people suffered a lower incidence of depredation by monkeys. Moreover, monkeys evidently prefer areas with weeds (Sussman and Tattersall 1981). Thus, greater predation from monkeys, possibly compounded by competition from weeds, may explain why the trees we studied in nonweeded areas still had 0% germination despite growing next to tracks.

Density, Regeneration, and Reproductive Output

Sixty-five years ago, Vaughan and Wiehe (1941) found 33 Tambalacoques in a 1 ha plot in Macchabé (33 trees/ha). In 2003, we surveyed a total of 0.7 ha (70 random quadrats of 100 m^2) in Macchabé, Mare Longue, and Brise Fer and found only 4 Tambalacoques (5.7 trees/ha). This gives an indication of the dramatic decline of the species over the past 65 years. Another study of 1 ha in Macchabé forest indicated a threefold drop in tree density over 60 years (Motala 1999), compared to the results of Vaughan and Wiehe (1941). Moreover, when tree size classes were compared, it was found that the larger individuals showed the greatest decline in density.

However, population declines were less severe in plots that had been weeded. In a plot of Vaughan and Wiehe (1941) that was resurveyed after 49 years, 63% of individuals above 10 cm dbh had died, but these losses were partially offset by recruitment of 48.5% (Strahm 1993). This plot had been weeded at least twice between the two surveys. Weed removal also had a positive effect on other native plant species. From random plots in nonweeded areas, Brise Fer was better preserved (table 11.2), and Macchabé more degraded, having fewer native species (60 vs. 66 for Brise Fer), with aliens accounting for 89.8% of all stems (Ramlugun 2003).

The degree to which the Tambalacoque shows a "healthy" J-shaped size distribution is apparently related to the degree of forest degradation, with a closer approximation to a J-shaped curve in the most well-preserved forest than in the most alien-invaded forest (table 11.3, fig. 11.3). Flowering and fruit production were also correlated with forest degradation, with the smallest proportion of flowering or fruiting trees, and the lowest number of fruits per tree, in the most degraded of the three forests (table 11.4).

DISCUSSION

We have shown that a prevailing hypothesis for the apparent failure of Tambalacoque regeneration—a putatively obligate mutualism with the extinct dodo (Temple 1977)—fails to account for the tree's decline. Several other leading hypotheses (e.g., Cheke et al. 1984; Wyse-Jackson et al. 1988) also failed to gain support. An unfortunate elaboration of appealing-sounding hypotheses based largely on suppositions—what might be termed a "biopoetic" approach—misdirected and delayed conservation efforts, allowing further degradation of the tree's population. A population survey

Table 11.2 Number of native tree species, percentage of alien stems, and basal area of native and alien species in the three study sites on Mauritius

	Brise Fer	Mare Longue	Macchabé
Number of native species	66	67	60
Percentage of alien stems	84.6	84.8	89.8
Basal area of native (m²/ha)	53.0	32.9	36.1
Basal area of alien species (m²/ha)	16.0	32.3	27.3

Table 11.3 Number of individual Tambalacoque trees, separated by height class, in three upland forests of Mauritius

	≤20 cm	≥130 cm			All
	Brise Fer	Brise Fer	Mare Longue	Macchabé	All
Nonweeded	—	70	31	37	138
Weeded	27	132	22	3	184
Total	27	202	53	40	322

and some experiments showed that, in forests invaded by alien plants, Tambalacoque trees produce very few flowers and therefore have low reproductive output. Furthermore, alien animal species kill most of the few fruits that do develop, virtually halting regeneration. Because plant populations are usually seed-limited (Mueller-Landau et al. 2002), the drastic reduction in seed number results in a low density of saplings and young trees.

Several other tree species on Mauritius are known to have difficulty regenerating. Natural regeneration was either scarce or virtually nil in four other *Sideroxylon* species endemic to the various islands of the Mascarenes (two on Mauritius and one each on Reunion and Rodrigues; Friedman 1981). Coode (1979) noted a similar situation for the Mauritian-endemic canopy tree *Canarium paniculatum* (Burseraceae); and germination failure

Table 11.4 Number of mature-sized (dbh ≥10 cm) Tambalacoque trees flowering and fruiting and average number of fruits produced in three upland forests of Mauritius

Type of forest	Brise Fer	Mare Longue	Macchabé
Percentage flowering			
All	59.3	54.8	16.7
Nonweeded	40.8	39.3	24.1
Weeded	59.3	64.7	0
Percentage fruiting			
All	35.3	20.0	2.4
Nonweeded	21.7	7.1	2.6
Weeded	42.6	33.3	0
Average fruits per tree ± SE			
All	36.4 ± 8.9	2.4 ± 1.3	0.3 ± 0.3
Nonweeded	9.3 ± 7.1	0.07 ± 0.05	0.3 ± 0.3
Weeded	52.7 ± 13.5	5.3 ± 3.0	0

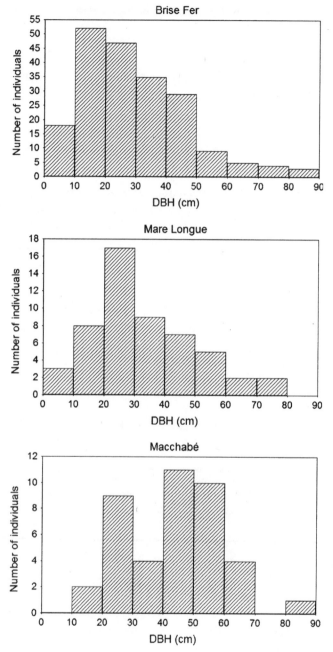

Figure 11.3. Size-class distributions of Tambalacoque trees for the Brise Fer, Mare Longue, and Macchabé study areas

in sowed seeds of *Elaeocarpus integrifolius* (Elaeocarpaceae), another tree endemic to Mauritius, has been reported (Coode 1987).

In islands such as Hawaii, New Zealand, and Mauritius, in situ conservation efforts often involve weeding alien plants and fencing out deer and feral pigs. Conservation measures for invaded forests are showing both positive results and some problems. In a dry forest of Hawaii, 40 years of fencing followed by recent weeding resulted in increased diversity and biomass of native plant species and regeneration of understory plants. However, canopy species still failed to regenerate (Cabin et al. 2000), even with supplementary watering and weeding (Cabin et al. 2002). Tambalacoque trees, however, increased their fruit production after weeding, with seedlings establishing within 7 to 8 years after weeding. A very similar situation was found in the Mauritius-endemic canopy tree *C. paniculatum* (Auchoybur 2003).

Apart from Mauritius, long-tailed macaques have been introduced only in Tinjil, Indonesia (ISSG 2003), and thus it is difficult to generalize about their ecological impact. However, other alien species can have similarly important ecological effects as predators on fruit, flowers, or seeds. In Mauritius, rats are the main predators on seeds of *C. paniculatum*, destroying more than 36% of seeds produced (Auchoybur 2003). Rats are the most widespread of all invasive vertebrate species, reaching 82% of the world's islands and island groups (Brockie et al. 1988). In the Hawaiian dry forest, many native species have their seeds depredated by rats (Cabin et al. 2000), as well in the Canary Islands (García 2002), and New Zealand (García 2002 and references therein).

Some alien weeds are also very problematic on islands. For example, *Psidium cattleianum* is one of the commonest invasive species of wet and cloud forests in several tropical regions, with a tendency to form monospecific stands (Huenneke and Vitousek 1990; Meyer 2000). It has been shown that *P. cattleianum* competes with native species for germination substrates (Huenneke and Vitousek 1990).

It is almost certain that the kinds of ecological alterations we have documented are occurring elsewhere, given the similarities of Mauritius with other invaded tropical islands and fragmented continental ecosystems. For example, Musil (1993) reported 40%–80% of failure in recruitment in the Cape Fynbos ecosystems, which has been severely fragmented and degraded by invasive plants. However, to our knowledge no studies have previously shown a decline in plant reproduction as a direct effect of invasive species, although such an effect could be expected.

Aggressively invading species pose critical threats to many island biotas. These threats are frequently poorly understood in tropical islands, where resources for conservation are usually scarce and where at least

30% of the world threatened plant species are found (Maunder et al. 1997). Research and conservation initiatives are direly needed in these ecosystems to prevent a large-scale loss of tropical biodiversity.

CONCLUSIONS AND IMPLICATIONS

1. The combined effects of numerous invasive alien species are causing continued forest decay in Mauritius. Similar situations also exist in other tropical forest ecosystems, both on islands and continents.

2. Vertebrate extinctions often lead to a breakdown in ecological interactions on islands. However, the major causes of the continued decline of native species in highly fragmented and alien-invaded forests, such as those on Mauritius, appear to result more from detrimental interactions with alien species than from lost interactions with extinct species.

3. The case of the Tambalacoque tree provides a good example of how insufficient experimentation and anecdotal observations, coupled with considerable supposition, can mislead biologists and hinder conservation efforts. Appealing-sounding hypotheses can be very wrong, highlighting the critical need for further empirical studies to help guide conservation efforts.

ACKNOWLEDGMENTS

We thank Danielle Florens and Jean-Claude Sevathian for help during initial stages of fieldwork, and Mr Pavaday for providing access to the urban tree. Dominique Strasberg, Christophe Thébaud, Anthony Davy, and two anonymous referees provided useful comments on the manuscript. We are also grateful to Yousouf Mungroo, Director of the National Park and Conservation Service in Mauritius, for permission to carry out our research and to Jean-Claude Autrey, Director of the Mauritius Sugar Industry Research Institute, for his encouragement and for reviewing the manuscript. Part of the study was made possible with funding from the British Ecological Society through its Overseas Bursary Scheme of 2003.

PART IV

Insidious and Poorly Understood Threats

Carlos A. Peres and William F. Laurance

INTRODUCTION

In a world experiencing astonishing growth in its populace and economies, tropical forests in most regions are shrinking to mere shadows of their former distributions. This has been documented with increasingly accurate remote-sensing probes, which can distinguish the harsh contrast between what is forest and what is not. In this age of satellite imagery, time-series observations—together with underlying laboratory, theoretical, and modeling research—reveal an alarming pace of environmental change. A great challenge for scientists is to understand the ecological, climatic, and socioeconomic consequences of such large-scale changes, and the complex forces that drive them.

To what degree will the surviving remnants of tropical forests retain their ecological integrity? These remnants can range from pallid collections of trees where many of the pieces of the puzzle are already missing, to a fully intact biota, complete with even the most disturbance-sensitive vertebrates. Because many environmental changes are relatively subtle—not causing a gross loss of forest cover—answering this question often goes far beyond the comfort of the image-analyst's armchair and requires hard-won fieldwork. Although technologies are rapidly improving, the spectral and spatial resolution of the most widely available remote-sensing methods are still too myopic to differentiate many types of forest-canopy disturbance, including low-impact selective logging.* Moreover, many forms of structural disturbance occur beneath the forest canopy, effectively hidden from even the highest-resolution and least-affordable sensors in satellites or airplanes. Even an intrepid field biologist may have difficulty perceiving some nonstructural disturbances, including nontimber resource extraction, although such changes might be important ecologically.

Quantifying the extent and intensity of many cryptic disturbances can require Herculean efforts even at moderate spatial scales—and for this reason, such changes are almost certainly underestimated and underappreciated. For example, the cascading effects of population declines from game hunting or nontimber resource removal can have prolonged lag times that may range from months to decades. Yet these subtle disturbances can spell the distinction between truly pristine forests, characterized by the full richness of their biological diversity and functioning, and those that have been corrupted in insidious ways. Quantifying the spatial extent of seemingly intact forests affected by cryptic disturbances is a great challenge for those attempting to monitor the integrity of the world's forested wildlands.

In addition to cryptic disturbances, few threats to tropical forests are more inherently insidious than poor governance and human corruption. Of course, corruption and malfeasance afflict all societies—one need look no further than the financial scandals that engulfed U.S.-based corporations such as Enron and WorldCom, where unscrupulous company executives defrauded investors out of billions of dollars, to be reminded that this is so. But some societies suffer more from corruption than do others. Sadly, political corruption, where individuals abuse public office for financial or other private gain, is pervasive in many developing countries. This occurs because government officials in such countries are often poorly paid, increasing the likelihood of bribery, especially when officials control valuable natural resources, such as timber, oil, minerals, and gemstones. Other societal phenomena that are common in developing countries, such as weak political institutions with poorly developed checks and balances, and widespread nepotism and political patronage, can also promote corruption.

Although corruption and weak governance can sometimes reduce environmental pressures by hindering development activity, they are usually considered a threat to sustainable development. Corruption can have large impacts on nature conservation, by promoting overexploitation of

---

* However, disturbances such as recent selective logging and surface fires can now be roughly quantified, at least for brief windows of time, using high spatial-resolution satellite sensors and spectral mixture analysis.

forests, wildlife, fisheries, and other resources, and by reducing the effectiveness of conservation programs. Many ecosystems worldwide, such as species-rich tropical forests and coral reefs, which are largely confined to developing countries, and temperate and boreal forests in Asia, South America, and Africa, are highly vulnerable.

The five chapters in this part highlight a diversity of cryptic, poorly understood, or insidious threats to tropical forests. Recent studies suggest that all could or have become serious threats to tropical forests and their ecological functioning.

Cryptic Surface Fires

In chapter 12, Jos Barlow and Carlos Peres describe the first major study of the ecological impacts of surface fires on neotropical plant and animal communities. This is by definition an embryonic science because major wildfires, such as that described in this study, were unprecedented in living memory in undisturbed Amazonian forests. Indeed, tropical moist forests were once considered virtually immune to wildfires because of their high moisture content, despite having substantial and continuous fuel loads. Intense droughts in recent decades triggered by strong El Niño events have changed all that, initially in Southeast Asia and later in the Neotropics. Even intact seasonally dry forests can now become flammable—even in areas well above the rainfall threshold between semi-deciduous and evergreen forests—especially with the rapid proliferation of human ignition sources.

As shown in this study, an initial, low-intensity understory fire, with blaze heights that rarely exceeded 40 cm, had profound effects on forest structure, composition, fruit production, and vertebrate communities, in large part because it greatly facilitated subsequent burns that were much higher in intensity. Because fires have been very rare in tropical forests over evolutionary timescales, even an unimpressive, slow-moving fire can kill even the largest emergent trees. Moreover, once burned, canopy-tree mortality increases drastically, resulting in a heavily punctured forest canopy and an accumulation of dry, flammable fuels that then paves the way to far more intensive, recurrent fires.

These recurring fires can drive a phase-shift, converting a complex rainforest into a degraded scrub or savanna dominated by fire-adapted plants and generalist animals. In the central Amazonian study area of Barlow and Peres, fires were aggravated by a high sand fraction in the soil, which reduced the water-retention capacity of the forest and considerably retarded forest regeneration during post-burn intervals. As a consequence, twice- and thrice-burned forests became a skeletal relict of previously unburned stands, both in terms of species diversity and the value of their ecosystem services.

Critical Role of Seed Dispersal

Humans have always hunted vertebrate prey but the scale of this harvest in tropical forests has reached crisis proportions in recent decades. Current patterns of large-game depletion in tropical forests are almost invariably unsustainable and can reach into the heart of even the largest and least-accessible tropical nature reserves, such as those of lowland Amazonia. Understanding the higher-order consequences of extirpating large-bodied frugivores will become increasingly important in managed tropical forests, where the residual vertebrate fauna may predominantly consist of small-bodied species. For many plants, especially those with large seeds, the surviving small frugivores may be poor or ineffective seed dispersers.

What impact will this have on plant communities? Successful seedling recruitment in higher plants involves a sequence of nested regeneration stages, from the production and pollination of fertile flowers to dispersal and germination of viable seeds, and early seedling emergence, growth, and survival. A disruption in any of these processes, or a reduction in the transition probability between consecutive stages, could result in a seedling-recruitment bottleneck and eventually the collapse of a population. In chapter 13, John Terborgh and Gabriela Nuñez-Iturri present a rare glimpse at what could happen to an otherwise undisturbed tropical tree assemblage should specialized seed dispersers be subtracted from this complex equation. On the basis of nearly 7000 trees and saplings mapped in a 4 ha plot of mature forest in Manu Park, southern Peru, they show that strong spatial constraints operate on successful recruitment. In so doing, they confirm at a community level the textbook-famous zone of low recruitment around adult trees, as postulated by the Janzen-Connell model. More than 90% of the saplings of the most common tree species, and all saplings of uncommon species, were located well beyond the projected crowns of conspecific adults, implying that they arose from dispersed seeds. This is possible only because high levels of dispersal for even large-seeded plants have occurred over millennia in this faunally intact Amazonian forest. Nonetheless, we are only beginning to understand how dispersal limitation will operate in vast tracts of overhunted forests elsewhere. The long-term impacts of such changes remain a challenge for future studies, but understanding the baseline conditions, such as those revealed in this study, is a key starting point.

Insidious Effects of Roads

Roads have long been a watershed event in the frontier expansion of tropical forests. By greatly increasing physical access to forests, they can open a Pandora's box of spontaneous activities—forest colonization, logging, hunting, mining, land speculation—that lead to large-scale forest loss or degradation. Yet thousands of kilometers of new highways are being constructed in pristine tropical forests each year, often with the financial backing of international aid and lending organizations.

In chapter 14, Susan Laurance synthesizes available studies on the myriad ecological impacts of roads and highways in tropical forests. Their effects are remarkably diverse and insidious. They cause important edge effects on forest habitat, facilitate the invasion of weeds and diseases, and may create important barriers to wildlife movements. Even narrow, unpaved roads can have significant but poorly understood effects on complex rainforest systems. For example, a narrow road with little traffic in central Amazonia had surprisingly strong impacts on the community structure of understory birds and disrupted local movements of many species. Many other forest-interior vertebrates shy away from road clearings. Speeding vehicles along roads are also a significant cause of road-kill mortality; one study in tropical Queensland found that around 3000 animals were killed each year along a 2 km stretch of highway. The direct and indirect road effects are, however, likely to be most severe where relatively intact forest cover has been stable on an evolutionary time scale. This chapter concludes with recommendations designed to ameliorate the effects of roadworks if they must be built.

Effects of Fragmentation versus Habitat Loss

The correlated processes of habitat loss and fragmentation are among the most critical of all threats to natural ecosystems. Among those who study fragmented ecosystems, an enduring controversy—with important implications for forest management—is the degree to which the spatial configuration of habitat fragments is important. Some studies suggest that the effects of habitat loss are paramount, and that the main determinant of ecological impacts in any landscape is simply the percentage that has been destroyed. Others, however, suggest that the spatial arrangement of the surviving fragments—their configuration, shape, and degree of connectivity—is also important. The former result implies that reducing forest loss is overwhelmingly the most important goal for ecosystem conservation, whereas the latter suggests that actively managing fragmented landscapes to reduce the myriad effects of habitat isolation and edge effects is equally important.

In the tropics, few regions have been as severely reduced and fragmented as the Atlantic forests of Brazil, where more than 93% of the original cover has vanished and less than a fifth of the remaining forest is under strict protection. In chapter 15, Pedro Develey and Jean Paul Metzger provide a pioneering attempt—the first ever in the tropics—to tease apart the relative importance of habitat loss and fragmentation on forest-bird communities. Their data-rich study (never presented previously) was based in the Brazilian State of São Paulo and involved collecting hard-won field data during three consecutive breeding seasons in a large number of fragmented and continuous-forest sites.

Their results clearly indicate that the spatial organization of the remaining forest cover is important and cannot be ignored in conservation plans. They conclude that the survival of diverse avian communities in the Atlantic forest and perhaps elsewhere in the tropics will require renewed conservation efforts, especially in landscapes in which less than 30% of the original forest cover survives—the approximate extinction threshold for many bird species, below which fragmentation effects increase markedly. In landscapes with more than 30% forest cover, they believe, source–sink dynamics and other processes that can rescue local populations will help to offset local extinctions of sensitive species, such as canopy frugivores and ground insectivores, regardless of habitat fragmentation patterns. Unfortunately, in the drastically degraded Atlantic forests, most landscapes have precious little forest remaining, and the remedial management of surviving fragments in these areas should be a top priority.

A Collapse of Forest Governance

Few threats to tropical forests are more vexing than widespread corruption and poor governance. Such problems can become even worse when a country is destabilized politically (see W. F. Laurance 2004b). In chapter 16, Kathy MacKinnon presents a compelling essay on the Gordian knot of policy issues that can influence the fate of tropical forests in developing countries. She draws on her decades of experience in Indonesia, a megadiversity country that is facing a "megacrisis" as it struggles to resolve deeply entrenched social, political, and economic discord. She paints a balanced picture between, on the one hand, rampant corruption and a widespread failure of government institutions to enforce environmental legislation; and, on the other hand, more sanguine prospects for innovative conservation opportunities outside park boundaries. The Indonesia of today—the most species-rich and forest-endowed country in Southeast Asia—is in the throes of economic, political, and social upheaval, which is creating a chasm between sustainable development and the disheartening reality of gross forest mismanagement.

MacKinnon argues that resolution of such regional crises will depend on the interplay between political constraints, government policies, and the needs and aspirations of local people. The fate of the largest remaining tropical wilderness areas in Asia, including Sumatra and large parts of Borneo and New Guinea, are in the hands of Indonesians, a nationality that encompasses diverse ethnic groups. Many parks and protected areas have received considerable donor assistance and international funding, yet they continue to be assaulted by myriad threats, exacerbated by widespread corruption and mismanagement. Forest conservation in tropical countries such as Indonesia will only be implemented through better understanding of the many benefits and ecosystem services provided by forests, and a demand for greater accountability from policymakers and government officials.

SYNTHESIS

The chapters in this part illustrate two key themes. The first is that, for all but the most strikingly obvious changes in forest cover, we often have surprisingly poor information about the distribution and nature of the many threats to tropical forests. For example, in many tropical countries, accurate maps of selective-logging concessions are lacking or very difficult to obtain. Far worse is the problem of illegal logging, which is rampant in many tropical regions. In the Brazilian Amazon, for example, the government estimates that 80% of all timber is stolen, with absolutely no form of environmental regulations or payment of government royalties. Illegal logging may be even more pervasive in Indonesia. At present, even the most intensive remote-sensing efforts can provide no more than fragmentary glimpses of logging operations, and even then, only for very recent timber cutting (but see the pioneering effort of Asner et al. [2005]).

Other widespread environmental insults, such as illegal gold mining—which leads to mercury contamination of forest soils, fish, people, and waterways, to extensive stream sedimentation, and to local decimation of fauna through unregulated poaching by miners—are also very poorly documented. We know, for example, that illegal gold miners are invading national parks and indigenous reserves throughout the Guianas, northern Amazonia, and the Congo Basin, but beyond such vague generalities we have little idea of their actual numbers and distribution. The pervasive effects of overhunting are even more poorly documented, and in the absence of detailed local studies can only be surmised indirectly by assessing the distribution of population centers and of rivers and roads that provide physical access to forests.

Perhaps most damaging of all are cryptic surface fires, which appear deceptively unimpressive while creeping slowly across the leaf litter but can exact a terrible toll on rainforest plants and animals. Because the fires are usually hidden beneath the canopy, we have very little direct information on their spatial distribution, aside from a few local studies and specialized remote-sensing analyses (e.g., Cochrane and Laurance 2002). We can only guess at the actual distribution of fires, by evaluating local climatic and soil conditions and the distribution of degraded lands and anthropogenic ignition sources. Only when the fire-degradation process has proceeded so far that the forest is in imminent danger of becoming fire-prone scrub or savanna, can we readily perceive its extent and progress.

The second major theme in this part is that the ecological consequences of many environmental threats in the tropics are poorly understood. A recent discovery dramatically illustrates this point. It had long been assumed that surface fires cause widespread mortality among small trees, lianas, and forbs, but larger trees (those greater than 30 cm in diameter) were thought to be nearly immune, their sheer size and thicker bark affording adequate protection from scorching. A series of short-term studies (usually no more than one year in duration) supported this interpretation. However, the first longer-term study of surface fires, conducted over a 3-year period, revealed a very different picture: big trees often survived the first year after a fire but then began to succumb dramatically to their injuries in subsequent years. As a consequence, it now appears that even cool-burning surface fires have serious impacts on large trees, with important implications for forest-carbon storage and ecosystem functioning (Barlow et al. 2003b).

This is but one example of scores of environmental mysteries that those alarmed about the fate of tropical forests are struggling to solve. For example, what will be the long-term effects of pervasive defaunation, from habitat disruption and overhunting, on tropical ecosystems? Will populations of plants that rely on chronically overhunted seed-dispersers eventually wither and collapse? Will competitive balances among plant species be fundamentally disrupted? Could such ecological changes ramify throughout the rainforest biota? Future efforts to manage and protect tropical ecosystems may ultimately hinge on present attempts to comprehend such ecological unknowns.

CHAPTER 12

Consequences of Cryptic and Recurring Fire Disturbances for Ecosystem Structure and Biodiversity in Amazonian Forests

Jos Barlow and Carlos A. Peres

Tropical evergreen forests have long been considered virtually immune to sustained fires largely because of excessively moist fuel conditions (Uhl 1998). Over the past two decades, however, this view has been gradually dispelled ever since large uncontrolled fires linked to severe El Niño Southern Oscillation (ENSO) events affected forests in east Kalimantan, Borneo in 1982–1983 (Leighton and Wirawan 1986). More intensive and widespread recurrent fires also followed the 1997–1998 ENSO event with large-scale wildfires occurring throughout many seasonally dry forests of Southeast Asia (Guhardja et al. 2000), Mesoamerica (Anon. 1998a), and Brazilian Amazonia (Barbosa and Fearnside 1999; Nepstad et al. 1999b; Peres 1999b).

As a result, tropical forest fires have been gradually attracting more attention from the science community (e.g., Wuethrich 2000; Cochrane 2003; W. F. Laurance 2003b), and there is now a growing consensus that an unprecedented dynamic of frequent fire incursions and increasingly severe wildfires can become rapidly established in many previously unburned tropical forests (Cochrane et al. 1999; Goldammer 1999; Nepstad et al. 1999b; Siegert et al. 2001). As the causes of fires are discussed elsewhere in this book (W. F. Laurance, chap. 5 in this volume), we make no attempt to describe them in more detail here. Instead we focus on the consequences of these fires, first examining levels of tree mortality across the Amazon basin and the wider tropics. We then assess the severity of effects of single "cryptic" low-intensity fires and more severe recurrent fires, examining changes in forest structure, and using three indicators of forest health and integrity: fruit production, the understory avifauna, and large vertebrates. Much of the discussion on the ecological effects of wildfires focuses on studies from the Brazilian Amazon, which reflects the current paucity of information from other tropical forest regions rather than any relative geographical importance.

EFFECTS ON TREE MORTALITY

Despite the importance of ground fires as catalytic agents of change in tropical forests (Cochrane et al. 1999; Nepstad et al. 1999a, 1999b; Cochrane 2003), studies documenting post-fire rates of tree mortality have been restricted to a relatively limited number of regions (table 12.1). These include the seasonally dry forests of Roraima and Bolivia near the phytogeographic limits of Amazonia (IBAMA 1998; Santos 1998; Barbosa and Fearnside 1999; Pinard et al. 1999), the heavily logged or fragmented forest landscapes in Malaysian Borneo (Woods 1989) and at the eastern flanks of the Brazilian Amazon (Holdsworth and Uhl 1997; Cochrane and Schulze 1999; Gerwing 2002), and lightly logged but seasonal forests of central Brazilian Amazonia (Haugaasen et al. 2003; Barlow and Peres 2004b). Despite the limited geographic replication, several hypotheses concerning the impact of fires on tree mortality have become established, and are summarized in the following sections.

Tree Mortality and Burn Severity

Tree mortality is strongly related to burn severity in many of these regions (Uhl and Kaufmann 1990; Pinard et al. 1999; Barlow et al. 2003a), with the number of live stems surviving the fire being strongly linked to its initial severity (fig. 12.1; Barlow and Peres 2006). As a result, more severe recurrent fires reduce the live-tree density much more than do initial low-intensity fires. Although the magnitude of this loss depends on the number of times a forest has burned, the degree of additional mortality is very similar across the studies conducted to date in the eastern (Cochrane and Schulze 1999; Gerwing 2002) and central Brazilian Amazon (Barlow and Peres 2004b) and in Borneo (Slik and Eichhorn 2003) (fig. 12.2).

There is also an interaction between burn severity and short-term differences in susceptibility according to stem size. For example, short-term

Table 12.1 A summary of studies documenting the percentage of trees (dbh ≥10 cm) dying in the aftermath of wildfires in humid tropical forests

Study	Region	Annual rainfall (mm)	Burned area sampled (ha)	Pre-fire logging intensity	Fire intensity	Time since most recent fire (months)	Tree mortality (%)
Cochrane and Schulze 1999[a]	Tailândia, eastern Amazonian Brazil	1500–1800	0.96	Moderate	Low	12	38
			1.26	Moderate	Moderate	12	64
			1.5	Moderate	High	12	90
Holdsworth and Uhl 1997	Paragominas, eastern Amazonian Brazil	1700	2.5	Moderate	Low	1 and 18	44
Kauffman 1991	Paragominas, eastern Amazonian Brazil	1700	—[b]	Moderate	Moderate	?	36–54
Gerwing 2002[c]	Paragominas, eastern Amazonian Brazil	1700	1.5	Moderate	Low	?	38
Peres 1999b	Rio Arapiuns, Central Amazonian Brazil	2200	1	Moderate	High	?	84
			1	Very light	Low	<1	11
Haugaasen et al. 2003	Rio Arapiuns, Central Amazonian Brazil	2200	2	None to very light	Low	9–15	36
Barlow and Peres 2004b	Rio Arapiuns, Central Amazonian Brazil	2200	5.5	None to very light	Low	ca. 36	42
			1.5	Light	High	ca. 36	74
IBAMA 1998	Roraima, Brazil	1000–2300	?	?	?	<6	8

Continued

Reference	Location						
Barbosa and Fearnside 1999	Roraima, Brazil	1000–2300	0.525	None to very light	Low	ca. 4	8
Santos et al. 1998	Roraima, Brazil	1800	?	?	?	<6	16
Pinard et al. 1999[d]	Guarayos forest reserve, Northern Bolivia	1300–400	—[d]	Light	Low to moderate	12	23
J. H. Vandermeer, pers. comm.	Southern Nicaragua	?	?	?	?	?	44–54
Slik and Eichhorn 2003[a]	East Kalimantan, Indonesian Borneo	2000–2500	0.9	Light to moderate	Low	36	59
			0.9	Light to moderate	High	36	79
Woods 1989	Sabah, Malaysian Borneo	ca. 2700	5.9	Heavy	Low to high	1 and ca. 24	53

[a] Excluding pioneers.
[b] 500 trees were sampled in an undetermined area.
[c] Excluding pioneers. Mean stems per hectare from moderately logged forest was used as the background tree density.
[d] Nearest trees to 50 points in burned forest.

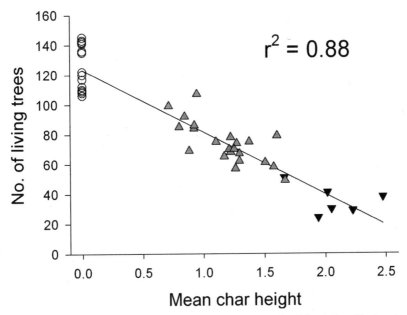

Figure 12.1. Relationship between burn severity (mean char height score) and the number of live trees in quarter-hectare plots ($F_{1,42} = 313.2$, $R^2 = 0.88$, $P < 0.001$). Unburned, once-burned, and twice-burned forests are indicated by open circles, gray triangles, and filled triangles, respectively. (Adapted from Barlow and Peres 2004b.)

mortality is markedly size-dependent following initially light burns, with smaller stems (i.e., 10–30 cm in diameter) having disproportionately higher levels of mortality than larger stems (Woods 1989; Holdsworth and Uhl 1997; Cochrane and Schulze 1999; Peres 1999b; Pinard et al. 1999; Haugaasen et al. 2003). However, the survival advantage of the larger stems does not persist following more severe recurrent fires, in which all stems become equally vulnerable (Cochrane and Schulze 1999; Barlow and Peres 2004a).

Temporal Increases in Tree Mortality

A temporal increase in tree mortality has been widely reported following fire events, within 1 year of fires (Holdsworth and Uhl 1997; Kinnaird and O'Brien 1998; Haugaasen et al. 2003), between 1 and 2 years (Cochrane et al. 1999), and from 1 to 3 years after fire disturbance (Barlow et al. 2003b). Although all these studies reported a marked temporal increase in mortality among smaller stems (<30 cm diameter at breast height [dbh]), in the

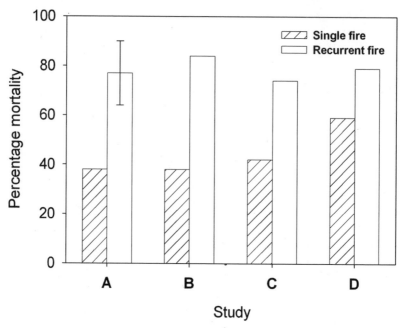

Figure 12.2. Percentage of tree mortality following single and recurrent wildfires in four studies; (A) Cochrane and Schulze 1999; (B) Gerwing 2002; (C) Barlow and Peres 2004b; and (D) Slik and Eichhorn 2003. Error bars in study A indicate upper and lower estimates of mortality.

longest-term study to date the greatest increase in mortality (relative to their original abundance) was found in the largest diameter class (≥50 cm dbh; Barlow et al. 2003b). Because a relatively small number of large trees account for a disproportionate amount of forest biomass (D. B. Clark and Clark 1996), the delayed mortality of an additional 14% of trees from 1 to 3 years after fire resulted in the loss of an additional 107 Mg/ha of live biomass, more than doubling the overall losses of live-tree biomass, from 23% after 1 year to 51% after 3 years (Barlow et al. 2003b).

Although this striking increase in large-tree mortality comes from a single study, its generality is supported by the high mortality rates suffered by large trees following recurrent fires (Cochrane et al. 1999; Slik and Eichhorn 2003; Barlow and Peres 2004b), droughts (Leighton and Wirawan 1986; Swaine 1992; Condit et al. 1995), and forest fragmentation and edge effects (W. F. Laurance et al. 1997, 2000a). The vulnerability of large trees to fire may be explained by greater hydraulic stress (Midgley 2003) or increased vulnerability to windthrow or pathogen attack.

Whatever the causes, a temporal increase in large tree mortality across tropical forests affected by surface wildfires would substantially increase the degree of biomass loss and committed carbon emissions, even without recurrent burns (Barlow and Peres 2004b). In one extreme scenario, the burning (by a low-intensity fire) of the 270,000 km^2 of forests in Brazilian Amazonia that were estimated to have become highly vulnerable to fire in December 1998 (Nepstad et al. 1999a) could result in as much as 2.2 Pg of carbon to be committed to the atmosphere through the initial burn and subsequent decomposition of dead biomass (Barlow and Peres 2004b). This would exceed the 1.7 Pg of carbon estimated to result from all tropical deforestation each year (Malhi et al. 2002a). Although burn coverage is highly unlikely to be as extensive as this in a single year, estimates of carbon losses can still be high. Diaz et al. (2002) estimate that 0.036 to 0.47 Pg of carbon could have been committed to the atmosphere as a result of the fires that burned 26,200 km^2 of forest in the "arc of deforestation" in the Brazilian Amazon in 1998, an amount equivalent to 5% of annual global human-induced emissions (Houghton 1999). However, these figures do not include the post-fire biomass accumulation resulting from forest regeneration, which in one study restored forest biomass to just 6% below primary forest levels, 3 to 5 years post-fire (O. Carvalho et al. 2002). It is clear that current levels of uncertainty remain high, despite the potential importance of carbon emissions from fires in tropical forests. A more accurate quantification will depend on additional long-term studies of post-fire tree mortality and regeneration, and more precise estimates of annual fire coverage.

Mortality Rates and the Historical Influence of Fire

It is likely that forest history has an influence on the vulnerability of tree species to fire. The detected post-fire mortality of 36% to 64% of all trees greater than or equal to 10 cm dbh in areas that can be described as core Amazonian forest is noticeably higher than 8% to 23% mortality at forest sites in Roraima and Bolivia along the fringes of the Amazon basin (table 12.1). It seems reasonable to hypothesize that these differences result from the repeated history of fires along the phytogeographical periphery of the Amazon forest, which may have exerted stronger selective pressure on the evolution of fire tolerance, or traits conferring some mechanism of fire resistance (Barlow et al. 2003a). However, differences in mortality rates between sites and regions remain difficult to interpret in light of the relatively limited geographical spread of study areas, and differences among studies in fire intensity, the timing of post-fire sampling, and the pre-burn history of logging (table 12.1), all of which can affect mortality rates (e.g., Holdsworth and Uhl 1997; Siegert et al. 2001; Barlow et al. 2003b).

EFFECTS ON FOREST STRUCTURE AND SPECIES COMPOSITION

Fires act in a similar manner to other forms of forest disturbance such as selective logging and natural tree-fall gaps, with the mortality of canopy trees increasing irradiance levels in the midstory and understory. These increased light levels shift primary production to lower forest strata and, combined with the physical perturbation of the forest floor, result in the rapid regeneration of many species of woody-stemmed pioneer species (Haugaasen et al. 2003; Barlow and Peres 2004b). However, the successional path following more severe recurrent fires is very different, resulting in most cases in the rapid colonization or regeneration of nonwoody species such as bamboos, other grasses, aggressive sedges, and nonwoody vines (Fearnside 1990b; Kinnaird and O'Brien 1998; Cochrane et al. 1999; Barlow and Peres 2004b). These plants dominate the understory in twice-burned forest, and may even suppress the regeneration of woody species, which are much less abundant following a second fire than after an initial low-intensity fire (fig. 12.3; Barlow and Peres 2004b). The dominance of these pyrophytic (fire-adapted and fire-promoting) plants can also encourage the recurrence of future fires (Fearnside 1990b), hastening the transition from forest to scrub and facilitating the persistence of an alternative low-biomass stable-state ecosystem (Nepstad et al. 1999b; Cochrane et al. 1999; Barlow and Peres 2004b).

Mounting evidence suggests that fires can cause long-term changes in forest species composition. For example, some morphological features of trees, such as bark thickness and buttresses, evidently affect survival of trees after low-intensity fires (Nieuwstadt and Sheil 2002; Barlow et al. 2003a). It can be predicted that tree families with thin bark and a high prevalence of buttresses will be negatively affected by fires, a prediction borne out by the significantly greater-than-expected declines of trees in the families Burseraceae and Sapotaceae in central Amazonia (Barlow and Peres, in press). However, future forest composition will depend on recruitment and regeneration as well as differential rates of mortality. While short-term studies confirm the expectation that pioneers will initially dominate the post-fire regeneration (Woods 1989; Holdsworth and Uhl 1997; Kinnaird and O'Brien 1998; Cochrane and Schulze 1999; Barlow and Peres 2004b), a study conducted in southeast Asia indicates

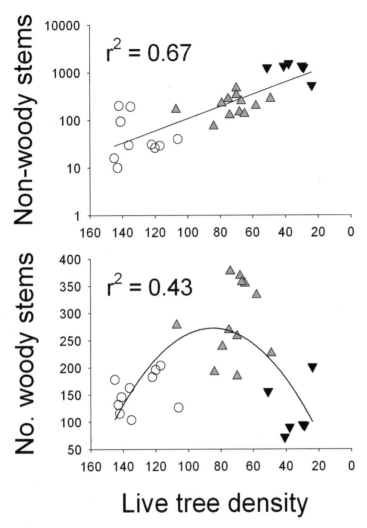

Figure 12.3. Changes in the density of woody and nonwoody stems (<10 cm dbh) 3 years after fire dis-turbance in relation to local burn severity (indexed as the number of live trees per plot). The sharp increase in the density of nonwoody stems appears to inhibit the regeneration of woody stems following intense and recurrent fires. Unburned, once-burned, and twice-burned forests are indicated by open cir-cles, gray triangles, and filled triangles, respectively. (Adapted from Barlow and Peres 2004b)

that they may also continue to persist much longer, with the pioneer genus *Macaranga* spp. dominating seedling recruiting in burned forest up to 15 years post-fire (Slik et al. 2002).

FRUIT PRODUCTION

Because frugivory and animal-mediated seed dispersal are important processes in tropical forests (Howe and Smallwood 1982; Howe 1986; M. F. Wilson 1989; Jordano 1992; Peres and Roosmalen 2002), forest disturbance events affecting fruit production, or fruit-frugivore interactions, could have serious consequences for forest wildlife and future floristic composition. However, despite studies documenting assemblage-wide tree mortality (table 12.1), and speculation about its effects on fruit production (e.g., Kauffman and Uhl 1990; Nepstad et al. 1999a), only a single study (conducted 3–4 years after fires in central Brazilian Amazonia) has assessed the effects of fire disturbance on fruit production in the humid tropics (Barlow and Peres, in press; but see Kinnaird and O'Brien 1998).

To some extent the paucity of information on this topic may reflect the inherent difficulty of studying burned habitats: without the advantage of an a priori experimental design, the low live-tree densities and dense regeneration in burned forest make commonly used methods, such as species-specific comparisons of fruiting activity or fruit traps, extremely difficult. Barlow (2003) used terrestrial surveys of fruitfall along 16 km of census trails, which allow the overall rates of fruit production in burned and unburned forest to be quantified, and provide results that are comparable with those from other methods (S. Y. Zhang and Wang 1995; Stevenson et al. 1998). However, terrestrial surveys of fruitfall do not provide the kind of information on individual species-specific responses that are available for other disturbance regimes (e.g., W. F. Laurance et al. 2003, 2006b).

One would predict that the fire-induced mortality of many trees and lianas should result in large reductions in fruit production, and this is supported by evidence from central Amazonia, where total fruiting tree abundance ($\geq$10 cm dbh) in once- and twice-burned forest was 83% and 39%, respectively, of that in unburned forest (Barlow and Peres 2006). However, these declines were far lower than the overall reduction in tree densities in each habitat, supporting previous assertions (e.g., Kauffman and Uhl 1990) that tropical trees would respond similarly to those in temperate zones, with many surviving understory and midstory trees having enhanced flower and fruit production in response to higher light levels in burned habitats (Barlow 2003). Nevertheless, large fruiting trees ($\geq$40 cm dbh) were much less abundant in burned than unburned forest (fig. 12.4; Barlow 2003), reflecting long-term patterns of tree mortality (Barlow et al. 2003b). This mortality resulted in substantial declines in mean fruiting-tree basal area, which in once- and twice-burned forest comprised only 71% and 38% of that in unburned forest, respectively (Barlow and Peres 2006). Because fruiting-tree basal area is a good proxy for overall stem fruit production (Leighton and Leighton 1982; Stevenson et al. 1998), these results indicate that substantial reductions in fruit production occur in burned forest, with the degree of change closely reflecting the loss of aboveground forest biomass in once- and twice-burned forest.

Lianas appear even more vulnerable to fires than trees (Gerwing 2002). Fruiting lianas accounted for only 10%, 7%, and 5% of fruiting stems in unburned, once-burned, and twice-burned forest, respectively (Barlow and Peres 2006). The high mortality rate of large woody lianas contributed to the collapse in mean liana-crown coverage with, respectively, 89% and 97% of this being lost in once- and twice-burned forest. This dramatic decline of liana crowns will have important implications for forest diversity: Not only do lianas contribute an average of 25% to stem diversity in tropical forests (Gentry 1991), but they are also important for many ecosystem-level processes (Schnitzer and Bongers 2002)

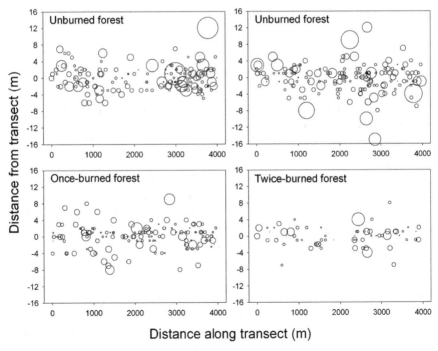

Figure 12.4. Fruit patches on the forest floor along four transects (1 m wide and 4 km long) censused between February 2001 and April 2003. Tree diameter is indicated by the size of each circle.

and many animals rely on them for fruit or nectar resources (Putz and Windsor 1987; Morellato and Leitao 1996; Palombit 1997; Davies et al. 1999; Peres 2000b; Meehan et al. 2002; Peres and Roosmalen 2002).

RESPONSES OF LARGE VERTEBRATES TO WILDFIRE

It is hardly surprising that terrestrial vertebrates with poor climbing abilities and low mobility should be affected by surface fires, and indeed reports of dead and injured animals (including tortoises *Geochelone* spp. and agoutis *Dasyprocta* spp.) are not uncommon in the immediate aftermath of fires (Peres 1999b; Peres et al. 2003b). However, many other arboreal vertebrate species also appeared to succumb to smoke asphyxiation, including several primate, sloth, porcupine, arboreal echimyid-rodent, and bird species (Mayer 1989; Peres 1999b). Sub-lethal injuries of more mobile animals were also common following more-severe burns, and local hunters reported killing terrestrial mammals (including armadillos *Dasypus* spp., brocket deer *Mazama* spp., and pacas *Agouti paca*) with fire-induced scars up to 4 years after a large Amazon forest fire (J. Barlow, unpublished data).

After the fires, surviving animals must either emigrate into nearby unburned areas (if available) or remain within the burned forest matrix. Animals moving into unburned forest can expect to face elevated interference competition through territorial aggression from conspecifics (e.g., Bierregaard and Lovejoy 1989); density-dependent reductions in fitness through exploitative competition for food, mates, or other resources; and disadvantages from their poor familiarity with the spatial and temporal distribution of resources. Overcrowding can aggravate all of these effects.

However, in many cases, the presence of territorial conspecifics (O'Brien et al. 2003), the large spatial extent of fires, and the fragmented nature of many burned forest landscapes (W. F. Laurance, chap. 5 in this volume, Peres and Michalski, chap. 6 in this volume) will prevent or discourage animals from moving into unburned areas. Only where game hunting has artificially reduced populations in adjacent unburned forests, is it possible to envisage a relaxation of intraspecific competition and density-dependent effects. But even under these conditions, the life expectancy of emigrants may still be low if local hunting pressure persists or increases, and is aggravated by collateral damage to food crops and other forest resources (Peres et al. 2003b).

Individuals remaining in burned forests face a different set of problems. Initially, large-bodied animals may be hunted relentlessly by rural peoples desperate to compensate for losses of their food crops (Peres et al.

2003b). This is exacerbated by a lack of cover in the relatively open understory, the loss of midstory and canopy foliage (Haugaasen et al. 2003), and increased clumping of animals around remaining fruiting trees or unburned forest patches (Lambert and Collar 2002). Reports of large numbers of mid- to large-sized diurnal primates (including brown capuchin monkey *Cebus apella*, red-handed howler monkey *Alouatta belzebul*, and orangutan *Pongo pygmaeus*) being killed by hunters shortly after wildfires are not uncommon (Saleh 1997; J. Barlow, pers. obs.).

Those animals that escape or which are unaffected by hunting in burned forests may face severe food shortages, as many canopy trees abort their fruit crops and shed leaves following traumatic heat stress (Peres 1999b). In some cases, animals may compensate by switching to alternative dietary items. For example, primates such as howler monkeys (C. Peres, pers. obs.) and orangutans (Suzuki 1988) can become increasingly folivorous, resorting to the post-fire regrowth of young leaves, while pig-tailed and long-tailed macaques (*Macaca nemestrina* and *M. fasicularis*) and gibbons (*Hylobates muelleri*) can overcome a period of fruit and flower scarcity by taking advantage of insect outbreaks, such as after the 1982–1983 fires in Indonesia (Leighton 1983; Berenstain 1986).

However, dietary switching is not an option for all species, and what is apparent across studies and continents is that habitat specialists with specialized feeding requirements are also the most likely to be negatively affected after fires (Barlow 2003; O'Brien et al. 2003). In Southeast Asia, nocturnal primates (western tarsiers *Tarsius bancanus* and slow loris *Nycticebus coucang*) and the Malayan sun bear (*Helarctos malayanus*) were either locally extirpated or drastically reduced in numbers following the major fires of 1986 (Doi 1988; Boer 1989). Furthermore, the fire-induced loss of important forest resources, such as figs (*Ficus* spp.), was correlated with a reduction in group size and reproductive success of siamangs (*Symphalangus syndactylus*), suggesting that they would go extinct one to two generations after the fires (O'Brien et al. 2003). In Brazil, most vertebrate species that declined in abundance 3 years after a single-fire event were still recorded in once-burned forest, although they were often at very low densities. However, even these low-intensity fires apparently extirpated bearded saki monkeys (*Chiropotes albinasus*) and dark-winged trumpeters (*Psophia viridis*) (Barlow 2003).

Recurrent fires have far more pronounced effects than a single fire, resulting in the decline or extirpation of almost all forest species. In central Amazonia, only two species of small-bodied primate (marmosets *Callithrix humeralifera* and titi monkeys *Callicebus hoffmannsi*) and one species of terrestrial bird (little tinamou *Crypterellus soui*) were more abundant in twice-burned than unburned forest (Barlow 2003); all of these are known to be highly tolerant of or frequently associated with second-growth. Although these data show that wildfires clearly have a heavy impact on large vertebrates, most available data are restricted to short-term changes in abundance. Such studies overlook the long-term effects of fires on diet, territoriality, group size, and other traits defining the feeding, ranging, reproductive, and socio-ecology of frugivorous vertebrates, all of which can all be affected by changes in forest structure and resource availability (Leighton and Leighton 1982; Terborgh 1983; S. Zhang 1995; O'Brien et al. 2003). Moreover, the apparent short-term persistence of some species should not be taken as evidence of their long-term sustainability within fire-degraded habitats (e.g., O'Brien et al. 2003).

UNDERSTORY AVIFAUNA

Surface fires also have major impacts on understory birds. In the only detailed study on fires and tropical forest avifauna, the abundance and point diversity of birds in mist-net samples one year after a fire, in once-burned forest, were both lower than in unburned forest (Barlow et al., 2002). While species richness and capture success remained unchanged over time in unburned plots (Barlow and Peres 2004a), the local avifauna sampled in burned plots showed considerable changes between 1 and 3 years after fires, and displayed some signs of recovery in bird species richness. However, the species assemblage also became increasingly dissimilar to that in unburned forest over time, largely due to the increase in the abundance of species associated with secondary habitats. A similar temporal response was reported between 1 and 5 years after selective logging (Mason 1996), reflecting increased understory regeneration in the years following disturbance (Mason 1996; Barlow et al. 2002; Barlow and Peres 2004a).

Habitat structure can also account for differences in the understory avifauna among plots sampled during the same post-fire period. Changes in variables such as canopy cover and understory regeneration (e.g., fig. 12.3) accounted for the high levels of species turnover along the gradient of burn severity, in 28 plots where the avifauna was examined 3 years post-fire (Barlow and Peres 2004a). The average similarity (using the Bray-Curtis similarity measure) between mist-net lines in once-burned and unburned forest was merely 29%, compared to a mean background similarity of 54% among unburned plots. In twice-burned forest, similarity decreased markedly to only 6%, and virtually no bird species were shared between unburned and twice-burned sites (fig. 12.5).

Most insectivorous guilds declined as a result of fires, although arboreal-gleaning insectivores increased in abundance with greater burn severity as a result of invading species associated with second-growth (fig. 12.5). All dead-leaf gleaning and ant-following birds were extirpated from twice-burned forest, whereas arboreal nectarivores, granivores, and frugivores were most abundant in once-burned forest by 3 years post-fire (Barlow and Peres 2004a). The responses of different foraging guilds to fire are broadly similar to those from other structural disturbance in tropical forests, including selective logging (Wong 1986; Lambert 1992; Danielsen and Heegaard 1994; Mason 1996; Thiollay 1997; Putz et al. 2001a), habitat fragmentation (Stouffer and Bierregaard 1995a, 1995b; S. G. W. Laurance et al. 2004; S. G. W. Laurance, chap. 14 in this volume), gradients of increasing anthropogenic disturbance (Canaday 1996; Thiollay 1999; Pearman 2002), and natural treefall gaps (Schemske and Brokaw 1981). However, the effects of recurrent fires appear to be more severe than any of these, with community composition probably most resembling forest regeneration following clearfelling (Raman Shankar et al. 1998; Borges and Stouffer 1999).

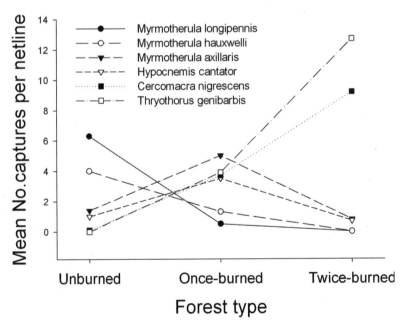

Figure 12.5. The three response types exhibited by the six most abundant arboreal-gleaning insectivorous bird species, clearly demonstrating species turnover along the gradient of burn severity (Barlow and Peres 2004a).

CONCLUSIONS AND IMPLICATIONS

1. Even a single, low-intensity surface fire is a considerable threat to tropical forests. Wildfires have severe consequences for trees, forest structure, fruit production, and the community structures of mid-sized to large vertebrates and the understory avifauna.

2. Despite being surprisingly strong, the effects of an initial, low-intensity fire are overshadowed by the devastating loss of biodiversity from recurrent fires. Recurring fires greatly increase rates of tree mortality, and lead to local extirpation of almost all understory bird and large vertebrate species typical of undisturbed forest.

3. Many questions remain unanswered about fire impacts in rainforests, in particular regarding their long-term effects, their effects in different regions of the humid tropics that vary in dry-season severity and soil characteristics, forest recovery rates in the absence of recurrent fires, and the synergistic influences of fires and other forest disturbances such as hunting, logging, and forest fragmentation.

4. Without major investments in education, increased governance, and the creation and protection of large reserves of undisturbed forest, fires can be expected to become increasingly prevalent in the humid tropics, representing one of the major emerging threats to forest wildlife.

ACKNOWLEDGMENTS

We are grateful to the political leadership of the Reserva Extrativista do Tapajós-Arapiuns and generous local support. Torbjorn Haugaasen and Bernard Lagan helped with data collection. Financial support for this study was provided by the Center for Applied Biodiversity Sciences of Conservation International and a National Environment Research Council Ph.D. studentship at the University of East Anglia. We are grateful to Reinaldo Barbosa for providing details for table 12.1, and to Bill Laurance, Torbjorn Haugaasen, Mark Cochrane, and Paulo Moutinho for their helpful comments on the manuscript.

CHAPTER 13

Disperser-Free Tropical Forests Await an Unhappy Fate

John Terborgh and Gabriela Nuñez-Iturri

Tropical forests are experiencing defaunation on huge scales. Hunting has reduced or eliminated the populations of large primates and terrestrial mammals over vast portions of Southeast Asia, Central Africa, and Amazonia, including remote regions that have so far escaped logging (Redford 1992; Robinson et al. 1999; O'Brien and Kinnaird 2000; Brashares et al. 2001; Fa et al. 2002; Milner-Gulland et al. 2003; Peres and Lake 2003; W. F. Laurance et al. 2006a). Large-bodied frugivores are especially important because they constitute a disproportionate fraction of the mammalian biomass of many tropical forests and disperse the seeds of large numbers of tree species (Andresen 1999; Peres and Roosmalen 2002; Poulsen et al. 2002). Yet, throughout the tropics, it is the largest frugivores that are most favored by hunters and consequently suffer the most systematic depletion (Peres 2000a; Robinson and Bennett 2000; Jerozolimski and Peres 2003).

Loss of dispersal function is likely to have complex consequences, because dispersal itself is a complex process in which different dispersers generate quite distinct kinds of seed shadows that may then be further modified by secondary dispersers (Smythe 1989; Forget and Milleron 1992; Levey and Byrne 1993; Forget 1996; Andresen 1999; Jansen et al. 2002). Careful studies have revealed that seeds are deposited by certain dispersers in particular kinds of sites: early successional stands (Howe 1977), "near objects" (Kiltie 1981), bat feeding roosts (Romo 1997), habitual latrines (Fragoso 1997; Julliot 1997), gaps (Wenny 2000), and, no doubt, other distinctive microsites not yet identified.

The importance of dispersal for successful reproduction of canopy trees has been widely demonstrated in focal species studies (de Steven and Putz 1984; Howe et al. 1985; Smythe 1989; Forget 1996; Cintra 1997; Fragoso 1997; Howe and Miriti 2000). Extrapolation from single-species studies has led many investigators to assume that dispersal is an essential process at the community level (Howe 1984; Chapman and Chapman 1996; Silva and Tabarelli 2000; Poulsen et al. 2002). Nevertheless, a full and convincing empirical demonstration of the role of dispersal at the community level is still lacking (but see Mills et al. 2006). What happens when seed dispersal is diminished or curtailed by the decimation of dispersers? This extremely important question is being addressed by a number of investigators, and preliminary answers are taking form. Already there are clear indications that defaunation alters parameters of tree recruitment in marked and even contrasting ways (Wright 2003). A paucity of herbivores can lead to mass recruitment of some species (Dirzo and Miranda 1981), whereas depletion of both seed dispersers and seed predators leads to increased clumping of recruitment (Wright et al. 2000; Cordeiro and Howe 2001; Wright and Duber 2001; Roldan and Simonetti 2001; Wyatt and Silman 2004).

Here we show at the community level, first, that the spatial patterning of tree recruitment in an intact forest is highly constrained, and second, that substantial alteration of the processes that give rise to these patterns is likely to have predictably negative effects on both the abundance and aggregation of tree populations. We begin by modifying the Janzen-Connell model (Janzen 1970; Connell 1971) to make predictions about the consequences of defaunation. Then, we present data on tree recruitment in an intact forest community and show that strong spatial constraints operate on recruitment. Next, as a pattern within a pattern, we show that much seed transport in this forest is directed in such a way that different tree species effectively synergize each other's recruitment via their ability to attract dispersers. And finally, we comment on the implications of our data for the maintenance of tree diversity in defaunated tropical forests.

THEORY

The Janzen-Connell (J-C) model provides a theoretical framework for making predictions about the consequences of defaunation for tree recruitment. The model holds that tree recruitment results from the interaction of dispersal (which declines with distance from a fruiting tree) and escape from predators and pathogens (which increases with distance).

Maximum recruitment of saplings is presumed to occur at some unspecified distance from reproducing adult trees.

The J-C model can be easily modified to make predictions about altered conditions by raising or lowering the dispersal and/or escape curves. In a forest depleted of dispersers (large frugivorous birds and/or mammals, such as primates), dispersal will be truncated and few seeds will fall beyond the projected crown of a fruiting tree. If dispersal is thus curtailed while the processes that determine the escape curve remain unaltered, two consequences can be anticipated: (1) overall recruitment of saplings will be reduced, and (2) the recruitment distance will contract, resulting in increased aggregation of adults in the next generation (fig. 13.1). If both seed dispersers and seed predators are reduced by hunting, as is often true in neotropical forests (Peres 2000a), then the escape curve may rise closer to adults. In such a case, recruitment will decline less acutely than in the previous case, but spatial aggregation of the population is nevertheless expected to increase. Unfortunately, the community-level data needed to test these predictions directly are not yet available, although efforts to acquire such data are under way.

Here we shall examine some other properties and predictions of the J-C model in an effort to determine the proportion of tree recruitment that resulted from dispersed seeds. Our strategy is to analyze the spatial distribution of small saplings in a faunally intact forest. Because nearly all tree reproduction in this and most tropical forests is via seed, it will be

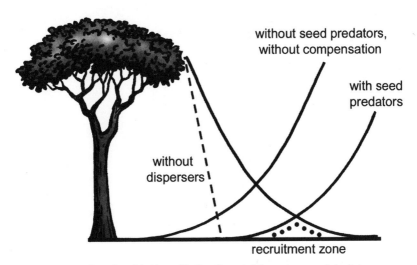

Figure 13.1. Janzen-Connell model with modifications (truncated dispersal curve and elevated escape curve) to show expected effects of decimating dispersers and/or seed predators via hunting

assumed that all saplings arose from seeds. The research was conducted at the Cocha Cashu Biological Station in Peru's Manu National Park, a tropical, moist forest site that supports a complete spectrum of large birds and mammals at ecologically regulated densities.

METHODS

We begin by examining the distribution of saplings greater than or equal to 1 m tall and less than 1 cm diameter at breast height (dbh) in relation to conspecific adult trees (those ≥10 cm dbh; Terborgh et al. 2002a). To locate saplings in relation to adults, we first mapped, marked and identified all trees greater than or equal to 10 cm dbh in a 200 × 200 m plot (4 ha) of mature floodplain forest ($N = 2336$). Then, using a common coordinate system, we mapped small saplings (all species, $N = 4347$) within a 90 × 90 m (0.81 ha) block at the center of the larger 4 ha adult tree plot. This arrangement ensured that any conspecific adult within at least 55 m of any sapling would have a known location (fig. 13.2).

Using an Excel macro, we then calculated the distances of the saplings of 109 tree species to each conspecific adult in the encompassing adult tree plot. Inclusion in the data set required that both small saplings and adults were present in the respective mapped areas. A lack of saplings of several canopy emergents and gap pioneers excluded these species from consideration. Also excluded were palms and any understory species suspected of

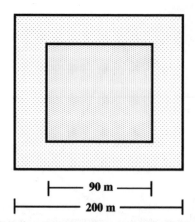

Figure 13.2. Sampling design for determining recruitment distances of saplings. All adult trees, defined as those ≥10 cm dbh, were mapped and identified within a 200 × 200 m plot (4 ha). Saplings (≥1 m tall and <1 cm dbh) were mapped and identified within the 90 × 90 m inner square, ensuring that any conspecific adult within at least 55 m of a sapling would have a known location.

John Terborgh and Gabriela Nuñez-Iturri

attaining reproductive status at dbh less than 10 cm. Among the 109 species, numbers of saplings per species ranged from 1 to 213, and numbers of adults ranged from 1 to 169. Once we had obtained the distances of saplings to all conspecific adults in the encompassing 4 ha plot, we took the minimum distance to nearest conspecific adult to be the recruitment distance for that sapling. In all, we tallied recruitment distances for 1803 saplings of 109 species, or a mean of 16.5 saplings per species. (Note that "recruitment distance" as operationally defined here is distinct from the zone of maximum recruitment hypothesized by the J-C model.)

The J-C model predicts that recruitment will be partly to completely suppressed near and under the crowns of conspecific adults. To assess whether this prediction held in our tree community, we asked whether the number of saplings located under the crowns of conspecific adults was more or less than the number that would occur if saplings were distributed at random. To perform this test, we needed data on the radii of adult crowns of the various tree species. Data on the crown dimensions of adults of 44 (out of the 109) species were available in a database compiled for a different purpose. We were thereby able to assess whether saplings occurred at less than or equal to one crown radius of adults of these 44 species more or less often than expected, under the assumption of random dispersion of saplings.

In the second part of our work, we tested the hypothesis that saplings of primate-dispersed tree species would be more likely to occur under the crowns of primate dispersed heterospecifics than elsewhere in the forest. Numbers of saplings of 14 primate-dispersed tree species were tallied in 100 m^2 sample areas in three situations: (1) under the crowns of 9 species of primate-dispersed trees (included within the 14, $N = 65$; saplings conspecific with the focal tree were not included in the counts), (2) under the crowns of two wind-dispersed and one autochorous tree species ($N = 24$), and (3) at random points on the forest floor ($N = 32$). Information on the dispersal modes of the 14 tree species was taken from Terborgh (1983), Symington (1987), and Peres and van Roosmalen (2002).

RESULTS

Sapling Recruitment Distances

Aggregated recruitment distances of the 1803 small saplings show a modal value of 5 to 10 m (fig. 13.3). However, the median recruitment distance is 19 m and 93% of all saplings lie 5 m or more from the nearest conspecific adult. To put this into better perspective, we note that the mean crown radius of 36 of the 44 species for which we have such data is 5 m. The crown radii of 5 of the remaining 8 species are between 5 and 10 m, and only those of the last 3 species are greater than 10 m. Thus, among all saplings, those growing beneath conspecific adults constitute a small minority.

We now ask whether the saplings of less common species are as likely as those of common species to occur under conspecific adults. Such a pattern could be implied by the J-C model because neither the dispersal nor the escape curve is assumed to vary systematically with adult tree abundance, although Janzen (1970) did mention that the seed shadows of common trees will tend to overlap. We begin by classifying the saplings into three groups based on the density of conspecific adults: 10 or more per ha, between 2 and 10 per ha, and 2 or fewer per ha (fig. 13.4). Median recruitment distance for the 7 most abundant species (out of 109) is 10.4 m; for the next most common 69 species it is 23.5 m, and for the 33 least common species it is 51 m. We refrain from referring to the latter as "rare" species, because most, if not all, have abundance ranks within the top half of a floodplain tree community containing more than 700 species.

The distributions of recruitment distances leave the overall impression that saplings are "repulsed" from the vicinity of adults, and more assertedly so for less common species, but more conclusive evidence is

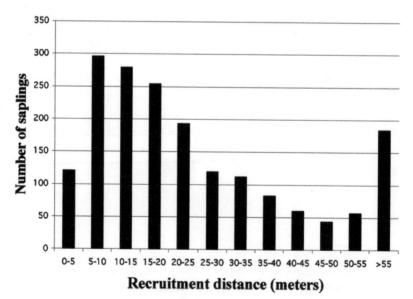

Figure 13.3. Recruitment distances of 1803 saplings of 109 tree species. Because the buffer surrounding the central plot is 55 m wide, recruitment distances beyond 55 m are not determined accurately and are therefore aggregated.

John Terorgh and Gabriela Nuñez-Iturri

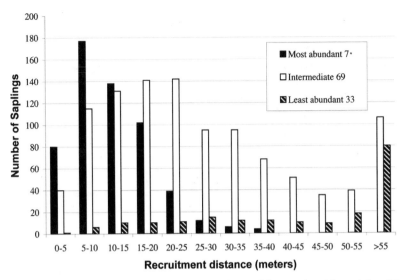

Figure 13.4. Recruitment distances of saplings grouped by density of the respective adult populations (10 or more per ha; between 2 and 10 per ha; and 2 or fewer per ha)

needed. For that, we turn to the 44 species for which we had data on the crown dimensions of adult individuals. For each species, we computed the projected area of an average crown and multiplied that by the density of adults to obtain an estimate for the fraction of the plot lying beneath crowns of that species. This gave an expected proportion of recruitment distances of the species in question that would fall within 1 crown radius of a conspecific adult, assuming that saplings were randomly dispersed on the forest floor. For each of the 44 species we then compared the expected number of saplings lying within one crown radius of an adult with the observed number (fig. 13.5). In 37 out of the 44 species, the expected number was greater than the observed number; in 4 species it was less (but only by a fraction of an individual in three cases), and in 3 species, both the expected and observed numbers were zero. The results were significant or highly significant, separately, for saplings representing each of the three adult abundance classes.

Synergistic Recruitment of Primate-Dispersed Tree Species

We tested the hypothesis that saplings (≥ 1 m tall and <1 cm dbh) of primate-dispersed tree species would be more likely to occur under the crowns of primate-dispersed heterospecifics than elsewhere. Our data con-

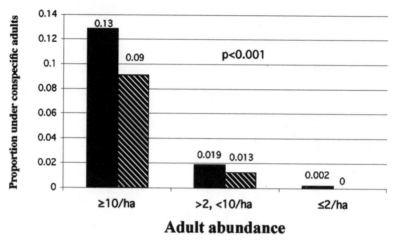

Figure 13.5. Expected (filled bars) and observed (hatched bars) proportions of saplings under the crowns of conspecific adults. Saplings were grouped according to the density of the respective adult populations.

sist of counts of saplings of 14 primate-dispersed tree species within 100 m² sample areas under target trees and at random locations on the forest floor (exclusive of saplings conspecific with the target tree). Log-transformed data were analyzed by one-way ANOVA and were highly significant ($F = 18.8$, df = 2, $P < 0.0001$). Post-hoc comparisons between all pairs were analyzed by the Tukey-Kramer procedure. The numbers of saplings under primate-dispersed crowns proved to be significantly higher (by about 2.5-fold) than under the crowns of wind- and auto-dispersed species or at random points. There was no significant difference between the numbers of saplings found at the latter two types of sites (fig. 13.6).

DISCUSSION

In the faunally intact forest at Cocha Cashu, saplings often appear at distances from adults that are multiples of the adult crown radius, and appear under conspecific crowns less often than would be expected if they were arrayed at random on the forest floor. These findings strongly support the J-C model. In particular, the zone of low recruitment around adults postulated by Janzen and Connell has been confirmed for the first time at a community level (but see Mills et al. 2006).

The finding that recruitment distances are strikingly different for saplings representing adults of high, intermediate, and low abundance

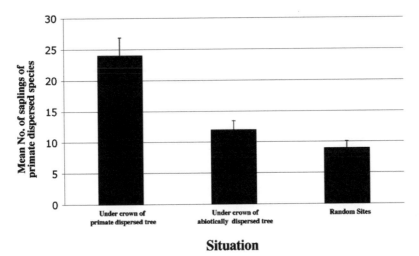

Situation

Figure 13.6. Numbers of saplings of 14 primate-dispersed tree species in 100 m² sample plots located (a) under the crowns of heterospecific primate-dispersed tree species, (b) under the crowns of abiotically dispersed tree species, and (c) at random points on the forest floor. Primate-dispersed species: *Ampelocera ruizii, Batocarpus costaricensis, Brosimum alicastrum, Brosimum lactescens, Buchenavia grandis, Copaifera reticulata, Duguetia quitarensis, Hirtella lightioides, Matisia cordata, Oxandra acuminata, Pseudolmedia laevis, Quararibea wittii, Ruizodendron ovale, Sloanea obtusifolia, Sorocea pileata, Spondias mombin*; abiotically dispersed species: *Ceiba samauma, Hura crepitans,* and *Luehea cymulosa.*

was not anticipated by either Janzen or Connell, but neither does it contradict their model. Of the relatively few saplings found under the crowns of conspecific adults (<5% overall), a large majority (88%) belonged to one of the 7 most abundant species in the community. In contrast, not one of the 136 saplings belonging to the least abundant set of species was under a conspecific crown, and at a median distance of 51 m, most were separated from conspecific adults by many crown radii.

Tree recruitment in this intact tropical forest community thus exhibits a high degree of spatial structure that appears to be driven largely by the interaction of dispersal with mortality factors, as hypothesized by Janzen and Connell. More than 90% of the saplings of the most common tree species and 100% of the saplings of uncommon tree species were situated beyond the projected crowns of conspecific adults, implying that they arose from dispersed seeds. These estimates of the proportions of new recruits arising from dispersed seeds are considerably higher than those reported by C. O. Webb and Peart (2001) for a forest in Borneo (46% of seedlings). The discrepancy between our results and theirs may be attributable to (1) their use of seedlings that might still be subject to strong density-dependent mortality (rather than saplings), and (2) a less precise mapping procedure that might underestimate the numbers of seedlings arising from dispersed seeds. In our data, the less common the species, the more recruitment appears to be "repulsed" by the presence of adults (Terborgh et al. 2002a). Additional spatial structure is overlain on this general pattern by the preferential recruitment of primate-dispersed species under the crowns of other primate-dispersed species. Undoubtedly the behavior of other animals that handle fruits and seeds, both dispersers and seed predators, generate additional "overlays" on the general pattern.

Much attention in the recent literature has been directed to the concept of dispersal limitation, defined as the failure of seeds to arrive at suitable sites for germination and subsequent development (Hurt and Pacala 1995; J. S. Clark et al. 1999; Hubbell et al. 1999; Muller-Landau et al. 2002). Defined in this way, dispersal limitation results from two at least partially independent functions, the number of seeds produced by a given species (fecundity), and the final distribution of those seeds after the completion of primary and secondary dispersal (Muller-Landau 2002). Given the low population densities of most tropical tree species and the small crop sizes of species bearing large-seeded fruits, dispersal limitation is likely to be widespread in tropical forests (Forget et al. 1999; Schupp et al. 2002). However, the consistency with which saplings of less common species appeared tens of meters from the nearest adults suggests that dispersal limitation in the forest at Cocha Cashu is more a product of limited fecundity than of inadequate distribution of seeds.

Both the magnitude of the recruitment distances of many species and the zones of low recruitment around adults imply that tree recruitment is highly sensitive to the diverse roles played by the many animals that handle fruits and seeds. We refer not merely to the presence vs. absence of animals, but to the specific details of their abundances and behaviors, such as directed transport of seeds. If one takes the J-C model at face value, then dropping the dispersal curve without raising the escape curve forecasts a general collapse in the populations of those species that are unable to recruit near adults. In the highly diverse forest under consideration, such species constitute a large majority. A slightly less drastic outcome would be predicted where hunting removed not only dispersers but some seed predators as well (e.g., peccaries). In this case, a somewhat enhanced escape from enemies might allow population persistence, but probably at lower densities. Greater clumping of adults would be another predicted consequence.

A somewhat indirect test of the latter case was conducted by Asquith et al. (1997), who compared the seed and seedling survivorship of several canopy tree species at a series of sites around Lake Gatun, Panama, and on its islands. The sites encompassed a diversity gradient of mammalian seed predators, ranging from a full guild of these animals on the mainland to only a single species, the spiny rat (*Proechimys semispinosus*), on tiny Lake Gatun islands. Smaller landmasses supported fewer vertebrate seed predators (and dispersers), but seed removal rates remained essentially constant across sites because spiny rats increased in density in apparent compensation for missing larger seed predators. If compensatory increases in smaller seed predators were to be a predictable consequence of overhunting their larger counterparts, the prognosis for the future of such forests would be bleak.

However, another study conducted on the Panama mainland suggests a less drastic scenario. Wright and Duber (2001) examined seed survival and seedling recruitment in the palm, *Attalea butyracea*, at a series of sites representing a gradient of hunting intensity. Densities of *Attalea* seeds dispersed more than 10 m from adult palms were reduced by an order of magnitude at the most intensively hunted site, as would be predicted by the J-C model, and suffered elevated attack by bruchid beetles. These results differed from those of the island study cited above in that there was no evidence of compensatory increases in spiny rats in the absence of peccaries and agoutis (Wright et al. 2000). A plausible explanation for lack of compensation in the mainland situation is that spiny rats may have been limited more by predators than by competition with larger granivores. Even where vertebrate seed predators are reduced along with dispersers, severe effects of diminished dispersal can be anticipated, because the repulsion of sapling recruitment away from adults may often result from the actions of invertebrates and fungi as well as those of vertebrates (Augspurger 1983a, 1983b; Howe et al. 1985; Givnish 1999).

CONCLUSIONS AND IMPLICATIONS

1. Tree recruitment in a mature floodplain forest of western Amazonia exhibits a high degree of spatial structure that appears to be driven largely by the interaction of dispersal with mortality factors, as hypothesized by the Janzen-Connell (J-C) model. More than 90% of the saplings of the most common tree species and 100% of the saplings of uncommon tree species were located beyond the projected crowns of conspecific adults, implying that they arose from dispersed seeds. This was particularly the case of gut-dispersed tree species.

2. Both the magnitude of the recruitment distances of many species and the zones of low recruitment around adults imply that tree recruitment

is highly sensitive to the diverse roles played by the many frugivores and seed predators that handle fruits and seeds. Significant declines or local extinctions of large-bodied dispersal agents is likely to lower the dispersal curve, which could result in a general collapse in the populations of those plant species that are unable to recruit near adults.

3. We conclude that evidence in support of the J-C model is compelling. We thus feel that the burden of proof should be on those who chose to ignore it or diminish its significance. Given the evidence presented, the J-C model clearly implies that tropical tree communities will gradually implode wherever dispersers are depleted. We should all take note and make this point emphatically in discussions and presentations on tropical forests and their future management.

ACKNOWLEDGMENTS

We extend our profound thanks to the many individuals who collectively invested more than two person-years in mapping, marking and measuring plants at Cocha Cashu: Manuel Sanchez, Percy Nuñez, Kyle Dexter, Fernando Cornejo, Pabla Ferreyra, Mailen Riveros, Martha Jarrell, Viviana Horna, and Patricia Herrera. We also thank the Instituto Nacional de Recursos Naturales (Peru's natural resources agency) and the administration of the Manu National Park for authorizing our research at Cocha Cashu. We are grateful to Adrian Tejedor for preparing figure 13.1. The research was supported by the Andrew W. Mellon Foundation and a Pew Conservation Fellowship to John Terborgh. Gabriela Nuñez-Iturri thanks Miguel Bravo for field assistance and the Lincoln Park Zoo, Wildlife Conservation Society, and the Canon National Parks Science Scholars Program for financial support.

CHAPTER 14

Rainforest Roads and the Future of Forest-Dependent Wildlife: A Case Study of Understory Birds

Susan G. W. Laurance

In frontier tropical regions, road construction is widespread and escalating, and is among the most important causes of rainforest destruction. Roads are the first step leading to many of the existing and emerging threats to tropical forests, such as deforestation, habitat fragmentation, edge effects, selective logging, surface fires, illegal mining, and overhunting (Fearnside 1990a; Chomitz and Gray 1996; W. F. Laurance 1998; Cochrane et al. 1999; Nepstad et al. 1999b, 2001). Today, large intact areas of tropical rainforest occur only in areas where there is little or no present-day human access. Once physical access is provided, rapid ecosystem change can occur at both a local and landscape level.

Most economists see road building on the frontier as crucial for social and economic development (Chomitz and Gray 1996). Roads provide the means for humans to access natural resources such as timber, land, and minerals, and for the movement of these resources to markets. The end result may be income-earning industries such as logging, farming, and mining, yet for the ecosystem there is widespread habitat disturbance and fragmentation (W. F. Laurance et al. 2001b, 2006a; Peres 2001a).

In addition to promoting human encroachment, roads themselves can have a number of direct impacts on wildlife habitats and populations. Although little information is available on the effects of roads in tropical forests, roads in temperate areas have been studied for more than 20 years. These studies have identified four major ecological effects of roads on wildlife populations: habitat loss and alteration, edge effects, wildlife mortality, and barrier effects (A. F. Bennett 1991; Forman et al. 2002). I assess each of these in a tropical context, with special emphasis on the Amazon and its diverse assemblage of understory rainforest birds. I then discuss the dramatic expansion of road networks in the Amazon and implications for forest conservation.

HABITAT LOSS

Habitat loss is the initial and most obvious impact of road development. The amount of land covered by roads worldwide is by no means trivial. The United States, for example, has 6.2 million km of roads, which cover 1% of the country and are equivalent to an area the size of Austria (Forman 2000). In the relatively pristine Brazilian Amazon, linear clearings account for about 48,000 km of paved and unpaved roads. These clearings have directly resulted in the removal of about 150,000 ha of rainforest habitat, totaling around 0.2% of all current forest loss in Brazilian Amazonia.

Road networks may destroy or alter scarce or fragile habitats (Forman and Deblinger 2000). While the planned construction of new roads and highways (ca. 7500 km) in the Brazilian Amazon, as outlined in the "Avança Brasil" proposal (W. F. Laurance et al. 2001b), may directly destroy up to 30,000 ha of rainforest, this impact is quite minor relative to the extent of remaining forest. However, in localities such as tropical Australia or Central America, which have already lost much of their rainforest, the creation of new roads may have greater consequences on already threatened species and ecosystems.

The amount of habitat loss caused by road construction is affected by topography and regional climate. Road construction in mountainous landscapes requires more forest clearing because roads meander backward and forward along ridges, and result in far longer roads than those across flatter landscapes. Furthermore, unsealed roads are very difficult to maintain in high-rainfall regions, particularly if shaded by overhanging trees. Commonly, such roads are accompanied by larger than necessary clearings to facilitate the drying of the road surfaces. Cyclones and hurricanes in some tropical regions can also have important interactions with roads. For example, roads may increase soil erosion by funnelling or impeding the large surface flows associated with cyclones, and this can cause significant disturbances to other habitats such as riparian zones (J. A. Jones et al. 2000) and offshore reefs (MacDonald et al. 1997). On a positive note, road construction in cyclone-affected areas requires many large culverts and bridges to allow for water movement, and these structures may facilitate the movement of wildlife under roads (Goosem 1997, 2004).

The landscape of central Amazonia, in comparison to other tropical areas such as Australia, Papua New Guinea, and Costa Rica, is only moderately undulating. The ancient Guianan Shield has eroded into a landform that ranges from small plateaus to stream channels that are deep and moderately spaced. In this region, highway construction follows a standard cut-and-fill pattern, where roads are cut through forested hills and gullies are filled with rock. These cuts and fills result in large, steep rock walls that support only a scattering of the hardiest pioneer shrubs and small trees. Culverts along these highways are frequently inadequate and the subsequent impediment of water causes extensive tree mortality in adjacent gullies, further adding to the loss of terrestrial and aquatic habitat associated with the road construction.

Although road construction results in habitat loss for forest-dwelling species, some plant and animal species can exploit these open and comparatively arid environments. Plants common to forest gaps and clearings, such as grasses, shrubs, vines, and exotic species, dominate the regeneration of cleared road verges (S. G. W. Laurance 2001; Goosem 2004). Forest and open-habitat wildlife species may also frequent road clearings. Canopy and gap-loving birds commonly follow canopy edges into road clearings (Cohn-Haft 1995), where higher light levels promote the production of flowers, fruits, and new leaves (Gentry and Emmons 1987) and attract herbivorous insects (Fowler et al. 1993). The constant sunlight and warm road surfaces will also attract basking reptiles (A. F. Bennett 1991). Finally, terrestrial and aerial predators appear to be regular users of low-traffic roads for hunting and movement (A. F. Bennett 1991; S. G. W. Laurance 2001).

EDGE EFFECTS

When a road is cut into the forest, a linear strip of vegetation is removed, creating an abrupt edge with the forest. Environmental conditions at the forest-road boundary can be markedly altered. The removal of forest canopy increases solar radiation, temperature, and wind turbulence in the road clearing and along its margins (Kapos 1989; Williams-Linera 1990; Turton and Freiburger 1997). Additional physical changes along forest edges can include soil compaction, dust, sedimentation of streams, and altered run-off (Trombulak and Frissell 2000). There may also be detectable changes in the chemical environment such as increased levels of heavy metals, herbicides, pesticides, ozone, and nutrients (Diprose et al. 2000; Trombulak and Frissell 2000).

Structural changes frequently occur in the vegetation adjacent to road clearings. In response to environmental changes such as wind turbulence and desiccation, tree-mortality rates immediately adjacent to road clearings can increase, resulting in more canopy gaps and higher tree-turnover rates (Lovejoy et al. 1986; W. F. Laurance et al. 1998; S. G. W. Laurance 2001). Permanent clearings can result in recurring disturbances that affect forest structure at all vertical strata, causing increased understory cover, woody debris, vines, and generally a lower canopy height (fig. 14.1) (S. G. W. Laurance 2001). Floristic changes are common along road boundaries and generally include an increase in pioneer and early successional species as well as the invasion of exotic plants (S. G. W. Laurance 2001; Goosem 2004).

Many of the biotic and abiotic changes described above are influenced by the width of the road clearing. Narrow roads that retain some canopy cover are less likely to be invaded by exotic species or infested by vines. Heavy vine infestations along the edges of wide roads can increase canopy damage and perpetuate canopy disturbance, resulting in continuing habitat disturbance (W. F. Laurance et al. 2001c). Edge effects may also be influenced by the topographical position of the road. Ridges and sun-exposed aspects may be more prone to climatic extremes, leading to greater disturbances than gullies or protected aspects (Turton and Freiburger 1997).

In a few cases, the response of tropical wildlife to edge-related habitat changes along roads has been assessed (e.g., Develey and Stouffer 2001).

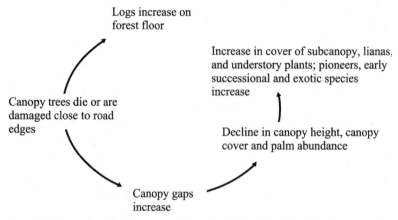

Logs increase on forest floor

Increase in cover of subcanopy, lianas, and understory plants; pioneers, early successional and exotic species increase

Canopy trees die or are damaged close to road edges

Decline in canopy height, canopy cover and palm abundance

Canopy gaps increase

Figure 14.1. Structural and floristic responses of vegetation in the vicinity of road clearings in tropical rainforest

For example, I investigated edge-related changes in understory-bird communities along roads in central Amazonia. Based on a 2-year mist-netting study, bird abundance and composition were examined at three distances (10, 70, and 130 m) from the forest-road edge. The degree to which regrowth vegetation along forest borders reduced edge effects was also assessed, by comparing bird communities near roads with varying levels of regrowth (S. G. W. Laurance 2004).

In my study, total bird captures and insectivores increased with distance from road edge, whereas the captures of nectarivores and frugivores did not vary significantly. Among insectivores, six of eight foraging guilds showed significant responses to edge distance; five declined significantly in abundance near roads (army ant followers, mixed-species flock-core members, mixed species flock-regular members, solitary understory species, and terrestrial species) whereas one guild (edge/gap specialists) increased significantly (S. G. W. Laurance 2004). Forest regrowth alleviates edge effects to some degree for three insectivore guilds, with captures of army ant-followers, mixed species flocks, and understory solitary birds increasing on or near edges with tall regrowth (S. G. W. Laurance 2004).

Factors related to the volume of road traffic, such as vehicle noise, movement, and chemicals, could cause additional edge-related disturbances to forest wildlife. Road traffic in my Amazon study was restricted to just 6 to 8 passes per week. However, in a similar study on small rainforest mammals in tropical Australia, traffic volume reached 480 cars per day (Goosem 2000, 2002). Generalist species of small mammals increased in abundance near road edges whereas forest-interior species declined. Like my study in Amazonia, retention of an intact canopy over the road reduced negative edge effects on some forest-interior species, although two species still avoided road edges (Goosem 2000, 2002).

Changes in species composition of wildlife communities, such as increases in generalist, edge, gap, canopy, and exotic species in edge-affected habitats, could potentially have pervasive or functional impacts on forest communities. This is because species that do well in disturbed habitats may also function as nest parasites, predators of seeds or other animals, or competitors (Brittingham and Temple 1983; Wilcove 1985).

An emerging threat from roads is the introduction of new infectious and parasitic diseases to forest plants and wildlife. For example, roads in frontier regions, such as in South America, West and Central Africa, and Southeast Asia, can play an important role in the spread of human-borne diseases, such as malaria. Roadworks can provide favorable environments for pathogens and vectors (such as impeded, standing water) and facilitate the introduction of new or infected hosts. Although the risks to wildlife have not yet been assessed, human and domestic-animal activities on forest edges and inside the forest can lead to changes in vector ecology, behavior, and disease patterns, which could pose an important threat to vulnerable populations of hosts (Daszak et al. 2001).

WILDLIFE MORTALITY

Corpses of dead animals littering road sides are one of the most obvious effects of roads on wildlife. Although large animals are generally more noticeable, the frequency of deaths for small animals is probably greater. For example, in a 3-year study along 2 km of tropical highway, Goosem (1997) recorded about 4000 vertebrate roadkills, of which more than 3000 were amphibians.

For species that are widespread and relatively common, road mortality probably does not constitute a significant population sink (M. E. Jones 2000). However, for species where road-kill rates exceed natural deaths, road networks may be serious cause of population decline (Maeher et al. 1991; Fahrig et al. 1995). Such species may be rare (such as the Florida puma, *Felis concolor coryi*; Maeher et al. 1991), slow moving, nocturnal, or attracted to roadside habitats or resources (such as amphibians that mate or reproduce in roadside drains or wet roads; Goosem 1997).

Several features of roads can affect the frequency of wildlife collisions, including vehicle speed, traffic volume, night-time traffic, adjacent vegetation, visibility, and width of the road clearing. A number of temperate studies have found that wildlife collisions increase with vehicle speed and traffic volume (see reviews by A. F. Bennett 1991 and Forman et al. 2002). More collisions also occur at night, when many animals are active and visibility is low. In a Nepalese national park, where night-time driving was initially banned for 3 years and then later permitted, road collisions with wildlife abruptly rose sixfold (Rajvanshi et al. 2001).

Narrow roads may result in more road deaths for wildlife because they maintain higher canopy cover and are perceived as less of a habitat discontinuity than are wide roads, thereby tempting more wildlife to cross. Goosem (1997) found that road-kill rates declined with increasing road width for all rainforest vertebrate groups (amphibians, reptiles, birds, mammals). Road-kill frequency has also been found to increase in areas where riparian habitat is traversed by roads, evidently because such habitats are hotspots for certain species. In heavily fragmented ecosystems, remnant riparian vegetation can form important wildlife corridors (S. G. W. Laurance and W. F. Laurance 1999), yet where these habitats are intersected by roads wildlife mortality can become a serious drain on a limited population (Goosem 2004).

In addition to identifying which species are at risk of car collisions, road-mortality data can also be used to assess which species do not perceive road clearings as a barrier (A. F. Bennett 1991; Forman et al. 2002). Wide roads lead to fewer road-kills because animals often choose not to cross the larger clearings, but such roads may become a barrier to wildlife movements resulting in the isolation of populations.

BARRIER EFFECTS

For many forest species, a road cut through rainforest is a major habitat discontinuity: It is brighter, hotter, drier, frequently polluted by noise and chemicals, and offers very little habitat or cover. Some wildlife species may see road clearings as natural boundaries for the delineation of home ranges or territories (Burnett 1992; Develey and Stouffer 2001). If individual home ranges are established on one side of the road or the other, then there will be little opportunity or motivation for territorial individuals to cross roads, at least on a routine basis and, as a result, the natural continuity of a population can be disrupted (Forman et al. 2002).

To date, little is known about the effects of roads and other linear clearings on the population continuity of tropical wildlife. However, from our understanding of the responses of rainforest wildlife to other habitat disturbances—particularly habitat fragmentation, logging, and edge effects—we can tentatively identify traits that could make a species susceptible to the potential barrier effects of linear clearings. Such traits may include a tendency to avoid forest edges and treefall gaps, low mobility, strict arboreality, and foraging or dietary specializations for forest-interior microhabitats.

Although highly mobile, understory rainforest birds in central Amazonia are a diverse and specialized community, with some members being highly sensitive to habitat disturbances such as forest fragmentation (Stouffer and Bierregaard 1995b), logging (Thiollay 1992), and edge effects (S. G. W. Laurance 2004). In a study of road clearings that varied in their width and amount of regrowth along the road verge, I showed that roads significantly disrupted the local movements of many bird species and guilds (S. G. W. Laurance et al. 2004). At road sites that were 30 to 40 m wide and lacked forest regrowth, movements of four guilds of understory insectivores were significantly inhibited (fig. 14.2), whereas frugivores and edge/gap specialists were not affected. At road sites with intermediate amounts of regrowth (vegetation 3–8 m tall), four bird guilds showed no significant reduction in local movements whereas movements of two guilds declined significantly (fig. 14.2). Finally, at sites where tall regrowth extended across the road clearing, only one guild, solitary understory birds, had significantly reduced road-crossing movements (fig. 14.2) (S. G. W. Laurance et al. 2004).

Small rainforest roads can clearly limit the local movements of some understory birds, but do they affect actual dispersal movements? Studying the natural dispersal movements of highly mobile species such as birds is very difficult, particularly in dense rainforests. I used translocation experiments with radio-tracked or banded adult birds to assess their ability to traverse highways and large clearings. Because it has been removed from its established territory and mate, a translocated bird has an intense motivation to return to its original territory. Although local movements of all study species were significantly reduced by 30 to 40 m wide road clearings, translocated birds of all species were all able to cross a highway clearing of 50 to 75 m in width. This result suggests that small (<75 m wide) road clearings are a behavioral rather than physical barrier to bird movement. Larger clearings (250 m wide), however, were not crossed by translocated birds, even though control translocations in intact forest revealed that birds readily traversed much larger distances to return to their territories (S. G. W. Laurance and Gomez 2005).

Other tropical studies have demonstrated that roads and powerlines can inhibit the movements of some rainforest mammal species (Burnett 1992; R. Wilson 2000; Goosem 2001). Roads of 8 to 20 m width caused a limited reduction (ca. 20%) in movements of small (<500 g) terrestrial mammals, but had no effect on larger species. Powerline clearings (60 m wide), however, inhibited the movements of both small and larger mammals (Goosem 2004). Strictly arboreal species, such as the lemuroid ringtail possum (*Pseudocheirus lemuroides*), never descend to the ground and have only been observed crossing roads via continuous canopy connections (R. Wilson 2000; S. G. W. Laurance, pers. obs.).

In tropical regions where roads frequently lead to human colonization, the original road-clearing width inevitably increases over time. The significance of such road-barrier studies (Goosem 2004; S. G. W. Laurance 2004), particularly those that have demonstrated the physical limits to wildlife movements (S. G. W. Laurance and Gomez 2005), is that they predict the disruption of wildlife populations by relatively small clearings (>250 m wide). In many degraded landscapes, such as those created by Brazilian settlement schemes (fig. 14.3), clearings of this width are established long before the landscape becomes highly fragmented. Hence, wildlife and plant

CLEARED SITES

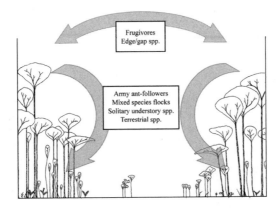

INTERMEDIATE REGROWTH SITES

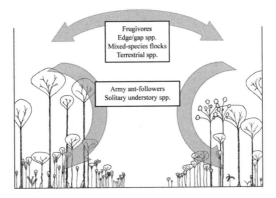

TALL REGROWTH SITES

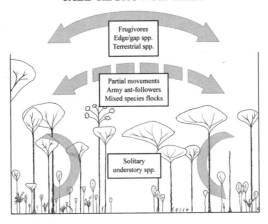

Figure 14.2. Movements of understory bird guilds across small rainforest roads in central Amazonia. Arrows depict movements, and full or partial inhibitions at three types of road sites: cleared sites, intermediate regrowth sites, and tall regrowth sites.

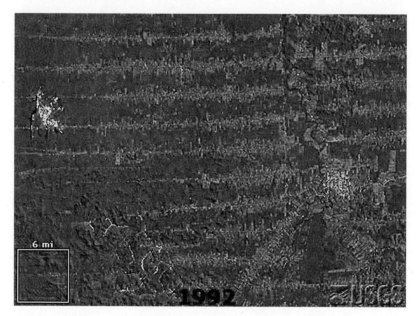

Figure 14.3. Satellite image (USGS 1992) of the deforestation patterns adjacent to roads in a colonization scheme in Rondônia, Brazil. Remnant forest habitat occurs in long parallel strips and its wildlife communities may be fragmented by only relatively small clearings (<250 m).

populations can become effectively isolated by widening roads even when more than 50% of primary forest cover still remains in a landscape.

Tropical biologists are now investigating ways to increase the permeability of roads to wildlife movement. In tropical Australia, the movement of wildlife through large culverts, beneath bridges, and on overpassing structures (such as ladders) is currently being studied (Goosem 2004). The effectiveness of these human-made structures for wildlife largely depends on the degree to which they mimic the natural environment, with the most successful being a tall bridge, which allows a continuous strip of natural or little-disturbed vegetation to occur below (Goosem 2004).

THE ROAD-EFFECT ZONE

The road-effect zone is an estimate of the area of land surrounding a road that is affected to some degree by its presence. This measure is useful as a planning tool to summarize the detrimental effects of roads (fig. 14.4) and to quantify the extent of affected habitat. In central Amazonia, for example, a small, unpaved rainforest road (10–35 m width) appears to have a zone of impact on understory rainforest birds of about 180 m in width (S. G. W. Laurance 2001). Wider, paved highways will likely have larger-scale effects. Assuming that the road-effect zone I measured was typical—which is a highly conservative assumption, given that the road I assessed was narrow, unpaved, protected from hunting, and had very little traffic— then about 864,000 ha of understory-bird habitat in Brazilian Amazonia has been degraded so far (based on an estimated 48,000 km of paved and unpaved roads in the region; S. Bergen, pers. comm.). Clearly, however, the width of the zone is not constant and will change in response to clearing size, traffic volume, the taxa or processes examined, and the compounding effects of local road density (Forman and Deblinger 2000).

The road-effect zone is a practical tool for examining the effects of land-use practices that can significantly increase road density. Commercial logging and population-settlement schemes, for example, cause significant increases in road density in tropical frontier areas, yet they create very different road patterns. Logging operations require extensive networks of small roads to remove commercially valuable trees, and these roads tend to follow a dendritic (hierarchical branching) pattern of construction (Forman 1995). Planned settlement schemes, however, can be distinguished by a main road, which is then traversed by many parallel smaller roads—leading to a rectangular road pattern commonly referred to as "fishbone."

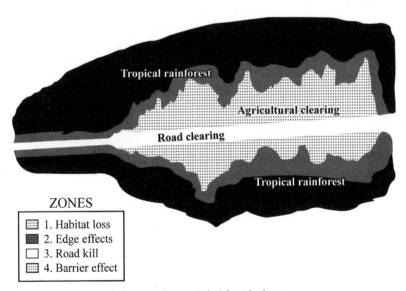

Figure 14.4. Predicted road effect zones for a tropical rainforest landscape

These two contrasting types of road-building patterns can have very different effects on wildlife. Logging roads lead to a significant decline in suitable forest-interior habitat and hence a decline in the abundance of many rainforest birds (Thiollay 1992), arboreal mammals (W. F. Laurance and S. G. W. Laurance 1996), and other sensitive wildlife (Fimbel et al. 2001). Logging disturbance can be widespread and ranges in severity depending on harvest intensity and collateral damage (Putz et al. 2001b). These disturbances can be relatively short term if the logging operation is a single harvest and the roads are closed afterward and the area left to recover. However, if logging is ongoing and road clearing is followed by settlements, then there will be a far greater and longer-term effect. In the first case, sensitive wildlife populations may have declined significantly, but the area may still permit wildlife dispersal across the landscape, especially as the forest recovers structurally. In the second scenario, dispersal may be precluded for sensitive species, eventually leading to the extirpation of populations.

In the Amazon, many settlement schemes are government-sponsored colonization programs that seek to provide small land holdings to farmers (Fearnside 1986). Land clearing is concentrated along roads and the effect is far more severe and localized than with traditional selective logging operations. The deforestation and habitat fragmentation patterns of settlement schemes in Brazil have been examined in several studies (Fearnside 1986; Skole and Tucker 1993; Dale and Pearson 1997). Dale and Pearson (1997) modeled how different land-use practices affect fragmentation patterns in settlement schemes in Rondônia, but may have underestimated the impact of smaller clearings on forest-interior wildlife. The recent translocation study by Laurance and Gomez (2005) suggests that even narrow (ca. 250 m wide) clearings can fragment populations of disturbance-sensitive birds and probably other species. When road clearings become wider (fig. 14.3), as is frequently the case in colonization schemes, fragmentation effects are likely to be severe.

ROADS AND THE FUTURE OF AMAZON RAINFORESTS

The extent to which roads can lead to dramatic landscape changes in tropical habitats was examined in two predictive studies on the future of the Brazilian Amazon (G. Carvalho et al. 2001; W. F. Laurance et al. 2001b). Both studies assessed the effects of a $40 billion Brazilian government initiative titled Avança Brasil (Advance Brazil). The program encompasses major infrastructure projects such as highways, railway and gas lines, and hydroelectric schemes, all intended to accelerate economic development in timber, mining, and industrial agriculture (W. F. Laurance et al. 2001b).

In their analysis, W. F. Laurance et al. (2001b) included optimistic and nonoptimistic scenarios, which were based on the amount of predicted deforestation and habitat degradation near roads, highways, and other infrastructure. While both scenarios predicted that the Amazon will be drastically altered over the next 20 years, the models suggested that deforestation rates would decline by 269,000 to 506,000 ha per year if the many new infrastructure projects under Avança Brasil did not proceed. Furthermore, if the projects are canceled, the rate of forest degradation (conversion of pristine or lightly degraded forests to moderately or heavily disturbed areas) is predicted to decline by 1.53 to 2.37 million hectares per year (W. F. Laurance et al. 2001b). Although such analyses are too large in scale to address the local responses of different wildlife species to roads, they highlight the critical role of planning to reduce future road impacts on forests. Future studies should consider the ecological effects not only of major highways but also of smaller roads, which are far more abundant in the Amazon.

DESIGNING ROADS TO MINIMIZE THEIR IMPACTS

Although relatively little research has been conducted on tropical-forest roads, it is apparent that roads have important direct effects on wildlife habitat and populations (fig. 14.4 and table 14.1). It is possible at this stage to make several general recommendations about roads with respect to the future protection of tropical rainforests. Key priorities are as follows:

1. Plan roads very carefully and conservatively. Once a road has been constructed, it frequently initiates a spontaneous process of forest colonization,

Table 14.1 Summary of the direct effects of roads on tropical wildlife

	Habitat loss	Edge effects	Road mortality	Road inhibition
Amphibians	X		X	
Arboreal mammals	X		X[a]	X[a]
Terrestrial mammals	X		X	
Forest-interior mammals	X	X		X
Forest-interior birds	X	X		X

[a]Some species are stricly arboreal and avoid ground movement.

logging, mining, hunting, and other activities that are nearly impossible for governments to control, especially in remote frontier areas. Hence, governments can generally control where roads are located, but after that the roads open a "Pandora's box" of activities that governments rarely can control.

2. Maintain large roadless areas to ensure cost-effective forest conservation.

3. Control road proliferation in logging concessions and block road access immediately after logging has concluded.

4. Minimize road building inside protected areas and wildlife corridors, because this can disrupt population continuity and reduce habitat quality.

5. Increase the permeability of roads for wildlife by the following measures, where road construction must occur:

   a. Building wildlife underpasses and overpasses (culverts, bridges, and rope ladders of appropriate size and design to promote wildlife movement).

   b. Keeping roads narrow and maintaining overhead canopy cover wherever possible. This may increase wildlife mortality from road-kills but the increased population connectivity is usually more important. Furthermore, such measures ensure that population connectivity occurs across the landscape and is not localized to just a few points, as will occur with under- and overpasses.

CONCLUSIONS AND IMPLICATIONS

1. The rapid proliferation of paved highways and unpaved roads is among the most important emerging threats to tropical forests. In the Amazon, as well as in other tropical regions such as Southeast Asia, West and Central Africa, and Central America, the expansion of roads and highways is one of the most important proximate drivers of forest loss and degradation.

2. In addition to facilitating greatly increased physical access, which leads to forest colonization, logging, and hunting, the roadworks themselves can cause significant changes in forest ecosystems, by promoting invasions of weeds and diseases, causing edge effects, and inhibiting movements of disturbance-sensitive wildlife.

3. Roadworks should be greatly minimized in tropical protected areas, which contain a large number of forest-interior specialists that are sensitive to habitat alteration. The ecological impacts of roadworks can be reduced by limiting road widths, maximizing forest-canopy cover over the road, minimizing impediments to natural streams and rivers, and by

providing culverts, overpasses, and other structures to facilitate animal movements across roads.

4. Very careful planning must be conducted prior to road construction. Once a road has been constructed, it frequently initiates a spontaneous process of colonization and other activities that are nearly impossible for governments to control, especially in remote frontier areas.

ACKNOWLEDGMENTS

I thank W. Laurance, A. Bennett, K. Sieving, and C. Peres for comments. The Biological Dynamics of Forest Fragments Project (BDFFP), the Australian Postgraduate Awards, University of New England, and Wildlife Conservation Society provided support for the Amazonian research. This is publication number 445 in the BDFFP technical series.

CHAPTER 15

Emerging Threats to Birds in Brazilian Atlantic Forests: The Roles of Forest Loss and Configuration in a Severely Fragmented Ecosystem

Pedro Ferreira Develey and Jean Paul Metzger

Each year, vast expanses of tropical forest are reduced and fragmented. Fragmentation and habitat loss provoke dramatic transformations in plant and animal communities, leading, among myriad consequences, to local extinctions of many vulnerable species (Willis 1979; Bierregaard and Lovejoy 1989; Klein 1989; Malcolm 1997; W. F. Laurance et al. 1998, 2006b; Tabarelli et al. 1999). Species in fragmented forests are susceptible not only to the effects of habitat reduction and isolation, but also to other anthropogenic impacts—such as climate change (Williams and Hilbert, chap. 2 in this volume), fire (W. F. Laurance, chap. 5 in this volume), overhunting (Peres and Michalski, chap. 6 in this volume), and logging and fuelwood gathering (Oluput and Chapman, chap. 7 in this volume)—that can interact synergistically with fragmentation. Collectively, the interrelated processes of habitat loss and fragmentation are probably the greatest single threat to global biodiversity (W. F. Laurance and Bierregaard 1997).

In a strict sense, habitat fragmentation is defined as the subdivision of continuous areas of habitat (Fahrig 1997, 2003; McGarigal and Cushman 2002). But other changes in land-cover configuration also typically occur in fragmented landscapes—over time, forest fragments become progressively smaller and ever more isolated, the amount of abrupt habitat edge increases, and the total area of original habitat declines. Thus, as deforestation proceeds, the spatial configuration of forest fragments (which includes fragment size), as well as that of other landscape elements such as corridors and secondary forest, also evolve. Despite the ubiquity of fragmented landscapes in the tropical world, it is unclear to what degree the spatial configuration of fragmented forests is important to biodiversity (Fahrig 2003).

Does the configuration of fragments have a large effect on biodiversity, or can most of the deleterious effects of fragmentation be explained largely as a consequence of the proportion of original habitat lost in the landscape? The distinction between these two concepts is quite vital, because if the former is correct then managing degraded or developing landscapes to minimize fragment isolation and edge effects, and to maximize fragment size, is a key priority (e.g., W. F. Laurance and Gascon 1997); whereas if the latter is correct, then we should simply maximize the total amount of habitat and not concern ourselves with the spatial configuration and size of surviving forest remnants.

Several authors have argued that habitat loss has a greater impact on species survival in fragmented landscapes than does habitat configuration. From a review of 17 empirical studies that attempt to separate effects of habitat loss and configuration, Fahrig (2003) found that 10 showed that habitat cover had a stronger effect. For instance, a study independently comparing the effects of habitat cover and configuration in a temperate forest in Oregon, USA (McGarigal and McComb 1995), concluded that habitat area had a greater effect on overall bird abundance and that only 2 of 15 species examined were clearly affected by habitat configuration. Likewise, a study of forest birds in 94 Canadian landscapes also concluded that habitat cover had stronger effects on species occurrences than did habitat configuration (Trzcinski et al. 1999). For 31 of the species in the Canadian study, 25 showed a positive, significant relationship between habitat area and species presence, whereas only 6 showed a significant relationship with habitat configuration.

However, other studies have highlighted the importance of habitat configuration for species survival. Villard et al. (1999), for example, suggested that habitat configuration and cover were both important determinants of species occurrences for Canadian birds in landscapes with 4% to 68% forest cover.

Based on spatially explicit simulation models and a review of fragmentation impacts on birds and mammals, Fahrig (1997) and Andrén (1994) suggested that a critical threshold should occur in fragmented landscapes at about 20% to 30% of habitat cover. Once total forest cover falls below this percent range, the relative importance of habitat configuration for species persistence is predicted to increase. This is because, below this threshold, forest isolation tends to increase sharply. Despite much discussion about fragmentation thresholds, however, there is still no convincing empirical evidence for this prediction, and no general consensus about its existence (Fahrig 2003; Lindemayer and Luck 2005; Radford et al. 2005). One key problem is that most studies are based on theoretical models, and empirical studies are scant (Flather and Bevers 2002). Furthermore, thresholds in landscape structure likely vary according to species characteristics, such as their sensitivity to fragment size, isolation, and edge effects (McGarigal and McComb 1995; With and Crist 1995; Villard et al. 1999), further complicating efforts to search for generalities. Extinction thresholds, defined as the minimum amount of habitat below which a population of a particular species goes extinct, are also difficult to define. Because species differ in their ecological requirements, there is no single threshold common to all species (Fahrig 2001), and the extinction threshold may vary even for the same species in different landscape types.

Insofar as we are aware, no study in the tropics has yet assessed the effects of *both* habitat cover and configuration on species survival. Fahrig (2003) suggests that the effects of habitat configuration may be greater in tropical than in temperate systems because of the large number of habitat specialists in this biome that would presumably be sensitive to factors like fragment size, shape, and isolation, but this prediction is yet to be tested. The Brazilian Atlantic Rainforest provides a vital opportunity for such a study, because it is a recognized global biodiversity hotspot (Myers et al. 2000) with only 8% of the original forest remaining (SOS Mata Atlântica/INPE 1993), and because most of the surviving forest fragments are small (Metzger 2000). In this biologically diverse and severely threatened environment, understanding the typical responses of species to habitat cover and configuration, and identifying the existence of any fragmentation or extinction thresholds, is essential for viable conservation planning and habitat restoration.

Here we focus on the avifauna of the Brazilian Atlantic forest to test the relative roles of forest cover and configuration on species persistence. We evaluate the relative importance of forest cover and habitat configuration separately at the community, functional guild, and species levels. In doing so, we hope to gain new insights into one of the most poorly understood consequences of land-cover change in the tropics.

METHODS

Study Sites

We conducted fieldwork on the Atlantic Plateau of the southeastern State of São Paulo, Brazil, between the counties of Tapiraí (23°50′ S, 47°30′ W),

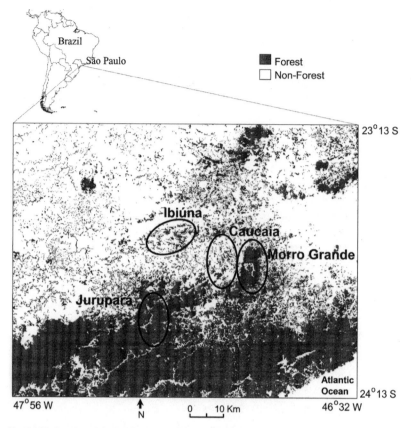

Figure 15.1. Location of the four landscapes considered in this study

Ibiúna (23°39′ S, 47°13′ W), and Cotia (23°55′ S, 46°50′ W) (fig. 15.1). The entire region was once covered by Atlantic forest, classified as "dense montane ombrophyllous forests" (*sensu* Veloso et al. 1991). We selected two fragmented landscapes in Caucaia and Ibiúna, with 31% and 26% habitat cover, respectively (fig. 15.2). The two landscapes are each about 10,000 ha in area, share a similar vegetation type, and are characterized by disturbed forest and second-growth patches. Pastures, agricultural crops, horticulture, and farmhouses comprise most of the local matrix. We also chose two forested landscapes to evaluate the composition of bird communities in areas of continuous forest with more than 90% habitat cover: Jurupará State Park (26,000 ha), predominantly a mature forest, and the Morro Grande Reserve (10,000 ha), a predominantly secondary-growth continuous forest.

These four landscapes have similar environmental conditions, and are located at the high slopes of the Serra do Mar, at 800 to 1000 m above sea

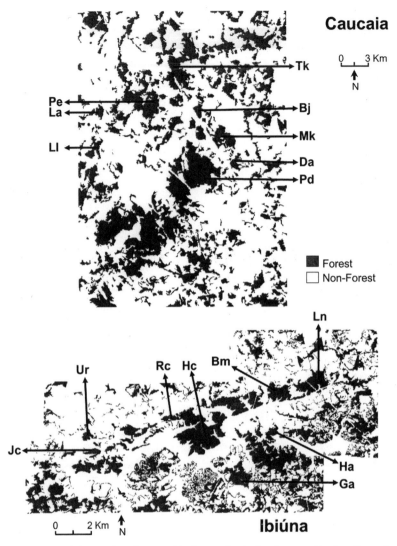

Figure 15.2. Detailed views of two fragmented landscapes, showing forest cover and configuration, and the locations of study fragments. Fragment codes are given in table 15.5

level. The climate at these sites is subtropical and rainy, with an average annual precipitation of 1340 mm and mean temperatures of 17°C in the coldest months and 22°C in the warmest months.

Sampling Design

In each of the two fragmented landscapes (Caucaia and Ibiúna), we surveyed birds in eight forest fragments. Half of these fragments were in areas with a high proportion of forest cover, and the remainder had a low proportion of forest cover. In the forested landscapes, we surveyed four plots within secondary forest (Morro Grande Reserve) and four plots within mature forest (Jurupará State Park). Replicate plots within continuous forests were located at least 1 km apart. In each of these 24 sites (16 fragments and 8 continuous-forest areas), we established a study plot of approximately 20 ha.

Birds were sampled using point-counts (Reynolds et al. 1980) during the breeding seasons of 2000, 2001, and 2002. We established four point-counts in each 20 ha study plot and, during 10-minute intervals, recorded all bird species detected by sight or vocalizations within an estimated 100 m radius. Counts were done between 0600 and 1000 hours and avoided rainy and windy weather. Point-counts were located at least 100 m away from forest edge and were separated from one another by at least 200 m. We visited each point-count site five times during the study period, for a total of 20 counts per site (yielding 160 counts at each fragmented landscape and 80 counts at each forested landscape). To minimize observer biases and problems in species identification, one observer (PFD) conducted all counts, and the calls of most species were tape-recorded using a semi-directional microphone. In cases of ambiguity, calls were compared with those in reference collections.

Five functional guilds of birds were defined using a combination of the scheme developed by Willis (1979) and our field observations. These included (1) medium to large canopy frugivores, (2) edge species, (3) ground insectivores, (4) mixed-species flocks, and (5) nectarivores.

Landscape Analysis

Our landscape-structure analysis was based on visual interpretation of available aerial photographs from the year 2000 (Caucaia, 1:10,000 scale; Ibiúna, 1:20,000 scale). We used only two land cover classes—forest and non-forest—that were easily identified at both spatial scales. All identified polygons were digitalized in ArcView 8.3, and the aerial photographs were also georeferenced and mosaiced.

For each fragment, we quantified forest cover and forest configuration within an 800 m radius from the center of each fragment. This spatial scale seems reasonable for species with intermediate home range sizes (5–10 ha), which is likely the case of most understory birds (see Develey 1997 for mixed-species flocks). This scale of measurement also avoided a high degree of overlap (i.e., >50%) between the neighborhoods of adjacent fragments. The sizes of fragments ranged from 14 to 492 ha, and fragment area was highly correlated with the proportion of forest cover remaining within an 800 m radius of the fragment center ($r = 0.936$, $P < 0.001$).

While forest cover could be measured with only one index (percentage of forest cover [PF]), several indices were used to measure different aspects of forest configuration, such as degree of forest fragmentation, proximity of other forest, fragment connectivity, fragment size, and fragment shape (table 15.1). All landscape metrics were measured in grid images with a 5 m resolution using Fragstats 3.3 software.

Because most landscape indices were intercorrelated, we used a Principal Component Analysis (PCA) to extract independent factors. Whenever necessary, log-transformations were used to normalize the

Table 15.1 Landscape metrics used to quantify forest cover and configuration in an 800 m radius from the center of each fragment examined (see McGarigal and Marks 1995)

Acronym	Name	Unit	Description
PF	Percentage of forest cover	%	Percentage of forest area
AREA_MN	Mean fragment area	ha	Mean area of fragments
GYRATE_AM	Area-weighted mean radius of gyration	m	Area-weighted mean radius of gyration (i.e., average distance of all fragment pixels to the central pixel) of forest fragments
NP	Number of forest fragments	unitless	Number of forest fragments
LPI	Largest forest fragment	%	Percentage of the landscape that is covered by the largest forest fragment

Continued

ED	Edge density	m/ha	Sum of the contact areas between forest and nonforest (m) divided by the landscape area (ha)
CPLAND	Percentage of core area	%	Percentage of the forest core area considering edges of 50 m
CLUMPY	Clumpiness index	unitless	Degree of forest clumpiness
PROX	Proximity index	unitless	The sum, for all neighboring fragments within an 800 m radius of the focal fragment, of fragment area divided by the square of the edge-to-edge distance between that fragment and the focal fragment
C_F50	Forest connectivity	m	Radius of gyration of a subgraph composed of the focal fragment and all forest fragments within 50 m of that fragment

data. The first component of the PCA explained 71% of the variance and was significantly correlated with almost all landscape metrics (table 15.2). This first axis was most strongly correlated with the percentage of remaining forest (PF) and fragment area (AREA_MN), although it was also significantly associated with two indices of forest fragmentation (NP, LPI). The second PCA axis explained an additional 16% of the variance and can be considered a landscape configuration axis that is almost independent of forest cover (table 15.2).

To obtain an index that exclusively represents forest cover, we extracted from PCA axis 1 the information related to fragmentation, using the residuals of a linear regression between this axis and the number of fragments in the landscapes; this metric is hereafter termed COVER. The same procedure was used to derive an index of forest configu-

Table 15.2 Factor loadings of the two first principal components. Significant correlations are shown in bold.

Index		PCA1	PCA2
PF	Forest proportion	**0.93**	0.06
Log(AREA_MN)	Mean forest patch area	**0.97**	0.00
GYRATE_AM	Area-weighted mean radius of gyration	**0.96**	−0.01
NP	Number of fragments	**−0.92**	0.14
LPI	Large patch index	**0.96**	0.00
ED	Edge density	−0.46	**0.84**
Log(CPLAND)	Core area percentage	**0.99**	−0.04
CLUMPY	Clumpiness index	**0.77**	−0.51
Log(PROX)	Proximity index	0.57	0.66
Log(C_F50)	Connectivity index	**0.78**	0.18

ration, by removing the nonsignificant correlation between PCA axis 2 and forest cover; this metric is hereafter termed CONFIG. These two indices are not correlated ($r = 0.10$, $P = 0.704$), indicating that they represent statistically independent effects of forest cover and configuration (table 15.3).

Data Analysis

We conducted analyses at both the landscape and individual-fragment levels. At the landscape level, we first compared species richness in

Table 15.3 Pearson correlation coefficients between indices of forest cover and configuration with the original landscape metrics. *$P < 0.05$; **$P < 0.01$; ***$P < 0.001$.

Index		Forest cover	Configuration
PF	Forest proportion	0.58*	0.00
Log(AREA_MN)	Mean forest patch area	0.32	−0.05
GYRATE_AM	Area-weighted mean radius of gyration	0.38	−0.06
NP	Number of fragments	0.00	0.18
LPI	Large patch index	0.53*	−0.05
ED	Edge density	−0.05	0.87***
Log(CPLAND)	Core area percentage	0.43	−0.09
CLUMPY	Clumpiness index	0.06	−0.54*
Log(PROX)	Proximity index	0.42	0.63**
Log(C_F50)	Connectivity index	0.25	0.14

fragmented and forested landscapes to determine the occurrence of species according to forest cover and quality (mature or second-growth forest). We then used Detrended Correspondence Analysis (DCA) to examine the composition of bird communities in fragmented sites. We excluded species not present in at least five study sites to reduce the weight of rare species. A Mann-Whitney U-test was used to compare the bird abundance between the two fragmented landscapes, based on the mean number of bird detections in the eight fragments of each landscape.

At the level of fragments, the relative importance of COVER and CONFIG in explaining species richness and variation in bird abundance was determined by forward stepwise linear regression models.

RESULTS

Forested versus Fragmented Landscapes

A total of 150 bird species were detected across all study areas. Jurupará, the primary forest area, had the highest species richness, with 33 species being found only there. Bird richness declined gradually as one moved to the continuous second-growth area (Morro Grande) and the two fragmented landscapes. Considering only the two continuous landscapes, 37 species recorded at the primary forest site were missing at the second-growth site. For these two continuous forests, differences in species richness are probably due to local vegetation conditions.

Another 29 species occurring at the continuous second-growth site were absent from the two fragmented landscapes (each with less than 31% of suitable habitat). Taking into account that sampling effort in the fragments was twice that in continuous-forest sites, the loss of species is even more evident. Large-bodied canopy species, such as the parrot *Triclaria malachitacea*, the toucanet *Selenidera maculirostris*, and the cotinga *Carpornis cucullatus*, disappeared from the fragmented sites. Other species recorded only in continuous forests included the ground insectivore *Chamaeza campanisona*, the mixed-flock follower *Philydor atricapillus*, and a species sensitive to hunting, *Tinamus solitarius*. Some species were observed only in the fragmented landscapes, but these typically used forest edges and were only sporadic visitors of forested areas.

These data indicate that, in our study area, the local extinction threshold for 66 species (44% of the community) is above the 31% forest-cover value. Hence, nearly half of the bird community could not persist in fragmented landscapes with this level of forest loss.

Although the two fragmented landscapes have a similar proportion of forest cover (31% at Caucaia and 26% at Ibiúna), the forest configuration at Caucaia appears to be more fragmented and less connected (table 15.4). These landscapes differed significantly only in the number of fragments and their connectivity, but Caucaia had a higher edge density, closer proximity among forest fragments, and smaller forest fragments, when compared with Ibiúna (table 15.4).

There also was a clear separation in bird-community composition between Ibiúna and Caucaia (fig. 15.3), as demonstrated by differences between these two avifaunas along DCA axis 1, which explained 30% of the total variation in the data set. A total of 58 species occurred in at least 5 of the 16 fragments we sampled within the two landscapes (appendix 15A). Nonetheless, forest fragments at Ibiúna were also well-separated from one another, with the four most isolated sites (Jacu, Rocha, Sr. Hélio, and Urubu) having high values on axis 1. DCA axis 2, which explained 10.7% of the total variation in the data set, was closely related to fragment size (r_s = −0.84, $P < 0.01$), with large fragments having smaller values on axis 2.

Although the ordination of 58 common bird species clearly separated the two landscapes (fig. 15.3), only 18 of the species differed significantly in abundance between the two landscapes. Ten species were more abundant at Ibiúna, whereas 8 were more abundant at Caucaia. Among the 10

Table 15.4 Main differences between the fragmented landscapes of Caucaia and Ibiúna. Significant (P < 0.05) relationships are indicated in bold (based on Mann-Whitney U-tests).

| Index | | Mean values | | P |
		Caucaia	Ibiúna	
PF	Forest proportion	38.2	48.5	0.323
Log(AREA_MN)	Mean forest patch area	1.7	2.6	0.073
GYRATE_AM	Area-weighted mean radius of gyration	263.0	344.5	0.198
NP	Number of fragments	**13.1**	**7.4**	**0.015**
LPI	Large patch index	28.8	41.3	0.304
ED	Edge density	80.3	67.5	0.136
Log(CPLAND)	Core area percentage	2.3	3.0	0.166
CLUMPY	Clumpiness index	0.955	0.965	0.054
Log(PROX)	Proximity index	5.6	5.7	0.935
Log(C_F50)	Connectivity index	**5.7**	**7.1**	**0.003**

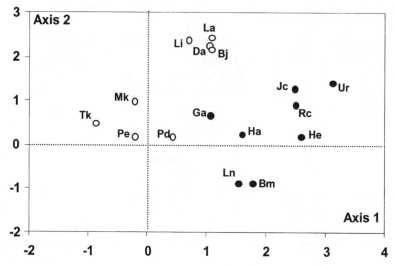

Figure 15.3. Ordination of bird communities in 16 forest fragments, using 58 bird species that were detected in at least five sites (filled circles, Ibiúna fragments; open circles, Caucaia fragments). Fragment codes are given in table 15.5.

most common species at Ibiúna, 5 (*Columba picazuro, C. cayennensis, Dryocopus lineatus, Picumnus cirratus,* and *Turdus rufiventris*) are common in disturbed forests and can use nonforested areas; 4 are typical of forests but can benefit from secondary forests (*Automolus leucophthalmus, Syndactyla rufosuperciliata, Conopophaga lineata,* and *Turdus albicollis*); and *Penelope obscura* is a canopy frugivore that usually disappears in heavily hunted sites. The 8 most common bird species at Caucaia (*Lepidocolaptes fuscus, Dendrocolaptes platyrostris, Batara cinerea, Herpsilochmus rufimarginatus, Hemitriccus diops, Attila rufus, Schiffornis virescens,* and *Hylophilus poicilotis*) are more dependent on forests for survival.

Bird Communities at the Fragment Level

The proportion of forest cover within 800 m radii of fragment centers ranged from 20% to 83%, and was therefore above the 20% fragmentation threshold proposed by Fahrig (1997). Under these conditions, forest cover should be the main determinant of species richness and persistence. However, both percent forest cover and the configuration indices had high standard deviations (table 15.5), particularly the proximity index (PROX), which allowed us to evaluate the effects of these factors on species richness and abundance.

Table 15.5 Sixteen study fragments and main spatial characteristics of the landscapes surrounding those fragments. AREA: area of the studied fragment; PF: percentage of forest; PROX: proximity index; NP: number of fragments; ED: edge density; AREA_MN: mean area of the fragments.

Focal fragment	Initials	AREA (ha)	PF (%)	PROX	NP	ED	AREA_MN (ha)
Caucaia							
Lacerda	La	14.00	25.24	75.69	20	85.04	2.54
Beto Jamil	Bj	14.08	28.67	214.76	18	91.66	3.20
Dito André	Da	18.78	25.25	65.98	17	102.01	2.99
Lila	Ll	28.88	20.48	22.51	15	65.85	2.75
Mioko	Mk	52.17	32.56	219.63	10	69.51	6.55
Pedro	Pe	53.08	47.22	2,696.48	8	85.10	11.87
Takimoto	Tk	99.39	51.56	2,787.48	8	84.09	12.96
Pedroso	Pd	175.10	74.60	851.59	9	59.48	16.66
Mean		56.93	38.20	866.77	13.13	80.34	7.44
Std dev		55.71	18.35	1,186.89	4.91	14.24	5.60
Ibiúna							
Jacu	Jc	24.41	33.41	78.27	8	72.96	8.40
Urubu	Ur	27.74	22.04	13.02	11	54.79	4.03
Rocha	Rc	32.23	28.51	113.23	12	78.45	4.77
Haras	Ha	42.63	34.13	6,246.81	9	75.18	7.62
Linhão	Ln	102.76	63.82	2,760.55	7	71.71	18.33
Boi	Bm	115.00	58.05	196.97	5	59.90	23.31
Gavião	Ga	136.77	65.57	1,457.98	4	93.42	32.95
Sr. Hélio	He	491.71	82.47	113.23	3	33.73	55.26
Mean		121.66	48.50	1,372.51	7.38	67.52	19.33
Std dev		155.88	21.73	2,196.41	3.25	17.96	17.71
All fragments							
Mean		89.29	43.35	1,119.64	10.25	73.93	13.39
Min		14	20.48	13.02	3.00	33.73	2.54
Max		491.71	82.47	6,246.81	20.00	102.01	55.26
Std dev		117.92	20.14	1,725.37	5.00	17.00	14.10

Species richness in fragments was significantly affected by fragment configuration ($R^2 = 0.42$, $P = 0.006$, $N = 16$; table 15.6). In contrast, overall bird abundance was neither related to forest cover nor to forest configuration. In terms of functional guilds of birds, species richness of mixed-flock followers was related to fragment configuration ($R^2 = 0.37$, $P = 0.013$, $N = 16$), but no functional group was related to forest cover.

Table 15.6 Most-significant relationships between forest cover or forest configuration and bird species richness and abundance, using multiple regression models (N = 16 fragments)

Response variables	First step				Second step				Final model				Shapiro-Wilk of residuals
	Predictor	Slope	R^2	P	Predictor	Slope	R^2	P	R^2	Adj. R^2	F	P	
Richness													
Total	CONFIG	+	0.42	0.007	COVER	+	0.06	0.228	0.48	0.40	$F_{2,13} = 6.10$	0.014	ns
Mixed flocks	CONFIG	+	0.37	0.013	COVER	+	0.10	0.144	0.47	0.39	$F_{2,13} = 5.80$	0.016	ns
Abundance													
Ground insectivores	COVER	+	0.29	0.034					0.29	0.24	$F_{1,14} = 5.62$	0.034	ns

Considering the overall abundance of species in different functional groups, only ground insectivores showed a significant relationship ($R^2 = 0.29$, $P = 0.03$, $N = 16$), being positively affected by forest cover. Regression models relating the abundance of each species to indices of forest cover and configuration indicated that five species were mainly affected by fragment configuration, two by total forest cover, and three by the interaction between cover and configuration (table 15.7). For seven species, configuration was more important than forest cover in explaining changes in abundance.

DISCUSSION

In the highly endangered forests of the Brazilian Atlantic Region, we contrasted the poorly understood effects of forest loss and fragment configuration on bird communities. Relative to continuous primary forest, continuous secondary forest showed a severe reduction in bird species richness, especially for those species that are highly sensitive to environmental changes (Stotz et al. 1996). This decline was strongly related to reduced habitat quality, rather than changes in landscape structure (habitat loss and/or fragmentation). This information is crucial for conservation purposes because, in most fragmented landscapes of the Atlantic forest, only secondary or disturbed forests survive. Only large, continuous tracts of primary forest can sustain the full complement of vulnerable bird species in this region. In landscapes with little remaining forest, sensitive species such as canopy frugivores and ground insectivores often disappear, regardless of fragmentation patterns. The loss of canopy frugivores, in particular, could have important ecological ramifications, as they are important seed dispersers (see Terborgh, chap. 13 in this volume; W. F. Laurance et al., chap. 23 in this volume).

Effects of Forest Cover and Configuration

We also observed substantial differences in bird abundance and composition between the two fragmented landscapes, which had roughly similar amounts of forest cover (31% in Caucaia and 26% in Ibiúna). Species that were more abundant at Caucaia than Ibiúna were usually those that are negatively affected by fragmentation, as evidenced by studies in other Atlantic forest sites (Willis 1979; Aleixo and Vielliard 1995; Anjos 2001; Marsden et al. 2001). However, many of the most abundant species at Ibiúna occurred in more isolated forest fragments. This pattern suggests that effects

Table 15.7 Most-significant relationships between forest cover or configuration and abundance of individual bird species, using multiple regressions (N = 16 fragments)

Species	First step				Second step				Final model				Shapiro-Wilk of residuals
	Predictor	Slope	R^2	P	Predictor	Slope	R^2	P	R^2	Adj. R^2	F	P	
Batara cinerea	COVER	+	0.41	0.008	CONFIG	+	0.14	0.068	0.55	0.48	$F_{2,13} = 7.81$	0.006	ns
Columba cayennensis	CONFIG	+	0.46	0.005	COVER	+	0.24	0.007	0.69	0.65	$F_{2,13} = 14.64$	<0.001	ns
Elaenia mesoleuca	CONFIG	–	0.33	0.021					0.33	0.28	$F_{1,14} = 6.80$	0.021	ns
Herpsilochmus rufomarginatus	CONFIG	+	0.45	0.005					0.45	0.41	$F_{1,14} = 11.23$	0.005	ns
Platyrinchus mystaceus	CONFIG	–	0.45	0.005	COVER	+	0.08	0.160	0.53	0.46	$F_{2,13} = 7.35$	0.007	ns
Saltator similis	CONFIG	–	0.41	0.009	COVER	–	0.15	0.057	0.56	0.49	$F_{2,13} = 8.13$	0.005	ns
Trichothraupis melanops	CONFIG	+	0.28	0.037					0.28	0.22	$F_{1,14} = 5.33$	0.037	<0.05
Thraupis sayaca	COVER	–	0.25	0.052	CONFIG	–	0.15	0.099	0.39	0.30	$F_{2,13} = 4.22$	0.039	ns
Grallaria varia	COVER	+	0.34	0.019					0.34	0.29	$F_{1,14} = 7.07$	0.019	ns
Pyriglena leucoptera	CONFIG	+	0.26	0.044					0.26	0.21	$F_{1,14} = 4.91$	0.044	ns

of forest fragmentation and/or habitat loss at Ibiúna are more intensive than at Caucaia, and may indicate that spatial configuration of forest remnants is also important in influencing bird community structure.

However, the relationship between forest configuration and structure of the avian community at the landscape level was unexpected. This is especially so if we consider that overall fragmentation at Ibiúna is less intense than at Caucaia and that the fragments at Ibiúna are larger and better connected. This becomes clearer when data are broken down for each fragment. At this scale, the four fragments in the northern portion of the Ibiúna landscape are more isolated and fragmented than other fragments in this landscape. These four fragments were separated from other Ibiúna fragments in the DCA plot (fig. 15.3), having a lower abundance of species typical of core forest habitats, such as *Sclerurus scansor* and *Schiffornis virescens*, as well as higher abundances of forest-edge species such as *Saltator similis*.

In comparison to studies in temperate regions (e.g., McGarigal and McComb 1995; Trzcinski et al. 1999), we found that the configuration of forest fragments had a relatively pronounced effect on species richness and abundance. This is so despite the fact that the two fragmented landscapes we studied still had more than 20% remaining forest habitat, which is higher than the fragmentation threshold suggested by Fahrig (1997). Notably, landscape configuration alone explained about 40% of the variance in avian species richness and number of species attending mixed-flocks (table 15.6). Among the seven species whose abundance was most affected by landscape configuration, three showed reduced abundance with increasing edge density and forest proximity. These are species that benefit most from fragmentation (*Platyrinchus mystaceus*, *Elaenia mesoleuca* and *Saltator similis*) and are rare in continuous forests (Machado and Da Fonseca 2000; Marsden et al. 2001). Moreover, two of these species can inhabit nonforest habitats. Four other forest-dependent species (*Columba cayennensis*, *Herpsilochmus rufomarginatus*, *Trichothraupis melanops*, *Pyriglena leucoptera*) appeared to be negatively affected by fragmentation.

All species except *Thraupis sayaca* and *Saltator similes*, which are common in nonforested habitats and live even in urban areas, were positively affected by increasing forest cover. The Giant Antshrike (*Batara cinerea*) and Variegated Antpitta (*Grallaria varia*) are largely dependent on forest habitats, and tend to disappear in small and isolated fragments (Willis 1979; Aleixo and Vielliard 1995; Ribon et al. 2003). The Variegated Antpitta and its congeners also commonly disappear from small forest fragments in the Amazon (Stratford and Stouffer 1999) and the sub-Andean zone (Renjifo 2001). Among the species considered in this study, those that were most forest-dependent showed a positive response to forest cover, whereas those that were least dependent (i.e., edge species) responded to forest configuration, becoming less abundant in more fragmented and isolated areas.

Comparisons with Other Studies

Our findings contrast with those of most similar studies. Among these, only Villard et al. (1999) found that forest configuration was as important as forest area, for 6 of 15 bird species considered in their study. However, they used three configuration indicators and a single indicator of forest cover (Fahrig 2002), which may have led to an overestimation of the configuration effect. In our study, forest cover and configuration were each quantified by a single factor, eliminating this potential statistical problem. Hence, our results clearly suggest that forest configuration has an important impact on bird community composition, at least for avifaunas in the direly threatened Atlantic forest of Brazil. It should also be emphasized that earlier studies considering effects of forest cover and configuration (e.g., Trzcinski et al. 1999; Villard et al. 1999) have not attempted to dissociate landscape-configuration information that is usually embedded statistically in habitat-cover data (see table 15.2). This suggests that the relative effects of reduced forest cover have been overestimated in previous studies.

Our findings also suggest that the effects of forest configuration are unlikely to increase as habitat cover declines. This argues against the existence of a simple fragmentation threshold for this bird community and suggests that forest cover and configuration act independently within the range of forest-cover values considered here (20%–83%). It further implies that species responses are individualistic, being more strongly related to their individual characteristics and behavior, than to general landscape features such as forest cover and configuration.

The absence of similar studies in tropical forests makes comparisons difficult. This is especially so because, in contrast to temperate forests, plant and animal communities in tropical forests are far more diverse, a characteristic that can result in different response patterns. Fahrig (2003), for instance, suggests that the effects of configuration in tropical systems should be greater than in temperate systems, because tropical systems sustain many habitat and microhabitat specialists than can be sensitive to changes in habitat configuration. Our results in the Brazilian Atlantic forest support this view.

In terms of the 58 relatively abundant bird species considered here, about 80% (46 species) showed no significant changes in abundance in relation to habitat structure (primary vs. secondary forests). This number of apparently insensitive species is high when compared to other studies. The difference probably occurred because forest fragments in this study were mostly secondary forests, where many sensitive species had already disappeared (66 species, using Jurupará as a baseline). Moreover, most other species were very rare and thus were excluded from our analyses. The spatial scale we used to measure fragment configuration might also influence the results, particularly for species with large home ranges, such as canopy species. In those cases, forests near but outside our study areas could have affected our results (McGarigal and McComb 1995).

Our results indicate that the most sensitive bird species in our study region had disappeared from landscapes with less than 30% remaining habitat area. This suggests that, if a fragmentation threshold does exist in this region, it is probably above the 30% level. Unfortunately, many landscapes in the Brazilian Atlantic forest have already suffered far worse forest loss than this. In such drastically fragmented landscapes, the effects of forest configuration may be relatively important in determining bird community composition.

CONCLUSIONS AND IMPLICATIONS

1. We assessed the influence of forest cover and configuration on bird communities in the highly endangered Brazilian Atlantic forests. Although some studies suggest that conservation efforts should focus on preserving the largest possible area of forest regardless of how it is spatially distributed, our results show that the spatial arrangement of forest is also crucial and therefore cannot be ignored in conservation plans.

2. Our findings suggest that the local maintenance of a species-rich avian community in the southern Atlantic forest will require focusing conservation actions on landscapes with more than 30% remaining forest cover. This relatively high percentage of forest appears to be an extinction threshold for many bird species.

3. Landscapes with small percentages of forest cover can still maintain many forest-bird species. It is imperative that habitat-restoration efforts increase in such heavily degraded areas, and these should consider the effect of forest spatial arrangement on species persistence.

4. Our results lead us to conclude that both total forest cover and spatial configuration should be considered important when tropical habitat

conservation and restoration programs are implemented. This conclusion is consistent with the idea that tropical forests contain many habitat and microhabitat specialists that are relatively sensitive to the spatial characteristics of fragmented landscapes, such as the proportion of habitat edge, fragment size and shape, and fragment isolation.

5. Focusing only on total habitat area, and ignoring landscape configuration, could be a dangerous strategy for the conservation of tropical biotas. This is especially the case for severely fragmented ecosystems like the Brazilian Atlantic forest, where the effects of landscape configuration appear to be relatively important for the persistence of forest birds and possibly other wildlife species.

APPENDIX 15A

Table 15A.1 Total number of records of the 58 bird species occurring in at least 5 out of 16 forest fragments. See table 15.5 for fragment codes.

Species	Fragments															
	Caucaia								Ibiúna							
	Tk	Mk	Pe	Pd	La	Li	Bj	Da	Bm	Ln	He	Ga	Ha	Rc	Ur	Jc
Crypturellus obsoletus	2	2	4	4	0	3	0	0	6	2	4	1	2	1	0	0
Crypturellus tataupa	1	0	2	0	0	0	0	0	0	3	1	1	0	1	0	0
Penelope obscura	0	1	0	0	0	0	0	0	0	1	3	0	0	5	1	1
Columba picazuro	0	0	0	0	0	0	0	0	2	0	3	0	3	5	0	1
Columba cayennensis	0	0	0	0	0	2	0	1	2	2	4	0	1	2	8	7
Leptotila verreauxi	3	1	0	2	1	1	3	2	0	6	3	1	2	1	6	5
Leptotila rufaxilla	3	3	0	1	1	1	4	7	0	1	4	2	0	1	1	1
Piaya cayana	1	3	5	4	1	1	2	3	1	4	2	5	1	4	3	1
Phaethornis eurynome	2	3	1	1	2	2	0	4	0	3	1	3	7	3	0	4
Thalurania glaucops	4	2	1	3	2	0	3	2	2	0	1	0	2	2	1	2
Picumnus cirratus	0	0	2	1	0	1	2	0	6	7	1	4	4	1	6	1
Celeus flavescens	4	0	1	6	1	0	6	3	6	2	6	4	3	2	2	7
Dryocopus lineatus	1	0	1	0	0	0	0	0	2	2	1	3	3	1	0	0
Veniliornis spilogaster	5	1	2	0	0	0	0	1	2	4	0	2	1	0	0	2
Batara cinerea	5	1	4	6	2	2	4	1	3	3	0	3	1	0	0	0
Thamnophilus caerulescens	6	3	12	6	7	8	9	5	12	4	10	8	9	8	11	14
Dysithamnus mentalis	8	7	7	3	6	16	9	6	10	10	7	7	9	10	8	9

Species																
Herpsilochmus rufimarginatus	16	6	11	3	9	10	10	8	2	0	0	12	8	0	0	0
Pyriglena leucoptera	3	0	10	0	0	3	1	3	10	4	2	4	5	4	0	0
Grallaria varia	8	3	7	9	3	3	1	0	5	9	0	1	0	0	0	0
Conopophaga lineata	3	2	3	5	5	5	7	3	16	11	4	5	9	6	11	16
Synallaxis ruficapilla	2	2	5	1	2	6	2	3	7	10	3	3	7	1	0	3
Syndactyla rufosuperciliata	3	3	6	4	6	1	5	4	1	2	0	1	3	0	0	4
Automolus leucophthalmus	1	0	5	0	4	4	0	0	3	1	8	4	5	5	12	2
Xenops minutus	1	2	3	2	0	0	1	1	0	2	0	0	0	0	0	0
Sclerurus scansor	7	4	7	4	1	1	2	0	1	2	1	1	0	2	0	0
Lochmias nematura	0	0	0	2	1	0	0	3	1	1	1	0	2	1	0	4
Sittasomus griseicapillus	9	3	10	7	0	0	2	1	0	0	0	4	0	0	0	0
Dendrocolaptes platyrostris	0	2	1	0	1	1	3	1	0	0	0	1	0	0	0	0
Lepidocolaptes fuscus	15	9	14	8	6	9	13	6	5	3	0	10	8	0	0	0
Camptostoma obsoletum	0	4	1	2	1	0	0	3	6	3	1	1	1	0	1	3
Elaenia mesoleuca	1	0	0	1	0	0	2	0	0	0	9	2	0	2	5	2
Leptopogon amaurocephalus	1	1	11	3	3	8	4	2	2	4	0	4	1	1	0	4
Hemitriccus diops	1	0	1	1	3	7	1	2	0	0	0	0	0	0	0	0
Tolmomyias sulphurescens	7	8	15	6	9	8	3	7	9	7	6	11	9	3	4	12
Platyrinchus mystaceus	2	4	0	6	2	3	1	4	3	3	9	2	3	3	5	3
Lathrotriccus euleri	7	4	6	6	6	1	8	2	5	2	4	4	2	2	3	3
Attila rufus	2	0	1	1	0	3	1	2	0	0	0	1	0	0	0	0
Myiarchus swainsoni	0	0	3	0	1	0	2	0	3	2	2	0	2	0	3	5
Pitangus sulphuratus	1	0	3	1	3	0	5	1	0	0	1	1	1	1	1	1
Myiodynastes maculatus	0	3	1	3	0	0	0	2	0	1	2	1	1	1	2	0
Pachyramphus polychopterus	3	2	5	4	1	2	3	0	2	5	4	1	7	0	0	0
Chiroxiphia caudata	22	15	18	12	13	16	6	12	21	21	10	9	9	5	3	7
Schiffornis virescens	6	6	2	6	0	5	2	2	0	0	1	1	0	0	0	0
Pyroderus scutatus	0	0	2	4	0	0	1	1	2	0	1	0	0	0	3	0
Turdus rufiventris	2	3	3	5	1	4	6	4	8	13	8	8	5	7	8	9
Turdus albicollis	9	3	6	10	6	2	4	1	11	11	7	5	2	7	9	9
Cyclarhis gujanensis	6	4	13	9	4	13	10	8	2	8	3	11	8	2	8	11
Vireo chivi	8	3	7	9	7	9	7	5	7	6	10	7	6	7	12	13
Hylophilus poicilotis	6	2	8	9	1	9	2	4	0	1	0	5	4	0	2	0

Continued

Parula pitiayumi	0	3	8	1	2	0	0	0	0	3	0	3	3	2	1	1
Basileuterus culicivorus	11	6	9	3	13	16	9	10	11	9	9	6	5	14	14	15
Basileuterus leucoblepharus	10	4	17	16	3	7	7	8	22	18	26	10	16	0	8	9
Coereba flaveola	1	0	1	0	0	0	0	1	2	0	1	1	0	0	0	0
Tachyphonus coronatus	2	0	0	0	0	0	1	1	5	3	0	0	0	0	0	2
Trichothraupis melanops	1	0	5	0	1	2	1	1	0	4	1	2	3	2	0	0
Thraupis sayaca	0	0	1	0	0	0	2	0	0	1	1	0	1	1	6	3
Saltator similis	1	1	0	0	1	0	1	2	6	0	8	0	4	1	9	9

ACKNOWLEDGMENTS

We thank William Laurance and Carlos Peres for in-depth editing, Carla Morsello for translating and revising the manuscript, and Ana Maria Godoy for assisting with the figures. Financial support was provided by the State of São Paulo Research Foundation (no. 00/03457-1), in the framework of the Research Program on Characterization, Conservation and Sustainable Use of the Biodiversity of the State of São Paulo (BIOTA/FAPESP) project no. 99/05123-4, Biodiversity conservation in fragmented landscapes of the Atlantic Plateau of São Paulo.

CHAPTER 16

Megadiversity in Crisis: Politics, Policies, and Governance in Indonesia's Forests

Kathy MacKinnon

TROPICAL FORESTS AND BIODIVERSITY LOSS

Since the Earth Summit in Rio de Janeiro in 1992, biodiversity conservation in general, and tropical rainforests in particular, have received an unprecedented amount of attention and funding (Castro et al. 2000; World Bank 2002a, 2006). There is no indication, however, that the rate of rainforest loss or degradation is slowing either globally (Achard et al. 2002; Rudel 2005) or regionally (J. MacKinnon 1997; Wikramanayake et al. 2002). Of the three major tropical forest regions, Southeast Asia has the highest relative rate of forest destruction and logging (W. F. Laurance 1999). Estimated forest loss in Indonesia, a recognized megadiversity country, has almost trebled over the past 15 years, rising from 900,000 hectares annually in 1990, to at least 1.7 million hectares by 1997 (World Bank 2001; Holmes 2002), and to perhaps 2.5 million hectares by 2003, according to some recent estimates (World Bank 2005a). Clearly, such rates of forest loss will have an enormous impact on biodiversity and future development options.

Tropical lowland rainforests support some of the richest concentrations of plant and animal species on Earth. More than half of the mammals of Peninsular Malaysia are confined below 350 m elevation and 81% are restricted to altitudes below 660 m (Stevens 1968). On Borneo, 61% of the resident birds (244 species) are confined to mixed lowland forests, including many Sunda endemics (D. R. Wells 1984; K. MacKinnon et al. 1996). These forests serve as colonizing sources for higher elevations; thus the loss or fragmentation of lowland forests can lead to impoverishment of nearby montane-island avifauna (J. MacKinnon and Phillipps 1993). From a biodiversity perspective, therefore, it is a particular concern that the old-growth lowland tropical forests of the Philippines, Thailand, Vietnam, Sumatra, and Borneo, all rich in endemics, have already been reduced to less than 10% of their original areas (J. MacKinnon 1997; Terborgh and van Schaik 1997). Yet, forest loss still continues. Holmes (2002) predicted that all lowland forests outside protected areas (excluding swamp-forests) would be lost by 2005 in Sumatra, and by 2010 in Kalimantan, Indonesian Borneo. Although a few fragments of lowland forest remain, international comparisons show that Indonesia's rate of forest loss is among the highest in the world (World Bank 2005a).

Logging plays a major role in forest destruction. In some areas, forests are clear-felled prior to land conversion for agriculture, plantations, and transmigration settlements. Usually, however, it is not the initial logging activities per se that cause forest loss but the land clearance afterward when logging roads and poor concession management provide access for illegal loggers and migrant farmers. Destruction and degradation of tropical forests lead to habitat fragmentation with plants and animals becoming restricted to "islands" of habitat, often too small to support viable breeding populations. A much quoted rule of thumb holds that a 90% reduction of habitat will result in the eventual loss of about half of the species found in that area (E. O. Wilson 1992). Species loss continues as fragments become isolated from sources of colonization. Very small fragments lose species rapidly, especially larger mammals and predators at the top of the food chain (W. F. Laurance et al. 2002b), but even relatively large patches of forests lose a significant number of species (Lambert and Collar 2002). The relationships between animals and plants, and their role in forest dynamics, are complex and poorly understood but the loss of key species from an ecosystem can have far-reaching effects (Crooks and Soulé 1999; Terborgh et al. 2001).

Forest fragmentation also results in a dramatic increase in the amount of habitat edge and magnifies the impact of edge effects, with such impacts altering ecological processes over large areas and even deep inside apparently undisturbed forest (W. F. Laurance 2000). Studies in the 90,000 ha Gunung Palung National Park in western Borneo showed a year of reproductive failure and no recruitment for canopy trees when there was a dramatic increase in seed predation, mainly as a result of bearded pigs flooding into the park from surrounding degraded areas (Curran et al. 1999). Moreover, large mammals such as tigers, elephants, and rhinoceros are disproportionately affected by habitat reduction as they tend to avoid forest boundaries where human activities increase disturbance and human-wildlife conflict is greatest (Kinnaird et al. 2003).

Forest destruction is proceeding so fast that this decade is probably the last chance to protect viable areas of tropical forests; indeed for some countries in tropical Asia it is already too late. Very few countries in Southeast Asia have an adequate and effective protected-area network (J. MacKinnon 1997). Lowland forests are generally under-represented in national protected-area networks and are under the greatest pressure from agricultural expansion and logging. The concept of sustainable forest management outside protected areas, although widely promoted, is rarely even attempted. The fate of Asian forests will thus depend on reconciling the often-conflicting needs and agendas of people, parks, and government policies and promoting economic development and growth consistent with wise natural-resource management and maintenance of forest cover. It is obvious that the security of forests and the quality of forest management depend on good governance (Environmental Investigation Agency 1998; Jepson et al. 2001; Forest Watch Indonesia/ Global Forest Watch 2002).

PARKS AND PEOPLE

Protected areas are the cornerstones of biodiversity and species conservation (Kramer et al. 1997; Bruner et al. 2001). Indonesia was one of the first countries in Southeast Asia, and indeed the world, to use the best principles of conservation biology to plan a national protected-area system representing all habitats in seven biogeographic regions; many of these areas became national parks (J. MacKinnon and MacKinnon 1986). Moving from conservation planning to realistic and effective park management on the ground, however, has proven to be a major challenge, even with substantial investments in conservation (K. MacKinnon 2005).

While parks are created for their biological values, it has become apparent that their long-term survival, protection, and management depend on a whole host of other factors: political, social, and economic (Brandon 1997; K. MacKinnon 1997; Brandon et al. 1998; Oluput and Chapman, chap. 7 in this volume; Fagan et al., chap. 22 in this volume). Some neighboring communities may have derived parts of their livelihoods from the conservation area and now have their rights restricted; others may be recent immigrants, exploiting forest resources and clearing land for agriculture. In both cases, their needs for land and livelihoods must be addressed. The philosophy that wildlife conservation and protected areas in poorer countries were doomed unless local communities participated in, and benefited from, conservation efforts led to a whole generation of integrated conservation and development projects (ICDPs). The concept of reconciling the needs of conservation and social and economic development is appealing but often ICDP initiatives have met with only limited success, satisfying neither the conservation nor rural development agenda (Wells and Brandon 1992; Noss 1997; Brandon et al. 1998; Hackel 1999; Oates 1999; Wells et al. 1999; K. MacKinnon 2001; Terborgh et al. 2002b; McShane and Wells 2004). A case study from Sumatra illustrates the challenges of attempting conservation in a changing development and political environment.

KERINCI-SEBLAT NATIONAL PARK

Kerinci-Seblat National Park (KSNP) is one of the largest conservation areas in Southeast Asia (fig. 16.1). Extending south along the Barisan Range, the park straddles four provinces and covers more than 1.4 million hectares, encompassing a range of habitats from lowland and hill forests to unique highland wetland systems, montane forests, and subalpine habitats on Mount Kerinci, Sumatra's highest mountain (3805 m). The park harbors more than 4000 plant species, 350 bird species (including 14 of the 20 Sumatran mainland endemics), and 144 mammal species (73% of the Sumatran mammal fauna and one thirtieth of the world total). It protects some of the last viable populations of rare and endangered mammals such as the endemic Sumatran hare *Nesolagus netscheri*, small Sumatran rhinoceros *Dicerorhinus sumatrensis*, clouded leopard *Neofelis nebulosa*, Sumatran tiger *Panthera tigris*, Malay tapir *Tapirus indicus*, and elephant *Elephas maximus*. Many of the wide-ranging large herbivores and predators require expansive areas of lowland forests to maintain viable populations. The integrity of the park, and its biodiversity values, are threatened by agricultural encroachment and cinnamon plantations, mining concessions that overlap park lands, poaching of tigers and rhinos, and legal and illegal logging in the lowland and hill forests (K. MacKinnon 1997; Jepson et al. 2001).

The area was declared a park in 1982 but almost immediately the Ministry of Forestry excised the dipterocarp-rich lowland forests for logging concessions, a policy decision with major implications for conservation efforts. The ICDP began in 1997, financed with a World Bank loan and a grant from the Global Environment Facility (GEF), with the stated objectives to protect forests and biodiversity both within and beyond park boundaries. In 1999, after lengthy consultations with adjacent communities, the boundaries were agreed and the park was legally gazetted, the first

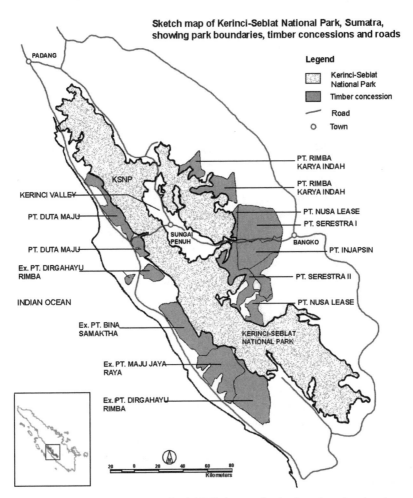

Sketch map of Kerinci-Seblat National Park, Sumatra, showing park boundaries, timber concessions and roads

Legend

- Kerinci-Seblat National Park
- Timber concession
- Road
- O Town

Figure 16.1. Map of Kerinci-Seblat National Park (KSNP), Sumatra, showing the main roads and provincial boundaries

national park in Indonesia to achieve this status. The project provided financing to strengthen park management, spatial planning tools for development planning, and grants to adjacent communities to support alternative livelihoods and rural development in return for commitment to conservation agreements. Considerable resources were provided for village facilitators to work with local communities to designate boundaries, map resources, and develop reciprocal conservation commitments. More than 75 boundary villages eventually received some kind of development assistance under the ICDP after entering into conservation agreements with local government and conservation authorities.

After 6 years and more than US$18 million of investment, the ICDP failed to achieve its conservation objectives. Today KSNP is still under threat from agricultural encroachment and illegal logging. It was probably never realistic to expect that providing alternative development options alone would induce local communities to reduce their impact on Kerinci's forests, especially because much agricultural expansion is not for subsistence but to plant a valuable cash crop, cassiavera (*Cinnamomum burmanni*). Although it was clear that strong enforcement and protection measures would be needed, park staff have been unable to deter land clearance and illegal logging, even with generous resources for patrolling, training, and equipment. There was no clear linkage between development activities and conservation and few of the beneficiary communities maintained their reciprocal commitments to respect park boundaries (see also Oluput and Chapman, chap. 7 in this volume). Some beneficiaries continued incursions into the forest, and were often a greater threat than villages that received no benefits. At least one local bupati (regent) still complains of the park as an opportunity cost, even though in recent months his constituents have suffered floods, landslides, loss of rice crops, and severe hardship as a result of heavy rains washing away logged hillsides.

The substantial investments spent on expensive technical assistance, facilitation, and development grants could probably have been more effectively used on strengthened enforcement, including prosecution of known offenders, and a more aggressive outreach campaign to raise local awareness of the park's values and linkages to ecosystem services, such as flood control. A major NGO was funded to provide community facilitation because of its long-term commitment to the area but walked away at the end of the project, just like other contracted technical assistance.

Despite the establishment of an inter-provincial planning committee, regional development strategies continue to threaten park integrity. Local governments press for roads that would bisect the park, fragment forest habitats, and provide further access for agricultural expansion and poaching. Three mining companies have exploration concessions that overlap park boundaries. Embattled park staff continue to challenge and oppose transport and mining plans that conflict with park-management objectives.

Recognizing the need for a landscape approach to conservation, the project provided resources for biological surveys and improved forest management in neighboring concessions around park boundaries. Plans to provide training for reduced-impact logging were abandoned because of extensive illegal logging within concessions. Rapid ecological-assessment surveys pinpointed areas of high biodiversity and of importance for wildlife or watershed protection. Proposals have been prepared to return these high value forests to the park but the Ministry of Forestry has been slow to process these requests, even though the concessionaires are in flagrant violation of logging contracts and, in some cases, no longer active. The current forest anarchy within Sumatra, with illegal logging operations extending even within national parks, further complicates the situation with local authorities unwilling to close down illegal sawmills that often have the backing of high political and military figures (Jepson et al. 2001).

Both the donor and the government agree that the ICDP was too ambitious, with too many objectives and too much funding over too short a time period (World Bank 2003). Would the park have been in the same difficult condition without the project? Although illegal logging is a serious threat and park patrols are met with increasing violence, KSNP appears to have suffered less deforestation, both within the park and in the broader ecosystem, than other Sumatran parks such as Barisan Selatan and Gunung Leuser (Kinnaird et al. 2003; Y. Robertson, pers. comm.). The biologically rich Sipurak Hook area was finally incorporated into the park in 2005, and there is still hope that other important areas, such as the RKI Finger, will be given conservation status and that adjacent concessions will be closed to stop logging roads reaching park boundaries. Dedicated tiger and rhino anti-poaching patrols continue to work effectively, with NGO support. At least, partly because of the project, a newly formed consortium of local NGOs has become an effective force for conservation and change in the region. KSNP is still one of the most important conservation areas in Southeast Asia, supporting populations of many of Asia's large mammals as well as providing watershed protection for two of Sumatra's major rivers, the Batanghari and Musi, which in turn service millions of hectares of downstream farmlands.

Although the global conservation community complains about inadequate funding for parks and protected areas (see Whitten and Balmford, chap. 17 in this volume), the Kerinci story illustrates that even generous budgets will not ensure success where there is no real political commitment and local support for conservation (K. MacKinnon 2005). The economic crisis in Indonesia, combined with political upheaval, decentralization, and breakdown of law and order are leading to unprecedented pressures on Indonesian forests and national parks. Illegal logging is rampant. Other globally important national parks such as Tanjung Puting in Kalimantan, and Gunung Leuser in Sumatra have also suffered badly (Environmental Investigation Agency 1998, 1999; Environmental Investigation Agency and Telapak 2000). Both are relatively well resourced, with the Leuser ecosystem receiving substantial donor assistance under a project financed by the European Union. The lowland forest ecosystem around Leuser is further threatened by the Acehnese government's plans for a network of new high-ways (Ladia Galaska) to criss-cross the province and open up new development opportunities as part of the Aceh peace process. Rebuilding in Aceh after the devastating tsunami will also put the park's forests under pressure as an accessible, and poorly protected, source of timber. The aftermath of the dreadful tsunami and reconstruction efforts in Aceh will put further pressures on Aceh's forests for timber for rebuilding. Without real political support at the provincial and local levels, the future for Indonesia's parks and forest conservation looks bleak.

POLICIES AND DISINCENTIVES: ROOT CAUSES OF BIODIVERSITY LOSS

The threats to Indonesia's parks are not unique. Many parks and reserves in Southeast Asia, and elsewhere in the tropics, face similar challenges. Most conservation efforts seek to reduce local pressures on parks and species but there is increasing awareness that it is not always the small-scale illegal activities of local communities that are the greatest threat. Forest degradation and loss are often the result of actions and development agendas driven by policies generated far away from the site of conservation concern (Brandon 1997; Rudel 2005).

A variety of policies are affecting the rate of tropical forest loss in Southeast Asia: land use, resettlement, and transmigration policies to colonize frontier regions; provincial and national transport and communication policies that encourage road building through primary forests; energy policies that promote the flooding of lowland valleys for hydroelectric power schemes; pricing policies for timber and agricultural products; subsidies for agricultural plantations and wood- and pulp-processing ventures; and land-tenure policies that encourage colonists to settle frontier areas, promote land uses that lead to soil degradation, and increase demand for new farming land. In many countries, agriculture and forestry policies deliberately encourage the opening up of remote forest areas; construction of new roads then leads to further agricultural encroachment and greater hunting and harvesting efforts. Even global markets can influence rates of forest loss. Recent studies in Barisan Selatan Park, Sumatra, show that rates of forest clearance in the park increase when global coffee prices rise as local farmers expand their holdings to plant more coffee (Kinnaird et al. 2003). Rapid population growth and economic development will continue to fuel the expansion of the agricultural frontier into Asia's forests.

The serious liquidation of Indonesia's forest resources began under Suharto's government (1966–1998) with the declaration of all Indonesian forests as state property and the allocation of logging and mining concessions to powerful conglomerates and politico-business families. The short concession period, lack of regulation, levy system, and other perverse incentives all encouraged poor logging practices (Repetto 1988). Weak concession management failed to prevent pioneer farmers from illegally clearing logged stands. During the 1990s, forest loss in Indonesia was fueled by the timber industry's rapacious demand for wood, a growing pulp and paper industry, the surging demand for land on which to establish plantation crops, and plans to increase the mining of the vast deposits of coal, gold, and other minerals that lie beneath the archipelago's forests (Barber 1998; Forest Watch Indonesia/Global Forest Watch 2002). Agricultural expansion and development of oil-palm plantations exploded in the 1990s, with the total area under plantations in Indonesia growing from 1.8 million hectares in 1994 to an estimated 2.4 million hectares in 1997. More than 6 million hectares of forest in Sumatra and Kalimantan have been allocated for oil-palm plantations. Plantation owners have profited from clearing and selling the timber, yet to date only one-third of this area is planted with oil palm. Even so, companies are still demanding that more forest be converted for plantations. In late 2005, NGOs and donors alike protested with the Indonesian government over widely advertised proposals to clear large tracts of hill forest for oil-palm plantations along the Kalimantan-Malaysian border in Borneo, despite the fact that much of the land is known to be unsuitable for oil palm. Hopefully the new Heart of Borneo Initiative, recently launched by the governments of Indonesia, Malaysia, and Brunei, will change the emphasis to conservation and sustainable management of forests rather than large-scale conversion for plantation agriculture.

The devastating fires that have periodically raged through Indonesia's forests can also be attributed to policy failures. The 1982 and 1983 fires in east Kalimantan were an ecological and economic disaster. The most visible culprits were the shifting cultivators and pioneer farmers who lit the fires, but the conditions that made the forest vulnerable were the direct result of two decades of deforestation, encouraged by government land-use policies to open a frontier region to large-scale commercial exploitation (Mackie 1984). Kalimantan and Sumatra have suffered subsequent large-scale fires, especially when the monsoon rains came late. The 1997 and 1998 fires are some of the worst yet, with more than 6.5 million hectares of forest lands damaged in Kalimantan and 9 million hectares across the archipelago (Holmes 2002). These fires were the result of deliberate burning by logging concessionaires, plantation owners, and small farmers, all intent on clearing more land for agricultural production (and often subsidized by government to do so). One of the affected sites was the much-criticized scheme to convert one million hectares of central Kalimantan peat-swamp forests for rice production, on lands known to be unsuitable for agriculture. These vast wildfires triggered profound ecological changes (see Barlow and Peres, chap. 12 in this volume); as well as impacting degraded lands, the fires nibbled into more than 17 reserves and national parks, destroyed fruit trees, and increased threats to endangered species such as orangutans *Pongo pygmaeus* and other forest primates and large mammals (Kinnaird and O'Brien 1998). The economic costs of the 1997 and 1998 fires have been conservatively estimated at $10 billion in terms of burned timber, lost tourism revenue, and human health in Indonesia, and billions more, if one considers the whole of Southeast Asia where 20 million people were blanketed in smoke for months on end (Barber and Schweithelm 2000).

GOVERNANCE AND POLITICAL COMMITMENT

Under the current policy environment, many conservation efforts in Southeast Asia are unlikely to achieve their long-term objectives—most will, at best, simply hold the line. The economic crisis that rocked the region in the late 1990s put additional pressures on the region's forests. So too did the devastating floods in China that led to a moratorium on logging in that country's upper watersheds as well as an increase in imports of timber from neighboring countries to feed China's explosive economic growth, both for domestic use and trade. While Chinese wood imports have risen dramatically—from $6 billion in 1996 to $16 billion in 2005—and are expected to double again by 2015, so too have its exports. The volume of wood going into exports is roughly equal to 70% of the volume imported. Many of these exports (furniture, plywood, and other forestry products) are going to the United States and Europe, with the United States alone accounting for 35% of China's wood-based exports (A. White et al. 2006). Not surprisingly, logging companies have expanded their activities to meet this global demand, and all accessible forests in the region, from Indochina to New Guinea, are under threat from logging, even where timber stocks are low.

The Indonesian economic crisis led to political upheaval and the downfall of the Suharto regime but the exploitation and unsustainable management of forests continued and expanded. The interim government of President Habibe (1998–1999) responded to demands for reform by pushing through legislation on regional autonomy, with decentralization and devolution of decision making to the district (*kabupaten*) level.

However, decentralization has proved disastrous for Indonesia's forests (W. F. Laurance 2004b). With little capacity for development planning but a need to generate their own revenues, many *kabupaten* governments have encouraged forest clearance for plantations. Forest conversion provides revenues as well as an excuse to harvest natural forest to feed the pulp and paper industries (Forest Watch Indonesia/Global Forest Watch 2002). Threats to remaining forest are further compounded by new exploitation permits for mineral deposits and pressure from the Ministry of Mining to release watershed-protection forests and even national parks for mining.

Today, Indonesia is a society in transition, driven by economic, political, and social crises, and the gap between sustainable development and the reality of current forest mismanagement could hardly be wider. Illegal logging networks have seized control of Indonesia's forests (McCarthy 1999; Jepson et al. 2001). Timber plunder, followed by forest clearance, is rampant in the lowland forests of the Sunda Shelf. Corruption is pervasive with civilian and military officials involved in harvesting and marketing illegal timber (J. Smith et al. 2003). Companies buy illegal timber from their own concessions to exceed their legal annual allowable cut. Conflicting laws on land and cutting rights, as well as failures in enforcement, allow both legal and illegal parties to log without hindrance.

Illegal logging has become the de facto institutional arrangement governing Indonesia's forests. In 1998, less than 26 million of the 58 million m^3 of Indonesia's timber supply came from legal production (D. G. Brown 1999). By 2000, even the Indonesian Plywood Association (APKINDO) was complaining that illegal sources, from Sumatra and Kalimantan, were supplying at least 1 million of the 7 million m^3 of Indonesia's timber market share in China (Forest Watch Indonesia/Global Forest Watch 2002). Illegal logging is occurring throughout Indonesia, even within national parks, with networks of "entrepreneurs" and state bureaucrats apparently immune from prosecution (Environmental Investigation Agency and Telapak 2000; Jepson et al. 2001). The lack of regulation in forest management reflects a general breakdown in governance and a paralysis of central authorities to enforce contradictory national laws in an era of decentralization. The resource exploitation model of the Suharto regime is therefore being replicated at the local level.

So far, the central government has proved unable, or unwilling, to stem this illegal harvest, in spite of repeated commitments to donors to improve forest management, stop further conversion of forest to oil-palm plantations, reduce mill over-capacity, and stop illegal logging in national parks. Early in 2002, the Navy seized three ships exporting illegal timber from central Kalimantan, but these were later released after considerable internal and external political pressure. The destruction of forest ecosystems and globally important biodiversity continues, with no coordinated strategy to stop the losses, even though indiscriminate logging will result in long-term damage to watershed forests and increase fire risks. The Department of Forestry, with the support of the WWF–World Bank alliance, has developed a ten-step program to stop illegal logging, improve monitoring, and increase transparency and accountability in the forestry sector. However, these commitments, like the previous export ban on Indonesian timber, will be meaningless without concerted action to enforce closure of illegal sawmills and stop illegal logging operations. Whereas the political and economic situation may rebound, the forests will not.

FUTURE PERSPECTIVES

Economic progress in developing countries is often based on exploitation and conversion of natural capital. By the mid-1990s, the area of tropical forests worldwide allocated to logging concessions exceeded the area in reserves by a factor of 8:1 (Johns 1997). In Indonesia, half of the total forest estate (more than 35% of the country) is scheduled to be logged, while protected areas, many of them "paper parks," cover less than 10% of the land area. With increasing population and pressure on land, it is probably unrealistic to hope that many more large, protected areas will be established. Conservationists must, therefore, look for innovative opportunities for conservation outside park boundaries.

Production forests in Indonesia could play an important role in conservation. Selectively logged forests, especially when large and only lightly disturbed, can support a high proportion of mature forest species including most mammals and many species that are unable to survive in isolated forest reserves (Johns 1988, 1997). Production forests can also serve as useful buffers and corridors between strictly protected areas. They have the potential to effectively supplement the conservation estate if (but *only* if) strict management regimes eliminate or minimize clearance, fire, and penetration by settlers and hunters so that forests remain as forests (Lambert and Collar 2002). This may require policy changes but in most cases stricter application and enforcement of existing regulations would greatly improve forestry practice. Habitat corridors through agricultural lands and protection of watersheds can all benefit conservation. So too can development decisions that draw farmers away from marginal lands or route roads away from wilderness areas. Slowing agricultural expansion into forests, mountains, and other habitats that are marginal for agriculture, and rehabilitating degraded lands for sustainable production, should become a priority for the future.

Indonesia is a megadiversity country in megacrisis. The same governance and enforcement failures apply to harvesting of wildlife as well as timber, with the escalating wildlife trade for markets in China and Vietnam leading to export of large numbers of Indonesian mammals and reptiles and further "emptying" of Indonesia's remaining forests (World Bank 2005b). If the current state of resource anarchy prevails, the lowland forests of western Indonesia will be totally destroyed within the decade. Indonesia's government and the international community face some hard choices if they wish to stem this biological catastrophe. This is a regional and global emergency but resolution depends on the interplay between political realities, government policies, and the needs and aspirations of local people. Protection and sustainable management of forest lands and resources will be a key determinant of Indonesia's short-term economic recovery and the long-term future of its peoples. From a conservation perspective, immediate priority must be given to stopping illegal logging in Tanjung Puting N. P. and Gunung Palung N. P. (Kalimantan), and the major Sumatran parks (Gunung Leuser, Kerinci-Seblat, and Barisan Selatan). As Indonesia's lowland forests vanish, it also becomes increasingly important to secure conservation status for remaining tracts, such as the Sipurak Hook, newly repatriated to Kerinci-Seblat National Park in Sumatra, and for the Sebuku-Sembakung area in east Kalimantan, where irreplaceable and globally important biodiversity is under imminent threat from proposed oil-palm plantations. Such action would send a clear message that decentralized government can also mean responsible government.

The situation is not entirely gloomy. The Indonesian Ministry of Forestry has a process under way to reassess the status of current logging concessions. Two national parks in Java have been expanded: Gunung Gede has been extended by 6000 ha to now encompass 21,975 ha and Gunung Halimun by 70,000 ha to now span 113,357 ha. The extension forests are mostly hill and montane areas, with approximately 60% natural habitat remaining. They have been captured from Perhutani, the government logging company, and from watershed forests on Gunung Salak (Supriatna, pers. comm.). These parks protect the water supplies of major cities such as Jakarta, Bogor, and Sukabumi; their economic value in provision of domestic and agricultural water supplies is estimated conservatively at US$1.6 billion annually. A new national park, Sembilang, has been established in south Sumatra to adjoin Berbak N. P., Indonesia's first Ramsar site and a wetland of international importance. Sembilang N. P. will protect some of the last remaining lowland swamp and mangrove forests of south Sumatra, areas recognized as important fish spawning grounds. In Riau, a new national park, Tesso Nilo, has been established to protect some of the last remaining lowland forest in Sumatra; this area has previously been logged but still harbors good numbers of large mammals, including elephants. In Sungai Wain, east Kalimantan, an enlightened bupati and local government are supporting and enforcing conservation of watershed forests with high biological value. These are all positive steps for forest conservation, but such success stories are rare and already foreign companies are hustling for logging tracts in Indonesian Papua, one of the last great remaining forest wildernesses.

Protected areas are often perceived as opportunities forgone because their ecological and environmental services are poorly recognized. Yet the benefits of many conservation areas, in terms of watershed protection, flood control, genetic reservoirs, research, and recreation potential, can far outweigh their costs in terms of staff salaries, operational costs, and loss of agricultural opportunities. Indeed many protected areas can be justified according to traditional cost-benefits criteria (J. MacKinnon et al. 1986; McNeely 1988). A consortium of Indonesian NGOs is undertaking a concerted media campaign, INFORM, to promote public debate and raise awareness about the forestry crisis. The intent is to mobilize civil society to understand the benefits of forest conservation and to demand action to stop logging in parks and watershed forests. Much more will need to be done, however, if Indonesia wishes to retain a permanent and sustainably managed forest estate. Unless determined action is taken soon to curb the current state of resource anarchy (see J. Smith et al. 2003; W. F. Laurance 2004b), the majority of Indonesia's lowland forests could be lost forever.

CONCLUSIONS AND IMPLICATIONS

1. Indonesia is one of the most forest-rich and biologically important countries in Southeast Asia. The Asian economic crisis, ensuing political upheaval, and decentralization, accompanied by a need for local governments to raise their own revenues, have greatly increased pressure on Indonesia's forests. The crisis is exacerbated by lack of governance and no effective enforcement to protect the country's forests.

2. The most species-rich lowland forests are easily accessible and particularly vulnerable to agricultural clearance and timber exploitation. At present rates of exploitation, it has been estimated that the lowland

forests of the Sunda Shelf—some of the richest forests on Earth—will be almost totally destroyed outside protected areas in Sumatra and Kalimantan within the next decade.

3. Illegal logging is occurring in all forests, even watershed-protection forests, conservation areas, and national parks. National and local governments seem unable, or unwilling, to stop these illegal activities.

4. Many parks and protected areas have received considerable donor assistance and international funding. Even generous budgets and international pressure have failed to stem illegal logging and continuing forest loss within these parks. Conservation and sustainable forest management cannot succeed without good governance and political commitment.

5. Forest conservation in Indonesia will only come about through better understanding of the multiple benefits and ecosystem services provided by forests and a demand from civil society for greater accountability from policymakers and government officials. Resolution of the Indonesian forestry crisis is tied more broadly to politics, policies, and governance issues.

ACKNOWLEDGMENTS

I am grateful to Bill Laurance and Carlos Peres for their invitation to submit this chapter and to the many colleagues and friends with whom I have worked on forest conservation issues in Indonesia and elsewhere in Southeast Asia.

PART V

Solving and Mitigating Emerging Threats

William F. Laurance

INTRODUCTION

How Do We Save Tropical Forests?

The sixteen preceding chapters document the daunting diversity of threats to tropical forests and their biota. How can we counter, or at least reduce, these threats?

Clearly, this is an enormous challenge. Pressures on tropical forests will only intensify in coming decades, in part because of burgeoning population pressures (human numbers are likely to exceed 9 billion by the year 2050, with most of this increase in the developing world), and in part because of growing per capita consumption, as developing countries strive to attain the lofty economic standards of wealthy nations. Moreover, efforts to promote conservation in developing nations are hindered by inadequate funding, limited educational and institutional capacities, and frequently pervasive corruption. Increasing globalization—with its growing demands for open markets, international competitiveness, and foreign investment—also intensifies pressures on forests.

Developing nations are unlikely to conserve tropical forests on a large scale unless they benefit from doing so. Tropical forests provide a diversity of natural ecosystem services, but who gains from these services? Some of the benefits accrue to local populations. Many of these rewards are obvious—food, fuel, medicines, timber, and myriad other natural products; the protection of erodible soils; reductions in landslides and flooding; watershed protection; and homes for some indigenous peoples. Other local benefits, such as the role of forests in helping to maintain local rainfall regimes, are less apparent but nonetheless important.

But many benefits of tropical forests are global, rather than local. Tropical forests play a major role in the global carbon cycle, sequestering billions of tons of carbon in their living biomass and soils, and thereby helping to slow global warming. The forests also act as massive heat engines, driving global patterns of atmospheric circulation and transporting vast amounts of moisture to higher latitudes. In addition, forest products, including some pharmaceuticals, crops, and other commodities, have global benefits. It is clear that the global community should, therefore, pay some of the costs of forest conservation.

The six chapters that follow focus on strategies for tropical forest conservation. The first chapter provides an insightful economic perspective on forest conservation, contrasting the costs and benefits to local, national, and global communities. The second describes a new strategy, conservation incentive agreements, for providing direct payments for conservation. The third chapter describes how carbon-offset funds and other international mechanisms might be used to promote conservation in the Amazon, whereas the fourth evaluates the benefits and costs of ecotourism for tropical conservation. In the fifth chapter, the authors assess a multidisciplinary program for reducing the impacts of hunting pressure in Central Africa. The final chapter proposes a bold vision for expanding protected-area networks in the tropics.

Who Should Pay for Forest Conservation?

In chapter 17, Tony Whitten and Andrew Balmford argue persuasively that tropical forest conservation is being seriously underfunded. Furthermore, they ask who, in principle, should pay for forest conservation, distinguishing three groups of stakeholders: local people, the national community, and the global community. They then examine the various costs and benefits of forest conservation accruing to each group. Notably, they find that many important benefits of conservation benefit all three groups of stakeholders, but that most of the direct and indirect costs fall upon local communities.

This situation is not only unfair, but it also creates important disincentives that erode local support for conservation. The only viable solution, they argue, is for the global community of governments, nongovernmental organizations (NGOs), and international donors to bear a much larger fraction of conservation costs. This can be achieved by various means, many of which they detail and most of which involve direct compensation to local or national bodies. How much money is needed to achieve an adequate minimum level of tropical forest conservation? According to Whitten and Balmford, around $10 billion per year—a number that might initially seem large but actually pales in comparison to that expended each year on many other human endeavors.

Conservation Incentive Agreements

In chapter 18, Eduard Niesten and Richard Rice critique many of the existing approaches for generating sustainable income in the tropics, such as ecotourism, integrated conservation and development projects, carbon-offset funds, sustainable agroforestry, and reduced-impact logging. Most of these strategies, they argue, have been disappointing in terms of revenue generation and achieved only limited long-term forest conservation. Moreover, they have consumed large amounts of conservation funding, and some have even had negative environmental impacts on forests.

The fundamental problem, they argue, is that conservation rarely can pay for itself. They advocate a fundamentally different approach, which they term the conservation incentive agreement. Under these agreements, a donor (such as a conservation organization) directly pays a natural-resource owner (such as a government or local community) to conserve the resource they control. This approach has been used successfully in Guyana, where Conservation International (CI) has purchased from the national government the rights to a large logging concession. Instead of logging, however, CI protects and manages the forest and provides funds to local communities for promoting sustainable economic activities. For a modest investment (around US$0.64 per ha per year), a sizable tract of forest is being protected from logging, at least for now.

In theory, the CI approach is highly promising, but in reality it faces some tall hurdles. First, it will be challenging to find enough well-heeled donors to fund conservation incentive agreements on a scale anywhere close to that needed for large-scale forest protection. Second, for such agreements, the devil is clearly in the details. To be effective, the agreements must be managed by highly motivated individuals with a long-term commitment to the project, and who ensure that local communities are effectively consulted and compensated. Individual differences among projects, cultures, and the personalities of key players could largely determine the success or failure of any agreement. For these reasons, conservation incentive agreements are likely to be but one of many arrows in the conservationist's quiver.

Mitigating Amazon Deforestation

In chapter 19, Philip Fearnside shows that forest loss and degradation in Amazonia make a significant contribution to global greenhouse gas emissions, and also have large effects on regional hydrology, heat transport, and aerosols. When deforested, each hectare of Amazonian forest spews forth 200 tons of carbon in the form of greenhouse gases. Because extensive tracts of Amazonian forest still survive, policy changes that slow deforestation could have large climatic benefits. Moreover, much deforestation in Amazonia is for low-productivity cattle ranches owned by wealthy landowners, rather than small-scale farmers, so forest loss could be reduced with limited social costs.

Unfortunately, efforts to mitigate climatic change in Brazilian Amazonia have so far focused largely on exotic-tree plantations, such as *Eucalyptus*, that have little value for biodiversity. Fearnside argues, however, that slowing or avoiding deforestation—as might be achieved under provisions of the Kyoto Protocol—could potentially have far greater climatic benefits. An enormous advantage of this strategy is that it could provide serious funds for conserving primary forests, in the form of carbon-offset funds.

This idea seems simple enough, but the national and international politics involved are positively Byzantine. For example, Brazil fears that accepting funds for avoiding deforestation might result in a loss of sovereignty over some of its lands, whereas some Europeans oppose it because it might allow the United States to simply buy its way out of its international obligations—by paying tropical countries to retain their forests, rather than cutting its own carbon emissions. Significant practical difficulties in implementing avoided-deforestation projects also exist. Fearnside synthesizes the various arguments for and against the avoided deforestation and argues, convincingly, that this golden opportunity to promote large-scale forest conservation must not be lost.

Tourism and Forest Conservation

In chapter 20, Stephen Turton and Nigel Stork assess the opportunities and potential threats posed to tropical forests by the rapid worldwide growth of ecotourism. The impacts of tropical tourism have been evaluated in depth at only a few locations, and the authors draw heavily on recent work in north Queensland, which has seen a large increase in ecotourism following designation in 1988 of the Wet Tropics World Heritage Area.

In general, the authors believe that ecotourism can provide important benefits for nature conservation. Tourism contributes to local and national economies, provides incentives for governments to invest in natural-resource management, and can give a major boost to conservation of "flagship" species, such as mountain gorillas. However, not all tropical countries will find it feasible to establish thriving ecotourism industries. Ecotourists require adequate transportation infrastructure, attractive hotels and facilities, and trained (often multilingual) staff and tour leaders, which may require years and considerable investment to develop. Potential tourists are also very sensitive to perceptions of physical danger; in Central Africa, for example, current efforts to promote ecotourism in Gabon could be hindered by political instability in neighboring countries.

Nevertheless, in at least some tropical countries, nature-based tourism will grow substantially in the near future. As this occurs, the potential for negative impacts on tropical forests and indigenous populations will also grow. Turton and Stork suggest that many of these impacts can be reduced by existing management techniques, such as concentrating visitors in a few high-profile areas, properly designing and constructing visitor sites, and managing visitor use through education or regulations. They also believe that a priority is to monitor the impacts of visitors on key components of the environment, so that baseline data can be used to visitor management. Such monitoring may be beyond the means of some developing countries, but can likely be achieved via partnerships with tourist operators, universities, and NGOs.

Reducing Threats from Hunting

Industrial logging is one of the most rapidly proliferating land uses in the tropics. When forests are infiltrated by logging roads, the impacts of hunting can rise dramatically. In chapter 21, Paul Elkan and his colleagues assess an ambitious program by the Wildlife Conservation Society (WCS) and local partners to limit the impacts of hunting in several large logging concessions in the Republic of Congo.

In most tropical logging operations, hunting pressure rises dramatically, for several reasons: forests suddenly become much more accessible to hunters; the local population grows rapidly from the influx of job-seekers and commercial hunters; and the loggers themselves are avid hunters, using their vehicles for transporting hunters into forests and for transporting bushmeat out of forests. For targeted wildlife species, which are usually large in size with slow reproductive rates, the logging-hunting synergism can have devastating impacts.

The WCS approach demonstrates that the impacts of hunting on sensitive wildlife species can be reduced substantially, provided that the logging company is supportive and the program is managed effectively. Among the most important measures was a zoning system that reduced the influx of outsiders, limited hunting in key wildlife areas, and fostered a sense of responsibility for game management by the local communities. Monitoring of hunting regulations by law-enforcement teams, and the closure of disused logging roads, were also important in ensuring compliance. Another critical measure was that the logging company provided timber workers and their families with alternative sources of protein, such as fish, chicken, and other domestic animals, to replace bushmeat.

As the WCS study shows, limiting the impacts of hunting in logged forests is achievable, but it is not cheap or easy to implement. For these reasons, careful regional planning—for example, locating sawmills as far from forests as possible, to help reduce population pressures in forests—is vital. It is unrealistic to expect logging companies to absorb the costs of reducing hunting; this can only be achieved with pressure from timber consumers and with financial and logistical support from international donors.

A Bold Protected-Area Strategy for Tropical Forests

In chapter 22, Chris Fagan, Carlos Peres, and John Terborgh propose what, at first blush, seems like an overwhelmingly pie-in-the-sky strategy for forest conservation: each tropical nation should sequester 50% of its original forest cover in protected areas. This number is obviously far higher than guidelines established by the World Conservation Union–IUCN, which advocated having more than 10% of the land area of each nation set aside for protection, or the ca. 12% of the Earth's land cover that is presently contained in various kinds of reserves. Shouldn't we be satisfied with 12%?

On closer inspection, the rationale underlying the Fagan et al. strategy becomes clear. First, the species–area relation suggests that a reduction of habitat area by nearly nine-tenths should lead to a sharp decline—by perhaps half—in species diversity. This, they argue, is simply unacceptable. Second, even such alarming predictions could be overly optimistic, as the 12% of habitat in reserves will almost certainly be fragmented and thus likely to lose even more species over time. Third, the 12% figure overstates the case: much less than half of this land area is fully protected under law, with the remainder being in "soft" reserves that are open to various forms of exploitation. Finally, many existing reserves are merely "paper parks"—receiving minimal protection and suffering from myriad forms of illegal encroachment. The authors vividly illustrate this point with an analysis of the numerous threats to protected areas in Central America.

For these reasons, the authors argue, conservation biologists must fundamentally reevaluate their views about the minimum extent of protected areas. The 50% figure they advocate includes not only strictly protected areas and soft reserves, but also multiple-use lands, such as sustainably managed production forests. Landscape-design principles, such as embedding strictly protected areas within matrices of soft reserves and production forests, could help to minimize the impacts of habitat fragmentation on reserve biota. Improving the management of reserves, to reduce the corrosive effects of illegal encroachment, is also a high priority. Thus, what initially seemed like a wildly unrealistic scheme quickly begins to sound eminently defensible. Given the great complexity and poorly understood nature of biogeographic patterns in the tropics, current strategies that involve protecting just a small fraction of the original forest area will likely result in serious losses of tropical biodiversity.

SYNTHESIS

Several common themes—and a few notable differences of opinion—pervade the chapters in this part. The first theme is that funding for tropical conservation is demonstrably and seriously inadequate. In most tropical countries, nature conservation relies upon protected areas that are too small, too few, increasingly isolated from other areas of forest, and degraded by a multitude of threats, including illegal logging, hunting, gold mining, and other illicit activities. Further insidious threats, such as those arising from climatic change, emerging pathogens, and invading species, might also degrade protected areas. Moreover, most parks and reserves are poorly situated with respect to key centers of biological diversity and endemism, the distribution of which is often poorly known in the tropics (however, a few notable exceptions exist, such as the incomplete Amazon Region Protected Area program and the recently expanded network of national parks in Gabon, both of which are significantly based on biogeographic criteria). Timber-production forests and other multiple-use areas are often so poorly managed that they increase, rather than slow, forest destruction; for example, the proliferation of timber roads in tropical frontier areas greatly facilitates forest encroachment by hunters, loggers, miners, and slash-and-burn farmers. As a consequence of these grim realities, it is apparent that a far bolder, less minimalist approach to tropical nature conservation is needed.

A second important theme is that developed nations must shoulder a much larger share of responsibility for funding tropical forest conservation. The global community receives vital benefits from tropical forests, albeit often indirectly, and thus it is only fair that they should help to pay for it. Although international donors presently fund a variety of tropical conservation programs, the lion's share of costs—including substantial losses in opportunity costs of forgoing development—is being borne by local communities and tropical countries.

It is one thing to argue for greater investment in conservation, and a second to achieve it. Perhaps the greatest challenge we face as conservation advocates is to convince wealthy nations to commit a much larger share of their foreign-aid allocations to tropical forest protection than is presently the case. A few large conservation funds do exist, such as the Global Environment Facility and the Pilot Program to Protect the Brazilian Rainforest, but these involve annual budgets of just a few hundred million dollars. As asserted persuasively by Whitten and Balmford, we should be thinking in far bigger terms—on the scale of roughly $10 billion per year—which might be feasible given the generous levels of the public's willingness to pay at present.*

How should payments from wealthy nations be made? A menu of potential mechanisms is available, including carbon-offset projects, integrated conservation and development projects, conservation incentive agreements, direct assistance from international donors and development agencies, and, somewhat less directly, ecotourism. Among conservation practitioners there is no consensus about which mechanisms are most promising—nor, perhaps, should there be. Different nations and circumstances will demand different solutions, so it is naïve and overly simplistic to think in terms of a one-size-fits-all solution for the tropical biodiversity crisis.

* For example, average tax payers in industrialized countries such as Italy and the United Kingdom are willing to pay US$45 per household per year to fund a protection program covering 5% of Brazilian Amazonia, and US$59 per household per year to fund a program protecting 20% of the region (Horton et al. 2003).

A final theme is that critical opportunities for conservation funding are in danger of being lost from petty infighting or mulish obstinance. A disturbing example of this is the recalcitrance presently being exhibited by Brazil's foreign ministry, which has fought vociferously against provisions under the Clean Development Mechanism of the Kyoto Protocol that would harness carbon-offset funds directly to slow tropical deforestation. Not only is the ministry declaring that Brazil should forgo this opportunity, but they are arguing with equal fervor that all other tropical nations should do likewise. Their reasons? Simple xenophobia: they fear losing control of their natural resources to foreign powers, should they accept funds linked to avoiding deforestation. However, most well-informed observers agree that international agreements can be framed that would allow Brazil and other countries great latitude in developing their sovereign natural resources while simultaneously using carbon-offset funds to help promote forest conservation.

In a similar vein, the stance of some high-profile European NGOs, who have opposed the avoided-deforestation provision under Kyoto on the grounds that it fails adequately to "punish" the United States for its indisputably excessive carbon emissions, seems equally myopic. Should the United States be criticized? Absolutely. Should perhaps the greatest single opportunity to gain major international funding for tropical forests be abandoned for reasons of misconstrued political correctness? Absolutely not. Let us hope that better-informed heads prevail in ongoing efforts to use carbon-offset funds to slow the alarming pace of tropical forest destruction.

CHAPTER 17

Who Should Pay for Tropical Forest Conservation, and How Could the Costs Be Met?

Tony Whitten and Andrew Balmford

Attempts to conserve tropical nature are reaching a crisis point. On the one hand, more species and more habitats are at risk in the tropics than elsewhere—due to a combination of rapidly rising human populations, acute poverty, increasing per capita consumption (Myers and Kent 2003), and the higher densities in tropical areas of both species in general and intrinsically vulnerable species with small range sizes in particular (BirdLife International 2000; Hilton-Taylor 2000). On the other hand, while in overall terms conserving what remains of wild nature makes striking economic sense (Balmford et al. 2002), conservation is under-funded almost everywhere, with the shortfall in required resources being particularly extreme in developing countries. For example, while developed-world expenditure on terrestrial reserves runs at only around one-third of the estimated requirement for an effective network (estimated as 15% of the Earth's land area, counting World Conservation Union–IUCN Protected Area categories IV–VI as well as the more conventional I–III [James et al. 1999a]), current expenditure in the developing world is reckoned, very roughly, to be less than one-twentieth of that needed (James et al. 1999a; Balmford et al. 2003; fig. 17.1).

Bridging the funding gap represents an urgent challenge to those interested in tropical conservation. How can we bring about a substantial and sustained increase in funds available for developing-country conservation in ways that are both ethically acceptable (with those who benefit most from conservation paying the most for it) and pragmatic (with

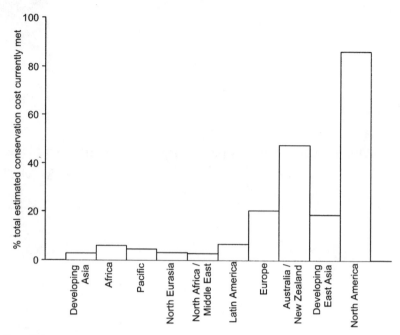

Figure 17.1. Variation in the extent to which the total direct costs of terrestrial-reserve networks (effectively conserving 15% of different regions) are currently met, versus mean per capita income (from Balmford et al. 2003)

the identified sources being plausible)? Money is by no means the sole obstacle to achieving conservation and not all of the money already available is being used as efficiently as it could be. Nevertheless, we contend that many opportunities for better conservation are lost through inadequate funding (for example, see Leader-Williams and Albon 1988) and that many of the poorest and biologically richest countries require greatly increased funding for sustained conservation of their often large, diverse, and highly threatened protected areas.

This chapter explores two aspects of how we might respond to this challenge. We start by considering how, in principle, the shortfall in support for tropical conservation ought to be met. Building on earlier arguments of Bell (1987), Wells (1992), and Kramer and Sharma (1997), we examine how the costs and benefits of conservation are presently distributed across local, national, and global stakeholders. Who pays for conservation at the moment, and where do the different types of benefit accumulate? We argue that, in principle, each constituency should meet needed increases in funding in rough proportion to the value of benefits it receives from conservation (also bearing in mind the constituencies' differing abilities to pay). In the second part of the chapter, we use these insights into the distribution of conservation benefits and current conservation spending as context for exploring where the extra investment might come from. What do recent developments tell us about the scope for increased support from private donors, business, and taxpayers? Might additional means help to narrow the gap between conservation needs and current support?

We do not attempt to tackle other major hurdles to tropical conservation, such as the chronic shortage of trained conservation professionals, the need to expand in-country public and political support for conservation, the impending impacts on natural habitats and species distributions of climate change (see Lewis et al., chap. 1 in this volume, Williams and Hilbert, chap. 2 in this volume, Avissar et al., chap. 4 in this volume), and the need to develop institutions capable of delivering conservation benefits effectively and equitably on the ground (see Elkan et al., chap. 21 in this volume). These are all extremely important issues, but beyond the scope of this chapter. In addition, we concentrate here on protected areas of forest. However, while the detailed distributions of conservation costs and benefits will be different for nonreserved land, and will vary with habitat types and across individual reserves, we believe that our overall conclusions are broadly applicable across the terrestrial tropics.

WHO IN PRINCIPLE SHOULD PAY?

Consider three groups of stakeholders (after Wells 1992): *local* people,* living in or near an area targeted by a conservation intervention such as a park; the *national* community, which includes locally based commercial elites but mainly more-distant stakeholders; and the *global* community of concerned individuals, businesses, nongovernmental organizations (NGOs), governments, and intergovernmental organizations. How much does each of these constituencies currently pay for conservation?

WHO PAYS FOR CONSERVATION NOW?

In addressing this question, it is helpful to consider two classes of cost: the immediate costs of conservation activities, including that from acquiring

---

* The term *local* is used to denote long-term and often indigenous inhabitants who, although of varying socio-economic levels, depend significantly on resources from nearby forest areas.

or leasing land, managing or restoring habitats and populations, and enforcing restrictions on land use, which we term *active costs*; and the indirect costs of conservation, which we term *passive costs*, and which include the opportunity costs that arise when harvesting wild populations or converting wild habitats is restricted, as well as the cost of damage incurred by animals originating in the conserved habitats. Although some active and passive costs (such as the budget for running a national park's head office, or the opportunity costs to international consumers of reduced harvesting of protected species) are located at national or global levels, most costs are located in or near the areas targeted by a conservation project. The question is, who currently pays these costs?

Thinking first about the active costs of conservation programs in developing countries, these are generally met by state or national agencies and, to a lesser extent, NGOs, funded by a combination of national and international taxpayers and donors (fig. 17.2, top row). According to surveys of national protected-area agencies conducted by the UNEP–World Conservation Monitoring Centre in 1993 and 1995, international donors funded only ca. 20% of the total expenditure on developing-country nature reserves (James et al. 1999b). However, the report's authors suggest this was a very substantial underestimate—a point confirmed by comparison with a top-down analysis of international-donor investment in Latin America and the Caribbean (LAC) from 1990 to 1997 (Castro et al. 2000). Castro and colleagues report that international donors spent at least $180 million annually on LAC-protected areas at

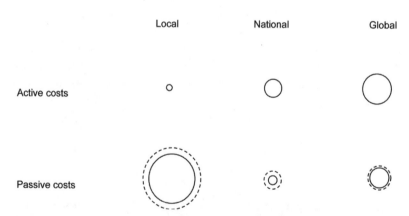

Figure 17.2. A rough schematic of the current distribution of the costs of conserving protected areas in developing countries. In the top row, the area of the circles describes the approximate relative contribution of local, national, and global communities to current expenditure on the direct costs of tropical reserves (estimated at ca. $750 million annually). In the bottom row, the area of the circles describes the current distribution of gross passive costs; actual costs (solid lines) may be lower than perceived (dashed lines) because current levels of use of natural habitats may not be sustainable.

a time when equivalent government expenditure was probably less than $150 million per year (James et al. 1999a, 1999b; note that these and all other costs have been updated to 2000 U.S. dollars). Although data are limited, for LAC at least, it appears that international funding for the active costs of conservation is probably greater than national-level outlays. Active payments by local communities, however, are generally limited to tax contributions, which because of widespread rural poverty are low in absolute terms.

In contrast, we believe that local communities probably bear the brunt of the gross passive costs of developing-country conservation (fig. 17.2, bottom row). The private benefits forgone as a result of the establishment of a protected area—through restrictions on harvesting certain wild species, lost opportunities to convert wild habitats to farms or plantations, or reduced prospects of development of new infrastructure (such as roads or electricity)—can be very substantial (M. Wells 1997; Balmford et al. 2002). In Madagascar, for example, the gross-opportunity costs of two parks to villagers that harvest wild resources have been estimated at between $39 and $125 per household per year (Kramer and Sharma 1997; Ferraro 2001), whereas abandoning slash-and-burn agriculture and harvesting would cost households between $93 and $191 per year (Brand et al. 2002); these costs probably represent more than 10% of household income (Ferraro 2001). In Kenya, the gross opportunity costs of the country's approximately 60,000 km² of parks and reserves have been estimated at $270 million annually (Norton-Griffiths and Southey 1995). For developing countries as a whole, one upper estimate (based simplistically on the value of land in strictly protected areas) puts the gross-opportunity costs of existing reserves at more than $5 billion each year—approaching an order of magnitude more than the ca. $750 million currently spent, by all agencies combined, on meeting their direct costs (James et al. 1999a, 2001; see table 17.1).

Most of these opportunity costs are met by local people (Bell 1987; Wells 1992; Kramer and Sharma 1997; Ferraro 2001; Brand et al. 2002). Added to this, local communities in some areas can bear significant costs when animals from nearby conservation areas damage crops, kill livestock, or attack people (Karanth and Madhusudan 2002). Conservation initiatives can also impose opportunity costs at national scales—through lost tax revenue from logging operations (Kremen et al. 2000)—and at an international level, through reduced exploitation of species in international trade. In addition, passive costs that are locally incurred may be partially transferred to national or, more commonly, international levels by the provision of compensation schemes or alternative development programs.

Table 17.1 Rough estimates of the likely annual total costs of a protected-area network covering about 15% of the tropical land area. Note that these figures refer to non-forest as well as forest biomes, and they become increasingly imprecise from top to bottom (all figures are in 2000 U.S. dollars; for estimation details, see James et al. 1999a, 1999b, 2001).

Costs	$ million/year
Active	
Current expenditure on existing reserves	~750
Shortfall in current expenditure	~1,500
Management costs of additional reserves needed to reach 15% target	~2,000
Passive	
Opportunity costs of existing reserves	~5,000
Opportunity costs of additional reserves	~6,500

Three other points need highlighting at this stage:

1. While the passive costs of conservation are often very significant at the local level, they are sometimes perceived to be even greater than they actually are (see dashed circles in fig. 17.2), because some particularly destructive uses of natural habitats may not be sustainable even in the short term. The opportunity costs of conservation are also sometimes inflated by incentive schemes that subsidize otherwise uneconomic habitat conversion (Myers 1998; van Beers and de Moor 1999; Myers and Kent 2001).

2. Nonetheless, the local communities most strongly impacted by the passive costs of developing-country conservation are generally among the poorest of the poor; it is both inequitable and impractical to expect them to continue to bear these costs into the future (Bell 1987; Wells 1992; Norton-Griffiths and Southey 1995; Kramer and Sharma 1997; Ferraro 2001).

3. Figure 17.2 is based on the current costs of existing terrestrial reserves. Despite some successes (Bruner et al. 2001; Nepstad et al. 2006), many tropical protected areas are deteriorating (van Schaik et al. 1997; Brandon et al. 1998; Oates 1999; Terborgh 1999; Terborgh et al. 2002b), and reserve managers estimate that roughly another $1.5 billion is needed annually to meet the full active costs of these reserves (James et al. 1999a, 1999b; table 17.1). Added to this, the total extent of existing reserves is far below the approximately 15% of land area considered a minimum safe

standard for conserving a representative sample of species, habitats, and ecosystem services over the medium to long term (IUCN 1993, 1998). Expanding developing-country reserve networks to meet this target has been very roughly estimated to cost about an extra $2 billion each year in active costs, plus about $6.5 billion annually to offset local opportunity costs (James et al. 1999a, 2001; table 17.1). These additions total some $10 billion annually.

Given the enormous inequities in the current distribution of conservation costs, and the need to spend a great deal more on tropical conservation if it is to succeed, how can we substantially increase conservation investments, and do so fairly? We believe the key is to examine the current distribution of conservation benefits, and the potential for these being increased in the future.

WHO BENEFITS FROM CONSERVATION?

Here we consider five classes of benefits that may arise from tropical conservation: (1) sustainable consumption of conserved resources (for food, timber and other fibers, and medicines); (2) nature-based tourism; (3) localized ecological services (such as regulation of water supply, prevention and reduction of storm and flood damage, and erosion and sedimentation control); (4) more widely dispersed ecological services (such as nutrient and climate regulation, and carbon storage); and (5) option, existence, and bequest values. Where do these benefits currently accrue, and can they be expanded to better offset the costs of conservation?

We contend that where wildlife and wildlife products are not commonly marketed, the benefits are generally greatest at the local level (fig. 17.3, first row). However, for the subset of such products that are marketed, the benefits at national scales can be substantial, and are generally less significant at the international level (although nevertheless important in some cases). Efforts to make existing harvesting regimes sustainable, to expand harvesting to other species, and to identify new markets lie at the core of many recent attempts to offset the local opportunity costs of conservation and simultaneously achieve development goals (IUCN/UNEP/WWF 1980, 1991; Reynolds et al. 2001). However, there is a growing view that, while the "use it or lose it" approach can work in some situations, it will frequently lead to overexploitation of wild resources, if not immediately, then as human populations and demands rise (Redford 1992; Robinson 1993; Barrett and Arcese 1995; Brandon 1997; Kramer

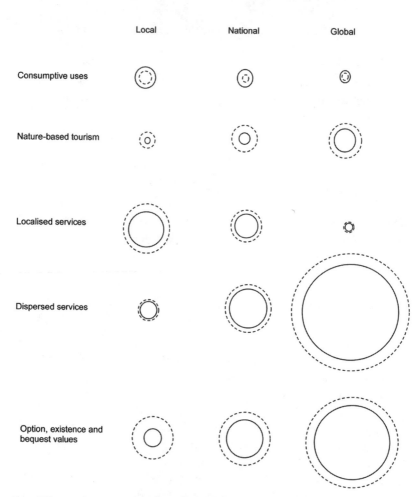

	Local	National	Global
Consumptive uses			
Nature-based tourism			
Localised services			
Dispersed services			
Option, existence and bequest values			

Figure 17.3. A rough schematic of the current distribution of five kinds of benefits from developing-country conservation. In each row, the area of the solid circles describes current benefits, whereas the area of the dashed circles describes potential, sustainable benefits in the future. The total size of a row's circles reflects that benefit's roughly estimated total value, relative to other benefits. (From Costanza et al. 1997)

et al. 1997; Brandon et al. 1998; Newmark and Hough 2000; van Schaik and Rijksen 2002). For this reason, we suggest that current levels of consumptive benefits from tropical reserves may not be sustainable, and will inevitably become lower in future, as stocks are depleted or permitted harvests are reduced (fig. 17.3, dashed circles in first row).

Nature-based tourism (i.e., tourism focused on enjoying wildlife or wild places) is also frequently advocated as a promising means to derive substantial benefits from conservation (Boo 1992; Goodwin 1996; Davenport et al. 2002). However, in most cases, the benefits of tropical nature-based tourism accumulate largely at national and especially international levels, rather than offsetting the opportunity costs at the local level (Brandon 1996; Wells 1997; fig. 17.3, second row). One study from Royal Chitwan National Park in Nepal, for example, reported that only 6% of nearby households obtain any income directly or indirectly from the 50,000 to 100,000 people that visit the park each year (Bookbinder et al. 1998). Likewise, only 0.2% of the total expenditure by tourists visiting Komodo National Park in Indonesia has, until recently, accrued to local villages adjacent to the park (Walpole and Goodwin 2000). Tourism is also extremely sensitive to periods of political instability. In Central Africa, for instance, lucrative gorilla tourism in Rwanda and the Democratic Republic of Congo plummeted following civil wars in both countries (Butynski and Kalina 1998), and Uganda's tourist industry has yet to recover from its collapse during Idi Amin's rule, which ended more than 20 years ago. Such impacts of war can be longer lasting than those affecting core conservation activities (Hart et al. 1996). Clearly, it is evident that the scope of the benefits of nature-based tourism must be increased, both in general—tourism is currently the world's fastest growing industry, with nature-based tourism believed to be its fastest growing sector (Davenport et al. 2002; Turton and Stork, chap. 20 in this volume)—and through targeted efforts to increase revenue-sharing, especially at the local level (Walpole and Goodwin 2000; Walpole and Leader-Williams 2001). Nevertheless, many biodiverse habitats and wild destinations are simply too remote, too dangerous, or insufficiently charismatic to attract large numbers of high-paying tourists (McClanahan 1999; Davenport et al. 2002). We therefore believe that the potential for expanding nature-based tourism and its benefits for conservation is more limited than sometimes suggested (Boo 1992; Goodwin 1996; Davenport et al. 2002; Turton and Stork, chap. 20 in this volume; fig. 17.3, dashed circles in second row).

The conservation of wild habitats can generate substantial (though commonly under-appreciated) benefits through the provision of localized ecological services (fig. 17.3, third row). For example, retaining forest cover can reduce the risks of downstream flooding, erosion, and sedimentation, while maintaining dry-season water flows through evapotranspiration and cloud interception. Likewise, coral reefs and mangroves act as nurseries for offshore fisheries, and absorb storm energy, thereby protecting coastal communities (for examples, see Kumari 1994; Sathirathai 1998; Becker 1999; A. T. White et al. 2000; Yaron 2001; R. K. Turner et al. 2002; Carret and Loyer 2003; Philipp and Fabricius 2003; but see Chomitz and Kumari 1998). These benefits mostly accrue at the local level, although in many developing countries, the provision of such services to major urban centers (such as Dar es Salaam, Tegucicalpa, and Quito) is dependent on the maintenance of upstream forest cover (McNeely 1988; Burgess et al. 2002; Spergel 2002; Dudley and Stolton 2003). At all scales, the value of these services is likely to increase as human populations grow, become wealthier, and disperse into previously unoccupied areas near patches of intact habitat (fig. 17.3, dashed circles in third row).

Dispersed ecological services are those whose benefits can be enjoyed at a considerable distance from the conserved habitat. For example, because atmospheric carbon circulates globally, the contribution of a conserved wetland or forest to carbon sequestration or storage benefits everyone. While often underappreciated, such services can be tremendously valuable (Myers 1996; Costanza et al. 1997; Daily 1997; Pimentel et al. 1997; W. F. Laurance 2006a), especially at the global scale, where they contribute to the welfare of large numbers of relatively wealthy peoples (fig. 17.3, fourth row). The value of dispersed ecological services is likely to grow as human populations increase and per capita incomes rise (fig. 17.3, dashed circles in fourth row).

We turn last to an array of nonuse values (fig. 17.3, fifth row): those arising from retaining the possibility of use in the future (option values), those that describe the value of simply knowing a habitat or species is still extant (existence values), and those that derive from being able to pass those benefits on to future generations (bequest values). These values have affected traditional views of the relationship between people and nature in many parts of the world, but are notoriously difficult to capture in dollar terms (OECD 2002). However, their lower bounds are represented by the donations to conservation NGOs, or by the contributions of developed-country governments to the Global Environment Facility (GEF)—currently in excess of $1 billion annually for biodiversity. In absolute terms, these benefits are greater at national than at local scales, and greatest of all at the global level. They can be expected to grow as people become wealthier and more numerous. They may also increase as natural habitats become scarcer, and if people become more aware of, and concerned about their natural heritage.

SO WHO SHOULD PAY?

Comparing these distributions of benefits and costs yields several broad insights into who ought to pay for tropical conservation:

1. A simple "fortress-and-fines" approach for imposing conservation on local people without due compensation or opportunities for participation is, in

our view, not only immoral (in suggesting that sizable opportunity costs should be met by the rural poor), but also unworkable in the long term if used alone because, as populations grow, the rising costs of enforcement would further increase the largely unmet active costs of conservation.

2. Solutions that meet local opportunity costs may also fail insofar as they rely on expanding already unsustainable resource exploitation, or on substantially increasing and redistributing income from nature-based tourism.

3. The unmet passive and active costs of tropical conservation will instead often have to be met from other benefit streams (see also Wells 1992; Kramer and Sharma 1997; Sinclair et al. 2000; Terborgh and Boza 2002). The needed increase in funding is so great that many new funding sources need to be identified at all levels (R. K. Turner et al. 2002). Moreover, because for some habitats the local costs of conservation may exceed the local benefits, cross-subsidies between scales may sometimes be necessary (J. Kellenberg, pers. comm.; see fig. 17.4 for an example). Spreading responsibility for meeting the costs of conservation across beneficiaries should also buffer conservation activities against economic fluctuations in individual countries or sectors.

Figure 17.4. Lowland rice farming below upland forests in eastern Madagascar. Lowland farmers stand to benefit if upstream forest clearing is reduced, because of reduced flooding and sedimentation in their rice paddies. However, a contingent-valuation study suggests that lowland farmers' willingness to pay for upland conservation is far lower than the opportunity costs to upland farmers of abandoning slash-and-burn agriculture. Other beneficiaries of conservation must meet these costs if Madagascar's upland forests are to be conserved effectively and equitably.

4. In general, the most promising sources of increased support will be those constituencies that already gain the most from conservation (i.e., the columns with the largest solid circles in fig. 17.3), and those whose benefits are likely to grow most in future (i.e., those with dashed circles much larger than solid circles). Looking from left to right across figure 17.3 at the relative magnitude of the overall benefits enjoyed by each group of stakeholders, the greatest contribution to meeting the currently unmet costs of tropical conservation should come from the global community, followed by national and then local stakeholders. Because the developed world and (to a lesser degree) urban communities of developing countries gain most from tropical conservation, it is only equitable and practical that they should pay the bulk of the costs for it—at present they do not (see also Wells 1992; Norton-Griffiths and Southey 1995; Kramer and Sharma 1997; R. K. Turner et al. 2002).

5. If we now examine the rows in figure 17.3, we see that the largest conservation benefits accrue not (as is sometimes supposed) from direct consumption or nature-based tourism, but from localized and dispersed services (e.g., services such as water flow in a river and absorption of carbon dioxide by plant matter), and from nonuse values (the value for things that we do not use but would feel a loss over if they were to disappear; see also Costanza et al. 1997). Accordingly, it is these benefit streams that best justify expanded support for tropical conservation, and which might be most readily tapped to provide new conservation funding. However, because such benefits are largely nonrival (i.e., they can be consumed by many people simultaneously) and nonexcludable (i.e., they can not be denied to anyone), persuading beneficiaries to invest in conservation to secure such intangible benefits over the long term will commonly require government intervention.

6. The idea that national and global beneficiaries of ecosystem services and existence values would pay local communities for their continued delivery raises several potentially difficult issues. We would argue, however, that most are soluble and none is unique to this model of conservation funding. For example, payment to not harvest species or not convert natural habitats raises worries about welfare dependency. However, communities can be required to be active in ensuring compliance with conservation objectives (Ferraro and Kiss 2002). Performance-based community support could be in-kind (for example, through the provision of health care or educational opportunities), rather than in cash (Ferraro and Kramer 1997). A related problem is that the provision of payments may stimulate immigration from elsewhere, increasing both costs and pressures (for examples, see Campbell and Hofer 1995; Merlen

1995; Oates 1999). But such a "honeypot" effect can be a problem for any scheme that seeks to address, rather than ignore, the passive costs of conservation, and can only be addressed through the early establishment of who does and does not have rights to compensation (Ferraro and Kramer 1997). Finally, mechanisms for delivering compensation need to be both equitable and effective. Payments should reach all those incurring opportunity costs, and should probably be delivered not as a lump sum but in a continuous stream (for example, see Niesten and Rice, chap. 18 in this volume), in direct exchange for ongoing production of conservation benefits (Ferraro and Kiss 2002; Ferraro and Simpson 2002).

In sum, our cost-benefit comparison suggests that a great deal of the increased support needed for tropical conservation should come from global stakeholders, in exchange in particular for the continued delivery of both dispersed ecological services and existence values (see also Wells 1992; Hardner and Rice 2002; Ferraro and Simpson 2002). The central challenge will be finding ways to bring these less-tangible benefits to the attention of decision makers.

HOW CAN WE BRIDGE THE GAP IN PRACTICE?

Having examined the principles of who ought to pay for tropical conservation and why, we now consider where in practice the extra funds might be raised. Of course the magnitude of the shortfall in funding means that we need to cast our net widely. We must look for increased support from local and national as well as global communities, via a mix of mechanisms: increased individual donations, bringing the market to bear, and expanding direct contributions from governments.

Increased Donations

At one end of the spectrum, increasing numbers of private individuals are joining NGOs (fig. 17.5). The emergence of new, tropically based NGOs is encouraging (although their most important contributions may be political, educational, and practical, rather than financial). At the other end of the spectrum, major contributions to tropical conservation from private individuals and foundations have risen dramatically over the past 5 years (for instance, see http://www.conservation.org/xp/CIWEB/news room/press_releases/2001/120901.xml for a single initiative totaling $261

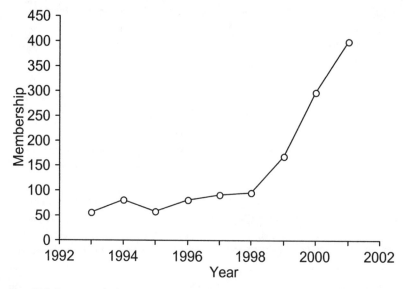

Figure 17.5. Recent growth of Nature Uganda, a local-membership NGO (data provided by NatureUganda)

million). Likewise, large corporations, mostly involved in primary industries, have made a number of extremely significant donations in recent years (for one new $50 million partnership, see http://www .investinginnature.com/index.htm). These are extremely encouraging moves, and it is hoped that they in turn will catalyze further contributions. Nevertheless, even these unprecedentedly large donations can provide only a fraction of the total resources needed, and all NGOs are currently suffering from reduced donations as a direct result of the unstable stock market.

Bringing Markets to Bear

An array of exciting new initiatives for funding conservation is also emerging from the commercial sector (Chichilnisky and Heal 1998; Daily and Walker 2000; Daily and Ellison 2002; Sandor et al. 2002). Some are essentially extensions of existing markets, and operate through global-level consumers choosing to invest in environmentally responsible companies, or paying premium prices for certified products that have been sustainably harvested (see, for example, the website of the Forest Stewardship Council [www.fscoax.org]). Again, these developments are extremely welcome, but they may only ever capture a fraction of the developed-world market (and less of the developing-world market). Their main potential probably lies in the sustainable management of natural resources outside reserves, rather than in financing protected areas (Hardner and Rice 2002).

Other market-based initiatives involve creating entirely new markets through which beneficiaries pay producers for the provision of ecosystem services. In many cases, the prompt for beneficiaries to pay for what they have previously received for nothing has come from new legislation. The most developed such market is for watershed protection. In Colombia, Ecuador, and Laos, hydroelectric companies are handing over a substantial portion of their revenues for upstream-forest conservation; similar payments are made by downstream water consumers in Ecuador, and by municipal authorities in El Salvador (see Kiss et al. 2002; Spergel 2002).

Many ambitious programs are now under way to try to secure funding for conservation through carbon-credit schemes (see Fearnside, chap. 19 in this volume). These have considerable potential to generate very large sums for conserving tropical forests, particularly as developing countries generally have a comparative advantage, in being able to achieve credible emissions reductions at far lower marginal costs than can developed nations (Kiss et al. 2002; Niesten et al. 2002; Niles et al. 2002). However, major hurdles remain: although reducing ongoing conversion of natural forests could make a large contribution toward meeting lowered carbon dioxide–emission targets agreed under the Kyoto Protocol (Malhi et al. 2002a), concerns about confirming compliance and about the validity of forest conservation as a carbon-sequestration instrument mean that, at present, habitat retention is not eligible for carbon credits under the Clean Development Mechanism of the United Nations Framework Convention on Climate Change (Bonnie et al. 2002; Niles et al. 2002; see Fearnside, chap. 19 in this volume, for a detailed discussion).

However, other means of funding conservation through paying for carbon storage are possible. Web-based initiatives run by organizations such as Climate Care and Future Forests enable individuals or organizations to make voluntary payments in proportion to their carbon emissions, with the revenues funding, among other things, tropical conservation projects (see www.climatecare.org and www.futureforests.com). The World Bank has just launched the $100 million Biocarbon Fund (www.biocarbonfund.org) with the triple goals of reducing greenhouse gases in the atmosphere, reversing land degradation and the loss of biodiversity, and improving local livelihoods in poor countries. In Costa Rica, a national tax on gasoline has funded annual payments to landowners of about $40 per ha for retaining and managing natural forests (Castro et al. 1998). Since 1997, 320,000 ha have been brought into this program. This funding base is now being expanded through contributions from hydroelectric companies (for hydrological services) and tourism operators (for provision of biodiversity and scenic beauty; Chomitz et al. 1999; Kiss et al. 2002; see fig. 17.6).

Expanding Direct Government Contributions

Despite encouraging and very significant growth in funding for conservation from private donors and markets, we believe that general tax revenues, raised by governments, will continue to be the principle means by which tropical conservation is financed, and by which the funding gap might be bridged (James et al. 2000). Although the total costs of effectively conserving a representative sample of tropical wild nature are vast—on the order of $16 billion annually for terrestrial reserves in the tropics (James et al. 1999a, 2001)—they represent only a tiny proportion of the global gross domestic product and tax revenues (~0.05% and ~0.2%, respectively; World Bank 2002c). Whereas donor governments

Figure 17.6. Costa Rican forests like this deliver a range of localized and dispersed services. New initiatives are now capturing some of these values and making payments to landholders who retain native-forest cover. (Photo courtesy of J. Kellenberg)

already make major contributions to tropical conservation, there is little evidence that this support has grown significantly since the first commitments to the GEF in the early 1990s (Horta et al. 2002; World Bank 2000, 2002a). The GEF has been a major new source of financing for conservation but its remit has been extended to include land degradation and persistent organic pesticides, with the result that the funds available for conservation and the other original programs are inevitably reduced. Of course, external funding is complemented by sizable cofinancing contributions from the tropical governments themselves, in addition to the provision of recurrent costs (World Bank 2004). Under debt-for-nature swaps, increased conservation investment can relieve developing-country governments of some of their debt repayments (see Spergel 2002 for a recent review), although such initiatives could be undermined by debt-relief programs in the poorest countries.

For donor governments, very sizable increases in conservation funding could be achieved through various means. Although it may appear naïve, we believe it is worth reiterating that all conservation needs could be readily met through only moderate reductions in military spending—a pertinent comparison given that environmental security is likely to be at least as important an issue during the twenty-first century as is national security (Raven 2002). Indeed, a globally effective conservation program could be funded for less than President G. W. Bush's increases in annual U.S. military expenditure alone. Alternatively, the necessary funding could come from eliminating just a small fraction of the almost $1 to $2 trillion currently spent each year on "perverse" subsidies, which simultaneously harm the environment and encourage economic inefficiency (Myers 1998; van Beers and de Moor 1999; Myers and Kent 2001). This would have the added benefit, in many cases, of reducing pressures on remaining habitats and lowering the local opportunity costs of conservation (see above). One other possibility is for northern governments to fund conservation by raising entirely new taxes (Spergel 2002); a recurring suggestion along these lines is a so-called Tobin tax on the approximately $2 trillion traded each day by currency speculators. A tax of 0.1% to 0.25% might help suppress harmful currency speculation while raising perhaps $100 to $300 billion annually for international environmental and poverty-related issues (see http://www.globalpolicy.org/socecon/glotax/currtax/2001/0913uk.htm and http://www.ceedweb.org/iirp/factsheet.htm). An unavoidable challenge, however, for significantly increasing donor-government expenditures on developing-country conservation is that (unlike military spending, perverse subsidies, or even development aid) such a commitment would require the north-south transfer of real financial resources.

The response to this challenge, of course, is to point out that such a transfer is entirely justified, given the very substantial flow of conservation benefits in the opposite direction (fig. 17.3).

It should be noted that all countries are now committed to achieving the eight Millennium Development Goals, which were agreed by the United Nations General Assembly in 2000 (http://www.un.org/millenni umgoals). While only one of these goals refers explicitly to the environment, the Water, Energy, Health, Agriculture, and Biodiversity framework used at the World Summit on Sustainable Development reveals the interaction between biodiversity and the other Millennium Development goals (which cover poverty, hunger, health, water, sanitation, education, and the means of executing development; see http://www.iisd.ca/link ages/sd/sdund). This should mean that biodiversity receives more attention and is "mainstreamed" in the course of development, especially in productive landscapes.

FOUR FINAL CONSIDERATIONS

This review of the practicalities of meeting the present gap in funding tropical conservation raises four other issues that are crucial, but which cannot be dealt with at length here:

1. If the recurrent costs of conservation cannot be adequately met though
    sustainable resource exploitation or nature-based tourism, but instead
    require support for local communities (such as through direct payments
    and other outside support), those contributions must be made on a recur-
    ring, long-term basis. While fixed-term projects are appropriate for build-
    ing a bridge, constructing a dam, or even subduing an outbreak of
    infectious disease, they are entirely inappropriate for conserving nature
    (Terborgh and Boza 2002, building on Wells et al. 1999; see also Janzen
    1986b; Sinclair et al. 2000; Ferraro and Kiss 2002). Ongoing, market-based
    payments offer one avenue for recurrent funding. A second route is via
    the establishment of conservation trust funds, which may pay for conser-
    vation simply from the interest on their endowment, or by drawing down
    a large capital sum over time (see Spergel 2002). Although it can prove
    difficult to attract donors for the heavily front-loaded support that these
    schemes require, they have been successfully established in Uganda,
    Malawi, Bhutan, Bolivia, Brazil, Peru, and Mexico (World Bank 2002a).
2. There is a perceived gap between the funding of conservation NGOs and
    the funding of conservation action on the ground. Both the NGOs and

donors are sensitive to this, and cooperation is growing among the main conservation NGOs to assess the effectiveness of their programs and thereby to improve their delivery. This requires better setting of targets, making meaningful measurements of change, and learning from the results (Salafsky et al. 2002). In addition, a scoring system has been developed to assess the effectiveness of protected areas (Hocking et al. 2000), a version of which is being applied to all World Bank, GEF, and WWF forest/biodiversity projects (Stolton et al. 2003).

3. While we have focused here on increasing the available funds to meet conservation costs, other initiatives are also important. For example, research and dissemination of results on the local delivery of ecosystem services by natural habitats (e.g., Carret and Loyer 2003), and the growing awareness of connections between the major faiths and conservation (Prance 1996; Ramakrishnan et al. 1998; Posey 1999; Goldsmith 2000; Biodiversity Project 2002; Palmer and Finlay 2003), can greatly increase nearby communities' awareness of the imperatives and benefits of conservation. This can offset their perception of the opportunity costs that conservation incurs, and in turn lower the active costs of enforcing conservation regulations (for an encouraging example, see Becker 1999). Likewise, locally relevant environmental education, the growth of local NGOs, and the development of nature-based tourism aimed at local and national communities (helped, for instance, by local-language-speaking field guides and lower park entry fees for nationals) can all increase the nonuse and localized-service values of conservation areas. This in turn can raise in-country support for conservation and thereby lower the active costs of conservation (Bell 1987; Davenport et al. 2002; Dourojeanni 2002).

4. Finally, there is a need to tackle the thorny issues of accountability, institutional weaknesses, and corruption in some host-country institutions— none of which is easily dealt with by money alone (R. J. Smith et al. 2003; W. F. Laurance 2004b; R. J. Smith and Walpole 2005).

CONCLUSIONS AND IMPLICATIONS

1. The gaps in funding tropical conservation are enormous, and the costs of failing to do so are all too clear. It is clear that nations of the developed world could fund effective conservation of the tropics if they chose to make it a high priority.

2. There are encouraging developments in tropical conservation, across several fronts, but an order-of-magnitude gearing-up of current

support is still needed. There is also a need to make better use of present conservation resources to assure donors that their money is making a real difference.

3. Effective conservation will be achieved only through greatly increased funding from developed donor countries, largely via its governments. This must be coupled with greatly increased efforts in working with global, national, and above all, local communities to better quantify, understand, and disseminate the benefits of conservation. This agenda underscores the global benefits of tropical conservation, and our international responsibilities for meeting its costs.

ACKNOWLEDGMENTS

We thank Monique Borgerhoff-Muller, Agi Kiss, Kathy MacKinnon, William Laurance, Norman Myers, Madhu Rao, Matt Walpole, and an anonymous referee for comments, and the editor of *Oryx* for permission to reproduce material from an earlier version of this chapter (Balmford and Whitten 2003).

CHAPTER 18

Conservation Incentive Agreements as an Alternative to Tropical Forest Exploitation

Eduard T. Niesten and Richard E. Rice

The sheer diversity of threats to tropical forests undermines the effectiveness of conservation interventions that focus on individual sources of pressure. Moreover, as contributors to this volume have discussed, new and unanticipated negative environmental impacts continue to emerge as a consequence of human activity in tropical forests. Many responses to anthropogenic habitat loss reflect a "use it or lose it" perspective, as seen in approaches that pursue a balance between low-impact resource use and conservation. Many such approaches, ranging from carbon-sink projects to sustainable forestry to eco-friendly agriculture, may in fact create a new source of pressure rather than reliably mitigate existing threats. Mechanisms to minimize habitat disturbance in general are needed. Such mechanisms must depart from the notion that "use" necessarily involves extraction, cultivation, or other forms of anthropogenic pressure on natural habitat.

The most successful models to date for achieving biodiversity-conservation objectives are formal protected areas, such as national parks, nature reserves, and so on (Bruner et al. 2001). Even in contexts where legal provisions and enforcement capacity are less than robust, national parks have achieved concrete, measurable conservation relative to their immediate surroundings. However, the motivation to search for other conservation tools is strong. Hesitation on the part of developing-country governments to dedicate more land areas for conservation, reluctance to increase the burden on strained environmental protection institutions, and concern over resource access and economic opportunities for local communities all suggest a limit to the land area that will receive formal protection in the tropics, and this limit may fall well below that required to maintain critical reservoirs of biodiversity (Fagan et al., chap. 22 in this volume).

To overcome limits to formally protected land area, practitioners have explored tools such as integrated conservation and development projects, green certification of timber and other forest commodities, and sustainable forest management (Wells et al. 1999; Rice et al. 2001; Hardner and Rice 2002). Despite the commitment of vast resources to these and other approaches, the number and scale of successes have yet to demonstrate robust, replicable conservation results. The challenge remains in presenting to decision makers in countries that sustain substantial shares of global biodiversity viable alternatives for extending direct protection beyond existing formal protected-area systems. In this chapter, we describe an alternative mechanism, conservation incentive agreements, which provide governments and local stakeholders in the tropics with direct incentives to forgo exploitation and development of forested areas. Such agreements comprise an explicit quid-pro-quo arrangement in which an area is zoned for conservation management, and in return, a biodiversity investor provides a structured stream of compensation. Instead of accepting the biodiversity sacrifices inherent in conventional "use it or lose it" approaches, the rationale behind conservation incentive agreements posits that "use" should include payments for the provision of biodiversity and other environmental services through habitat maintenance.

APPROACHES TO PROTECTING TROPICAL FORESTS

The sections that follow describe three approaches within the forestry sphere that have been forwarded as a means to promote biodiversity conservation: carbon sequestration through forestry, sustainable forest management, and agroforestry. Over the past decades, enormous sums of money have been spent on research, policy advocacy, and project implementation for these three strategies, yet there are still reasons to doubt their potential for biodiversity conservation. Indeed, these approaches may contribute to dynamics that, in theory, can create new threats to habitat rather than a solution to biodiversity loss.

Several contributors to this volume (Avissar et al., chap. 4; W. F. Laurance, chap. 5; Fearnside, chap. 19) analyze linkages between deforestation and climate change. Where such analyses demonstrate deforestation to be a contributor to climate change at local, regional, and global scales the conservation response is relatively straightforward; these findings create yet another rationale for measures aimed at slowing or halting deforestation. Analyses that demonstrate the deleterious impact of climate change on tropical forests (Lewis et al., chap. 1 in this volume; Williams and Hilbert, chap. 2 in this volume) reinforce this rationale, insofar as forest conservation can contribute to global efforts to mitigate climate change.

The most ambitious effort to date in the arena of climate change has been the Kyoto Protocol process, in which industrialized nations (the so-called Annex I countries) committed themselves to reductions in annual greenhouse-gas emissions relative to their baseline emissions in 1990. The protocol explicitly acknowledges the role of forest in provisions regarding Land Use, Land Use Change, and Forestry (LULUCF). Article 12 of the Kyoto Protocol establishes the Clean Development Mechanism (CDM), intended to facilitate developing-country participation in efforts to stem global warming. In principle, the CDM achieves reductions in global greenhouse-gas emissions through investments in environmentally sound development in developing countries. Because of low costs for land and labor, developing countries in the tropics offer some of the lowest-cost carbon-offset opportunities (Hardner et al. 2000). Annex I countries benefit from the low costs of projects in non-Annex countries, while developing countries benefit from increased investment. However, in 2001 the Parties to the Protocol elected to bar forest conservation from the CDM, restricting forest-based projects to afforestation and reforestation (to understand the political undertones driving this decision, see the extensive discussion by Fearnside, chap. 19 in this volume). Project activities designed to slow deforestation remain ineligible for crediting until at least 2013 (but see Santilli et al. [2005] for a proposal to credit earlier reductions in deforestation).

As a result, the Kyoto Protocol has become a new source of potential pressure on forests in the tropics (Niesten et al. 2002). First, by restricting LULUCF investments in developing countries under the CDM to afforestation and reforestation, the Kyoto Protocol raises the possibility that CDM crediting can act as an incentive to clear natural forest to make way for carbon plantations. For example, in February 2000, the Rio Foyel

S.A. timber company sought to replace 4000 ha of native forest with a pine plantation in southern Argentina, in the hope of obtaining future carbon credits. This project was blocked by local authorities, communities, and environmental organizations, but demonstrates how carbon credits can motivate destruction of natural habitat (Biodiversidad en América Latina 2000). Although projects with questionable net carbon benefits should not qualify for crediting under CDM guidelines, and reforestation is only creditable on areas cleared before 1990, the Rio Foyel example suggests that some timber companies may nevertheless seek to exploit a perceived opportunity. Additionally, land that might be suitable for natural-forest restoration now faces competition from plantation efforts subsidized by carbon values. These dynamics raise biodiversity concerns and undermine the carbon-sequestration benefits from plantation efforts.

The second potential impact on native forests in developing countries stems from the decision to credit forest-management activities in industrial (Annex I) countries under Article 3.4 of the Kyoto Protocol. This results in a carbon premium for standing natural forests in industrialized countries that, in effect, increases the opportunity cost of harvesting their timber. Therefore, timber in developing countries becomes relatively cheaper, potentially redirecting timber harvests from industrialized to developing countries. That is, if carbon premiums alter forest-management systems and harvest patterns such that timber supplies from Annex I regions decline, developing regions may respond by increasing their share of global timber markets. Carbon benefits from LULUCF measures in Annex I countries would then be negated by accelerated harvests in developing countries, with substantial biodiversity impacts should these harvests be directed to natural forests. Thus, rather than supporting habitat conservation, the new rules governing carbon-sequestration efforts in the realm of forestry generate incentives that may, in theory, pose a new threat to natural habitats.

Sustainable Forest Management

A second type of conservation intervention that may aggravate rather than slow biodiversity loss is sustainable forest management (SFM). SFM seeks to offer an alternative to conventional logging as widely practiced in the tropics, which involves the rapid extraction of high-value species with little regard for the condition of the remaining stand or long-term wood flows (Rice et al. 2001). SFM generally involves harvesting guidelines designed to encourage the growth of marketable timber and efforts to limit the damage to unharvested trees. Such measures seek to ensure a long-term, renewable supply of timber from a given area, which precludes outright clearing of forested areas. Proponents of SFM as a conservation strategy argue that, outside formally protected areas such as national parks (which are necessarily limited in extent), tropical forests are doomed to agricultural conversion unless they generate revenue through continual exploitation.

The enthusiasm for SFM rests on the conviction that a long-term, sustained flow of timber from an area generates greater profits than rapid, short-term mining of timber resources. However, basic renewable-resource economics demonstrates that this is unlikely to be the case in tropical developing countries that sustain much of the world's biodiversity (Pearce et al. 2000). Simply stated, SFM is financially unattractive because the returns from investments in future timber production commonly are lower than those earned by rapidly harvesting marketable trees and investing the profits elsewhere (Bowles et al. 1998; Gullison et al. 2000). Given slow timber-price appreciation and commercial timber growth over time, the high discount rates that characterize developing countries mean that income from logging in the future is negligible (Fearnside 1989b). The appropriate discount rate for project assessment or financial analysis in most developing countries typically is between 8% and 15% (Gittinger 1982; OECD 1995). Rice et al. (2001) show that assumptions of 3% annual timber-price appreciation and 4% annual timber growth are generous. Under these assumptions, the overall value of a stand increases by slightly more than 7% per year. Thus, typical discount rates in developing countries present profit-maximizing operators with a clear incentive to immediately harvest all commercially viable trees.

Efforts to enhance the financial viability of SFM include a variety of policies, but few of these measures enhance the competitiveness of SFM vis-à-vis conventional logging (Reid and Rice 1997). Instead, many policies that seek to buttress the profitability of SFM benefit all modes of timber extraction (see discussion that follows). As a result, even when they succeed in boosting the profitability of SFM, conventional logging often remains the more attractive option. Thus, many policies intended to facilitate SFM may have the undesired effect of also promoting conventional logging, and thereby increase the possibility of accelerating habitat destruction.

The effort to promote commercial exploitation of a greater number of tropical timber species is one example of such a policy (Toledo 1997). In most tropical forests, only a small number of tree species have established commercial markets. Efforts to expand the range of marketable species seek to make long-term management more attractive (Johnson and Cabarle 1993). However, this measure addresses neither the muted prospects for timber-price appreciation nor the low regeneration rate of commercial timber, and therefore does not encourage long-term management; instead, it may simply encourage rapid harvesting of a greater number of species, thereby intensifying conventional logging pressure. Similarly, policies to reduce waste and inefficiency in logging and timber processing have been proposed as a means to promote broader adoption of SFM (Johnson and Cabarle 1993; Gerwing et al. 1996; Holmes et al. 2000). Even when efforts to enhance processing efficiency lead to greater profitability (which is not necessarily the case), they may amplify incentives to harvest in the short term instead of enhancing relative profitability of future as compared to current harvests.

Compared to conventional logging, SFM may convey some conservation benefits (Johns 1997; Chazdon 1998). However, even SFM causes a major disturbance of the ecosystem and a significant change in habitat composition, with a variety of negative impacts on biodiversity (Frumhoff 1995; Bawa and Seidler 1998). Moreover, like conventional logging, SFM frequently increases the threat of hunting and forest conversion by providing road access to frontier areas, and, even when conducted carefully, logging can increase the intensity and frequency of forest fires (Holdsworth and Uhl 1997; Auzel and Wilkie 2000; Nepstad et al. 2001). Therefore, it is not clear that SFM would achieve a satisfactory level of conservation and biodiversity maintenance even if it were financially viable (fig. 18.1). Indeed, in many areas, the selective harvest of high-value timber that typically takes place in a first cut under conventional logging may be more conducive to forest recovery than the higher-intensity harvests and continual, long-term disturbance under SFM (Horne and Hickey 1991; Howard et al. 1996; Rice et al. 1997).

Agroforestry

Throughout the tropics, cultivation of agricultural commodities drives habitat conversion and biodiversity loss, motivating a search for low-impact cultivation methods (McNeely and Scherr 2001). Agricultural production techniques with reduced environmental impacts fall under the general rubrics of agroforestry (Schroth et al. 2004) and ecoagriculture. In areas where formal protection of land is difficult, environmentally compatible agriculture may preserve some degree of connectivity between protected areas. Because agricultural expansion will inevitably continue

Figure 18.1. Even if sustainable forest management were viable, there are reasons to doubt its benefits for biodiversity. (Photo from Conservation International/Haroldo Castro)

in many regions, efforts to mitigate its environmental impacts through changes in farming and land-use techniques offer a valuable contribution. A role clearly exists for targeted interventions to decelerate and mitigate the impacts of agricultural expansion, and, in many cases, agroforestry is a preferable alternative to clear-cuts and monocultures.

Where habitat conversion already has occurred, the ecological benefits of agroforestry systems, relative to land used for annual crops, cattle pasture, or monoculture plantations, justify efforts to promote agroforestry. Insofar as habitat conversion is inevitable (or already has taken place), agroforestry systems can serve as corridors or buffer zones, with benefits including lower usage of agrochemicals and reduced soil erosion, nutrient leaching, and watershed degradation; and, depending on selection of species included in the system, enhanced nitrogen fixation and carbon sequestration (Schroth et al. 2004). Perhaps the greatest biodiversity benefit of certain agroforestry practices is the potential for stabilizing the spatial proliferation of production systems, thereby creating buffer zones to protect forest borders (Angelsen and Kaimowitz 2004). Any degree to which agroforestry can halt or slow shifting agricultural frontiers is a valuable contribution to biodiversity conservation.

However, despite efforts to minimize its ecological footprint, agriculture in any form constitutes a major manipulation of the natural environment, rather than an alternative to habitat conversion; areas under agroforestry or ecoagriculture harbor far less biodiversity than does natural habitat. Thus, from a conservation perspective, agroforestry systems are a compromise rather than a solution (Terborgh and van Schaik 1997). Although agroforestry initiatives can create corridors or buffer zones in a patchwork of forest and production areas, they nevertheless impose serious disturbances on ecosystems; given a choice, biodiversity protection is better served by intact habitat than the fragmentation inherent in agricultural land mosaics (W. F. Laurance and Bierregaard 1997). Moreover, agroforestry systems may or may not be financially or ecologically sustainable in the medium to long term, and therefore offer uncertain outcomes even where adopted as a conservation strategy (Fearnside 1995a).

In the tropics, conventional agriculture commonly drives deforestation by exhausting soil resources, which in turn creates pressures for farmers to clear additional forest to find non-depleted soils (Angelsen and Kaimowitz 2001). However, agroforestry also can induce a similar dynamic, as in the case where declining productivity of cocoa trees leads farmers to establish new farming plots in pristine forests (Petithuguenin 1995; Ruf 1995). Even if agroforestry supports greater biodiversity than other cultivation systems, the greatest ecological benefits from agroforestry are derived if production is concentrated on already-cleared lands.

Although potentially less detrimental than other forms of land use, agroforestry systems are an environmental disturbance that does not necessarily result in a stable forest frontier (Ruf and Schroth 2004). The robustness of agroforestry systems over time in the face of changes in soil quality, water availability, and weather patterns as well as economic conditions remains an open question. Moreover, the more financially attractive sustainable agroforestry becomes, the stronger the incentives to convert remaining natural habitats to agroforestry (Angelson and Kaimowitz 2004). Fundamentally, agroforestry rewards farmers for increasing physical demands on ecosystems—perhaps less so than under other land uses, but certainly relative to conservation of natural habitat. This dynamic is particularly well illustrated in areas with deteriorating economic prospects for agricultural commodity production, such as coffee-producing regions under conditions of low coffee prices (Sanchez 2002). In such situations, the driver of continued habitat loss is not so much market incentives as a lack of viable exit options for farmers, as many farmers living at tropical forest margins have little choice but to rely on destructive agricultural practices for their survival.

Thus, despite its potential environmental benefits, agroforestry can sometimes become a driver of deforestation. Agroforestry can thus contribute to new and emerging threats such as those described in this volume, including the introduction of diseases, alterations in natural fire regimes, and other biodiversity impacts of sustained human presence in and near forest areas. As is the case for carbon-sequestration projects and sustainable forest management, the potential for achieving habitat conservation through agroforestry is limited, warranting a search for alternatives that provide more direct biodiversity protection.

CONSERVATION INCENTIVE AGREEMENTS

The Concept

The approaches described above reflect the notion that biodiversity conservation must be achieved as a corollary benefit of some other self-financing activity. This reflects a "use it or lose it" perspective for tropical forests. However, the dynamics that characterize human resource use inevitably create potential threats, even if the resource uses are predicated on principles of sustainability and minimizing impacts. Therefore, instead of seeking to minimize the impacts of human activity, the most reliable conservation strategies must minimize human activity itself in areas of conservation interest. Indirect approaches to biodiversity conservation such as carbon sequestration, sustainable forest management, agroforestry, and even ecotourism and integrated conservation and development projects, fail to address the root cause of biodiversity loss, namely the missing market for biodiversity values (Ferraro and Kiss 2002; Niesten et al. 2004a). Conservation incentive agreements seek to address this missing market by providing direct compensation in return for biodiversity conservation services (Hardner and Rice 2002). These agreements expand the notion of "use" to include such services, and avoid the tension between conservation and habitat pressure that characterizes indirect approaches.

A conservation incentive agreement directly compensates local stakeholders and relevant government bodies for conservation services. National resource authorities and local resource users forgo destructive exploitation of areas of habitat in return for a steady stream of structured investments, under a negotiated agreement between an investor, the host government, and local stakeholders. Thus, the agreement comprises an explicit quid pro quo of regularly scheduled payments in return for conservation performance based on measurable indicators. The range of parties to the agreement depends on the context. If the area to be conserved is privately owned, the agreement may involve only the conservation investor and the landowner. Government-owned land clearly requires an agreement with the relevant ministries and departments. Areas under traditional or customary land-tenure systems require agreements with indigenous communities. Often, areas of conservation interest fall under several overlapping ownership or use-right arrangements, indicating the need for thorough stakeholder analysis and consultation (Fearnside 2003a).

Negotiated elements of conservation incentive agreements include the amount of compensation for protection of the area in question, the investment portfolio where these payments will be directed, and provisions for the cost of management activities such as monitoring and enforcing natural resource protection. By thwarting the principal threats posed by destructive large-scale commercial logging and conversion to agriculture, and working with communities to rationalize subsistence resource use and possibly non-destructive, small-scale commercial extraction, a conservation incentive agreement preserves rights for use that are held, officially or by custom, by local stakeholders.

The conservation incentive agreement addresses the fact that many of the areas of interest for biodiversity conservation in the tropics are government lands in developing countries, often designated for industrial logging and conversion to commercial agriculture. This approach avoids or minimizes the need for new legislation by requiring similar agreements and contracts to those used to establish leases for logging, mining, or other resource development (Niesten and Rice 2004). Because land leases and resource concessions are in use all over the world, conservation incentive agreements are not a radical departure from the kinds of transactions with which governments are well acquainted. However, the fact that provision of biodiversity services can generate income directly through environmental service payments does represent a marked departure from indirect approaches in which biodiversity conservation is sought as a byproduct of other activities.

Direct payments under conservation incentive agreements may be extremely attractive to governments and local communities relative to extractive land uses such as logging or agriculture (Niesten et al. 2004b). Although extractive activities produce economic benefits (including employment, income, export earnings in foreign currency, and public tax revenues), for a broad range of resource-based products, economic prospects look less than promising and are highly sensitive to factors ranging from international market conditions to capricious local weather patterns. Additionally, government revenues based on taxing resource users are vulnerable to corruption and ubiquitous efforts to avoid taxation (W. F. Laurance 2004b). In contrast, a conservation incentive agreement can offer regularly scheduled, risk-free payments, denominated in stable foreign currency such as U.S. dollars. The underlying objective of a conservation incentive agreement is long-term habitat maintenance. Nevertheless, from the perspective of a host government, the expiration of an agreement's term presents an opportunity to revisit the issue of the best use of the area in question. Renegotiation and extension of the agreement may present an attractive option, as conservation incentive agreements can offer substantial, secure revenue potential for the host government and local stakeholders.

The benefits provided through conservation incentive agreements must outweigh returns from alternative uses of the target area. Where appropriate, this is accomplished by investing payments in economic activities that will provide alternative jobs and improve human welfare. Negotiated terms of a conservation incentive agreement can include, for example, the portfolio of activities to which annual payments will be directed. The conservation investor can supplement payments with health or education investments to benefit local stakeholders, particularly in remote communities that often lie beyond the effective reach of government services. For instance, a conservation incentive agreement might include bonuses to persuade teachers to serve in communities that face difficulties attracting qualified educators. Alternatively, payments could take the form of subsidies for medical supplies that otherwise would be inordinately expensive. The flexibility inherent in conservation incentive agreements allows them to be structured to include compensation that is appropriate at the community level, and thereby to generate trust and support for the agreement among local stakeholders.

An Example

A critical feature of direct incentives for conservation is that, unlike SFM or agroforestry, income no longer relies on habitat modification and natural resource extraction, but instead becomes a function of successful conservation (Ferraro and Kiss 2002). Perhaps the most basic illustration of the approach is a conservation concession that pays local stakeholders to desist from forest-clearing activity, and instead remunerates them for monitoring and enforcing habitat protection. In July 2002, Conservation International (CI) concluded such an agreement with the Government of Guyana. The process leading to this agreement began 2 years earlier, when the government issued to CI an exploratory permit for an area of roughly 80,000 ha in southern Guyana. This exploratory permit paralleled the first step taken by logging companies when applying for a Timber Sales Agreement in Guyana. During the exploratory phase, CI conducted a social impact assessment to determine the likely effects of the proposed concession on nearby communities, and developed a management plan specifying activities such as boundary demarcation, training of forest rangers, and collaborative engagement with local stakeholders. The social impact assessment and management plan formed the basis of the final application for the concession. Pursuant to this application, the government issued to CI a renewable 30-year lease over the area. Under the terms of the lease, CI pays the government annual acreage fees and royalties equal to those payable by timber concessionaires, amounting to about $30,000 and $11,000, respectively (Guyana Chronicle 2002).

The Guyana concession agreement also includes voluntary annual investments of $10,000 in community development projects, benefiting a population of about 670 people in three communities living near the concession. Outlays of the community investment fund are decided on an annual basis, and can target such things as health care, education, or materials purchases to support local economic activity. The concession is in a relatively remote, sparsely populated area that is not under immediate threat. Thus, this site serves as a pilot to explore the mechanics of designing and negotiating an agreement in a setting with low opportunity cost and with little risk of population pressure from in-migration. However, rapid infrastructure development on the Brazilian side of the Brazil-Guyana border, only 50 km away, suggests that logging pressure may intensify in the future in the absence of an alternative means to generate income and local benefits, as provided by the conservation concession.

Future Prospects and Challenges

The lack of market values for environmental goods and services remains the most serious threat to the world's dwindling biodiversity resources. Direct incentives for conservation, in the form of compensation in return for environmental services, offer the most promising means to address this missing market (see Whitten and Balmford, chap. 17 in this volume). For example, Gibbs (2001) recommends the purchase of large tracts of land as a habitat conservation strategy; in contexts where outright purchase is not possible, lease arrangements such as conservation incentive agreements offer an alternative. The World Bank (2002b) identifies the need for a mechanism to channel payments for conservation services, and

Nasi et al. (2002) similarly call for an international transfer mechanism to directly compensate countries that provide forest-ecosystem services through habitat protection. In a review of World Bank involvement in forestry efforts in developing countries, Lele (2002) also conclude that habitat conservation in forest-rich countries such as Brazil and Indonesia will require payments from developed countries for biodiversity services in the tropics. Deals like the conservation concession agreement concluded between Conservation International and the Government of Guyana in 2002 illustrate one promising mechanism for such payments.

Conservation incentive agreements require the determination of an appropriate level of compensation. The value of forgone extractive activity provides one indication. However, several other factors also should be considered in addition to this opportunity cost (Fearnside 1997a). For example, in the case of forgone logging, the economic value to the country of maintaining an intact ecosystem should be offset against the forgone logging revenues, as land clearing can increase soil erosion and vulnerability to flooding, and have other environmental costs. The employment impact of forgoing logging is also offset by new employment opportunities created in the conservation management sector; moreover, jobs created in the conservation management sector are likely to last substantially longer than those in the logging industry (fig. 18.2). Additionally, the host government

Figure 18.2. Conservation incentive agreements need to address the issue of forgone employment, but jobs in the logging sector are likely to be temporary, given the nature of the industry. (Photo from Conservation International/James D. Nations)

benefits when the compensation package includes investments in social services for local communities in or near the target area, because these investments relieve the government of the burden of providing such services. These are just some of the reasons that the appropriate level of compensation is not necessarily equal to the forgone profits from destructive exploitation.

In essence, conservation incentive agreements involve a payment for environmental services. As noted, the economic value of those environmental services may be a consideration when negotiating the appropriate level of this payment. However, just as compensation levels are not necessarily equal to forgone profits from extractive activity, the appropriate amount also is not necessarily equal to the full value of all environmental services. All that is needed is that the resource owners are made better off by participating in the conservation agreement than by engaging in destructive activity. In other words, once an area has been identified as a conservation priority, a conservation incentive agreement can be designed to offset the benefits forgone by desisting from destructive activity regardless of the economic value of environmental services. Indeed, some areas identified as priorities for biodiversity conservation may not be immediately threatened by large-scale destruction, in which case they carry only a low opportunity cost (Fearnside 1999a). In such areas, a conservation incentive agreement can be a low-cost vehicle for applying formal conservation status to the area to buttress enhanced management and secure it against future threats.

One concern that arises regarding conservation incentive agreements is whether directly competing against destructive exploitation will be inordinately expensive. There are at least two reasons why this concern is misplaced. First, in many areas the approach promises to be extremely cost effective. In a portfolio of conservation incentive projects that currently are at varying stages of implementation, the sizes of the areas in question range from 5000 to nearly 1,000,000 ha. The projected annual costs per hectare for these projects range from $0.05 to $15.00 per year. Weighted by size of the areas to be conserved, the average annual per hectare cost is about $1.30. Thus, while some projects may indeed carry dauntingly high recurrent annual costs, we believe that the average project will offer a readily affordable conservation opportunity.

Second, although the extent of global willingness to pay for conservation services is difficult to measure, the global community spends well over $1 billion each year on biodiversity conservation. Official development agencies spend on the order of $500 million per year on projects with a biodiversity conservation component (Hardner et al. 2000). Over the past decade, the Global Environment Facility has disbursed $4 billion, and leveraged an additional $12 billion, for environmental projects around the globe (GEF 2002). The annual budgets of international conservation NGOs reach a combined total in excess of $1 billion. Thus, willingness to pay (as indicated by funding currently allocated to conservation) is massive, and conservation incentive agreements are well within the means of existing budgets. Moreover, the introduction of a new conservation tool that can achieve clear, verifiable conservation results in terms of habitat protected may unleash new sources of demand, and thus funding, for biodiversity conservation. Indeed, many contexts may call for a combination of conservation approaches, and adding conservation incentive agreements to such combinations may increase the overall level of available funding while enhancing the probability of success.

Finally, the perception that other approaches are less expensive may in fact not be correct. For example, if conservation advocates persuade a developing country in the tropics to establish a national park, this does not mean that habitat protection has been achieved at no cost. Advocacy itself can be an expensive means of achieving biodiversity protection. Moreover, in the absence of financial support from the international community (which spends more than $1 billion per year on conservation), the opportunity cost of conservation and the burden of recurring costs of habitat management are simply shifted to the host country (which, in much of the tropics, is struggling to raise its citizens' standard of living with limited financial resources and therefore leaves protected areas unfunded). Similarly, the costs of designing, contracting, implementing, and monitoring carbon-sequestration projects, sustainable forest management, or agroforestry initiatives are also substantial.

CONCLUSIONS AND IMPLICATIONS

1. For several decades, conservationists, governments, and donors have sought ways to enable biodiversity conservation to pay for itself. Initiatives such as ecotourism, integrated conservation and development projects, and sustainable forest management are some of the results of this quest, but have proved disappointing both in terms of their ability to generate income and their success in achieving lasting conservation on a meaningful scale.

2. Efforts to achieve biodiversity conservation through carbon-sequestration projects, sustainable forest management, and agroforestry have consumed enormous amounts of conservation funding, without achieving commensurate conservation results. Moreover, by perpetuating

some anthropogenic pressure on forests, these approaches may, in theory, constitute or promote new threats to natural ecosystems.

3. Conservation incentive agreements explicitly accept that conservation, in many situations, *cannot* pay for itself. If natural habitat is to survive in the developing countries where most of the world's biodiversity resides, then the global community will have to channel funds to those governments and communities that are prepared to provide biodiversity services by setting aside conservation lands. Conservation incentive agreements offer a promising mechanism for such payments.

Mitigation of Climatic Change in the Amazon

Philip M. Fearnside

AMAZONIAN FORESTS AND CLIMATE CHANGE

Mitigation of climatic change in tropical forests has become one of the most controversial subjects in conservation. National governments and nongovernmental organizations (NGOs) have taken varying positions on mitigation measures such as planting trees and avoiding deforestation. These positions have also changed over time, sometimes abruptly, in response to political agendas. As I will argue here for Brazilian Amazonia (fig. 19.1), avoided deforestation has by far the greatest potential both for climatic benefits and for achieving other environmental objectives such as maintenance of biodiversity.

A clear distinction must be made between funding motivated by biodiversity concerns and that motivated by climatic-change mitigation. In this chapter, I argue that mitigation through avoided deforestation (which is entirely justifiable solely on the basis of society's willingness to pay for climatic benefits) can play an important role in maintaining Amazonian biodiversity—not that the much smaller pool of biodiversity funding should be hijacked for the benefit of climate-mitigation efforts.

The opportunity for climate mitigation to counter the powerful economic forces that threaten Amazonian forests lies in the much greater willingness of interested parties at present to pay for avoiding climate change as compared to avoiding biodiversity loss. As a binding international agreement, funding for climate-change mitigation via the Kyoto

Figure 19.1. Brazilian Amazonia and surrounding areas, with locations mentioned in the text

Protocol is expected to be much larger than could reasonably be expected from voluntary "public-relations" carbon projects financed by the private sector. For example, planners in the 1993–2001 Clinton administration were expecting that, over the 2008–2012 period, the United States would spend US$8 billion annually on purchasing carbon credits (J. Seabright, public statement, Brazil/U.S. Aspen Global Forum on the Kyoto Accords, Colorado, 9–11 October 1998). At that time, prior to the G. W. Bush presidency, the United States was expected to represent about half of the global carbon market in the 2008–2012 period. While the withdrawal of the Bush administration from Kyoto negotiations for the 2008–2012 period greatly reduces the potential carbon market on that time scale (as does the 2001 Bonn agreement, which eliminates avoided deforestation as a mitigation measure from 2008–2012), the magnitude of potential monetary flows on a longer time scale makes mitigation a major opportunity for conservation. Negotiation of key decisions at least got off to a symbolic start when the conference of the parties to the United Nations Framework Convention on Climate Change (UN-FCCC) met in Montreal in December 2005. The issue is therefore a very current one, and the decisions to be made cannot be taken for granted.

MITIGATION AND ADAPTATION STRATEGIES

"Mitigation" refers to measures to reduce the amount of climate change, as distinguished from "adaptation," which refers to protecting, moving, or changing human and natural systems to accommodate climatic changes with a minimum of disruption. Global warming is a major worldwide concern caused by net emissions of greenhouse gases such as carbon dioxide (CO_2), methane (CH_4) and nitrous oxide (N_2O). Most emissions come from burning of fossil fuels, but about 30% come from land-use change in the tropics, especially deforestation (Fearnside 2000b). Land-use changes release greenhouse gases from burning and from decay of biomass, as well as from soil changes, cattle, and hydroelectric dams.

In addition to its impact on global warming, deforestation also provokes climate change by diminishing the supply of water vapor from evapotranspiration, thereby reducing rainfall in Amazonia and in the heavily populated central-south portion of Brazil (Salati and Vose 1984; Fearnside 2004a; Marengo et al., 2004). Also, changes in the boundary layer above deforested areas in Amazonia can produce teleconnections that reduce summer rainfall in North America and elsewhere (Avissar et al., chap. 4 in this volume). Furthermore, aerosols in the smoke released by biomass burning impede rainfall formation by providing an excessive number of cloud-condensation nuclei, thereby forming water droplets that are too small to fall to the ground as rain (Rosenfeld 1999). Reduction of deforestation therefore mitigates a variety of climatic changes by avoiding atmospheric emissions and other land-use-change impacts.

In addition to avoiding deforestation, global warming can also be mitigated by planting trees in areas without trees. Atmospheric carbon is sequestered by being incorporated into tree biomass, and, depending on whether the wood is harvested and what products are derived, the carbon is maintained out of the atmosphere for variable amounts of time. Unfortunately, in current discussions about mitigation measures, a variety of land-use options have been lumped into the term "sinks," including temporary sequestration of carbon in biomass and wood products, permanent displacement of fossil carbon by substitution of coal or oil with wood or charcoal, and avoided emissions from slowing deforestation. Most criticism of "sinks" focuses on the first of these categories, silvicultural plantations.

TROPICAL FORESTS IN THE KYOTO PROTOCOL

Under the Kyoto Protocol, or under any alternative agreement that may take its place, credit for reducing emissions of greenhouse gases will be sold between nations, such that at least part of the emissions in highly industrialized parts of the world can be offset by reductions achieved at lower cost (and often also with greater collateral benefits) in other parts of the world. One way of reducing emissions that could generate such credits for sale is by reducing deforestation in Amazonia.

Avoided deforestation can be achieved in various ways. One way is at the project level, where specific activities can be shown to restrain or discourage clearing in an area. Protected-area establishment and defense is one type of such project, while efforts to implant licensing and inspection programs are another. Project-level approaches are subject to varying degrees of "leakage," or the negation of project benefits by changes that the project induces outside of its defined boundaries. Projects must include measures to minimize these effects, and to quantify and correct credit allocations for those that remain. Another set of options applies to programs, usually at the national level, rather than to individual projects (Fearnside 1995b). These measures are independent of most leakage effects, as any movement of deforestation activity that individual measures provoke within a country will not affect emissions totals at the national level. Program-level measures also escape the difficult task of showing a causal link to specific project activities, thereby greatly increasing the amount of credit that can be claimed. The downside is that these options require national-level emissions commitments, but, as will be discussed later, Brazil's national interests could be best served by embracing such a commitment and exploiting its advantages for much larger amounts of credit for avoided deforestation.

Prior to negotiation of the Kyoto Protocol in December 1997 (UN-FCCC 1997), there was wide agreement that both planting trees and avoiding deforestation were important measures in the fight against the greenhouse effect. In 1989, the German Parliament (Bundestag) held a series of hearings on tropical forests and global climate change (in which I twice testified), and produced a report that identified slowing tropical deforestation as a key priority for reducing global warming (Deutscher

Bundestag 1990). In 1992 a major new initiative, the Pilot Program to Conserve the Brazilian Rainforest (abbreviated as "PP-G7"), was approved by the G-7 industrial countries. Reducing emissions of greenhouse gases from deforestation was one of its principal purposes (e.g., World Bank 1992; Brazil MMA 2003). I served as a member of the PP-G7's International Advisory Group for 9 years (1993–2001), during which the G-7 countries donated more than US$250 million to the program; by far the largest contributions were made by Germany, followed by the United Kingdom. Major European environmental NGOs such as Greenpeace (Leggett 1990) and Friends of the Earth–UK (Myers 1989) published reports in which both planting trees and reducing tropical deforestation were forwarded as high priorities in the fight against global warming.

However, soon after the Kyoto Protocol was signed in December 1997, the European governments and European-headquartered NGOs abruptly turned against all forms of "sinks," including avoiding tropical deforestation. This anti-sink stance stemmed from a circumstance unique to the Kyoto Protocol's first commitment period (2008–2012), for which the emissions quota for each of the industrialized countries was fixed at the time of the Kyoto conference in 1997—before key decisions had been made, such as whether projects to reduce deforestation in tropical countries would be included in the Clean Development Mechanism (CDM). This circumstance presented a one-time opportunity for European countries to advance other agendas related to competition with the United States, where fossil fuel prices are approximately half of those in Europe (see Fearnside 2001d). If the doors could be effectively closed to purchase of significant quantities of carbon credits from projects in developing countries, then the United States would be forced to sharply increase its domestic fossil fuel prices in order to reduce emissions to the quota agreed in Kyoto, thereby leveling the competitive playing field with Europe.

A parallel logic underlay the attraction of European NGO members to opposing "sinks": resentment of the United States for its various sins in the world, including that country's role as the largest single emitter of greenhouse gases and its repeated obstruction of progress in climate negotiations. Environmental NGOs headquartered in Europe, such as Greenpeace, Friends of the Earth, World Wide Fund for Nature, and Birdlife International, split sharply over the issue with those headquartered in North America, such as Environmental Defense, Conservation International, The Nature Conservancy, the Natural Resources Defense Council, and the Union of Concerned Scientists. In Brazil, credit for avoided deforestation was supported by both grassroots and research

NGOs, including the Amazonian Working Group, the National Council of Rubbertappers, the Coordinating Body of Indigenous Peoples of Brazilian Amazonia, the Federation of Agricultural Workers, the Pastoral Land Commission, the Institute for Man and the Environment in Amazonia, the Institute of Environmental Research of Amazonia, and the Socio-Environmental Institute (Fearnside 2001b, 2001d). In both the United States and Brazil, branches and affiliates of European-headquartered NGOs, such as Greenpeace, Friends of the Earth, and the World Wide Fund for Nature, followed the European line in opposing credit for forests, with one important exception: Friends of the Earth–Brazilian Amazonia (e.g., Monzoni et al. 2000).

In the wake of the stunning withdrawal by U.S. President George W. Bush on 13 March 2001 from negotiations for the Kyoto Protocol's first commitment period (2008–2012), an agreement was reached by the remaining countries in Bonn in July 2001, ruling out avoided deforestation for credit (which could formerly have occurred under the Clean Development Mechanism [CDM] in the first commitment period; UN-FCCC 2001). Despite disagreements, prospects are much improved for agreement on avoided deforestation as a mitigation measure under the CDM beginning with the Protocol's second commitment period (2013–2017), with negotiations that began in 2005 (e.g., see W. F. Laurance 2006a, 2006b). Inclusion of forest is likely because the underlying motivation of the European opposition does not apply to the second commitment period, as the emissions quota for each country in the second period has yet to be negotiated. Because the net reduction in emissions to which each country's negotiators will agree is limited by the cost they foresee as needed to achieve the target, the existence of relatively inexpensive means of compliance means that negotiators will agree to deeper cuts in net emissions. Allowing a large source of low-cost credit from avoided deforestation therefore means that the countries will agree to reduce their emissions by more, and if forests are excluded the countries will simply agree to reduce by less.

Brazil's official opposition to crediting avoided deforestation stems from a completely different logic from that of the Europeans. Unlike the Europeans, who opposed all "sinks," the negotiating position of the Brazilian Ministry of Foreign Affairs was to oppose credit for avoided deforestation, but to argue for approval of credit for silvicultural plantations. The opposition to credit for avoided deforestation stemmed from a fear among Brazilian diplomats that deforestation is uncontrollable, and that Brazil could become subject to pressures that would jeopardize its sovereignty in Amazonia, if carbon credit were accepted and the country subsequently failed to control deforestation (Fearnside 2001c; see also Council on Foreign Relations Independent Task Force 2001). The individualistic nature of these opinions is clear from the generalized support for carbon credit in other parts of Brazilian society, including all of the state governments in the Amazon region (see IPAM 2000; Fearnside 2001d) and the environment minister, Marina Silva. Indeed, the former governor of the Brazilian state of Amazonas, Amazonino Mendes, even traveled to Chicago to attempt to negotiate sale of carbon benefits on the Chicago Board of Trade (*Amazonas em Tempo* 1999)—a gesture that is particularly telling given that the sovereignty concerns of Brazilian diplomats are the major obstacle to the country's adopting a favorable position on crediting avoided deforestation, and that Amazonino Mendes has long behaved as a vociferous defender of Amazonia against "foreign threats."

MITIGATION ACTIVITIES FOR AMAZONIA

Here I review six major types of climate-change mitigation activities that have been proposed for Amazonia, examining the pros and cons in economic, social, and political terms, as well as their value as global-warming countermeasures, and their conservation implications. These six options— plantations, agroforestry, soil sequestration, forest management, hydroelectric dams, and avoided deforestation—are summarized in table 19.1. It should be emphasized that the area to which a given option might expand (and consequently its potential contribution to mitigating global warming) is limited not only by Brazil's land area but also by the need for maintaining adequate areas in other uses, including food production.

Plantations

Brazil has one of the world's largest areas of silvicultural plantations—about 5 million hectares in 2000, mainly of *Eucalyptus* species (FAO 2001a). The country has been a leading diplomatic force in pushing for plantation expansion as a global-warming mitigation measure, beginning with the FLORAM Project (Ab Sáber et al. 1990) that foresaw an additional 20 million hectares of plantations for carbon in Brazil (mostly outside of Amazonia but also including two areas in the region, the Carajás railway in Pará and Maranhão and the former AMCEL/Champion plantations in Amapá). Although plans under the Clean Development Mechanism are more modest in scale, they could provide a significant impetus for expansion of plantations in Brazil (Meyers et al. 2000; Seroa da Motta and Ferraz 2000). The

Table 19.1 Comparison of mitigation options in Brazilian Amazonia

Mitigation option	Magnitude of potential climate benefit	Magnitude of financial costs per ton of carbon	Types of social and political costs and benefits
Silvicultural plantations For pulp	Modest due to short-term nature of sequestration	Relatively high cost per ton of carbon equivalent to permanent sequestration	Employment
For sawnwood	High on long term from logging displacement	High at present due to competition from low-cost timber from Amazonian forests	Employment
For charcoal	High, due to permanent nature of carbon displacement and Brazil's very large high-grade iron deposits	Low	Potential for strong negative impacts due to traditions of debt slavery and child labor in charcoal making
Agroforestry	Modest due to market limits	Moderate	Social benefits for small farmers
Soil sequestration Terra preta (black earth)	Substantial	Unknown	Social benefits if for small farmers
No-till agriculture	Low due to little additionality	Very uncertain: cost is low per hectare because no-till is often profitable in its own right; for same reason little carbon is additional	Mostly for large soy farmers
Pasture management	Low: despite large areas of pasture, results are slow limit extent to smaller areas	High, especially if time value given to carbon	Mostly for large ranchers

Table 19.1 Comparison of mitigation options in Brazilian Amazonia *(cont.)*

Mitigation option	Magnitude of potential climate benefit	Magnitude of financial costs per ton of carbon	Types of social and political costs and benefits
Charcoal amendments	High due to large areas of degraded pasture in which amendments might be applied for various land uses, although phosphates limit extent of pasture recuperation	Unknown	Low social benefits if for secondary forget (e.g., Carajás proposal)
Forest management			
Wood product sequestration	Very low or negative	Infinite cost if benefits zero or negative	Neutral
Reduced-impact logging	Substantial	Low (if all carbon is considered additional)	Modest benefits; little to small landholders
Avoided logging	High: current annual emissions approximately 60 million tons of carbon	Low	High loss on site; if substitution from plantations, then supply generated elsewhere
Hydroelectric dams	Much smaller than officially recognized in Brazil	Relatively high	Very high impacts
Avoided deforestation	Very large: current annual emissions approximately 450 million tons of carbon	Low financial cost (costs are political)	Beneficial; political cost in slowing deforestation

Bonn agreement of July 2001 allows credit under the CDM for plantations (afforestation and reforestation). Much of the future expansion of Brazil's plantations is likely to occur in Amazonia (Fearnside 1998, 1999e).

A recent initiative to plant 30,000 ha of *Acacia mangium* pulpwood plantations in Roraima (the Ouro Verde project) includes obtaining carbon credit under the CDM as a long-term goal (STCP 2002). Now approved for credit under the CDM is the 23,000 ha PLANTAR project in the non-Amazonian state of Minas Gerais. The PLANTAR project (PLANTAR 2003) would produce pig iron using charcoal from *Eucalyptus* plantations, although the claimed amount of climate benefit has been questioned (Van Vliet et al. 2003). Smelting pig iron with charcoal has long occurred in the Carajás area (Fearnside 1989a) and proposals for obtaining carbon credit for this activity continue to evolve. By replacing a fossil fuel (mineral coal and coke), charcoal use in smelting accumulates climatic benefits by permanently displacing fossil carbon, in contrast to pulpwood plantations where carbon in biomass and wood products returns to the atmosphere after a temporary period of sequestration (Fearnside 1995b). Displacement of fossil-fuel carbon is considered permanent because the avoided emission cascades forward in time: a ton of fuel not burned this year will be burned next year instead, the ton that would have been burned in year 2 passes to year 3, and so forth. Note, however, that some have argued that fossil-fuel displacement is not permanent because it lowers the cost of future extraction, thereby encouraging future use (Herzog et al. 2003). I have argued that, with more than 5 trillion tons of available fossil carbon on Earth, use will ultimately be limited by environmental (climatic) impacts rather than by extraction cost or physical availability (Fearnside 1995b). The Kyoto Protocol considers fossil-fuel displacement to be permanent.

Social impacts are a significant concern in promoting expansion of Brazil's charcoal industry, which is notorious for the degrading conditions under which the workers live, including "debt slavery," where families eternally indebted to patrons are not free to leave the charcoal-making camps (Sutton 1994; Fearnside 1996c, 1999c). Unfortunately, the requirement of "sustainable development" for CDM projects (under Article 12 of the Kyoto Protocol) has been interpreted to be the province of each host country to define (rather than being subject to an internationally standard set of criteria for what constitutes "sustainable development"). Countries can therefore obtain carbon credit for afforestation and reforestation projects that would not meet standards of bodies such as the Forest Stewardship Council, which must conform to international labor conventions whether or not the country in question has ratified those conventions.

Plantations in Brazil are primarily for pulpwood, followed by charcoal production. Longer-cycle plantations for sawnwood are rare. Wood production from plantations therefore does not displace logging in Amazonian forests. Brazil uses wood from Amazonian forests for virtually everything, including concrete forms, pallets, crates, plywood, and particleboard (Smeraldi and Veríssimo 1999). As long as Amazonian wood is available essentially for free, with only the cost of harvest and transportation to pay, one cannot expect to supply these products from plantations. The transition to plantation-wood sources will eventually occur, and it is in the country's advantage to provide mechanisms to achieve that transition while Amazonian forests remain standing, rather than waiting for resource depletion to work through "natural" market forces. Reducing Amazon logging, and the associated reductions in deforestation and surface fires, would have important climate-mitigation benefits.

Agroforestry

Agroforestry, or the combination of planted trees with annual crops, has many environmental and social advantages over predominant land uses in Amazonia, such as cattle pasture (Fearnside 1990c; Schroth et al. 2004). Provided that forest is not cut to make way for the agroforestry, trees in agroforestry systems will hold more carbon than would the vegetation otherwise occupying the site (S. Brown et al. 2000a). Avoided-deforestation benefits are sometimes also claimed, but great caution is needed to avoid exaggeration of these benefits (Fearnside 1999d). Market limits on the products of agroforestry systems, together with other limits, make expansion of these systems unlikely to significantly reduce the vast areas of degraded lands already present in Amazonia (Fearnside 1995a).

An initiative to subsidize agroforestry for its environmental services, particularly carbon benefits, is the PROAMBIENTE project (Mattos et al. 2001). This project would use funds (from Brazil's National Bank for Economic and Social Development) to finance small farmers, beginning with 13 pilot sites distributed among the Amazon region's nine states. Two arrangements are offered, one providing loans followed by payments for environmental services as determined by monitoring, and the other providing only the payments without loans. The creation of banking and organizational arrangements for integrating small farmers into carbon markets has already shown itself to be effective in stimulating agroforestry in Costa Rica and Mexico (Segura and Kindergard 2001; Nelson and de Jong 2003).

Managed secondary forest can provide a variety of products with lower labor and financial investment than agroforestry, provided that land is cheap. Plans for this type of management in degraded pasture lands along the Carajás Railway have been drawn up by the Companhia Vale do Rio Doce mining company. The plan is to use fertilization with powdered charcoal both to sequester carbon in soil and to increase the rate of biomass accumulation. The charcoal fertilization is based on very promising results with annual crops in experiments near Manaus (Glaser et al. 2002), although similar experiments with secondary forest trees have not yet been conducted.

Soil Sequestration

Sequestration of carbon in soil through changes in management has considerable potential for climate-change mitigation (Sombroek et al. 1993; Batjes and Sombroek 1997; Batjes 1998). However, the spatial extent of feasible management changes and the per-hectare benefits that these changes can provide are both limited. Because Brazilian Amazonia has an area roughly the size of France in cattle pasture (most of which has very low and declining productivity), the possibility of "recuperating" these vast areas through fertilization has often been raised (e.g., Serrão and Toledo 1990). The finite nature of phosphate deposits in Brazil (and the world) poses limits on this, as do market forces (Fearnside 2002b). In addition, many claims of increased carbon stocks in pasture soil are exaggerated by a failure to account for soil compaction (increased bulk density) under pasture. This compaction makes soil-carbon density appear to be greater than is actually the case, when the pasture soils are compared to natural soils in forests (see Fearnside and Barbosa 1998).

No-till agriculture (direct planting) maintains more soil carbon than does traditional tilling. Because no-till methods are often adopted on the basis of lower cost and greater profitability, independent of carbon benefits, meeting the CDM's requirement for "additionality" (demonstration that the carbon benefits claimed would not have occurred in a baseline no-project scenario) could be difficult. The primary focus of no-till agriculture is soybeans, which are rapidly advancing into Amazonia (Fearnside 2001e).

Soil amendments for agriculture and for recuperation of degraded areas often include application of lime. Use of lime, either as limestone ($CaCO_3$) or dolomite ($CaMg[CO_3]_2$), releases carbon dioxide when added to acid soil. These emissions must be considered in assessing the net benefit of the recuperation program. The same is true for carbon stocks in biomass, including underground biomass. Carbon losses from removal of woody plants, for example in recuperating degraded pastures, must be counted in assessing net benefits.

Carbon can be stored in soil in the form of charcoal (which decomposes very slowly), rather than increased organic carbon, which is the focus of most soil-carbon sequestration initiatives. Recent experiments showing dramatic yield increases when powdered charcoal is included as a soil amendment (along with modest amounts of fertilizer) have led to considering this as a part of soil-improvement proposals under the Terra Preta Nova project (Glaser et al. 2002; Sombroek et al. 2002). This initiative hopes to recreate the anthropogenic black earths (*terra preta do índio*) that modern inhabitants of Amazonia have inherited from pre-Colombian indigenous populations. The patches of black earth that dot the region today contain much more organic matter than do other soils, in addition to containing black carbon (charcoal). Artificial establishment of black earth offers the hope of more productive and sustainable agriculture and agroforestry, in addition to its climate-mitigation potential (Sombroek et al. 2003).

Forest Management

Sustainable forest management has often been suggested as a form of carbon sequestration because carbon in wood that is converted to long-lived wood products, such as fine hardwood furniture and construction timber, remains out of the atmosphere while the trees in the harvested location regrow and accumulate more carbon (e.g., Myers 1989). However, the fraction of the carbon stock that actually ends up in long-term products is miniscule, and short-term releases of much larger amounts of carbon take place from decay of the slash, stumps, and roots and from the many unharvested trees that are damaged or killed during the logging process. These losses more than outweigh the wood-product pools for many decades, and any value attached to time completely negates the very slow rise of carbon stocks in long-term products that can result from forest management (Fearnside 1995b). (Valuing time, most commonly done by applying a discount rate, converts the value of future costs and benefits to their present-day equivalents for the purpose of comparisons in decision making; invariably, future events have less weight than current ones, but the appropriate weighting is a matter of controversy; Fearnside 2002e.)

An additional key concern is that logging, even as a part of sustainable forest management, greatly increases the flammability of the forest and the risk of ground fires (Uhl and Buschbacher 1985; Uhl and Kauffman 1990; Cochrane et al. 1999; Nepstad et al. 1999a, 1999b; Cochrane 2003; W. F. Laurance, chap. 5 in this volume). These fires result in tremendous emissions and set in motion a recurring cycle of tree mortality and reburning that can degrade the entire forest (Barbosa and Fearnside 1999; Cochrane and Schulze 1999; Nepstad et al. 2001; Gerwing 2002; Barlow et al. 2003b; Haugaasen et al. 2003; Barlow and Peres, chap. 12 in this volume). Forest management plans, including those anticipating carbon benefits, virtually never consider the implications of increased ground-fire risk (Eve et al. 2000).

Reduced impact logging can have more immediate carbon benefits. Traditional logging practices in Amazonia, which can even involve wandering through the forest on bulldozers in search of logs, causes much more damage (and carbon emission) than does the loss of the harvested trees themselves (Uhl and Vieira 1989; Uhl et al. 1991; Veríssimo et al. 1992; Johns et al. 1996). The institution of known low-impact techniques therefore has immediate benefits for carbon, as well as for forest sustainability (Putz and Pinard 1993; Pinard and Putz 1996, 1997; Boscolo et al. 1997; Healey et al. 2000).

Avoided logging is another option with significant carbon benefits. The only example to date is in Bolivia, the Noel Kempff Mercado Climate Action Project (S. Brown et al. 2000b; Asquith et al. 2002). The project was negotiated by The Nature Conservancy, is financed by a consortium that includes American Electric Power and British Petroleum, and is owned and run by the Bolivian NGO Fundación Amigos de la Naturaleza (see Ellison and Daily 2003). The logging company that formerly exploited the area signed a "leakage agreement" to prevent re-investment of the funds in logging elsewhere. A system of monitoring tracks carbon stocks, as well as other features of the program such as the services and other benefits provided to surrounding communities.

Hydroelectric Dams

Hydroelectric dams are often promoted as climate-friendly energy sources, and credit for hydroelectric projects is permitted under the Kyoto Protocol for projects with a power density of more than 10 watts of installed capacity per square meter of reservoir area. Depending on how power density is calculated, the Belo Monte Dam, planned on Brazil's

Xingu River, could qualify, and the dam has often been mentioned in this context by Brazilian authorities. However, Belo Monte only reaches the very high power density of 10 W/m^2 if the calculation is made by ignoring the much larger areas of reservoir that would have to be created by additional dams upstream in order to regulate the flow of the Xingu River and make use of the full 11,000 megawatts of installed capacity planned for the dam (Fearnside 1996b, 2001a, 2006a).

An additional problem with using hydroelectric dams as a form of climate mitigation is that the dams themselves produce substantial emissions. Part of this comes during the first years of dam operation from the decay of trees that project above the water surface when the areas are flooded (Fearnside 1995c). Another large emission comes from methane produced by decay in the reservoir itself; much more important than the flooded wood biomass is the soft, rapidly decomposed organic matter in macrophytes (especially in the early years of a reservoir) and in the weeds that repeatedly grow and are flooded in the drawdown areas as the water level fluctuates. Only modest impacts are indicated if only the emissions from bubbling and diffusion through the reservoir surface are counted, as in the estimates currently being used for Brazil's national inventory of greenhouse gas emissions (Rosa et al. 2002, 2004, 2006). Unfortunately, if the much larger emissions from the water that passes though the turbines and spillway are considered, the emissions are approximately 10 times greater in the case of the Tucuruí Dam (Fearnside 2002a, 2004b, 2006b). Furthermore, hydroelectric dams have additional emissions from the concrete, steel, and other components of the dam construction itself, and these emissions occur years before any power is generated. Because emissions are greatest and generation is least in the early years of a dam, in contrast to electrical generation from fossil fuels, any value given to time in global-warming calculations weighs heavily against hydroelectric power (Fearnside 1997b).

Perhaps the greatest problem with hydroelectric dams as climate-mitigation measures is the tremendous environmental and social impact of these developments (Fearnside 1999f; WCD 2000). Although the CDM is supposed to be restricted to sustainable development, the decision to allow each country to define sustainable development for itself leaves the way open for projects with major impacts to gain credit. There could be no better example than Belo Monte: the upstream dams needed to regulate streamflow for Belo Monte would flood vast areas of tropical forest, almost all of it indigenous land, including more than 6000 km^2 for the Altamira Dam (formerly Babaquara Dam; Santos and de Andrade 1990; Fearnside 1999f, 2005a).

Avoided Deforestation

Avoided deforestation is the subject of debate on several different levels. One series of debates concerns the underlying data, such as the biomass of the forest and of the replacement vegetation. On another level are debates over the theoretical issues that determine how much climatic value avoiding a hectare of deforestation would have, while a third level involves the political interpretation of these results.

Forest biomass is a key measure, as carbon emissions are directly proportional to biomass, with only slight variations due to biomass effects on burning completeness and consequent trace-gas emissions. A wide range of estimates has been produced for the average biomass of Amazonian forest (see reviews in Fearnside et al. 1993; Fearnside 1997c, 2000c). If a low value from this range is picked, the result is a low estimate of deforestation emissions (and therefore of the benefits of reducing deforestation). Examples are provided by a series of Brazilian government estimates indicating little or even zero (!) emissions from Amazonian deforestation (see review in Fearnside 2000a). The choice of input parameters is often treated in a manner equivalent to picking a breakfast cereal in the supermarket, where one can pick whatever cereal one happens to like. Unfortunately, going into the literature to find a value for forest biomass is not the same as picking a breakfast cereal: some values are much better than others in terms of the underlying data and in the interpretation of those data. A recent debate over the biomass of Amazonian forests, and how to interpret it in terms of net emissions of greenhouse gases, illustrates this point (Achard et al. 2002; Eva et al. 2003; Fearnside and W. F. Laurance 2003, 2004). Great care must be taken that all components of the carbon stock are included, such as dead biomass, small-diameter trees, vines, palms, strangler figs, and other "non-tree" components, and belowground biomass. The full emission must include either the "committed emissions" after the year (or multi-year time period) used for the estimate, or the "inherited emissions" from decay or combustion of biomass that remains unoxidized from deforestation in the years prior to the year or period of interest. Regrowth in deforested landscapes of Amazonia is often overestimated by using data on secondary forests that are not derived from cattle pasture (which overwhelmingly predominates as a land-use history and which produces secondary vegetation that grows slowly; Fearnside 1996a; Fearnside and Guimarães 1996). To fully reflect the global-warming impact of deforestation, emissions of trace gases such as methane and nitrous oxide must be included, not only carbon (i.e., carbon dioxide). Inclusion of trace gases increases the impact of deforestation by 15.5 ± 9.5% over calculations that only consider carbon (Fearnside 2000c). All of the above factors are omitted in varying degrees from a number of widely used emissions estimates for Amazonian deforestation (see Fearnside and W. F. Laurance 2003, 2004).

The value of time is fundamental to the place of avoided deforestation in global-warming mitigation. Decisions on discounting or other forms of time-preference weighting (Fearnside 2002e) and on the time horizon for carbon accounting (Fearnside 2002f) make a tremendous difference in the credit assigned to avoiding deforestation, as compared to options at the two ends of the spectrum of permanence (the time that carbon remains out of the atmosphere): permanent displacement of fossil-fuel carbon and short-term sequestering of carbon in biomass in plantations. Heavy discounting of future costs and benefits favors plantations, while insisting on only "permanent" carbon (i.e., zero discount) favors fossil-fuel options (see numerical examples in Fearnside 1995b). The discussion of this is highly polarized, with groups opposed to all "sinks" (e.g., European NGOs) insisting that only "permanent" carbon be credited at all (e.g., Hare and Meinshausen 2000; Meinshausen and Hare 2000; WWF Climate Change Campaign 2000). However, the underlying philosophical position that a ton of carbon emission hundreds or even thousands of years in the future should be given the same weight in decision making as a ton of carbon emission today (i.e., zero discount) is completely at odds with the way human decisions are actually made (see Fearnside 1995b, 2002e). Because global warming is essentially a permanent shift in climate and associated probabilities of disasters, there is value to delaying global warming that is independent of questions concerning the pure time preference. A delay in global warming from time 1 to time 2 saves the lives that would have been lost to global-warming impacts between times 1 and 2. The question of what value should be assigned to time is a moral and political one, rather than a scientific one, and should be decided democratically after ample debate.

Various carbon-accounting frameworks have been proposed that establish an equivalence between carbon held out of the atmosphere for different lengths of time, including "permanent" displacement. "Ton-year" accounting methods represent one approach (Fearnside et al. 2000; Moura-Costa and Wilson 2000), but a method that is more likely to gain acceptance in international negotiations is Temporary Certified Emissions Reductions (T-CERs), based on what is known as the Colombian Proposal (Blanco and Forner 2000; Kerr and Leining 2000). This arrangement would allow market forces to determine the relative prices of certificates that correspond to carbon held out of the atmosphere for different lengths of time. When the certificates expire, they would have to be replaced either with a permanent fossil-fuel carbon displacement or with another temporary certificate, thereby solving the problem of "permanence" from the perspective of the climate. Means of limiting ("capping") the cost of these measures have also been proposed that alleviate a variety of diplomatic concerns (Schlamadinger et al. 2001). It is noteworthy that the European NGOs opposed the Colombian Proposal when it was first forwarded in October 2000, but abruptly reversed positions after the Bonn Agreement of July 2001. In 2002, 12 countries submitted views on the modalities governing these issues under the Kyoto Protocol, including refinements by Colombia and the European Union. Additional proposals from the academic community for transforming T-CERs into a system of "renting" carbon offsets (Marland et al. 2001), or to combine T-CERs with calculations based on the ton-year approach (Dutschke 2002), provide solutions to other perceived problems. The upshot is that if countries want to find solutions to the permanence "problem" they are quite capable of doing so, but if they want to seize on permanence as an excuse for excluding forests, they are also capable of pretending that the issue is insoluble.

"Leakage," or spillover effects outside of a project's boundaries that can negate the climate benefits achieved by the project, is one of the characteristics of project-based mitigation, including many energy-sector projects as well as forest-sector ones (Fearnside 1999d; S. Brown et al. 2000a). This can happen, for example, if farmers prevented from deforesting in a project area simply move elsewhere in the region and clear the same amount of forest at their new location. A variety of ways exists to design projects that minimize leakage effects, as well as for monitoring and compensating for the leakage that occurs (e.g., S. Brown et al. 2000a). A key assurance against leakage is carbon crediting that only pays for carbon benefits that have been achieved and verified, as opposed to mere plans or promises. The important thing is that leakage is a problem that can be minimized and adjusted for, and is not a justification for abandoning the effort to develop avoided deforestation as a mitigation strategy.

The same can be said for the difficulties in establishing an appropriate baseline or reference scenario for use in quantifying the "additionality" of the project effects (required for CDM projects under Article 12 of the Kyoto Protocol). "Additionality" refers to the need to demonstrate that the carbon benefits claimed would not have occurred in a baseline no-project scenario. Because the no-project scenario against which project results are compared is, of necessity, a hypothetical one, it carries uncertainty and a possibility of being "gamed" to falsely claim carbon credits. Again, various proposals exist to standardize procedures and minimize risks. The important thing is that uncertainty can be incorporated into the calculations and adjusted for in the credit granted. The fact that some mitigation options (such as avoided deforestation) have more uncertainty than others (such as plantations for charcoal) does not render the more-uncertain ones valueless. In fact, the large "jackpot" of climatic benefits from a successful program to slow deforestation is such that its expected value (the sum of the products of each possible result and its respective probability of being achieved) can be much higher than lower-risk options (Fearnside 2000d).

By insisting on very high levels of certainty, one effectively throws out the chance to make much more substantial advances in the fight against global warming. Uncertainty requirements represent a situation analogous to the problem of Type II error in statistics: by focusing all attention on reducing Type I error (the probability of mistakenly accepting a statement as true when it is not), one increases Type II error (the probability of not identifying a phenomenon that really exists), and can completely defeat the larger purpose of a study. In this case, it is the larger purpose of maximizing our reduction of global warming that is defeated by insistence on unrealistic levels of certainty for avoided deforestation measures (see Fearnside 2000d). Because the amount of carbon in each hectare of forest saved is so large, the effect of uncertainty can be more than compensated for by giving less carbon credit (Certified Emissions Reductions) than the amount of carbon actually present in the trees. Critics of avoided deforestation (e.g., WWF Climate Change Campaign 2000) virtually always make the unstated assumption that there is a one-to-one ratio between the amount of carbon credit given and the amount of carbon in the project's trees, such that any loss of biomass carbon represents a loss to the atmosphere. However, there is nothing in the Kyoto Protocol that specifies such a one-to-one ratio, and one can easily make deals that are advantageous to the atmosphere even in the face of impermanence, leakage, and uncertainty. Especially in the case of avoided deforestation, one can get substantially more real carbon than the face value of the credits that are given in exchange (Fearnside 2001d).

Climate change itself has become an excuse for rejecting avoided deforestation as a mitigation measure. The U.K. Meteorological Office Hadley Center's HadCM3 model (P. M. Cox et al. 2004) indicates climate change decimating Amazonian forests by 2080, while the dieback shown by a subsequent version of the model (HadCM3LC) is slightly less but still catastrophic (P. M. Cox et al. 2003). Early results of these models were seized upon by opponents of avoided deforestation as a justification for their positions (e.g., WWF Climate Change Campaign 2000). Needless to say, one might question whether such a finding, even if it were known with high certainty, would make it appropriate for environmental organizations to refuse to take up one of the most important potential weapons in the fight to save tropical rainforests. With more than 80% of Brazil's Amazon forest still standing, it is difficult to imagine throwing in the towel on the assumption that the forest is doomed anyway. But even if the forest is doomed, the proper place of environmental groups is to be fighting to save it tree by tree, rather than giving up in advance. If the forest only lasts for 80 years, then avoiding deforestation should be given a maximum of 80 years of credit rather than zero.

So, where do we stand in efforts to turn avoided deforestation into a mitigation measure on a scale that has significant benefits both for climate and for other conservation objectives? Much remains to be done. One area is the impact of planned infrastructure projects in Amazonia, which imply large increases in deforestation and greenhouse gas emissions (W. F. Laurance et al. 2001b; Nepstad et al. 2001; Fearnside 2002a; Soares-Filho et al. 2006). Were the decision-making process to take full account of the environmental costs of these projects, including their global-warming impacts, many would be seen as counterproductive and would not be undertaken. Progress has been minimal in incorporating such concerns into the planning process, despite frequent statements of intentions. Were credit for avoided deforestation a reality under the CDM, the motivation for such changes would increase dramatically.

Several kinds of strategies exist for reducing deforestation. One is to enforce the existing legislation (i.e., Brazil's "Forest Code") to reduce illegal clearing in private properties, particularly large properties. Because deforestation is largely for low-productivity cattle ranches belonging to wealthy landholders, the rate of forest loss could be substantially reduced without inflicting social costs (Fearnside 1993). An encouraging example is provided by a deforestation licensing and control program in the state of Mato Grosso, which showed strong indications of having a significant effect on clearing rates in the state over the 1999–2001 period (Fearnside 2003b). Unfortunately, deforestation surged upward in Mato Grosso, and throughout Amazonia, in 2002. (In Mato Grosso this may, in part, have reflected anticipation by large landholders of the October 2002 elections, when Blairo Maggi, the largest soybean entrepreneur in Brazil, was elected as state governor.) Notably, the estimate for 2001 deforestation in Mato Grosso produced by Brazil's National Space Agency (INPE 2003) was

inconsistent, both in magnitude and direction, with data from the Mato Grosso state government for clearing of rainforest and transitional forest (Fearnside and Barbosa 2004). Assuming that the decrease indicated by the state-government data is real, then the program's results are very important in demonstrating that deforestation is not beyond the control of government policies—a belief that lies at the core of the Brazilian Foreign Ministry's traditional opposition to recognition of carbon credit for avoided deforestation (Fearnside 2003b). It also implies substantial climate benefits over the 1999–2001 period (Fearnside and Barbosa 2003).

Enforcing legislation affecting private landholders is only one strategy for reducing and containing deforestation. Another is the creation and protection of various types of reserves. Most important of these are indigenous reserves, which have much larger areas and potential environmental significance than do the smaller areas designated as conservation units (Fearnside and Ferraz 1995; Fearnside 2003a; Nepstad et al. 2006). Negotiation with indigenous peoples has yet to begin and is an urgent priority. The satellite data from Mato Grosso show that, although most indigenous groups live up to their reputation as much better forest guardians than their nonindigenous counterparts, a few groups are allowing substantial clearings in their reserves (Fearnside 2002c, 2005b). This points to the urgency of making the as yet unremunerated environmental services of the forest a real source of income for indigenous groups. The best environmental results can be expected from direct payments for the services provided, rather than from indirect subsidies of activities like ecotourism or sustainable forestry (Ferraro and Kiss 2002). Indigenous peoples, as well as nonindigenous groups in Amazonian forests, must understand that their greatest asset is the environmental service of forest maintenance.

The Clean Development Mechanism of the Kyoto Protocol is only one way that Brazil could gain carbon credit for avoided deforestation. Were Brazil to join Annex I of the United Nations Convention on Climate Change (UN-FCCC) and Annex B of the Kyoto Protocol, credit could be gained through emissions trading (Kyoto Protocol Article 17), without such limiting restrictions as showing additionality and having detailed, georeferenced accounting for carbon stocks. Emissions trading under Article 17 is based on the much simpler National Inventories that are required of all signatories to the 1992 UN-FCCC. Because Brazil had a net emission of carbon from forests in 1990, Article 3.7 of the Protocol (the "Australia clause") guarantees that these emissions would be part of the country's assigned amount, and that any reduction below the 1990 level of emissions could be sold as carbon credit (Fearnside 1999b,

2001c). This could happen whenever the country decides to do so, even in the 2008–2012 commitment period, but would require that Brazil accept a cap on its national emissions of greenhouse gases. However, because the great majority of Brazil's emissions come from deforestation that produces little benefit to the country's economy and people, Brazil could limit or reduce its emissions more easily than virtually any other country in the world.

Whether Brazil takes advantage of its potential for climate mitigation through avoided deforestation is entirely up to the Brazilian government, or, more accurately, to the individuals who make up the responsible ministries within the government (Foreign Affairs and Science and Technology). Because opinions are so diverse on the issue, it is essentially a toss of a coin each time a new set of ministers is appointed. I believe that, sooner or later, individuals who support avoided deforestation will occupy these posts, and that once the country's negotiating position changes there will be no going back.

One must take a long-term view of the question of avoided deforestation. When I first began advocating forest maintenance for environmental services in 1985, the concept was essentially unknown (see Fearnside 1989b). Quantification of potential benefits prior to the Kyoto Protocol (e.g., Fearnside 1995b, 1997a) seemed highly theoretical at the time. Since then, there have been enormous advances, both in the science and in the diplomacy related to this question. The 5-year setback represented by the Bonn Agreement, although unfortunate given the pace of destruction in Amazonia, is minor on the longer scale of conservation efforts in the region. Further reducing the uncertainties associated with the benefits of avoided deforestation and the means of achieving them must remain a major priority for science. Pushing for acceptance of avoided deforestation both by the parties to the Kyoto Protocol and by the Brazilian foreign ministry must remain a major priority of conservation groups that defend Amazonian forests. Avoided deforestation cannot continue unrecognized for long: the arguments in favor of avoiding tropical deforestation as a major part of global efforts to mitigate climate change are too strong, and the benefits of tapping this source of value are too great to ignore (e.g., Moutinho and Schwartzman 2005; Santilli et al. 2005).

CONCLUSIONS AND IMPLICATIONS

1. Forest loss and degradation in Amazonia currently make a significant contribution to global greenhouse gas emissions, and the large areas

of surviving Amazonian forest mean that the potential for future emissions is greater than in other tropical areas where deforestation is more advanced. Policy changes that slow deforestation in Amazonia therefore have large potential climatic benefits. In addition to global warming, Amazonian deforestation also contributes to climate change through large effects on water cycling, heat transport, and aerosols. In the case of Brazilian Amazonia, where deforestation is largely for low-productivity cattle ranches belonging to wealthy landholders, the rate of forest loss could be substantially reduced without inflicting significant social costs.

2. Mitigation plans in Brazilian Amazonia have so far been concentrated on silvicultural plantations, such as *Eucalyptus* trees for charcoal production. The social and biodiversity benefits of these efforts are limited. Agroforestry (for example under the PROAMBIENTE project) is also planned, with greater potential for such benefits. Hydroelectric dams are often mentioned in this context of mitigation, but the social and environmental impacts (of which greenhouse-gas emissions are only one) make this a questionable option.

3. The agreement reached in 2001 regarding the Kyoto Protocol's first commitment period (2008–2012) rules out credit under the Clean Development Mechanism for avoided deforestation. However, inclusion of this option is likely from 2013 onward. Of all the mitigation measures, avoided deforestation could have the greatest potential benefits in Amazonia, in concert with other options such as avoided logging, reduced-impact logging, and forest-fire avoidance. An environmental licensing program in the state of Mato Grosso over the 1999–2001 period offers valuable and encouraging lessons on how deforestation could be reduced on a wider scale in Amazonia if the environmental services of the forest, such as in mitigating climate change, are properly rewarded.

4. Potential climate-mitigation measures in Brazilian Amazonia, especially avoided deforestation, could also be applied in other Amazonian countries and in tropical forests generally. Quantifying the costs and benefits of these measures and strengthening the institutional structures that assure their effectiveness should be major priorities in counterbalancing the growing list of emerging threats to tropical forests.

ACKNOWLEDGMENTS

My work is supported by the National Council for Scientific and Technological Development (CNPq: Proc. 470765/01-1) and the National Institute for Research in the Amazon (INPA: PPI 1 3620).

CHAPTER 20

Tourism and Tropical Rainforests: Opportunity or Threat?

Stephen M. Turton and Nigel E. Stork

Since the publication of the United Nations World Commission for Environment and Development Report (WCED 1987), commonly known as the Brundtland Report, the concept of sustainability has become the fundamental principle in decision making in the twenty-first century and has been adopted by governments throughout the world. The concept requires the integration of ecological, economic, social, and cultural considerations. Sadly, this principle does not appear to have been adopted in many countries with tropical forests where short-term financial gains have been made at the expense of these forests and the biodiversity and ecosystem services they sustain.

There are conflicting signs that one relatively new and emerging industry, nature-based tourism, in some locations is causing further damage to some of the most critical tropical forests around the world but, in other places, is helping rainforest conservation. Tourism has been associated with undesirable natural resource impacts in many parts of the world (Hunter and Green 1995), often because of the lack of adequate development controls. Like any industry that impacts on natural resources, tourism requires appropriate management strategies and systems to ensure both visitor enjoyment and resource protection for current and future generations. Nowhere is this more challenging a task than in tropical forests, where forest resource exploitation has led to a decline in the extent of native forests and the distribution of biodiversity, and is having severe socioeconomic impacts on host developing countries (Farrell and Marion 2001).

On a positive note, a number of international organizations, including the World Tourism Organization, United Nations Environment Program, and Conservation International (Christ et al. 2003), have recognized that this industry, if combined with effective protected-area management, has the potential to bring much needed ecological, sociocultural, and economic benefits to communities in developed and developing countries (Newsome et al. 2002). Tourism potentially is a much more sustainable use of tropical forest environments than other current uses of forest landscapes.

International and domestic tourists are continually seeking new experiences, particularly in ecologically complex and pristine environments such as coral reefs and tropical rainforests. By the very nature of these complex and sensitive environments, there is a real danger that over-visitation or inappropriate visitation, or the inappropriate construction of resort and associated facilities, may lead to their decline and become unsustainable (Davenport et al. 2002). Much of this has been brought on by the expansion of international tourism and the steady growth of domestic tourism in many tropical countries. In 2003, the world's travel and tourism is expected to generate US$4.5 trillion of economic activity (total demand), and over the next 10 years the industry is expected to achieve an annualized growth rate of 4.6% (World Travel and Tourism Council 2002). Because tourism has the potential to substantially improve the economies of many developing countries by bringing in large revenues, many countries have been prepared to allow the industry to grow very rapidly without putting in place safeguards and management protocols.

Tourism to natural areas has increased at an unprecedented rate over the last decade. In the late 1980s, nature-based tourism accounted for 2% of all tourism, rising to about 20% by the late 1990s (Newsome et al. 2002). This amounts to about US$20 billion per year from tourism activities in natural areas (WTO 1998, cited in Newsome et al. 2002). Along with tropical reef ecosystems, tropical forests are the most biologically diverse biome on Earth and this biodiversity is the primary draw card for tourists visiting this biome. While tropical forests have impressive structural and floristic diversity, it is the highly enigmatic animal life that dominates tourism interest in tropical forests at the international level (Newsome et al. 2002). Examples include gorillas in Central Africa, Orangutans in Indonesia, and bird watching in the Neotropics, Australia, and New Guinea.

In this chapter we assess both the positive and negative impacts of tourism and visitation in general on tropical rainforests at several spatial scales. Because tropical forest tourism is a relatively new industry, there have been few published studies to date of its impacts (Valentine and Cassells 1991; Farrell and Marion 2001). One exception to this is the growing body of literature associated with the so-called Wet Tropics region of northeastern Australia where the rainforest tourism industry has grown rapidly alongside that based on the Great Barrier Reef (Turton 2005). We draw heavily on our first-hand experience of the developing rainforest tourism industry in northern Australia, but also give examples of studies from other tropical countries that engage in rainforest tourism. We believe that the experience of rainforest tourism and how this is managed in northern Australia is an example of how it could be sustainably managed and developed elsewhere in the world. Much of our chapter looks at the local-scale and regional impacts of tourism and provides examples of how negative impacts can be mitigated. However, we also consider the global scale of tourism growth and how the benefits of this industry can be maximized.

THE GROWTH OF TROPICAL RAINFOREST TOURISM AND RECREATION IN NORTH AUSTRALIA

The Biological and Evolutionary Significance of the Wet Tropics

It is estimated that tropical, subtropical, and temperate rainforests today cover only 20,000 km^2 (~0.2%) of Australia (L. J. Webb and Tracey 1981). These remnant forests exist as a chain of discontinuous pockets extending more than 6000 km from northern Australia along the east coast to Tasmania. Some 7800 km^2 of tropical forests are located in north Queensland and of that about 90% are within the Wet Tropics bioregion, a narrow coastal strip of land some 20 to 80 km wide, which extends from Cooktown to Townsville (fig. 20.1). The Wet Tropics rainforests were the subject of much controversy in the 1980s as tropical biologists and conservationists highlighted the need to conserve them and to stop any further logging and adverse infrastructure development. The designation of a large part of these rainforests as a World Heritage Area was a major step toward their conservation. These rainforests represent 0.01% of the global tropical rainforest cover and yet, because of the unique nature of the biota and its evolutionary significance, the Wet Tropics region of north Queensland is one of the world's top biodiversity hotspots (Myers 1988b; Werren et al. 1995; but see Myers et al. 2000).

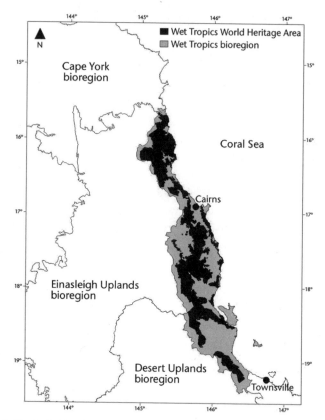

Figure 20.1. Wet Tropics bioregion in northeast Australia. The darker shade shows the Wet Tropics of Queensland World Heritage Area.

Although representing just 0.1% of Australia's land surface, the Wet Tropics contains the richest variety of animals and plants in the country, including two-thirds of the butterfly species, half of the birds, and a third of the mammals. Seventy vertebrate species are found nowhere else in the world (see Williams and Hilbert, chap. 2 in this volume). A very high proportion of the fauna and flora are endemic to the Wet Tropics (Commonwealth of Australia 1986). More than 400 plant and 76 animal species are officially listed as rare, vulnerable, or endangered (WTMA 1999). The Wet Tropics also provides an unparalleled living record of the ecological and evolutionary processes that shaped the flora and fauna of Australia over the past 400 million years when it was first part of the Pangaean landmass and then, later, the ancient Gondwana continent. For example, the rainforests of the Wet Tropics have more plant taxa with primitive characteristics than any other area on Earth. Of the 19 known families of primitive flowering plants (Takhtajan 1980), 12 are found in the Wet Tropics and 2 (Idiospermaceae and Austrobaileyaceae) are not found anywhere else in the world (Commonwealth of Australia 1986). For comparison, there are only 9 primitive flowering plant families in South American rainforests and none of these are endemic. Not surprisingly, the area has been described as a key to the origins and ancient habitats of primitive flowering plants.

The tropical rainforests of far north Queensland in Australia have been the focus of intense study with at least 1000 scientific articles focusing on this area published in the past 20 years. Many of these provide key insights into contemporary tropical forest issues such as habitat fragmentation, conservation genetics, and biodiversity conservation (see http://www.rain -forest-crc.jcu.edu.au/publications/publications.htm; also Williams and Hilbert, chap. 2 in this volume; W. F. Laurance and Bierregaard 1997; Bermingham et al. 2005). The threats to this area are as well understood as those of any area of tropical rainforest (WTMA 2002; Stork 2005) and it is informative to examine what we know about the impacts of tourism and recreation in this area and how these have been managed.

The Growth of Tourism in North Queensland and Its Economic Contribution

The Wet Tropics bioregion (fig. 20.1) contains several local-government areas that are among the fastest growing in Australia. As of June 1999, there were almost 200,000 people in the region, with an annual growth rate of 2.4% (WTMA 2002). This compares with the Queensland state average of 2.0%. By 2016, it is estimated that some 270,000 people will live in the region (WTMA 2002). The population growth is largely driven by the tourist industry, the largest employer in the region.

The Wet Tropics World Heritage Area (hereafter, WHA) is an outstanding tourist destination, and currently plays host to millions of tourists each year. Driml (1997) estimates that tourism in the WHA generates more than AU$750 million (US$515 million) each year. The Wet Tropics region has experienced significant increases in domestic and international tourism over the past 20 years, with some 2 million visitors per year in 1995 (WTMA 2002). Recent projections suggest that tourist number will be about 4 million per year by 2016, with an increase in international visitors being a major contributing factor (WTMA 2002).

Bentrupperbäumer and Reser (2002) have estimated that some 4.4 million visits per year occur to recognized Wet Tropics WHA sites, with 60%

of these being domestic and international tourists. The remaining 40% are local residents engaging in rainforest-based recreation activities. This emphasizes the importance of outdoor recreation activities to local residents in the region, a growing phenomenon throughout the developed world. In contrast, in developing nations with tropical rainforest, it would seem that international and domestic tourism contribute the greatest proportion of visitation compared with recreational use by local residents.

IMPACTS AT DIFFERENT GEOGRAPHIC SCALES

Impact Defined and the Importance of Scale

Although many consider the term *impact* to have a negative meaning, it is in fact a neutral term with many manifestations. Impacts of human activities in rainforests may be positive or negative depending on the magnitude and significance of the changes. Likewise, we must recognize that impacts may be biophysical, economic, social, cultural, and psychological in nature (Newsome et al. 2002). The degree of change resulting from an impacting process will depend on the *resistance* and *resilience* of the environment in question. *Resistance* may be defined as the ability of an ecosystem to withstand an impacting process, while *resilience* is its ability to recover following removal of the impacting process.

Table 20.1 shows human activities associated with pressures at a range of scales in regard to the Wet Tropics WHA (fig. 20.1). It is evident that the Wet Tropics WHA is subject to a wide range of impacting processes and phenomena at a range of scales, with many of these impacts being brought on directly or indirectly by tourism and recreation activities. While the activities shown in table 20.1 are explicit to the Wet Tropics WHA in Australia, many would apply to protected areas in other tropical rainforest regions (Farrell and Marion 2001). The consequences of the pressures for the natural values and processes of the WHA are described in table 20.2. It would be reasonable to state that tourism growth and increases in recreational demand would directly and indirectly contribute to many of these processes at a range of spatial and temporal scales.

Impacts of Tourism at the National and Global Scale

The focus of this chapter is on tropical forests. It is not an in-depth analysis of the global and national impacts of tourism, although it recognizes the importance of other, more fundamental impacts. Globally, the number of

Table 20.1 Human pressures at a range of spatial scales, arrayed along axes of the severity and extent of pressures (after E. Saxon, pers. comm.; see Stork 2005)

Extent of pressure	Severity of pressure ⟵⟶				
	Cumulative minor impacts		Interferes with natural processes		Transforms the landscape
Pervasive and/or permanent				Selective logging	Climate change
			Feral pigs	Power lines, roads, dams	Mountain top clearing
Widespread and/or long term			Fire		
		Small clearings	Water extraction		
Local and/or ephemeral	Casual camp sites	Walking tracks			

international visitors in all types of tourism is more than 500 million per year and each international or national flight a tourist makes contributes to the burning of fossil fuels. The total tree biomass required to compensate for the carbon emissions created by this travel is increasing each year and yet the area of tropical forests is diminishing. At the local level, the way that tourist infrastructure is designed and maintained also has a critical role in the global carbon budget. Curtis (2002) examined the carbon budgets of a range of hotels in northeast Queensland and demonstrated that the highest emitters were 2.5 times higher than the mean emission of 750 kg of carbon per bed per year. Hotels located within tropical forests or directly adjacent to rainforest were more ecologically driven in their design and as a consequence had much lower carbon usage (97 kg C/bed/year). Those in Cairns and other towns had much higher values (highest emission was 1785 kg C/bed/year). Of course, much of this is user-driven as many international and domestic tourists expect high levels of comfort, particularly air conditioning. There are also many positive benefits of tourism at the global and national scales, including wider public awareness of the significance of

Table 20.2 Consequences of pressures at various scales for natural values and processes of the Wet Tropics World Heritage Area (from E. Saxon, pers. comm.; see Stork 2005)

Extent of pressure	Severity of pressure			
	Cumulative minor impacts		Interferes with natural processes	Transforms the landscape
Pervasive and/or permanent			Loss of canopy cover and old growth habitats	Disintegration of biotic communities
		Disease reservoir	Decline in biodiversity	Destruction of unique habitats
Widespread and/or long-term			Attenuated natural fire patterns	
		Establishment of exotic weeds and pests	Reduced dry season stream flow	
Local and/or ephemeral	Site pollution and damage to local vegetation	Local disturbance of soil surface		

tropical forest conservation, protected-area financing, and local support for conservation due to incoming revenues from tourism to communities (Christ et al. 2003).

Impacts of Tourism and Recreation at the Regional Scale

While most studies of impacts of tourism and recreation in natural areas have focused on processes at the local or site level (e.g., Hammitt and Cole 1998; Newsome et al. 2002), we consider broader-scale impacts to be important in the context of the Wet Tropics bioregion (tables 20.1 and 20.2), and probably elsewhere in the tropics. These regional-scale impacting processes include (1) infrastructure development directly associated with tourism activities such as the creation of new resorts, new tourist attractions, or new airports; (2) associated linear infrastructure such as roads and walking tracks needed to transport tourists in the region or power lines and water lines to provide power and water (Goosem and Turton 2000; Goosem 2004; Turton 2005; S. G. W. Laurance, chap. 14 in this volume); (3) urban and peri-urban growth, particularly population increases in the areas adjacent to the Wet Tropics WHA; (4) water supply and increasing demand for more water storage; and (5) waste disposal in environmentally sensitive areas. Many of the regional-scale processes are interrelated, with the growth in tourism and associated service industries being the principal driver for all of them. This phenomenon has also been experienced in other tropical-forest tourist destinations, notably Hawaii and several Caribbean Islands (Newsome et al. 2002). As existing and new tropical-forest tourist destinations—many of which are in developing countries—grow in popularity, these regional-scale processes and associated negative environmental and social impacts are likely to become major issues for their governments and tourist industries.

Increases in the local-population and tourist numbers exert increasing demands for recreation and tourism activities, particularly in regard to road access, walking tracks, swimming holes, and visitor infrastructure. At the same time, there is a corresponding increase in demand for better and faster roads and highways, electricity, water supply, and waste disposal.

A considerable body of literature now exists examining impacts of linear infrastructure and service corridors on tropical rainforests, with the majority of this work being undertaken in northeastern Australia (Goosem 1997; Goosem and Turton 1998, 2000; Goosem 2004). Such infrastructure produces a multitude of biophysical impacts on rainforest ecosystems (fig. 20.2), including linear barrier effects on arboreal and ground-dwelling fauna, road kills, and edge effects, which may extend up to 100 m or more into the adjacent forest (Goosem 2004; S. G. W. Laurance, chap. 14 in this volume). Other effects include incursions of alien habitat along road verges and beneath powerline corridors, and conduit effects for feral animals and weeds. Roads also introduce contaminants, such as heavy metals, into rainforests where they may enter the food chain. Similarly, linear infrastructure disrupts the forest-canopy cover, leading to increased erosion and sedimentation. Access to water holes and fast-flowing rivers for recreation activities may result in adverse impacts on water quality at high-use sites, with corresponding effects on urban water supply. Butler et al. (1996) found that the recreational use of a number of local swimming holes they studied in the Wet Tropics caused

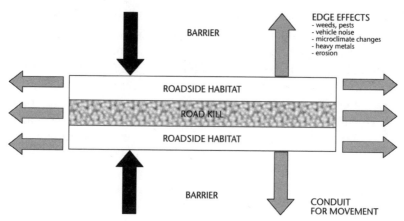

Figure 20.2. Ecological effects of roads and other linear infrastructure (powerline and service corridors) in forest ecosystems

substantial increases in, for example, the levels of suspended solids and nutrients, which had a negative impact on both the ecology of the rivers and visitor satisfaction. Many of these negative impacts could be addressed by providing adequate toilet facilities and signage to indicate the appropriate use of the rivers.

Mitigation of Tourism and Recreation Impacts at the National and Regional Scale

Management of tourism at the regional (and national) level requires the development of strategies that are agreed at various government levels and across numerous departmental portfolios. There needs to be agreements, for example, between the departments that manage biodiversity and natural resources and those that manage infrastructure developments, such as new tourist resorts, roads, power supply, and water extraction and use. Some decisions with respect to these resources are made at various levels, from national, state, and local government, and again joint strategies are needed. Many key areas for tourism are protected as national parks or world heritage areas and are therefore subject to protection at national and international levels.

It is dangerous for countries to see tourism alone as a panacea for their economic problems, as this industry is subject to the booms and crashes that other industries experience and in some instances is more susceptible to external forces than other industries such as agriculture. Most international tourists come from the USA, Europe, and Japan and these markets have proven to be particularly susceptible to issues such as security and disease. For example, in the last decade, events such as the crash of the Asian stock market, SARS, and the threat of international terrorism have had a devastating effect on nature-based tourism in Southeast Asia. All these events could have had a major effect on tropical-forest tourism and related tourism in northeastern Australia. The internal market for tourism, however, has grown.

Impacts of Tourism and Recreation at the Local Scale

Most studies on the impacts of tourism and recreation at the local scale have largely focused on walking trails, camp sites, water holes, and picnic sites (Hammit and Cole 1998), with the overwhelming majority of research being conducted in temperate ecosystems (Day and Turton 2000). Researchers have typically examined impacts of visitors on four main resource elements: fauna, flora, water, and soil (Hammit and Cole 1998; Newsome et al. 2002). Activities such as hiking, wildlife viewing, camping, and swimming tend to have localized impacts, but the cumulative effects may be significant (tables 20.1 and 20.2). Localized impacts from such activities may reduce functionality of facilities, increase safety concerns, reduce aesthetic enjoyment, contribute to visitor displacement, create conflict among visitor groups, and increase management costs (Bentrupperbäumer and Reser 2000; Farrell and Marion 2001).

There have been few studies on the impacts of recreation trampling in tropical rainforests, in contrast to the large volume of such work in temperate ecosystems (Jusoff 1989; Boucher et al. 1991; Stephenson 1993; Wallin and Harden 1996; Turton et al. 2000; Farrell and Marion 2001; Talbot et al. 2003; Turton 2005). The response of an ecosystem to recreation-related impact is mainly determined by the ecological characteristics of the biophysical system (Hammit and Cole 1998). As trampling in a tropical rainforest occurs in the understory, the main impacts occur at the lower vegetation stratum and soils. Tropical rainforests would be expected to have a low resistance to trampling because of moist, friable soils, year-round vegetative activity, and broad, thin leaves (Kuss 1986).

A recent, controlled experiment in upland tropical rainforest in northeast Australia on rhyolite and basalt soils (Talbot et al. 2003) confirms that rainforests have low resistance to trampling. Trampling, especially after 200 to 500 passes, reduced organic-litter cover. Soil bulk density increased with trampling intensity, particularly on basalt soils, while the permeability of both soil types decreased markedly with increased trampling intensity, even after just 75 passes. This study suggests that physical and hydrological changes may occur rapidly in tropical rainforest soils following low levels of trampling, particularly on basalt soils. It should be noted that the Talbot et al. (2003) study is limited to two soil types and does not provide any indication about recovery (resilience), but nonetheless confirms that tropical rainforest soils have low resistance to trampling.

Tropical rainforest soils and understory vegetation are expected to have moderate to high levels of resilience to trampling impacts (Day and Turton 2000), mainly because of high levels of primary production. A study of vegetation recovery of a closed trail in a Costa Rican rainforest (Boucher et al. 1991) found remarkable rates of regrowth within 12 months of closure. Studies of recovery of rainforest campsites in northeast Australia, following camp closure, have also demonstrated fairly rapid rates of recovery for soil, organic-litter cover, and seedling variables in comparison with non-impacted (control) campsites (R. A. Smith and Turton 1995). This finding suggests that tropical rainforest campsites should be rotated on a regular basis to ensure sustainable use of high-demand areas.

Impacts of tourism and recreation in tropical rainforests show distinct seasonal differences in northeast Australia (Turton et al. 2000), with the greatest impacts occurring during the so-called wet (rainy) season (November to April). At present, the wet season coincides with low levels of visitation from tourists and local residents, but changes in marketing strategies to increase tourism in these months may result in unsustainable tourism practices at some visitor sites. Ideally, tourism and recreation activities in rainforests should be concentrated at a few key sites and only during the drier months when the ecosystem has greater resistance.

Researchers have examined impacts of tourism on wildlife in tropical forests in Madagascar (Stephenson 1993), Indonesia (Griffiths and Van Schaik 1993; Kinnaird and O'Brien 1996), and Belize (Marron 1999). These studies have assessed a multitude of negative visitor impacts on small endemic mammals, clouded leopards, Sumatran rhinoceroses, orangutans, macaques, and howler monkeys. These studies suggest that visitor numbers and their activities at high-use sites must be closely monitored if wildlife tourism is to be sustainable in the long term. Poorly managed wildlife tourism has been shown to result in detrimental impacts on reproductive rates, foraging behaviors, and mortality rates of animals, especially for rare and threatened species. Such impacts may lead to higher-order changes in community composition. On the other hand, properly managed wildlife tourism in tropical forest regions has the real potential to provide economic, social, and cultural benefits to local (and often indigenous) people, as well as contributing to the conservation of habitats and endangered species. This is especially important in developing countries where government investments for management of protected areas are severely limited (Farrell and Marion 2001; Davenport et al. 2002).

Mitigation of Tourism and Recreation Impacts at the Local and Site Scale

Tourism and recreation impacts in tropical forests at the local (site) level require a combination of site- and visitor-management strategies and policies (Newsome et al. 2002). Site management in tropical forests is better focused on concentration of visitor activities at a limited number of sites, rather than dispersion across a larger number of sites (Bentrupperbäumer and Reser 2000). This strategy is based on the premise that tropical forests have low resistance to visitor impacts and respond best to site hardening or shielding to minimize negative impacts on soil, vegetation, and wildlife (Turton 2005). Bentrupperbäumer and Reser (2000, 2002) have examined visitor impacts in tropical rainforests in protected areas of northeast Australia and advocate managing visitors and their psychosocial behaviors to decrease negative biophysical impacts on rainforest ecosystems. Such an approach may reduce the need to apply more expensive site-management techniques, such as constructing board-walks, and compacting walking tracks at some visitor sites. Moreover, their approach has real value in managing low-use, long-distance walking trails in rainforest regions where extensive site management is aesthetically inappropriate and more difficult to apply. For example, prescribing appropriate low-impact practices and behaviors to users or promoting or enforcing the use of tour guides in ecologically sensitive areas can be very effective at reducing visitor impacts (Farrell and Marion 2001; Turton 2005).

WHAT IS THE RISK OF TOURISM TO TROPICAL FORESTS AND THEIR BIODIVERSITY, AND HOW CAN THE BENEFITS BE MAXIMIZED?

In the past 30 years, there has been growing concern about the fate of the world's tropical forests and the biodiversity they sustain. F. D. M. Smith et al. (1993), for example, examined the rate of change of threat status for species on global Red Data Book lists and concluded that 50% of all mammals, birds, and palms would disappear in the next 250, 350, and 70 years, respectively. Increasingly, attention has been focused on a few unique biodiversity "hotspots" (Myers 1988b). Myers and colleagues (Myers et al. 2000) suggest that 40% of the world's biodiversity is located in 20 tropical forest hotspots around the world. Many of these locations are prime targets for tourism, as international and domestic visitors want to witness their unique biodiversity, possibly before species become extinct.

In a very timely study of tourism and biodiversity, Christ et al. (2003) reviewed the growth of tourism in the world's biodiversity hotspots. They showed that, in the past 10 years, tourism has grown by more than 100% in 23 of the world's biodiversity hotspots located in developing countries, and that more than half of these each receive more than 1 million international tourists each year. More than half of the world's poorest 15 countries fall within biodiversity hotspots and, in all of these, tourism is already nationally significant or forecast to increase in importance.

Christ et al. (2003) also recognize that tourism development is a complex interaction among many players, with the private sector usually driving the process. The establishment of facilities is heavily dependent on multi- and bilateral development agencies. They recognize that the effective management of tourism and protection of the environment must be a partnership among the private sector, public sector, and society at large. Our experience with the sustainable development of rainforest in northern Australia concurs with that conclusion.

The benefits of tourism to developing countries are enormous. Tourism can be confined to relatively small areas, unlike less sustainable industries such as agriculture and forestry. Most visitors are happy to be guided to a few locations where they are guaranteed desirable views of nature or the adventure tourism they are seeking. The careful design of access to these sites and the design of on-site facilities such as roads, walking tracks, parking areas, toilets, and other facilities means that tourists can be directed and carefully managed and their impacts monitored and restricted. In this way visitors may access less than 1% of key rainforest sites but still be highly satisfied with their visits.

CONCLUSIONS AND IMPLICATIONS

1. Tourism is expanding rapidly in many of the world's biodiversity hotspots and particularly in tropical forests.

2. Despite this, there have been few studies of the development and management of tourism in tropical forests. The way that tourism devel-

opment impacts tropical forests, their biodiversity, and indigenous populations is likely to differ widely in different parts of the world. It is important that in-depth analyses and comparisons of a few key sites around the world are undertaken, to develop best-practice standards for others. These analyses should identify potential opportunities that tourism presents to local communities and to nations as well as looking at ways to manage the negative impacts. Some work undertaken in northeast Australia provides an ideal theoretical framework for application at tropical-tourist destinations around the world.

3. Tourism has the potential to provide new funding for management and protection of tropical forests and their biodiversity. Most tourists visiting new and exotic places are prepared to contribute to their safekeeping. As a result, much funding has already been contributed to conservation organizations and to management agencies. There is no doubt that tourism has made an enormous financial contribution to the conservation of flagship species such as the mountain gorilla.

4. Many countries looking to rainforest tourism to provide new socioeconomic benefits, will need to address the very different management issues that this presents. A number of techniques are available to minimize visitor impacts, including the proper location, construction, and maintenance of visitor sites and facilities, and managing visitor use through education or regulation.

5. Impacts of visitors on components of the biophysical environment need to be quantified at more sites in tropical forests so that baseline data may adequately inform visitor monitoring and management. Current barriers to monitoring in many tropical forest destinations, such as limited funding and staff resources, can be circumvented through partnerships with local tourist operators and with volunteers, regional universities, or nongovernmental organizations.

6. A comprehensive set of ecotourism tools has recently been proposed for parks in developing countries (Davenport et al. 2002). Suggestions include diversifying tourism infrastructure, revenue sharing, optimizing user-fee structures, increasing domestic tourism, and integrated conservation and development projects. We strongly urge the adoption and implementation of these simple, generally low-cost suggestions to improve management of nature-based tourism in tropical forests.

CHAPTER 21

Managing Threats from Bushmeat Hunting in a Timber Concession in the Republic of Congo

Paul W. Elkan, Sarah W. Elkan, Antoine Moukassa, Richard Malonga,

Marcel Ngangoue, and James L. D. Smith

Research over the past decade has shown that the hunting of wildlife in most tropical forests is occurring at unsustainable levels (Redford 1992; Robinson et al. 1999; Robinson and Bennett 2000; Fa et al. 2002). Commercial hunting associated with logging operations has depleted wildlife populations in forests of Asia, South America, and Africa (Bennett and Gumal 2001; Rumiz et al. 2001; Wilkie et al. 2001). A broad range of mammals, birds, and reptiles is harvested. Large-bodied species, many threatened and endangered, are often the first to be targeted and are slow to recover from population declines (Robinson and Bennett 2000). Investigations in areas where logging is selective have suggested that the greatest immediate threat posed to wildlife, is not the extraction of trees but the unsustainable hunting facilitated by the timber exploitation process (L. J. T. White 1992; Bennett and Gumal 2001; Wilkie et al. 2001).

Throughout the forests of Central Africa, commercial hunting facilitated by the timber industry is causing the systematic depletion of wildlife populations (Ape Alliance 1998; Wilkie and Carpenter 1999; Auzel and Wilkie 2000; P. D. Walsh et al. 2003). Of the 2.05 million km^2 of forests covering the Congo Basin, approximately 8% to 9% has been designated as protected area (Central Africa Regional Program for the Environment 2001). Much of the remaining forest is currently under commercial exploitation or is designated for future timber, petroleum, and mineral exploitation by private industry.

Commercial bushmeat and ivory poaching networks use timber-company infrastructure to penetrate remote forests (Ape Alliance 1998).

Logging activities have led to a domino effect of increased access to the forest, population growth, influx of capital, increased demand for bush-meat, and escalating commercial hunting (Wilkie and Carpenter 1999). Forestry camps often create markets and staging points for commercial hunting in previously remote, undisturbed areas. The revenue generated by timber-company communities attracts traders, commercial bushmeat traffickers, job seekers, and hangers-on and contributes to rapid demographic growth (Elkan 2003). Pressures on wildlife populations and cultural hegemony associated with logging activities impact local indigenous communities that depend on forest resources (Moukassa 2001).

To date, few field-based initiatives have been undertaken to attempt to minimize the direct and indirect effects of the forestry-exploitation process on wildlife populations (Robinson et al. 1999). Efforts to address the bush-meat crisis and promote wildlife management in tropical forests have thus far largely involved international media pressure, lobbying, and research. Policies to ban and/or manage hunting have been adopted in some countries (e.g., Sarawak and Bolivia) in an attempt to limit commercial hunting associated with logging (Robinson and Bennett 2000). In Africa, wildlife issues have only recently been introduced in the forest-policy reform processes (African Forest Law Enforcement and Governance 2003).

This chapter reports on efforts to promote landscape-scale wildlife management in a large timber concession in Central Africa. We describe conservation and management interventions and results obtained during 1999–2002 in the Kabo and Pokola timber concessions in northern Republic of Congo, adjacent to the Nouabale-Ndoki National Park.

CONTEXT

While rich in valuable mahoganies (*Entandrophragma* sp., Meliaceae family) and other hardwoods, the forests of the northern Republic of Congo also harbor some of the most important wildlife populations remaining in Central Africa, including forest elephants (*Loxodonta africana cyclotis*), lowland gorillas (*Gorilla gorilla gorilla*), chimpanzees (*Pan troglodytes troglodytes*), bongo (*Tragelaphus eurycerus*), buffalo (*Syncerus caffer nanus*), leopard (*Panthera pardus*), six species of duikers, and eight species of diurnal monkeys (Fay et al. 1990). The Nouabale-Ndoki National Park (NNNP), which is about 4000 km² in area, supports one of the most intact forests remaining in Central Africa and is contiguous with the Dzanga-Ndoki National Park in Central African Republic and Lobeke National Park in Cameroon (figs. 21.1–21.3). Four forest-management units surround the Nouabale-Ndoki

Park: Kabo (3000 km^2) and Pokola (estimated 5600 km^2) to the south have been exploited for timber since the 1970s. Timber harvest in Loundougou (3860 km^2) to the east, and Mokabi (3750 km^2) to the north began in 2002. Historically, timber exploitation in Kabo and Pokola has been largely selective, with harvests averaging two to three trees per hectare.

The logging concessions contain a large number of natural forest clearings, locally called "bais" and "yangas," which are a critical resource for large-mammal populations and support high densities of rare and endangered large mammals (Blake 2002; Elkan 2003). Although the region has relatively low human density (<1/km^2), over the past three decades, permanent settlements have been established along the Sangha River in association with the sawmills in Kabo and Pokola towns (Moukassa 2001). Development of a logging economy created a considerable export of game meat and uncontrolled hunting in many areas of Kabo and Pokola. In the mid-1990s, international conservation groups and some individuals, including Karl Ammann, brought attention to the poaching of great apes in Pokola and other timber concessions in the region (Ape Alliance 1998).

In early 1997 a timber company called Congolaise Industrielle de Bois (CIB) acquired rights to Kabo and the neighboring Loundougou concession in addition to Pokola, which it has been exploiting since the early 1970s. The expansion of CIB operations to Kabo compounded previous pressures on wildlife, such as revitalizing elephant-poaching networks, and greatly increased new ones, such as the systematic export of wildlife

Figure 21.1. Location of Republic of Congo in Central Africa

Tri-national parks and adjacent forest management units in northern Republic of Congo

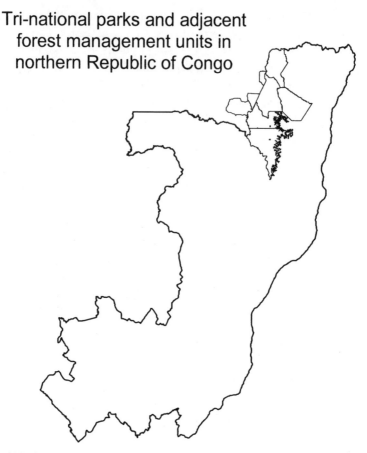

Figure 21.2. The northern Republic of Congo landscape and adjacent tri-national protected areas

products. CIB rapidly built a dike across the Ndoki River and created an advanced logging camp (Ndoki 2) 30 km south of the Park (fig. 21.3). A permanent road network was built to link the Ndoki 2 camp to Ndoki 1 camp in the Pokola concession and the sawmill towns of Kabo and Pokola. A major access road was also built to the Loundougou concession, passing just 5 km southeast of the Park border.

METHODS

Conservation

In 1991, Wildlife Conservation Society (WCS), with funding from the U.S. Agency for International Development (US-AID), began working

P. W. Elkan, S. W. Elkan, A. Moukassa, R. Malonga, M. Ngangoue, and J. L. D. Smith

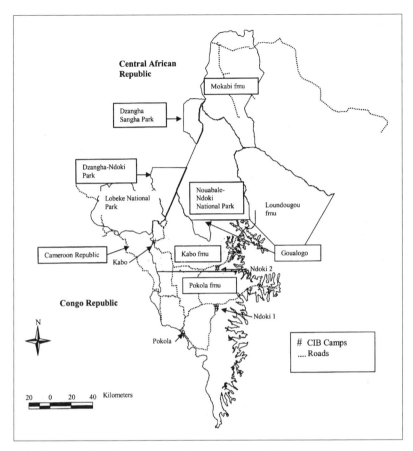

Figure 21.3. The northern Congo landscape including the Nouabale-Ndoki National Park, surrounding forest-management units, and adjacent protected areas in Cameroon and Central African Republic

with the Government of Congo to establish the Nouabale-Ndoki Park. Anticipating long-term management needs and increasing pressures on wildlife, WCS, CIB, and the Ministry of Forest Economy of the Government of Congo worked to finalize an agreement in June 1999 to develop a joint program titled Project for Ecosystem Management of the Periphery of the Park (PROGEPP, in French). The objectives of the project were to design, implement, and monitor sustainable wildlife-conservation and management systems in the Kabo, Pokola, and Loundougou concessions adjacent to the Park. Activities focused on education and awareness, wildlife management and hunting regulation, wildlife-law enforcement, and development of alternative activities and protein sources. Socioeconomic and ecological research and monitoring were used to inform strategy development, assess progress, and adaptively orient management interventions.

WCS was responsible for development and administration of overall project activities in collaboration with the Government of Congo and provided technical expertise and mobilized operational funding through its own and other international sources. The government assigned protection personnel, facilitated ecoguard training, and oversaw law enforcement. CIB initially provided support in the form of infrastructure (housing, guard posts, and a vehicle), diesel fuel, and other equipment for wildlife-protection activities, and later provided partial funding of antipoaching and alternative-protein activities. The company contribution amounted to 20% to 30% of the direct costs of alternative activities and wildlife-protection components in the first 3 years, with an increase to 30% to 40% of costs in 2002.

Research and Monitoring

Wildlife law enforcement. Government officers systematically collected information on return of law-enforcement efforts. Data were analyzed in terms of catch per patrol-unit effort (patrol unit = 1 Ministry officer accompanied by three to five ecoguards) and trends in violations (protected species, exports, zoning violations, obstruction of operations incidents, etc.) following methods adapted from Jachmann (1998).

Human demographics. Total-count demographic surveys were undertaken annually at the Kabo, Pokola, Ndoki 1 camp, and Ndoki 2 campsites with assistance from local guides familiar with the residents. Each household was visited and the members, ethnic group, and occupation registered.

Hunting. Observers recorded bushmeat entering at the Kabo, Ndoki 1, Ndoki 2, and Pokola sites on 10 randomly selected days per month, collecting information on species, age, sex, weight, origin, and price of hunted species. Two observers monitored the two principal entry routes to Pokola town because of its large size. Data collection was verified five to six times a month at random by a WCS researcher.

Protein consumption. Composition of meals was recorded at 20 randomly selected households on 10 randomly selected days each month at Kabo, Ndoki 1, and Ndoki 2. Twenty households were sampled on 10 randomly selected days twice a month at Pokola. Each household was visited in the morning and evening and asked to show the observer the type of meal they had prepared (bushmeat, fish, domestic meat, etc.). Information was recorded on protein composition, price, and ethnic group of each household.

Large-mammal abundance. We used ecological reconnaissance methods to generate information on large-mammal abundance and distribution and human activity in community hunting and nonhunting zones (Carrillo et al. 2000; Blake 2001). Seven different routes (16–40 km in length) were surveyed twice a year, in wet and dry seasons, following standard reconnaissance methods (L. White and Edwards 2000). A principal observer assisted by two local trackers recorded observations of all large mammal sign (tracks, dung, urine, hair, etc.) and the distance hiked was measured with a topofil and a GPS unit (Garmin 12XL). Animal-sign encounter rates per kilometer were determined.

RESULTS

Education and Awareness

A conservation-awareness campaign was the first step toward collaboration with the local communities and company employees on wildlife-management measures. Conservation awareness of local villagers, hunters, women's groups, company employees, and workers' unions was raised and a dialog established through individual contacts, films, meetings, and seminars. Over the course of the first 2 years of the project, every village in the Kabo and Pokola concessions participated in awareness meetings. New wildlife-management interventions and principles were described and discussed to promote community understanding, participation, and support. Nature clubs were established within local schools in towns and camps closest to the Park. Documentary films and a protected-species education program for primary schools, individuals, and target groups were used to improve conservation understanding on local, regional, and national levels.

Wildlife-Management Principles and Regulations

Private companies in the Republic of Congo are required by law to establish and register regulations governing employee conduct and use of company infrastructure. Violations of these regulations are subject to company-based penalties ranging from verbal reprimands to loss of pay and dismissal. To allow the company to punish employees for wildlife violations, steps were taken to legally modify the company's interior regulations to include wildlife-management principles. This process required more than 2 months of discussions between WCS, the government, and CIB representatives with employee labor unions (representing 1500 employees). An agreement was concluded integrating comprehensive wildlife-conservation and management regulations (table 21.1) and disciplinary measures for violations (scaled by severity of the violation). Unions agreed to these changes on condition of development of alternative protein sources (beef importation, fish, poultry farms, etc.) as a substitute for bushmeat.

Over the course of the first year a wildlife-management zoning plan was developed based on studies of traditional land-tenure systems of local communities, and was discussed and adopted with each village concerned (Moukassa 2001). Zoning provided for community hunting zones and no-hunting areas, and a prohibition of hunting in areas immediately sur-

Table 21.1 Wildlife-management regulations integrated into CIB logging-company policy in northern Congo

Regulations	Observations
1. Prohibition of snare hunting	Considered to be wasteful and unselectively kills nontarget species
2. Prohibition of hunting of protected species	Gorilla, chimpanzee, elephant, leopard, bongo, etc.
3. Prohibition of export of bushmeat from sites	Promotes local consumption only, discourages commercialization
4. Establishment of conservation zones within the concessions where hunting was not allowed	Particularly in forest clearings and other sensitive areas important for wildlife
5. Establishment of community hunting zones	Permitting legal hunting and local consumption
6. Development of a system of community-based hunter associations	Promoting controlled, legal exploitation
7. Restrictions on transport of hunters and wild meat in company vehicles	Hinders commercial hunting and export
8. Prohibition of transport of commercial hunters and traffickers	Restricts physical access
9. Specific regulations controlling subsistence hunting of forestry teams	Permits legal subsistence hunting while prospecting for trees but prohibits any export of bushmeat

rounding the Park. Committees of five major and six smaller villages (representing about 95% of the human population in the logging concessions) adopted regulations for community-hunting zones. These communities were receptive to the opportunity to reinforce their traditional rights, excluding access of nonlocal hunters to "their forest." Hunting committees were organized to act as focal points for resolution of wildlife-management problems and were responsible for member compliance of hunting regulations. These groups assisted in transforming an initially highly confrontational relationship (punctuated by at least two near-revolts in protest of wildlife controls) into a functional collaboration.

In late 1999, CIB began applying its regulations for wildlife management. Employee violations registered by the teams of government officers and ecoguards were documented and transmitted to CIB management (as well as the regional government authority). CIB then reviewed the charge and decided upon appropriate disciplinary measures depending on the seriousness of the violation. Penalties ranged from verbal reprimands for minor offenses to dismissal for more serious offenses, such as the poaching of elephants and apes. Key problems encountered included the following:

- Initially, CIB did not want to punish its skilled workers (particularly vehicle drivers) out of fear of reducing the efficiency of timber-production operations.
- Several CIB managers initially perceived wildlife-poaching problems as the responsibility of WCS and the government and not that of the company.
- Employee complaints regarding a perceived lack of protein availability and threats of labor strikes in protest of law-enforcement actions.
- Unions and employees contested disciplinary measures for wildlife violations in the regional judiciary system.

To address these problems, high-level CIB management clarified the company's commitment to improve wildlife management and communicated these principles directly to field managers and other concerned parties.

Enforcement of Wildlife Laws and Regulations

Law-enforcement efforts in the field were initiated in Kabo in 1999 with a small, mobile team of 2 Ministry officers and 8 ecoguards. Ecoguards recruited from local communities received training from the Ministry, the Congolese military, and WCS staff. Control posts were built at key vehicle-circulation points and mobile patrols surveyed sectors of the forest. Activities were progressively expanded geographically with increased rigor to apply hunting laws and company regulations. The protection unit was augmented as additional funding became available, reaching 6 Ministry officers and 25 ecoguards by late 2001 and being expanded to 40 guards in late 2002.

Wildlife law-enforcement activities in Kabo were initially focused on halting poaching of protected species, eliminating use of wire snares, and enforcing the zoning system. Hunters were encouraged to register their shotguns and purchase hunting permits from Ministry authorities, following Congolese wildlife laws. After enforcing these basic laws, a no-export rule was applied at the Ndoki 2 and Kabo sites. Pokola, with its large population center and long-established poaching networks, required a more staged approach. As in Kabo, initial efforts were prioritized to help stop protected-species poaching, reduce snaring, and enforce zoning. Because of pressure to feed the population and the potential for social unrest, a more gradual approach was used in controlling the supply of meat from source areas to Pokola town.

From 1999 to 2002, wildlife management and conservation systems were put in place over at least 600,000 ha of forest covering the entire Kabo concession and northern half of Pokola. During the first 3 years, a total of 3139 days of patrol-team effort resulted in the confiscation of 13 high-caliber rifles, 7 automatic rifles (AK47 and SKS, used for elephant poaching), 59 ivory tusks, 656 12-gauge shotguns, and 25,800 snares. Overall, 223 legal charges were filed against individuals for wildlife law violations (including 56 cases of protected-species poaching). Trends in return per unit effort demonstrated an overall reduction in poaching incidents, following an initial campaign in the Kabo concession (fig. 21.4, top panel).

Return rates for shotguns (per unit patrol effort) were significantly higher in Pokola (fig. 21.4, bottom panel) than Kabo ($W = 26$, $n = 3$, $P = 0.02$). Snare-return rates in Pokola were at least double those of Kabo in all years, and were 10 times greater in three 6-month periods ($W = 28$, $n = 6$, $P = 0.04$; Wilcoxon rank-sum tests; fig. 21.5). Personnel and financial limitations and conflicts with commercial-poaching networks slowed expansion of protection activities in the southern parts of Pokola.

Employees and nonemployees differed in the frequency of involvement in different types of violations ($\chi^2 = 82.98$, df = 5, $P < 0.001$). Employees took advantage of company infrastructure to transport hunters and bushmeat, whereas nonemployees were more often apprehended for elephant poaching (table 21.2). Differences in the frequency of violation categories were also found between the Kabo and Pokola concessions

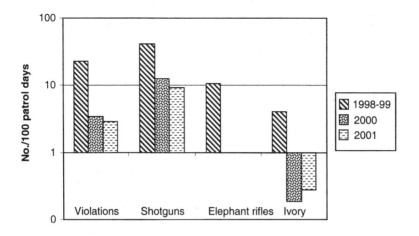

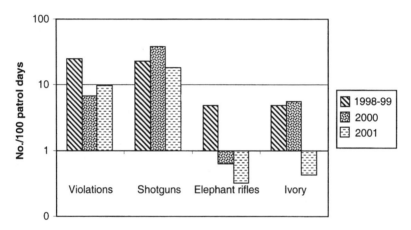

Figure 21.4. Violations and confiscations per 100 patrol days (on a log scale to facilitate interpretation) in the Kabo (*top*) and Pokola (*bottom*) forest-management units, 1999–2001

(χ^2 = 29.68, df = 5, P < 0.001; Chi-square tests). Eighty percent of all incidents of gorilla, chimp, and leopard (all protected species) poaching violations occurred in the Pokola concession.

In addition to CIB penalties, legal charges against violators of company regulations and wildlife laws were submitted to the Regional Direction of Forestry Economy for legal processing. Individuals charged with protected-species poaching were arrested and sent to the regional justice authority. Efficiency in processing of legal charges for wildlife violations varied with the level of motivation of the regional government administrators.

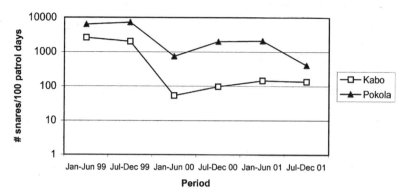

Figure 21.5. Snare-seizure rates per 100 patrol effort days (on a log scale) in the Kabo and Pokola forest management units, 1999–2001

Human Demographics

The population of Pokola town grew from an estimated 7200 inhabitants in 1999 to an estimated 11,400 in 2002, a 27% annual growth rate and 58% overall increase. Kabo village increased from 1406 in 1999 to 2333 in 2002, a 25% annual growth rate and 66% overall increase, associated with expansion of CIB activities. Seventy-nine percent of the population growth at the two CIB sites was caused by in-migration as a number of secondary industries (charcoal, wood theft, carpentry, etc.) developed in connection with the sawmills.

Demographic growth in the more-isolated company camps was slower than at industrial sites. At Ndoki 2 camp, after an initial 31% increase (from 616 to 806) from 1999 to 2001, population growth slowed to 6% to 7% per year (reaching 859 in 2002). This pattern is explained by initially high in-migration by employment seekers and hunting opportunists at the opening of the site in 1997. Over time, employment opportunities declined and wildlife-management activities discouraged commercial hunting, thereby contributing to the departure of some of the population.

Alternative Activities

Little domestic-animal protein was available prior to the initiation of the project, as local people in the area traditionally hunt, fish, and gather forest products (Moukassa 2001). We designed a program to develop sustainable systems to supply protein substitutes for bushmeat (with an

P. W. Elkan, S. W. Elkan, A. Moukassa, R. Malonga, M. Ngangoue, and J. L. D. Smith

Table 21.2 Frequency of employee and nonemployee incidences of different categories of wildlife violations in Kabo and Pokola logging concessions, 1999–2001

Violation category	No. of violations	Nonemployees	CIB employees
Elephant poaching	29	83%	17%
Hunting in a protected zone	19	21%	79%
Export of bushmeat	165	18%	82%

emphasis on the larger population centers) and to create revenue opportunities for local community members as an alternative to commercial hunting. We quickly learned that local communities would not easily modify their habits, unless it became a necessity.

Our program identified and worked with local people on vegetable gardening, fishing, and traditional animal-husbandry practices. Particularly well motivated individuals were targeted for introduction of new ideas, such as fish-farming, improved chicken farming, and beef importation. Fishing associations were organized early in the project to provide fishing supplies to traditional fishermen at low cost. Chicken vaccination and technical assistance supported development of local poultry farms. Guinea pig, porcupine, rabbit, and snail farming were trialed. Activities focused on both company sites (camps and towns) and traditional communities, with a direct emphasis on reducing hunting pressures. Regular meetings were held with CIB-employee unions to assess progress of alternative activities and to emphasize adherence to company wildlife regulations.

Over the 4-year project, company and local support for alternative protein increased dramatically. After relatively slow progress during the first 2 years, alternative sources of protein became a strong preoccupation for local people and the company, as protection efforts reduced the supply of bushmeat to larger towns. By early 2002, the project and company had established two large fish farms, four chicken farms with a capacity of 2000 chickens each, two butcher shops, one slaughterhouse, and five cold rooms to store imported produce. CIB further assisted local tradesmen by importing more than 30 cattle and up to 15 tons of frozen produce every 2 to 3 months. Material assistance was provided to local gardeners, chicken farmers, and fishermen, including 470 m of fencing, 22 kg of vegetable seed, 18,600 fishing hooks, and 11,600 m of fishing nets.

Protein consumption was monitored at the four project sites, with 19,207 household meals being examined in the Kabo and Pokola concessions in 2000–2002. Forty percent of all meals surveyed were made up of freshwater fish, 40% of bushmeat, and 7% of other species (snails, caterpillars, etc.). Only 2% of all meals included domestic meat, although this percentage increased noticeably in 2002. Fish was available year-round but quantities varied, with a peak in the dry season (February–March) and a scarcity in June–July. Bushmeat in the diet increased when fish availability was low. The frequency of domestic meat consumed at Pokola rose from 1%–2% of all meals in 2000–2001, to more than 5% in mid-2002, and greater than 25% in late 2002. Similar increases occurred at Kabo. These increases closely corresponded with greater importation of beef, chicken, and fish by the company, workers unions, and our project in 2002. However, domestic meat consumption was low (<1%) at Ndoki 2 and Ndoki 1 camps throughout the entire period.

Hunting Management

We recorded 10,178 mammals, reptiles, and birds (totaling an estimated 111,378 kg of wildlife biomass) entering Ndoki 1, Ndoki 2, Kabo village, and Pokola town during 1471 survey days in 2000–2002. Mammals were by far the most important hunted taxa, with 10 species of ungulates (including 6 duiker species, *Cephalophus* spp.) making up 74% of all kills and 87% of the biomass examined in the Kabo and Pokola concessions. *Cephalophus callypigus* accounted for 41% of total animal biomass and *C. monticola* represented 16% of biomass and was the most frequently killed species (37%). Primates constituted 16% of all kills, and rodents 4%. Reptiles were an important source of protein at the Pokola site, with dwarf crocodile (*Osteolaemus tetraspis*) constituting 4% of biomass and 3% of kills. Carnivores and birds were rarely taken (<1% of all kills).

The proportion of duikers, primates, and rodents differed among the four sites ($\chi^2 = 374$, df = 6, $P < 0.001$). A strong inverse relationship was found between the proportion of duikers and primates killed ($F = 39.0$, $R^2 = 80\%$, $P < 0.001$; simple linear regression). Rodents and small carnivores also tended to increase among kills as duikers decreased, although the relationship was weaker. These data indicate that hunters turned to primates and rodents when duikers become less available.

At the Ndoki 2 camp, where the history of hunting was more limited, duikers were consistently highest in importance (70% of all animals taken) with primates also being important (18%). In contrast, hunters reported that zones near Pokola town were generally depleted of wildlife. The relatively high frequency of rodents (4%) and even birds in the diet in Pokola indicated an increasing demand for animal protein, as bush-meat from nearby source areas became scarce because of increased law enforcement and human population growth.

Abundance of Large Mammals

From 1999 through 2001, data on the relative abundance of large-mammal sign and human activity were gathered over 1010 km of walking surveys traversing nine different wildlife-management zones of the Kabo and Pokola concessions. Elephants, gorillas, and chimpanzees were detected in all zones surveyed, excepting one. Bongo and buffalo sign was patchily distributed (with higher abundances being recorded than in the Nouabale-Ndoki Park and neighboring Loundougou and Mokabi concessions; Blake 2001; Elkan 2003). Human sign was lower in no-hunting zones than in hunting areas, suggesting a general respect of regulations ($P = 0.02$). Elephant-dung encounter rates were lower in hunting than protected zones ($P = 0.002$; Mann-Whitney U-tests), but tests for other species revealed no significant differences between hunted and nonhunted areas.

As wildlife-protection activities progressed, general observations showed increased large-mammal activity near roads and villages. Elephant sign, initially only observed more than 3 km from Kabo village, was regularly observed within 1 km of the village in 2002. Gorillas, which normally flee when encountering humans, exhibited curiosity behavior and approached humans on several occasions in the no-hunting areas of the Kabo concession, suggesting low disturbance levels. A solitary male gorilla visited the Ndoki 2 logging camp intermittently from 2000 to 2002, and remains unharmed by camp residents.

Level of Effort Required

Over the course of the project, financial and institutional support increased and allowed us to expand our activities, particularly with funding from the International Tropical Timber Organization (ITTO) in 2001. Level of effort grew progressively with $300,000, $550,000, and $800,000 in the first, second, and third years of the project, respectively, allowing expansion of activities to about 2000 km^2, 4000 km^2, and 6000 km^2 each year, respectively. Annual costs varied from $1.25 to $1.50 per ha with an increased effort required in Pokola brought on by greater hunting and human population pressures.

Not including the costs for initial establishment of project infrastructure, wildlife-protection costs were the most expensive activity (40%–50% of the annual budget), followed by research and monitoring (17%–19% of budget). Activities to provide alternative protein sources consumed about 10% of the project budget, and also received direct CIB support (such as use of cold rooms, transport of domestic meat, and construction of fish ponds). Conservation-awareness activities required a relatively low investment (10% of budget) but reached a broad geographical region. International technical expertise was fundamental to the development, training efforts, and administration of the program.

DISCUSSION

A critical, indirect impact of most tropical logging operations is a drastic increase in hunting pressure. Our program, designed to reduce the impacts of hunting on wildlife in logged forests, was a multifaceted approach involving local communities, private industry, and the Government of Congo, in a large area of forest adjacent to a protected area of global biodiversity significance. As a result, wildlife management was greatly improved in much of the project area and the process led to shifts in logging-company and community involvement after only 4 years of interventions. Although problems still persist in some areas and the sustainability of subsistence wildlife harvests are under investigation, the systems needed to address these problems have been firmly established.

Establishing personal relationships among organization representatives, investing considerable time in discussion of management issues, and engaging in frequent and transparent communication, were all critical elements of this program. From the beginning, high-level logging-company staff took a direct role in the process and communicated with employees and unions about the company's commitment to sustainable wildlife management. Moreover, direct and indirect involvement and support by the logging company increased greatly over the course of the project. Participation in the program has improved the company's image internationally as well its opportunities for eco-certification of its timber operation. The low human population density in the Kabo logging concession also facilitated law-enforcement activities and zoning of extensive no-hunting areas.

This project—an unusual collaboration among industry, government, and nongovernmental organizations—faced some major challenges.

Among the most difficult challenges were management issues that could potentially have had large impacts on logging-company revenues, such as establishing set-asides of high-biodiversity areas and altering road and logging-camp placement. Social conflicts caused by the reduction of the commercial bushmeat trade and pressures to feed the large town of Pokola also created important challenges for collaboration with local communities, and were the most complex to manage.

The wildlife-management efforts described in this chapter have been the subject of both praise and criticism. Criticisms include alleged corruption of law-enforcement units, concerns about restrictions of hunting rights, a perceived over-emphasis of the project on large mammals, and appeals that the private sector should cover all the costs. We agree that new conservation approaches involving private industry require a high level of rigor and should be critically examined, tested, and evaluated. However, such evaluations must objectively consider the biological context of the project, review and evaluate monitoring data, and take into account the complexities of the underlying ecological and societal problems.

In this project, detailed socioeconomic and ecological indicators and performance monitoring were used to assess progress in the logging concessions and to document logging-company involvement and improvements. Extensive internal control and personnel-management systems were developed to reduce corruption in wildlife-protection teams. Wildlife-management zoning reinforced traditional resource-tenure systems, and employment opportunities were created with the timber company and conservation projects for local communities. While much remains to be accomplished in the Kabo and Pokola timber concessions, the level of progress achieved thus far has substantially improved the conservation situation for all species of wildlife over an extensive area.

One of the strengths of this approach is that many of the costs of wildlife management can be passed on to the timber industry and, ultimately, to the timber consumer. Achieving this will require governments to adjust their policies and to develop viable eco-certification systems and other incentive mechanisms. The urgency of this task is highlighted by the fact that many protected areas in the tropics are surrounded by industrial-timber operations and are experiencing conservation threats requiring immediate intervention. The rapid escalation of these threats suggests that a stepwise approach, with international support being used in combination with funding and engagements from the private sector and Congo Basin governments, is direly needed.

Wildlife-Management Principles and Regulations

In this project, the integration of wildlife rules into logging-company regulations reinforced national laws and compensated for slow legal procedures within the country. Transport of game meat on vehicles is not technically illegal by Congolese law; however, the logging company, as a private owner, can regulate the use of its equipment and infrastructure as it deems appropriate. Wildlife-management zoning supported traditional land-use patterns in the region, reinforcing a sense of ownership by local communities and closed access to outsiders.

Although the more serious violations were strongly punished, the CIB Company needs to be more rigorous in its strategies to reduce employee involvement in commercial export of bushmeat. The employees most susceptible to trafficking bushmeat are vehicle drivers, who often greatly supplement their salaries by taxing hunters for transport on their vehicles. Taking disciplinary action or firing drivers could create temporary personnel shortages and interfere with logging operations. However, companies will need to accept such potential perturbations in operations as an inherent cost of improved environmental management. Additional benefits to the company from the controls we established were fewer thefts of company materials and reduced hunting by employees during work hours.

Law and Regulation Enforcement

The recruitment of ecoguards exclusively from local, traditional communities increased support for the initiative and provided alternative employment for some former poachers. The law-enforcement efforts limited the bushmeat supply to town markets and required local hunters to respect national legislation in relation to hunting permits. Changes have also been detected in hunting techniques: Hunters now make more use of shotguns than of metal cable snares. Law enforcement made it possible to extend protection to rare and endangered species, such as gorillas, chimpanzees, bongos, and forest elephants, over a large area in both community-hunting areas as well as no-hunting zones.

Persistent problems in this project include export of bushmeat to feed the Pokola population and elephant hunting for meat and ivory in southern areas. Elephant poaching has a higher financial return than is involved for other protected species but a similar penalty risk. Sustained elephant protection will require high-level government engagement, strict field-based law enforcement, and legal prosecution of poachers. Although enforcement efforts need to be expanded geographically, it was repeatedly demonstrated to us that without a minimum level of law-enforcement presence in an area, penalties and incentives are not yet strong enough for hunters to forgo short-term gains from commercial hunting.

Human Demographics

Sawmill sites and secondary-employment opportunities lead to demographic booms (more mouths to feed) and a socially volatile situation resulting in a difficult challenge for wildlife law enforcement. The majority of the human population in the Kabo and Pokola concessions was concentrated in and around the Pokola town and Kabo village where sawmills create employment and related opportunities, such as from wood theft, charcoal fabrication, and commerce. Additional attractions to Pokola for the regional populace are hospital facilities (the only functioning hospital in northern Congo), electricity, water, and television services provided by the logging company. Control measures, such as those employed in this project, can more easily slow rapid growth rates at logging camps that are geographically isolated, owned by the logging company, and have limited employment opportunities.

Alternative Activities

Allowing hunting to feed company employees has long been the norm in many tropical timber concessions (Robinson et al. 1999). Modifying company policy to require adherence to wildlife regulations creates a clear disincentive for employees to violate regulations, although sufficient alternative protein must be available for this to be effective in practice. Issues relating to alternative sources of protein were among the most socially complicated and sensitive of all management activities.

Efforts by project and company personnel led to a relative increase in availability of domestic-protein sources and fish in the Kabo and Pokola villages. Under pressure to feed the growing Pokola population, CIB, in mid-2002, began to import large quantities of frozen domestic meat to Pokola from Cameroon, a trend that was subsequently detected in household surveys. The seasonal availability of freshwater fish at all sites shows the need for development of freshwater-fisheries management.

Achievement of sustainable management of wildlife requires companies to help establish systems to provide alternative protein sources. The increased cost involved in purchasing domestic meat may also require companies to increase employee salaries.

Local communities have traditionally depended on wildlife as a renewable resource providing both protein and revenue (Davies 2002). Alternative opportunities for revenue generation were created through employment of ecoguards, efforts to provide alternative protein, and additional positions with the forestry company. While all of these have potential and should be pursued, employment opportunities in the timber industry have the greatest potential to provide an alternative to commercial hunting for local communities. At present, many companies recruit skilled labor from capital cities (and even other countries) for convenience. This takes opportunities away from local communities and inevitably increases secondary in-migration. It is therefore in the long-term interest of timber companies, local communities, and wildlife for the timber companies to invest in training and development of skilled, locally recruited workforces and to prioritize the recruitment of local indigenous peoples.

Hunting

While the sustainability of wildlife harvests in the Kabo and Pokola concessions is still being investigated, data indicate that significant progress has been made in the control and management of hunting pressures. A reduction in snaring has reduced waste and harvest of nontarget species. Designated no-hunting zones serve to protect important populations of game and key habitat, acting as population sources for neighboring hunting areas. Restrictions on bushmeat export helped to close the "open-access" nature of the trade, although community-based wildlife-management programs require the presence of semi-permanent law enforcement and monitoring in the short and medium term.

The ability to spatially manipulate hunting pressures, and potential for species-specific hunting regulations, show significant potential to aid sustainable-hunting programs in areas with small communities (McCullough 1996; Robinson and Bodmer 1999; Novaro et al. 2000; Eaton 2002; Elkan 2003; Milner-Gulland et al. 2003). Spatial manipulation of hunting can be achieved through zoning, restricting access, designated and organized hunts, and additional company regulations.

Wide-ranging rare and endangered species such as elephants and bongo received increased protection under our program as they moved across the logging concession and protected-area landscape (Blake 2002; Elkan 2003). Data suggest that, to conserve bongo and buffalo populations, efforts should be focused beyond park borders to the concessions that have higher populations of these species. In several logged areas that received conservation protection, both gorilla and chimpanzee sign were found in locally high numbers despite local hunting of small game (Elkan 2003). The presence of these species in hunting zones and increasing observations of calm behavior in the presence of humans support law-enforcement data, indicating an overall reduction in protected-species poaching over a large area.

Our efforts to improve wildlife management in northern Congo indicate substantial progress over a relatively short period of time, the potential for sustainable wildlife management in some areas, and growing threats to wildlife as logging expands into previously remote forests. Although inherently difficult, and at times controversial and socially contentious, this work has demonstrated the practical strengths of joint wildlife-management interventions based on collaboration among the Government of Congo, a nongovernmental organization, a private timber company, and local communities. In 2002, the Government of Congo, based on the initiative presented here, introduced policies requiring logging companies to finance wildlife-management programs. The approaches and lessons gained from the initiative in northern Congo should serve to promote and inform improved management efforts in logging concessions in the Congo Basin and tropical forests elsewhere.

CONCLUSIONS AND IMPLICATIONS

1. Our study demonstrates that hunting pressures can be controlled in logging concessions, which are rapidly proliferating in Central Africa. A multifaceted approach based on private-sector commitment and community collaboration is critical for the establishment of sound wildlife management.

2. A zoning system reduced in-migration of commercial hunters, helped to foster a sense of responsibility for game management among local communities, and decreased hunting in key wildlife-habitat areas.

3. Increasing the price of bushmeat, reducing its availability through law enforcement, and attaching "ownership" to wildlife by creating exclusive zones for local communities, all function to increase the value of hunted wildlife.

4. Semi-permanent monitoring by law-enforcement teams is essential to ensure continued compliance with wildlife regulations and also provides employment to local communities.

5. In logging concessions, closing roads after exploitation and minimizing the number of access routes is fundamental for reducing the costs and number of personnel needed to manage hunting.

6. The commitment of the logging company to provide an alternative source of protein for its work force is essential to wildlife-management programs.

7. Efforts to locate sawmills in existing populated centers (away from forests) should be promoted in regional-planning efforts and by organizations providing international assistance.

8. The human and financial resources needed to manage hunting are comparable to the costs of managing protected areas in the region. These costs vary depending on human population pressures, and there is an economy of scale as concession sizes increase.

9. Integration of the costs of wildlife-management programs into timber-business plans will be essential to sustain such efforts. Policy frameworks are needed to create incentives for private-sector investment in wildlife management.

10. Timber concessions with high biodiversity, important wildlife populations and habitat, and which act as buffer zones or corridors for nearby protected areas, are priorities for conservation intervention.

11. Timber-certification schemes must be clarified and consumers educated to take into account wildlife-conservation concerns specific to the forests of the Congo Basin.

ACKNOWLEDGMENTS

We thank the Government of Congo for its support of the work described here. Jean Marie Mevellec and Jacques Glannaz of CIB and Hinrich Stoll and Robert Hunink of TT Tropical Timber International worked constructively to integrate wildlife management in production forests. The ITTO, governments of Switzerland, Japan, USA, and France, United States Agency for International Development, Central Africa Regional Program for the Environment, United States Fish and Wildlife Service, United States Forest Service, Liz Claiborne Art Ortenberg Foundation, and Columbus Zoo partially funded this project. John Robinson, Mike Fay, and Amy Vedder helped to develop the initiative. John Robinson, Kent Redford, David Wilkie, and two anonymous reviewers provided helpful comments on the manuscript. Kimbembe Bienvenue, Calixte Makoumbou, and many other research assistants and ecoguards assisted with data collection. This chapter is dedicated to Frederic Glannaz, CIB Forest Management Planner.

CHAPTER 22

Tropical Forests: A Protected-Area Strategy for the Twenty-first Century

Chris Fagan, Carlos A. Peres, and John Terborgh

The 3000 conservationists present at the fifth World Parks Congress in Durban, South Africa (September 2003) rose in a standing ovation to the announcement that, as of 2003, the nations of the Earth had collectively incorporated more than 12% of the Earth's surface in formally constituted protected areas. While the announcement was met with evident jubilation from an international gathering of conservationists, we ask the following question: From a scientific standpoint, should this news evoke rejoicing or despair?

In a scientific context, the Durban announcement contains both good news and bad. The good news is that the portion of the Earth's surface placed under formal protection has tripled in the past 20 years; the bad news is that science suggests that it will be necessary to protect a far larger portion than 12%—and that the preserved area will need to be more effectively distributed among endangered biomes—if biodiversity is to be conserved in the long run. Another bit of good news to emerge from Durban is that some of the world's most esteemed leaders are recognizing that we must redouble our efforts to conserve biodiversity.

The congress opened with a ceremony that included the reading of a message from United Nations Secretary General, Kofi Annan. Annan began his speech by reflecting on the impressive growth in the number of protected areas since the fourth congress held 10 years earlier in Caracas, Venezuela. This positive note was quickly tempered by an admonition against complacency, as he warned that the world continues to lose biodiversity and ecosystem functions. Many protected areas, although ostensibly protected, still are not effectively implemented and, at a practical level, exist only on paper. The Secretary General concluded by emphasizing the importance not only of creating new protected areas, but also of ensuring that established reserves are fulfilling their function of conserving biodiversity.

As the congress unfolded, delegates from around the world described their experiences in weak or failing protected areas, as well as their profound concerns for the new and intensified threats that the future would bring. These concerns were given prominent voice in "The Durban Accord"—a vision statement for the twenty-first century based on the outcomes of the congress. The first page proclaims, "In this changing world, we need a fresh and innovative approach to protected areas and their role on broader conservation and development agendas." While applauding recent progress in establishing protected areas, the accord reminds us of the imminence of threats to tropical nature and their increasing severity—as outlined in many of the chapters in this volume. These threats highlight a pressing need for a new conservation strategy.

Expanding lists of endangered species in all quarters of the globe attest that existing conservation strategies do not adequately conserve biodiversity and the ecosystem functions that sustain it (e.g., Brandon et al. 1998; Stolton and Dudley 1999; Liu et al. 2001; IUCN 2003). For example, a recent study suggests that up to half of the world's plant species may qualify as threatened with extinction under the IUCN classification system (Pitman and Jorgensen 2002).

The challenge of stanching extinctions will inevitably continue to grow before it eases. As deforestation, forest fragmentation, invasive species, climate change, and other threats described in this book proliferate around the globe, the pressures on already-beleaguered protected areas will intensify. Such routine threats as poaching and logging will be exacerbated by continued population growth, globalization, and heightened competition for the Earth's finite resources. The challenge of conservation will be especially daunting in tropical developing countries where poverty and political instability often combine to make protecting nature a neglected afterthought. These harsh realities call for concerted measures to strengthen existing protected areas and to extend protection into priority areas, such as the remaining wilderness areas and "biodiversity hotspots" (Myers et al. 2000).

The prescriptions we propose to protect tropical biodiversity are based on four primary goals. The first goal is to create protected areas large enough to harbor intact biological systems complete with top predators and large-scale natural disturbance regimes such as fire and flooding. The focus should be on expanding protected areas into the large expanses of tropical forest wilderness that remain intact. The second goal is to improve the management of all categories of protected areas, both core-protection areas and extractive reserves. The third goal is for both the design and management of protected areas to be based on sound conservation science. For this to work, many scientists must cross into unfamiliar territory and become leaders in the conservation movement by engaging in the political discourse inherent to conservation. Conservation scientists can provide the answers to important questions needed to generate political will and public support for conservation. And the fourth goal, the management effectiveness and conservation status of protected areas, must be consistently and independently evaluated, with the resulting information shared with managers and other stakeholders through established systems of information exchange. Independent evaluations and information sharing will create transparency and guarantee that the people with the authority over protected areas are held accountable for their actions.

This chapter begins with a synopsis of the science that must be implemented if tropical biodiversity is to sustain itself through the coming century. Crucially, protected-area networks must be large enough to sustain keystone species—particularly large carnivores—as well as landscape-scale disturbance regimes. We also consider the prospects of extending the coverage of protected areas in frontier regions and encouraging the "rewilding" of degraded forestlands. We then continue by reviewing the latest data on the extent and conservation status of tropical forest protection. Finally, we conclude by recommending some guiding principles in our strategy for ensuring the protection of tropical forests in a future that will bring unprecedented global change and new challenges to conservation.

IF TWELVE PERCENT IS NOT ENOUGH, THEN HOW MUCH IS ENOUGH?

High contemporary extinction rates demonstrate that current conservation strategies, which rely heavily upon protected areas, are falling seriously short of what is needed. One reason extinctions continue to rise is that much of the Earth's protected-area system was created prior to the rise of conservation biology, and for reasons—such as protecting scenic landscapes, especially at higher elevations poorly suited for agriculture—that are unrelated to conserving biodiversity. This statement applies to nearly the entire U.S. national park system. For example, U.S. national parks have failed to prevent an exponential rise in the endangered species list because they are too few, too small, and (in many cases) located in regions containing few or no threatened species (Dobson et al. 1997). Such considerations suggest that the protected-area systems of many countries, including the United States, require a major overhaul. This is a tall order that will require both popular support from the bottom and political will from the top. Nonetheless, the impressive progress made over the past 20 years in establishing new protected areas is encouraging, and indicates the newfound importance the global community has attached to protecting nature.

Conservation of biodiversity is only one aspect of tempering human profligacy to bring society into balance with the finite resources of the Earth. A few thousand years ago, natural ecosystems dominated the planet. Thus, it is more than mildly paradoxical that we are thrilled to learn that nature is being given a mere 12%. Twelve percent is not enough; it is not even close. The most naïve interpretation of species–area curves yields the prediction that, with the destruction of 90% of the habitat, the number of species in a region will be reduced by half. But this simplistic prediction fails to take into account that the remaining natural habitat is likely to be fragmented into bits and pieces, each representing a small fraction of the remaining habitat area. Moreover, predictions derived from species–area curves apply only to the moment of reduction in area. Yet we now know from dozens of studies that habitat fragments, even quite large ones, invariably lose species over time (Willis 1974; Newmark 1995; Lynam 1997; Beier et al. 2002; W. F. Laurance et al. 2002b; Brashares 2003; Ferraz et al. 2003). By failing to account for this extinction "debt," which must be repaid over time, species–area curves overestimate the likelihood of long-term persistence for vulnerable species. Thus, both habitat loss and time are the enemies of species survival.

Furthermore, we know that the effect of time is inversely related to the size of a habitat fragment: species loss is fast in small fragments and slow in large ones (Terborgh 1975; Pimm et al. 1988; Ferraz et al. 2003). This knowledge lies behind the imperative of creating the largest possible protected areas and tying them together with corridors to make them, effectively, even larger. The science behind this model is well grounded and we regard it as highly unlikely that new findings will point in contradictory directions (Soulé and Terborgh 1999).

The loss of species with reduction in habitat area, and the loss of species over time in a given habitat isolate have both been extensively quantified and have entered into the realm of predictive science (Diamond 1972; Terborgh 1974; Ferraz et al. 2003). Yet there are still missing pieces of the puzzle. At a platitudinous level, it can be stated that species disappear from habitat isolates because area reduction disrupts the status quo ante. But, it is difficult to generalize further because at this point the analysis seems to break down into particularities resulting from the fact that *many* features of an intact ecosystem can be disrupted by area reduction. For example, disturbance regimes (such as the strength and frequency of fires and floods) and other fundamental processes are nearly always affected. On the biological side, large-bodied species, those high on the trophic ladder, and those with low initial population densities are vulnerable to stochastic population loss and, frequently, to poaching (Woodroffe and Ginsberg 1998; Peres 2001b). As species high on the trophic pyramid disappear or decline, the strengths of predation and herbivory often diverge from initial conditions, with cascading effects that can be devastating to both plants and smaller animals (Terborgh et al. 1999, 2001). Further consequences of fragmentation include the disruption of migrations and the exposure of interior habitats to harmful commuter species, such as brown-headed cowbirds in North America that fly kilometers into continuous forest from their open feeding habitat to parasitize songbirds (S. K. Robinson et al. 1995). Area reduction also invites the invasion of alien species, both plant and animal. In short, area reduction carries with it a multiplicity of disruptive effects, many of which have negative consequences for native species (W. F. Laurance and Bierregaard 1997).

Large, interconnected networks of protected areas offer the best defense against the ravages of habitat-area reduction and time (Soulé and Terborgh 1999). Unfortunately, the science behind this conclusion has matured only within the past two decades. Are we simply too late with the science needed to conserve nature? Certainly, in some countries, opportunities to protect adequate amounts of habitat have been preempted by "development." Where opportunities are limited, transnational protected areas may be an option. But there is opportunity in the fact that a sizable fraction of the Earth's terrestrial realm (estimates range between 30% and 50%) is still relatively "wild" (Mittermeier et al. 2003). It is within these remaining wildlands that opportunities still exist to create large, interconnected reserve systems that can retain their full species complements over time. What is needed now is the political will to make it happen (Kamdem-Toham et al. 2003).

To generate the political will, scientists must take the lead. We can provide the answers that make conservation more attractive and help to "market" the importance of biodiversity. And we must have convincing answers to the question certain to be posed by political decision makers: "If the 12% announced at Durban isn't enough, then how much is enough?" A recent study in the Cape Floristic Region of South Africa indicates that setting spatially explicit targets for biodiversity conservation is possible even in regions where few data on the distribution of biodiversity are available (Pressey et al. 2003). However, this approach to conservation planning is still relatively new and, in most cases, science cannot answer this question precisely. All we currently can say is, "the more the better"—although this answer will leave decision makers on shaky ground, as they like to have hard numbers that define concrete goals.

Although all numbers in this context have a certain arbitrariness, 50% is a much better figure than 10% or 12%. It is better because the science tells us that the lower figures are far from adequate, and it is better because—no matter how hard we campaign—we are unlikely to exceed 50% as a global figure. Thus, we can always insist that more is needed and always be confident that we are on solid ground in doing so.

Fortunately, there are still large areas of tropical-forest wilderness that remain intact, such as Amazonia and the Guyana Shield in South America, Gabon and the Congo Basin in Africa, and the island of Papua New Guinea. Large reserves can be created in these regions without engendering prohibitive social conflict or hardship. Where forests have already been degraded, the restoration of landscapes to a more natural character can be achieved through the process termed "rewilding" (Soulé and Noss 1998; Simberloff et al. 1999). Rewilding can be achieved by active management, such as by reintroducing top carnivores and other large mammals and by allowing forests to recover via succession. While it is of paramount importance to protect as much of the remaining wilderness areas as possible, fragmented or otherwise degraded forests have the potential to become valuable conservation assets if properly managed or simply left alone.

Those who would advocate protecting 50% of a nation's territory risk being dismissed as utopian and unreasonable, so some qualifications are in order. It certainly would be utopian to insist that half of the Earth's terrestrial, aquatic, and marine habitats should be conserved without development as strict nature preserves. But it is not utopian to insist that largely natural vegetation (including managed production forests) should prevail over half of the Earth's terrestrial realm. Natural and seminatural vegetation can serve many purposes in addition to conserving biodiversity. Such vegetation can protect watersheds, provide ecological services, produce timber and other natural products, nurture livestock, and, above all, enhance human quality of life. A vision that enhances the quality of life for humans is not a hard sell.

Clearly, we are not advocating "locking up" half of the Earth's natural resources; to the contrary, we are proposing a concept of sustainable development that includes strict nature preserves embedded in a matrix of economically productive land uses based on natural vegetation. The current allocation to strict nature conservation is far too low (see below). The relatively modest IUCN goal of 10% seems distant at this stage. Even 20% would be insufficient to bring the extinction rate near to zero.

As a comment on the feasibility of what we are proposing, let us consider a concrete example—that of the United States. About 40% of the United States is public land managed under various restrictions for multiple uses. By law, nearly all of this land remains under natural vegetation cover, but far too much of that has been overgrazed and clearcut, and some important ecosystems are missing entirely. Nevertheless, were it not for public land, the endangered species list would be far longer than it is. The fact that the United States has, in some fashion or another, conserved roughly 40% of its national territory indicates that an eventual goal of 50% is not beyond reach.

Other countries are beginning to realize that intergenerational interests are best served when a substantial fraction of the land remains in the public sector. For example, large portions of the Amazon forest remain in the public domain (in the form of *terras devolutas* or "unclaimed" public lands), as are comparably large fractions of the boreal forest in Canada and Russia. To be sure, most of this land is not gazetted for nature conservation, but under current law nearly all of it could remain under natural or seminatural forest cover. This represents a good beginning. The next step, both in the United States and elsewhere, is to promote biodiversity-friendly management of public lands. This means allowing natural-process regimes to organize the landscape, and controlling hunting to prevent the extirpation of top carnivores and other large animals. These are the most crucial measures that must be adopted to prevent the unraveling of nature.

A few countries have already embarked on such a far-sighted vision. Venezuela is perhaps the leading example, having already brought 46% of its national territory into its protected-area system. What convinced the politicians? Two arguments were paramount: preserving scenic landscapes for tourism, and protecting watersheds for agriculture and urban populations. Today, Venezuela benefits from a remarkable and perhaps unique status in the world as a country in which verdant mountains protected in national parks are visible from the streets of nearly every major urban center.

The specific objectives for which protected areas are managed vary over a broad spectrum from strict nature protection to the promotion of "sustainable" uses. Biodiversity protection is a principal objective only of areas within IUCN categories I through IV. These can be referred to as "hard" or "core" protected areas in which biodiversity conservation, ecological integrity, or wilderness takes precedence over other uses (Soulé and Terborgh 1999). "Soft" protected areas are those assigned to IUCN categories V and VI, in which management objectives such as recreation and resource extraction often take precedence over strict pro-

Table 22.1 IUCN protected-area categories, their primary management objectives, and the global number and extent of protected areas (from Chape et al. 2003)

IUCN category	Primary management objective	No. of sites	% of total no. protected areas	Area covered (km^2)	% of total area protected
Ia. Strict nature reserve	Scientific research	4731	4.6	1,033,888	5.5
Ib. Wilderness area	Strict wilderness protection	1302	1.3	1,015,512	5.4
II. National park	Ecosystem conservation and recreation	3881	3.8	4,413,142	23.6
III. Natural monument	Conservation of natural features	19,833	19.4	275,432	1.5
IV. Habitat/-species management area	Conservation through active management	27,641	27.1	3,022,515	16.1
V. Protected landscape/-seascape	Landscape/-seascape conservation and recreation	6555	6.4	1,056,008	5.6
VI. Managed resource protected area	Sustainable use of natural resources	4123	4.0	4,377,091	23.3
Unspecified category	Usually multiple objectives	34,036	33.4	3,569,820	19.0
Total		102,102	100	18,763,407	100

tection. Table 22.1 lists the principal management objectives of the six IUCN categories and includes the global number and extent of protected areas in each category.

According to the 2003 United Nations List of Protected Areas, there are currently 102,102 protected areas covering 18.8 million km^2, or 12.65% of the Earth's surface. Marine protected areas account for approximately 10% of the total ocean surface. Protection of the terrestrial realm covers 17.1 million km^2, or 11.5% of the Earth's land surface (including areas lacking IUCN categorization). As conservationists, we can rightfully be proud that these figures represent a tripling of the area protected since the IUCN Bali Congress of 1983, and a doubling since the Caracas Congress in 1993.

However, much of the gains in protected areas globally are in soft rather than hard protected areas. IUCN category VI (Managed-Resource Protected Area) is now the second largest category (23.3% of the global total), despite the fact that this category was created only in 1994. The popularity of soft protected areas reflects a widespread desire to link conservation with the sustainable use of natural resources. While category-V and -VI areas can have significant conservation values as buffer zones or corridors, their ability to sustain biodiversity as stand-alone conservation units is largely uncertain.

Globally, 44.8% of all protected areas and 47.9% of the total area protected are included in categories V and VI, or do not have a designated management category. Considering that nature is given primacy only in hard protected areas (categories I through IV), the portion of the terrestrial realm that qualifies as nominally fully protected drops from 12.65% to 6.05%—a figure that includes many "paper parks," especially in developing tropical countries that sustain much of the world's biodiversity. Thus, the 12% figure announced at Durban rather seriously overstates reality. Of particular relevance to this book is whether such a high proportion of tropical-forest reserves also have only limited protection.

BIODIVERSITY PROTECTION IN TROPICAL FORESTS

The most recent estimate of the percentage of the world's tropical and temperate forests (tropical moist, tropical dry, temperate broadleaf and mixed, and temperate needleleaf) under protection within IUCN management categories is 10.5%. This figure increases to 16% when those areas without a designated category are also included. When we examine the percentage of global forests protected within IUCN management categories, we find that tropical moist forest is the best protected, with more than 11% in reserves, and tropical dry forest is the least protected with less than 9% protected. Ten percent of temperate broadleaf and 11% of mixed and temperate needleleaf forests are protected (Chape and Sheppard 2003). To show the distribution of tropical forest protected areas by respective IUCN categories, we must look at the most recent estimates from the year 2000.

A total of 11.1% of tropical moist and dry forests (2,265,254 km^2 of the global total of 20,379,828 km^2) were considered to be protected in 2000, but only 7.2% of this total was included in the core IUCN categories I–IV (fig. 22.1). These figures are likely to be somewhat higher today, given a progressive increase in the number and extent of protected areas since 2000 in all biomes, including tropical forests. Nevertheless, as of 2000, only 7.2% of tropical forests were being managed with biodiversity protection as a primary objective, an amount considered to be wholly insufficient as a conservation target (Soulé and Sanjayan 1998; Rodrigues and Gaston 2001).

Are these protected areas located in the best places to maximize biodiversity protection? A recent analysis of lowland Amazonian Peru (below 500 m above sea level), which contains levels of local biodiversity

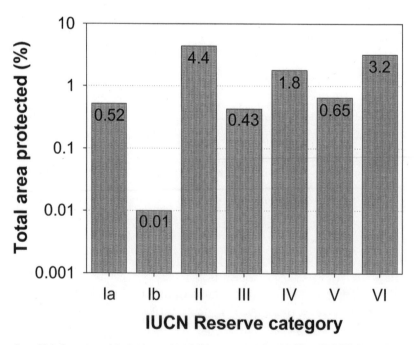

Figure 22.1. Percentage of the total area of tropical forest protected worldwide under IUCN Categories I–VI (UNEP-WCMC 2000), scaled to illustrate the dismally small percentages in several categories.

Chris Fagan, Carlos A. Peres, and John Terborgh

unmatched anywhere else on the planet, was informative. It revealed that, of the 14.9% of the region currently under protected status, 7.4% is without a defined IUCN category, and 4.6% is in category VI. Thus, a mere 2.9% of the region is within national parks (category II) (Llactayo and Pitman 2003). Remarkably, for a region renowned for its world-class biodiversity of birds, mammals, and trees, not even 3% is under strict protection. While this is not proof of a global trend, numerous other biologically rich regions in the tropics are similarly underprotected (e.g., Myers et al. 2000).

Many tropical countries are creating extractive and sustainable-development reserves in the conviction that nature can be used sustainably without damaging biodiversity—a belief that is maintained more by wishful thinking than by hard science. Even commercially important nontimber resources that justify the designation of numerous extractive reserves, such as the seeds of Brazil nut trees (*Bertholletia excelsa*, Lecythidaceae), may eventually face demographic collapse from chronic overharvesting (Peres et al. 2003a). It is thus important to question to what extent these areas are really protecting biodiversity in the long term. For example, current patterns of game hunting can have a disastrous effect on biodiversity, not only by depleting populations of large mammals and birds (Peres 2000a; Fa and Peres 2001), but by disrupting essential ecological functions, such as seed dispersal and herbivory (Redford 1992). A recent study in the Peruvian Amazon found that, unless hunting is restricted, large-bodied herbivores, carnivores, and primates are unlikely to persist in permanently settled multiple-use zones surrounding core protected areas (Naughton-Treves et al. 2003). Selective logging frequently degrades the structure and quality of forest-interior habitat (Fimbel et al. 2001), and opens the forest to subsistence and commercial hunters (Robinson et al. 1999; Walsh, chap. 10 in this volume). Even more detrimental is that logging greatly increases forest vulnerability to fire, by elevating the amount and continuity of desiccated fuel in the understory (Cochrane et al. 1999; Barlow and Peres, chap. 12 in this volume). Hunting, logging, and wildfires often co-occur in fragmented tropical landscapes, vastly aggravating vertebrate-species losses from area and edge effects (Peres 2001b; W. F. Laurance, chap. 5; Peres and Michalski, chap. 6 in this volume).

The trend to create extractive reserves as opposed to protected core areas in tropical regions is partly the result of increasing pressures on developing-country governments to provide for local people living in remote areas. For example, Brazil recently announced the creation of six new protected areas covering 3.8 million hectares in the State of

Amazonas, which brings that state's protected-area coverage to more than 40%. While we applaud the bold protection of such a large expanse of biologically rich forest, almost all of the area set aside is in sustainable development or extractive reserves (WWF-Brazil 2003). We assert that these areas cannot be relied upon to safeguard biodiversity indefinitely and should not be considered as substitutes for core-protection zones. Unfortunately, in the Brazilian Amazon and elsewhere in the humid tropics, the amount of tropical forest currently within core protection zones is entirely inadequate (Peres 2002). Moreover, many core areas designated for biodiversity protection are failing to achieve their primary management objective (e.g., van Schaik et al. 1997; Terborgh 1999; Oates 2002).

THE STATE OF TROPICAL PROTECTED AREAS

Forests in the tropics are currently being lost at much faster rates than those in temperate zones. According to the most recent estimates in the State of the World's Forests report published by the United Nations Food and Agriculture Organization (FAO 2001b), the annual rate of tropical deforestation between 1990 and 2000 was 14.2 million hectares, which translates into a loss of almost 1% of tropical forests each year. Globally, 0.38% of the world's forests were lost each year in the 1990s, but at the same time, large areas were converted to forest plantations, primarily in the temperate zone, resulting in a 0.22% net annual loss of forests (FAO 2001b). The absolute rate of tropical deforestation was highest in the Neotropics, followed by Africa and then Southeast Asia. Brazil, Indonesia, Sudan, Zambia, Mexico, Democratic Republic of Congo, Myanmar, Nigeria, Zimbabwe, and Argentina experienced the highest net loss of forest area between 1990 and 2000. The highest net gain of forest area occurred in the temperate countries of the Russian Federation, the United States, China, Belarus, and Kazakhstan (FAO 2001b).

Tropical forests are being exploited and degraded by road-building schemes, new settlements, and myriad extractive industries, including mining, petroleum developments, and logging; they are also being converted into other land uses such as monocultures and cattle ranches. For example, according to the Brazilian National Space Research Institute (INPE), the mean annual deforestation rate in the Brazilian Amazon from 1995 to 2002 was about 19,800 km^2—equivalent to 8 football fields per minute—a significant increase from the annual rate of 13,500 km^2 from 1990–1994. Alarmingly, the annual deforestation rate jumped to over 23,000 km^2—equivalent to 10 football fields per minute—from 2002 to

2005 (W. F. Laurance et al. 2004a; W. F. Laurance, chap. 5 in this volume). In order to truly evaluate the state of tropical forests, we need to consider not only net changes in forest cover, but also the scale and extent of degradation of standing forests. This poses a challenge beyond the resolution of low-cost remote-sensing products.

It is of key importance to understand how tropical forests are faring within established protected areas. If forests are being degraded and species being lost within such areas, then we can safely surmise that unprotected forests are deteriorating even faster. The literature is full of studies offering evidence that protected areas are failing to meet their basic management objectives (e.g., Wells and Brandon 1992; Kramer et al. 1997; Terborgh 1999; but see Nepstad et al. 2006). One such study found that most parks and reserves in the Amazon basin within Brazil are largely accessible to hunters and other extractive activities, and only 1.16% of this region is both strictly protected (at least on paper) and inaccessible to illegal resource extractors (Peres and Lake 2003).

Bruner et al. (2001) assessed the impacts of anthropogenic activities in and adjacent to protected areas in the tropics to determine whether parks are effective in protecting tropical biodiversity. The authors concluded that, in comparison to adjacent areas, the majority of parks are successful to varying degrees at mitigating land clearing, logging, hunting, fire, and grazing. Tropical parks thus do provide better protection than surrounding areas, but this is only a glass that is half full. The more crucial question is whether parks are maintaining the natural conditions that sustain biodiversity. Data collected by ParksWatch (www.park swatch.org) indicate that many neotropical parks are failing to resist a litany of anthropogenic threats.

Since 2000, ParksWatch has conducted on-site evaluations of 46 protected areas in Mexico, Guatemala, Peru, and Venezuela. Using a standardized questionnaire, field researchers reported the existence and intensity of various threats affecting primarily national parks (table 22.2). The values indicate the occurrence of threats, not the degree to which they are threatening a given park.

Overall, the data reveal the pervasive occurrence of threats to the parks of every country. For example, 100%, 86%, and 87% of parks evaluated in Guatemala and Mexico, Peru, and Venezuela, respectively, are threatened to varying degrees by hunting or fishing. In Mexico, five of the eight threats included in the analysis occur in 100% of the parks evaluated over the past 3 years. Such findings are probably typical of poorly implemented parks elsewhere in Latin America and throughout the tropics, despite distinct national political, social, and economic climates.

Table 22.2 Percentage of protected areas in four tropical countries affected by eight different classes of major threats (data collected by ParksWatch from 2000 to 2003)

Country	N^a	Agricultural encroachment	Hunting and fishing	Logging and fuel wood	Grazing	Mining	Fire	Road building	Hydro power	Mean no. of threats[b]
Guatemala	11	73	100	82	45	9	82	82	18	4.9
Peru	14	79	86	57	71	64	64	64	7	5.3
Mexico	6	83	100	100	100	50	100	100	17	6.5
Venezuela	15	60	87	60	40	20	87	7	7	4.0

[a] Number of protected areas sampled in each country.

[b] Average number of threats affecting each park, calculated using the eight major classes of threats in this study.

A study in Brazilian Amazonia, for example, found chronic understaffing in 25 of 26 federal reserves (Sá et al. 2000). Because financial constraints often prevent reserves created on paper from being adequately implemented, defensibility criteria against external threats should be explicitly considered to complement other reserve-design and siting criteria (Peres and Terborgh 1995). Furthermore, these results highlight the critical importance of conducting routine evaluations of protected areas in order to identify threats and management deficiencies.

THE IMPORTANCE OF POLITICAL WILL

A recurring pattern is that a few favored parks receive the lion's share of a nation's conservation budget, whereas the rest are underfunded. Favored treatment is afforded to flagship parks, which either generate substantial revenue from tourism, or serve another special purpose deemed important by the government (e.g., protection of the water supply of a major city). Uneven treatment of protected areas by governments arises out of the difficult decisions faced in allocating scarce resources to competing needs. Unfortunately, for parks, the criteria used to allocate budgets are rarely congruent with biodiversity values.

An example of this phenomenon is found in Guatemala's Maya Biosphere Reserve (MBR) where the country's flagship park, Tikal, a World Heritage Site and Guatemala's most important tourist destination, receives a disproportionate amount of financial and other support when compared with the reserve's other core zones. This support translates directly into stronger protection and fewer threats (table 22.3). Tikal has both the largest budget (US$ per ha) and the highest surveillance capacity (hectares per park guard) of all the reserve's core zones. Compared to Laguna del Tigre National Park, an exceptional wetland wilderness as recognized by Ramsar, Tikal enjoys more than six times the park-guard density and 16 times the budget. Regarding their ability to resist threats, Tikal is affected by approximately half the number of threats as Laguna del Tigre National Park (8 to 15) and almost one-quarter the number of "severe" threats, as defined by ParksWatch (3 to 11).

While surveillance capacity and budgets seem to influence the presence of threats and severe threats in the MBR, the existence of roads and the varying ease with which these parks are accessed is also significant. We argue that it is not coincidental that the two areas with the fewest number of threats and severe threats, Mirador-Rio Azul National Park and Dos Lagunas Protected Biotope, are those that cannot be accessed by a

Table 22.3 Number of threats, park guards, and budgets of the core zones of Guatemala's Maya Biosphere Reserve (MBR) (from Albacete 2003)

Protected area	No. of threats	No. of severe threats	Surveillance capacity (ha/park guard)	Budget (US$/ha)
Tikal National Park	8	3	1028	10.42
Laguna del Tigre National Park	15	11	6589	0.63
Sierra del Lacandón National Park	14	6	5800	3.21
Yaxhá, Nakum and Naranjo National Monument	12	5	1689	3.90
San Miguel La Palotada (El Zotz) Protected Biotope	11	6	2911	0.52
Laguna del Tigre Rio Escondido Protected Biotope	14	9	4590	0.33
Dos Lagunas Protected Biotope[a]	5	0	4950	0.28
Mirador–Río Azul National Park[a]	8	0	5845	1.05

[a]Not accessible by a road.

road. A multiple regression model indicates that the combined effects of accessibility and financial resources per unit area are strong predictors of the prevalence of severe threats ($R^2 = 85.3\%$, $F_{2,5} = 14.5$, $P < 0.001$), and a combination of accessibility and surveillance capacity accounted for an even higher proportion of the variation in incidence of threats ($R^2 = 87.7\%$, $F_{2,5} = 17.8$, $P < 0.001$).

The analysis clearly indicates that resource allocation (in the form of budgets and park guards) and physical access are both extremely important determinants of whether protected areas are effective. If physical access is high, then high levels of resources and commitment are needed to ensure park protection. For example, Tikal, the reserve with the third fewest threats, is accessed by a paved road but has an entrance gate that is monitored by armed guards.

Despite the heavily skewed resource allocation toward Tikal, the park is not without problems. Forest fires, the collection of nontimber forest products, unsustainable tourist levels, and occasional poaching do threaten the park to varying degrees. This occurs despite the fact that Tikal's guards carry guns—the only protected area in Guatemala where this is allowed. But relative to most of the other core zones in the Maya Biosphere Reserve, Tikal is in excellent shape and receives generous support, because it generates substantial tourist revenues. What is occurring in Guatemala is pervasive throughout the tropics: the lack of political will to protect certain parks is not always the result of a lack of capacity to do so (van Schaik et al. 1997).

CONCLUSIONS AND IMPLICATIONS

1. The Earth's nations have collectively "protected" more than 11% of the terrestrial realm. Less than half of the total consists of "hard" protected areas (IUCN categories I–IV). The remainder is unclassified or contained in "soft" protected areas (IUCN categories V and VI) open to various kinds of exploitation. If the current world commitment to terrestrial conservation remains unchanged, analyses of species–area curves suggest that roughly half of all terrestrial species will eventually be doomed to extinction.

2. For this reason, conservation biologists must press for a greatly expanded commitment to biodiversity protection. We propose a goal of 50% of land in protected areas as being biologically justifiable and politically attainable in many tropical countries. The 50% goal can be achieved by embedding strict nature preserves within an interconnected matrix of soft protected areas. The main obstacles to attaining this goal are widespread ignorance of the requirements of biodiversity conservation and a lack of political will.

3. We need to improve the data on habitat and spatial requirements of large predators and other keystone species, as well as endangered and endemic species for each country. IUCN Red Lists for endangered species are needed not only to prescribe inclusion of suitable habitat into protected areas but also to serve as a measuring tool for the effectiveness of conservation efforts.

4. Species-specific data must serve as the foundation for creating new protected areas, using conservation paradigms like reserve cores, buffers, and connectivity, as well as more recently developed approaches like representation, gap analysis, hot-spots, and ecoregional planning. This way we can meet the needs of key species while ensuring the perpetuation of ecosystem-level processes, including natural disturbance regimes.

5. Much of the battle to secure additional conservation areas will be won or lost in the next two or three decades. Priority should be given to the last remaining tropical wilderness regions, such as Amazonia and the Guyana shield, Gabon, the Congo Basin, and the island of New Guinea, which, although largely unprotected on paper, remain comparatively intact.

6. In these relatively intact regions, crucial opportunities are still available for rapidly expanding the existing system of reserves so long as

financial, sociopolitical, and institutional constraints can be overcome. Before additional parks are established, however, thoughtful planning will be required because the financial and political costs of creating new reserves can detract from competing investments in the near-term implementation and management of existing reserves.

7. In regions where expansion of protected areas is no longer feasible, conservation efforts should focus on improving management of all types of conservation areas. Core-protection areas, such as national parks as well as multiple-use zones, need to function as intended by law. This will enable the recovery of degraded forests currently under only nominal protection.

8. Information on park management problems, ranging from corruption to insufficient budgets and staff levels, should become public knowledge. This will help to prevent governments from neglecting certain protected areas in favor of those that are deemed more important financially or otherwise.

9. The design and official decree of protected areas is only half of the battle. The other half is ensuring that these areas are implemented, consolidated, and made viable in the long run. On-site evaluations of threats and management effectiveness should become mandatory for all protected areas. The evaluations should be conducted by independent groups to ensure transparency and that the results are accurate and unbiased. Just as we have Red Lists for endangered species, we need Red Lists for parks.

10. If protected areas are to withstand future threats and sustain biodiversity, park stakeholders (including government agencies in charge of parks, NGOs, individual managers, donors, and policy makers) need accurate information on the success and failures of protected areas. Evaluating the management effectiveness of protected areas identifies threats to ecological integrity and as well as management deficiencies. Information sharing needs to be mandatory so that park evaluation and monitoring is more than just a hollow exercise.

11. Most of all, scientists need to be seen and heard. We need to step out of our academic retreats and into the marketplace of ideas. Publishing in journals is important and necessary, but the transmission of ideas from the pages of journals to acts of congress is too slow a process for our fast-moving world. Scientists must engage the public in conservation issues by sharing their concerns. An informed public generates political will for conservation, and ensures that political leaders are held accountable for their actions. If our ideas are truly going to change the world, then we need to build bridges to the popular media and inspire people with a vision of how science can lead the way to a better world for ourselves and our children.

PART VI

Summary and Implications

CHAPTER 23

Emerging Threats to Tropical Forests: What We Know and What We Don't Know

William F. Laurance, Carlos A. Peres, Patrick A. Jansen, and Luis D'Croz

In this book we have defined emerging threats to tropical forests in four ways: (1) those that are relatively new, such as the emerging chitrid pathogen that is devastating some rainforest frog populations; (2) those that are unprecedented in scale or rapidly growing in importance, such as tropical surface fires; (3) those that are poorly understood, such as the environmental consequences of large-scale climatic and atmospheric changes; and (4) those that are synergistic in their effects, such as the devastating combination of habitat fragmentation and chronic overhunting on vulnerable wildlife.

Here we highlight key themes of this book, and introduce a few additional topics of importance, such as illegal gold mining and linkages between forests and oceans. We attempt to distinguish between what we presently know (or think we know) and what we don't know about emerging threats to tropical forests.

MANY THREATS ARE HARD TO DETECT

Many subtle anthropogenic impacts are obscured from view by the dense forest vegetation—a serious problem in tropical conservation. Remote-sensing methods provide a powerful means to quantify threats to forests at large spatial scales, but some important environmental changes are difficult or impossible to detect with remote sensors (table 23.1).

Table 23.1 Detectability of different threats to tropical forests using available remote-sensing techniques. Marginally detectable threats are those that can be detected, at least partially, using high-resolution methods or specialized detection algorithms that are expensive, technically challenging to implement, and available only for limited or specific areas.

Readily detectable	Marginally detectable	Not detectable
Deforestation	Recent selective logging	Hunting or defaunation
Habitat fragmentation	Surface fires	Harvests of many nontimber
Major forest fires	Effects of climate change	forest products
Major highways	on plant phenology	Effects of pathogens
	Small-scale gold mining	Compositional shifts in plant
	Wider roads (6–20 m width)	communities from climate
	Some invasions of exotic	change
	plant species	Nonrecent selective logging
		Narrow roads (<6 m width)
		Most secondary effects

Gross changes to forest cover, such as deforestation and forest fragmentation, can be quantified using readily available and affordable remote-sensing methods like the Landsat Thematic Mapper (Landsat TM) and Landsat Enhanced Thematic Mapper (Landsat ETM+), which have a spatial resolution of about 30 m. These approaches suffer from some technical challenges—for example, it can be difficult to distinguish primary forest from mature regrowth, and cloud cover obscures the ground in much of the imagery from humid tropical regions—but they generally function well for mapping land-use changes.

There is, however, a trade-off between the spatial resolution of satellite sensors and the frequency with which they cover any particular point on Earth (W. Turner et al. 2003). Older sensors, such as the Advanced Very High Resolution Radiometer (AVHRR), with a spatial resolution of 1 km, offer more frequent coverage of any particular point, and thus are useful for detecting shorter-term changes, whereas higher-resolution sensors such as Landsat provide less frequent coverage. Very high-resolution sensors, such as the Ikonos satellite, aerial photography, and airborne sensors like aerial videography and LIDAR (light detection and ranging), have a spatial resolution at or below 1 m and thereby can distinguish individual tree crowns and canopy gaps (fig. 23.1). However, these methods are expensive, sometimes technically challenging to use, and cover only small areas (Read et al. 2003; W. Turner et al. 2003).

William F. Laurance, Carlos A. Peres, Patrick A. Jansen, and Luis D'Croz

Figure 23.1. High-resolution remote-sensing methods, such as this Ikonos image with 1 m resolution, can detect individual tree crowns and canopy gaps. Such methods are useful for detecting and mapping some kinds of forest degradation. (Photo from the Biological Dynamics of Forest Fragments Project, Manaus, Brazil)

As remote-sensing technology improves, our ability to perceive more subtle changes in tropical forests will undoubtedly progress (e.g., Asner and Vitousek 2005). However, it may be many years before new methodologies are applied on a large scale for land-use planning and conservation. For example, it is currently possible to detect canopy damage from recent (<3-year-old) selective logging in tropical forests using satellite imagery from Landsat (Asner et al. 2004a, 2004b, 2005) and Ikonos (Read et al. 2003; fig. 23.1), but these methods are still under development and have only been applied to limited areas of the tropics. Likewise, sophisticated algorithms that assess changes in spectral patterns from remote-sensing imagery, such as linear mixture modeling, neural networks, and maximum likelihood classification, can be used to map recent surface fires with moderate accuracy (Cochrane and Souza 1998; Eva and Lambin 2000; Foody et al. 2001; Cochrane and Laurance 2002; Alencar et al. 2004). Again, however, these methods are currently being applied only in a few specific places and in a research context.

Among the greatest challenges at present is to improve the practical application of remote sensing and GIS (geographic information systems) to quantify tropical deforestation globally. This is less a research challenge than a practical one of committing financial resources to the task and making imagery, equipment, and standardized methods available widely. Landsat imagery, which is readily accessible, can and should be used to provide systematic annual estimates of primary-forest destruction in every tropical nation. This is vital because the principal data source on tropical deforestation, the decadal estimates from the United Nations Food and Agricultural Organization (FAO), is based on reports from individual countries (FAO 1993; Whitmore 1997) that are notoriously unreliable (Fearnside 2000b). For example, the FAO data suggest that tropical deforestation rates declined in the 1990s relative to the 1980s (this trend resulted in part from a change in the 1990s in how the FAO defined tropical forests), whereas more rigorously quantified data from remote sensing suggest that rates actually accelerated in the 1990s (Achard et al. 2002).

To date, reliable deforestation data are available for only a few countries. Brazil, for instance, deserves credit for producing and publicly releasing annual, remote sensing-based estimates of Amazonian deforestation since 1988, despite the criticism such estimates have attracted (e.g., W. F. Laurance et al. 2001b, 2004a) because they reveal accelerating forest loss. Other large-scale assessments of tropical deforestation based on remote sensing are beginning to appear, but these rarely encompass more than a small fraction of all tropical forests. For example, a recent study that estimated the mean annual deforestation rate for humid tropical forests worldwide from 1990 to 1997 (Achard et al. 2002) focused on "hotspots" of deforestation, but included only 6.5% of all tropical forest cover.

Many other threats to tropical forests are not detectable with even the most sophisticated, high-resolution sensors (table 23.1). For example, unless they substantially alter forest structure, activities such as hunting, harvests of many nontimber forest products, proliferating exotic pathogens, invading species, and many ecosystem-level responses to climatic change, are virtually imperceptible to remote sensors. Detecting such subtle changes is a key challenge for researchers, and one that will hopefully improve as new techniques and high-resolution imagery become more readily available. For example, South American forests have apparently increased in productivity in recent decades, possibly in response to rising carbon dioxide fertilization (Lewis et al., chap. 1 in this volume). These changes were detected using laborious plot-based studies, but such changes might now be discernable with satellites using spectral analyses of vegetation greenness (Paruelo et al. 2004).

A key message of this section is that many threats to tropical forests are currently obscured to us. Some of these threats can be detected using expensive remote-sensing imagery or specialized techniques that are currently being developed, but are not presently available on a systematic- or widespread-enough basis to have real relevance for conservation. Many other changes can only be detected by muddy-kneed scientists or observers working on the ground, whose efforts are necessarily limited in scale. In most cases, the closer we look, the more human-related environmental changes we find in tropical forests (Peres et al. 2006). Given these realities, it is essential that we are not misled into believing that all is well when we see a green-looking forest on images—the threats to tropical forests are far more pervasive and insidious than we often realize.

EVEN CRYPTIC DEGRADATION CAN HAVE DIRE EFFECTS

A growing body of evidence suggests that even relatively imperceptible forms of degradation can have serious impacts on tropical ecosystems. Here we describe some of the most important types of cryptic degradation, most of which are discussed in greater detail in the preceding chapters.

Defaunated Forests

An ever-increasing fraction of surviving tropical forests is being plagued by the "empty forest" syndrome—in which key animal species have been selectively removed. Selective defaunation is caused by overhunting, forest fragmentation, and other forms of habitat deterioration, sometimes operating in concert (Peres 2001b; Michalski and Peres 2005; Peres and Michalski, chap. 6 in this volume; Terborgh and Nuñez-Iturri, chap. 13 in this volume). Top predators and medium- and large-sized species targeted by hunters (fig. 23.2) are usually the most vulnerable (Peres 2000a; Wright 2003), and their population declines or local extinctions can have myriad consequences for the structure of ecosystems.

Few appreciate just how pervasive hunting is in the tropics. Recent estimates of the annual wild meat harvest are 23,500 metric tons in Sarawak (Bennett 2002), 67,000 to 164,000 tons in the Brazilian Amazon (Peres 2000a), and 1 to 3.4 million tons in Central Africa (Fa et al. 2002). Much of the Amazon basin (Peres and Lake 2003), including most of its protected areas (Peres and Terborgh 1995), is already physically accessible to hunters. Forests in Southeast Asia also suffer from chronic overhunting, in part because of the dramatic proliferation of logging roads in forests (Bennett and Gumal 2001; MacKinnon, chap. 16 in this volume; fig. 23.3). In Vietnam, twelve species of large vertebrates have been extirpated over the last four decades, largely from overhunting (Bennett and Rao 2002). Hunting pressure is even heavier in tropical Africa (Fa et al. 2002; Elkan et al., chap. 21 in this volume). In West Africa, at least one primate species

Figure 23.2. The extirpation of large-bodied vertebrates, such as forest elephants, can have important secondary effects on forest ecosystems (Photo by William F. Laurance)

has probably been driven to global extinction by overhunting, and game species have been so severely depleted that the commercial bushmeat trade in some regions has virtually collapsed (Oates et al. 2000).

One likely implication of overhunting is that plant species that formerly relied on large-bodied vertebrates as seed vectors may be poorly dispersed and thereby suffer increased density-dependent seed or seedling mortality (Jansen and Zuidema 2001; Wright and Duber 2001; Terborgh and Nuñez-Iturri, chap. 13 in this volume). Such changes may alter plant community composition, leading to declines in animal-dispersed plants and compensatory increases in abiotically dispersed species (Dirzo and Miranda 1991; Asquith et al. 1997; Tabarelli et al. 1999; Cordeiro and Howe 2001, 2003; Jansen and Zuidema 2001). Large-seeded plants are especially vulnerable. For example, two genera of large-bodied primates, woolly monkeys (*Lagothrix* spp.) and spider monkeys (*Ateles* spp.), are often

Figure 23.3. A proliferation of logging roads frequently leads to a drastic increase in hunting pressure in tropical forests (Photo by William F. Laurance)

driven to local extinction in overhunted forests across Amazonia and the Guianas (Peres 1999a). Many large-seeded tree and liana species rely on these primates for dispersal, and a number of these require passage through the primate gut to stimulate germination. A simulation study based on field data suggests that primate-dependent plants will suffer a 40% to 50% reduction in gut-dispersal in overhunted forests (Peres and Roosmalen 2002). Moreover, because dietary overlap between different guilds of frugivores can be very low (Gautier-Hion et al. 1985; Poulsen et al. 2002), a severe decline of a particular guild of frugivores could lead to a collapse of seed-dispersal services for dependent plant species.

Another likely consequence of selective defaunation is "mesopredator release," whereby populations of smaller omnivores explode following the extirpation of large, dominating predators that formerly regulated their abundance (Crooks and Soulé 1999). Hyper-abundant mesopredators, such as raccoons, opossums, and rodents, can increase predation pressures on nesting birds (Loiselle and Hoppes 1983; Sieving 1992), larger seeds, and seedlings (Asquith et al. 1997; Wright and Duber 2001). On small, predator-free islands in a large hydroelectric reservoir in Venezuela, arboreal folivores such as howler monkeys, iguanas, and leaf-cutter ants were many times more abundant in small (<12 ha) than in larger islands or the mainland. A herd of capybaras (giant rodents) stranded on one remote island proceeded to convert the forest understory into bare ground covered with capybara dung (Terborgh et al. 2001). Likewise, the disappearance of numerous species of insectivorous birds, bats, and army ants from Amazonian forest fragments may be at least partly responsible for elevated abundances of arthropods in fragments (W. F. Laurance et al. 2002b), which in turn could lead to increased herbivory pressure (e.g., van Bael et al. 2003).

These studies demonstrate that selective defaunation can potentially have pervasive, long-term effects on forests. However, much is still unknown. Whereas many remaining forests will suffer severe defaunation, others will be only partially defaunated, and the fate of these "half-empty" forests is uncertain (Redford and Feinsinger 2001). What is well established is that large species and those high up the trophic ladder often vanish or decline sharply in forests subjected to anthropogenic pressures. Much additional effort is needed to comprehend the kaleidoscopic changes in tropical forest structure and composition that occur as a consequence.

Nontimber Resource Extraction

In addition to widespread game hunting, other extractive practices can also have significant impacts on forests and their resources. For example, Brazil nut trees (*Bertholletia excelsa*) in extractive reserves may eventually face demographic collapse from chronic overharvesting. Tree stands in Amazonia with a long history of nut harvesting show a striking paucity of smaller individuals (Peres et al. 2003a). In the Peruvian Amazon, the fruits of *Mauritia flexuosa* palm trees, a long-lived forest emergent, are often collected only once by felling whole reproductive adults (Vasquez and Gentry 1989). In Uganda, harvesting of poles and associated disturbances from trampling and anthropogenic fires can depress tree populations near the margins of nature reserves (Oluput and Chapman, chap. 7 in this volume). Overharvesting of fuelwood can create enormous pressures on forests and greatly reduce wood debris and hollow-bearing snags that are essential for some birds, mammals, and other wildlife (Marcot 1992; Liu et al. 1999). In Borneo, traditional, low-intensity harvests of ironwood (*Eusideroxylon zwageri*) have intensified drastically with increased logging-road access into forests (Peluso 1992). If such practices are not managed effectively, populations of prized or commercially important species can decline sharply, sometimes with important impacts on wildlife.

Surface Fires

Intense fires used for deforestation, such as those lit by slash-and-burn farmers after felling trees and clearing the understory vegetation, are readily detectable both by satellites and on the ground. Surface fires, however, are far less intense and are obscured from sight by the dense canopy overhead. These unintentional surface fires are proliferating at an alarming pace throughout the seasonal tropics (Cochrane et al. 1999; Nepstad et al. 1999b; Cochrane 2003). A combination of increasing ignition sources, logging, habitat fragmentation, and strong droughts is evidently responsible (W. F. Laurance, chap. 5 in this volume), although even undisturbed rainforests can burn under some conditions (Barlow and Peres, chap. 12 in this volume).

Surface fires are deceptively unimpressive, creeping slowly through the understory and consuming only leaf litter and debris near the ground. But because rainforest species are poorly adapted for fire, surface fires cause heavy plant mortality and alter many aspects of forest ecology (Cochrane and Schulze 1999; Peres 1999b)—promoting a proliferation of weedy plants, altering forest microclimate, killing slow-moving animals, and reducing the abundances of many sensitive vertebrate and invertebrate species (Peres et al. 2003b; Barlow and Peres, chap. 12 in this volume). Until recently, large tropical trees were thought to be relatively resilient to surface fires, but recent studies have shown that many big trees die within 2 to 3 years after a surface fire, leading to a substantial loss of live tree biomass and generating considerable atmospheric carbon emissions (Barlow et al. 2003b).

The most alarming aspect of surface fires is that they can initiate a positive feedback cycle that eventually leads to complete forest destruction. Although the initial surface fire consumes only leaf litter and debris, it kills many trees and vines, thinning the forest canopy and leading to a rain of dead leaves and branches that accumulate on the forest floor like dry tinder. The forest is now ripe for a second surface fire, which typically produces an order of magnitude more heat than the first fire (Cochrane et al. 1999). In this manner, each fire creates more fuel than it destroys, progressively thinning the forest and making it drier and more fire prone. Rainforests cannot withstand this withering recurrence of fire; forest fragments have been observed to "implode" over time as their margins collapse inward (Gascon et al. 2000; Cochrane and W. F. Laurance 2002). Scorched landscapes that remain in the aftermath of recurring fires have little or no conservation value.

In much of the tropics, the incidence of forest fires is strongly influenced by the El Niño Southern Oscillation, which is associated with periodic droughts or rainfall deficits. Some evidence suggests that global warming could promote more frequent or more severe El Niño events, and that large-scale forest destruction could promote regional declines in rainfall (see discussion that follows). Such changes could potentially cause a major increase in the extent and frequency of destructive fires.

Climatic and Atmospheric Changes

Changes in the global climate and atmospheric composition are almost certainly affecting some tropical forests, although there is much uncertainty over the magnitude of these effects and the specific environmental drivers (Lewis et al., chap. 1 in this volume). For example, Amazonian forests have become increasingly dynamic in the past few decades, with higher rates of tree mortality and turnover (Phillips and Gentry 1994; Phillips et al. 2004). In addition, carbon storage (Grace et al. 1995; Phillips et al. 1998; Baker et al. 2004a) and productivity (Lewis et al. 2004b) in these forests are apparently increasing (but see Wright 2005). Significant shifts have also been detected in the composition of Amazonian liana (Phillips et al. 2002b) and tree (W. F. Laurance et al. 2004b) communities. One leading hypothesis is that increasing atmospheric carbon dioxide levels, which could potentially accelerate plant growth and alter competitive balances among species, are responsible.

At least some tropical ecosystems are likely to be altered by rising global temperatures. Most vulnerable are cool-adapted montane species, whose geographic ranges will contract and become increasingly fragmented with rising temperatures (Parmesan 1996; Peterson et al. 2002). Montane habitats may also become drier, because higher temperatures increase the typical basal altitude of the cloud layer, reducing moisture from rainfall and mist interception (Pounds et al. 1999; Still et al. 1999). For locally endemic species confined to high elevations, which can only migrate further upward as conditions heat up, the conservation prognosis is especially grim (Williams and Hilbert, chap. 2 in this volume). Moreover, it has recently been argued that global warming is already increasing the vulnerability of high-elevation amphibians to epidemic disease (Pounds et al. 2006). Rising temperatures are unlikely to threaten lowland species directly, but it has been suggested that they could accelerate the metabolic rates of plant species, reducing both plant growth and forest carbon storage (D. A. Clark et al. 2003).

Global warming is likely to alter large-scale patterns of precipitation, which should have an even stronger effect on tropical forests than do rising temperatures (Houghton et al. 2001; P. M. Cox et al. 2000, 2003, 2004). In general, rainfall variability and dry-season severity are expected to increase in the tropics (K. J. E. Walsh and Ryan 2000; Houghton et al. 2001). For any particular region, however, the specific nature of these changes, and their effects on local ecosystems and biota, are very difficult to predict (Mahlman 1997). Differences among regions will arise in part because the nature and concentration of atmospheric pollutants vary. In South and Southeast Asia, for example, the burning of wood, traditional biofuels (such as dried cow dung), and crop waste are an important cause of climate change (Lelieveld et al. 2001). This burning emits large amounts of atmospheric soot, which creates a dense brown haze that blankets South Asia and the Indian Ocean for much of the year. The soot absorbs large amount of heat from the sun, and thus has a significant impact on regional climate, reducing surface temperatures, evaporation, and summer monsoon rainfall (Ramanathan et al. 2001). Comparable effects are likely in other tropical regions where wood and biofuel burning are common.

A particular concern is that future warming of the climate might increase the frequency or severity of El Niño events, as suggested by some global circulation models (Timmerman et al. 1999; Houghton et al. 2001). In large areas of the tropics—the Neotropics, Southeast Asia, and Australasia—El Niño events cause droughts or rainfall deficits that lead to dramatic peaks in forest burning (Leighton and Wirawan 1986; Cochrane et al. 1999; Nepstad et al. 1999b; Barlow and Peres, chap. 12 in this volume). El Niño events also lead to strong oscillations in fruit abundance that impact populations of seed-dispersing animals (Wright et al. 1999). El Niño–related droughts may already be increasing in intensity, as some of the strongest droughts on record have occurred since the early 1980s (Trenberth and Hoar 1996, 1997; Dunbar 2000).

Finally, global warming may lead to a greater frequency and intensity of destructive storms—such as Hurricane Katrina and Cyclone Larry, which caused extraordinary damage to the southern United States and north Queensland, respectively—as a consequence of rising sea-surface temperatures (Mahlman 1997; Easterling et al. 2000; K. J. E. Walsh and Ryan 2000; Houghton et al. 2001; Emanuel 2005; Webster et al. 2005). Such storms are most likely in the cyclone, hurricane, and monsoon zones from about 7° to 20° latitude, and can have major impacts on forest dynamics in island, coastal, and sub-coastal areas (Whitmore 1975; Lugo et al. 1983). Fragmented forests may be especially vulnerable to increasing windstorm damage, which often causes major changes in forest structure, dynamics, microclimate, and floristic composition (W. F. Laurance 1997, 2004a; W. F. Laurance, chap. 5 in this volume). Strong storms can also lead to severe flooding, causing extensive tree mortality in low-lying areas and where water is impeded (Mori and Becker 1991). Devastating flash floods, such as those that killed tens of thousands of people over the last decade in China, Nicaragua, Venezuela, and the Caribbean, are most likely where the headwaters of streams have been deforested.

Gold Mining

Another growing but largely cryptic form of forest degradation is small-scale gold mining (fig. 23.4), which now involves more than one million people worldwide (De Lacerda 2003). Although they operate throughout the tropics (Uryu et al. 2001; De Lacerda 2003; Limbong et al. 2003), small-scale miners are especially active in the Guiana Shield of South America (Mol et al. 2001; Peterson and Heemskerk 2001) and in Central Africa (Van Straaten 2000). Miners use high-powered water cannons to dissolve river and stream sediments into mud, which is then passed over a sieve where the gold particles amalgamate with mercury; the mud is then released downstream. The gold is obtained by evaporating the mercury, although some mercury also escapes into streams and rivers.

Figure 23.4. Illegal gold mining, such as this in Brownsberg Nature Reserve, Suriname, can completely destroy creek beds. Forest recovery seems impossible here (Photo by Peter J. Van der Meer)

The miners, who often operate illegally near camps served by clandestine airstrips, have little regard for environmental concerns and can have severe local impacts on riverine systems, indigenous communities, and human health. Among the most serious threats is from mercury pollution (Nriagu et al. 1992; De Lacerda 2003), which accumulates in fish (Mol et al. 2001; Uryu et al. 2001) and in the wildlife and people eating those fish. Serious health effects, including neurological damage, memory loss, changes in personality, and birth defects, can occur among people living along rivers (Mol et al. 2001; De Lacerda 2003).

Gold mining instigates other kinds of environmental degradation (fig. 23.4). The torrent of dissolved sediment released by miners has serious effects on the aquatic fauna and flora downstream (Mol and Ouboter 2004). Heavy sediments may also threaten river navigability, water reservoirs, and hydroelectric plants, and creek beds can be destroyed forever. Large mud holes created by miners become breeding grounds for malaria-carrying mosquitoes. The soils in mined areas are so heavily disturbed that forest recovery is greatly impeded (Peterson and Heemskerk 2001). Miners who operate illegally also promote overhunting of wildlife, and corrupt, threaten, and ultimately destabilize local indigenous peoples who may safeguard large aboriginal territories (Nriagu et al. 1992; Mol et al. 2001). In the Guianas and northern Amazonia, indigenous lands and national parks have been overrun by tens of thousands of such gold miners (Nriagu et al. 1992; Mol et al. 2001).

Exotic Pathogens

Yet another form of cryptic environmental degradation is caused by exotic pathogens. Although among the least understood of all emerging threats, exotic pathogens have taken a catastrophic toll on some species (Cunningham et al., chap. 8 in this volume, Benítez-Malvido and Lemus-Albor, chap. 9 in this volume). Prominent examples include species extinctions and population collapses of stream-dwelling frogs in tropical Australia, Central and South America, and elsewhere as a result of rampant chitridiomycosis (Berger et al. 1998; Lips et al. 2003a; Pounds et al. 2006); the extirpation of Hawaiian land birds from malaria and avian poxvirus following the introduction of an exotic mosquito vector (Van Riper et al. 1986); drastic declines of many African ungulates from rinderpest (Plowright 1982); and epidemic "dieback" of Australian plant communities from introduced *Phytophthora* fungus (Shearer and Tippett 1989). Given the alarmingly high rate of exotic-species introductions worldwide, and the limited effort devoted to the study of emerging diseases in wildlife and native plants, it is almost certain that such examples represent a mere sliver of the entire problem.

Wildlife populations are not only vulnerable to "pathogen pollution"—the rapid movement of pathogens into new geographic areas and hosts. They can also be predisposed to infectious disease by stresses from environmental changes. Habitat fragmentation, for example, increases the likelihood of disease in some plant (Benítez-Malvido and Lemus-Albor, chap. 9 in this volume) and animal (Stokstad 2004) populations. As the catchphrase "A warmer world is a sicker world" so vividly conveys, global warming could play a major role in expanding the geographic ranges of many pathogens into new areas where they will encounter immunologically naïve hosts (Harvell et al. 2002; Pounds et al. 2006). Other human activities, including hunting (Walsh, chap. 10 in this volume), exotic species introductions, and an increased proximity of humans and their domestic animals to wildlife (Deem et al. 2001), can also increase wildlife vulnerability to infectious disease. Species declines or extinctions from exotic pathogens not only reduce forest biodiversity, but also can have pervasive impacts on ecosystem functioning (Cunningham et al., chap. 8 in this volume).

Invading Species

By escaping many of their natural enemies, introduced plant and animal species can increase dramatically in abundance in their new geographic ranges (Elton 1958; Rejmánek 1989, 1996). Globally, exotic species are increasing at an astounding rate and are making major inroads into many nature reserves (Usher et al. 1988; Usher 1991; G. W. Cox 1999).

In the tropics, different ecosystems vary greatly in their relative invasibility. Undisturbed tropical rainforests, and extreme environments such as harsh deserts and mangrove forests, have generally suffered few invasions (Whitmore 1991; Rejmánek 1996, 1999; Fine 2002). Other ecosystems, including disturbed and fragmented rainforests, dry and strongly seasonal forests, open habitats, and oceanic islands, are far more vulnerable to invaders (Elton 1958; Janzen 1983, 1988; Vitousek 1990; Rejmánek 1996; W. F. Laurance 1997). Island biotas, in particular, can be decimated by introduced species (Baider and Florens, chap. 11 in this volume).

What accounts for these differences? Initially, the high species richness of undisturbed rainforests was thought to account for their low invasibility (Elton 1958; Holdgate 1986). However, more recent studies suggest that two factors, the functional diversity of life forms in a habitat and propagule pressure (the number of well-adapted individual animals or plant seeds arriving in a particular community), collectively provide a better explanation for among-habitat differences in invasibility (Rejmánek 1996; Fine 2002). Ecosystems with high functional diversity, such as undisturbed rainforests, have few available resources, whereas those with low functional diversity, such as depauperate islands, have more resources that can be exploited by invaders. Hawaii, for example, lacks native successional trees that fix nitrogen, a limiting nutrient. This ecological niche has been filled by the introduced nitrogen-fixing tree *Myrica faya* (Myricaceae), which now dominates early successional habitats (Vitousek 1990). Successfully invading bird species in Hawaii are most likely to be those with bills that differ morphologically from already-resident congeners (Moulton and Pimm 1986). Similarly, subtropical south Florida has no native trees that are both flood tolerant and fire resistant; the exotic *Melaleuca quinquenervia* (Myrtaceae) is now overwhelming native grasslands there (Schmitz et al. 1997).

Propagule pressure is also a key determinant of biological invasions (Rejmánek 1989; Fine 2002). Undisturbed rainforests are rarely invaded because they receive few propagules from exotic species, especially from those well adapted for rainforest conditions. Most exotic plants in the tropics are ornamental species or agricultural pests, which are normally grown in full sun. This explains why most exotics are favored by disturbances that greatly increase light availability in rainforests. It also explains the lack of invaders in mangrove forests and deserts, because humans rarely introduce drought- or salt-adapted species (Fine 2002). When they are introduced in the tropics, plants that are shade tolerant and rely on generalist seed dispersers can and do invade undisturbed rainforests. In tropical Africa, for example, the exotic tree *Maesopsis eminii* (Rhamnaceae) has escaped forestry plantations and is now invading undisturbed forests (Viisteensaari et al. 2000). Native forests in Tahiti are being overrun by the exotic tree *Miconia calvescens* (Melastomataceae; Meyer and Florence 1996), while the shrub *Clidemia hirta* (Melastomataceae) is invading undisturbed forests in Malaysia and Hawaii (Ickes and Williamson 2000).

In the future, species invasions will almost certainly become an evergreater blight on tropical ecosystems. Anthropogenic disturbances, such as habitat fragmentation, logging, and fires, are increasing in extent. These disturbances promote invasions by exotic trees, shrubs, vines, and forbs that are having major impacts on tropical forests worldwide (e.g., Rejmánek 1996; Thompson 1988; Fine 2002; but see Teo et al. 2003). For example, exotic and disturbance-loving lianas often become hyperabundant in small forest fragments where they suppress tree survival, growth, and regeneration (Humphries et al. 1991; I. M. Turner et al. 1996; Viana et al. 1997; W. F. Laurance et al. 2001c). Invaders are especially problematic when they alter major ecosystem properties. Feral pigs, for instance, cause serious disturbances to rainforest soils and vegetation in Hawaii and north Queensland. In the monsoonal tropics of northern Australia, introduced water buffalo alter hydrological properties and facilitate weed invasions in forests and wetlands (Cowie and Werner 1993; G. W. Cox 1999). In tropical dry forests, human-lit fires suppress tree regeneration and promote invasions of exotic grasses, which in turn produce flammable litter that helps to carry future fires—a process that has contributed to a drastic decline in tropical dry forests worldwide (Janzen 1988; D'Antonio and Vitousek 1992; Shugart et al., chap. 3 in this volume).

The number of exotic species that can invade undisturbed tropical forests could rise considerably in the future (Fine 2002). Particularly worrisome are shade-tolerant tropical hardwoods, which increasingly are being grown in tree plantations (Grainger 1988). In recent decades, 45 species of woody legumes have been established in 42 tropical countries, and many are already becoming invasive (C. E. Hughes and Styles 1989). When exotic, shade-tolerant trees with generalist pollinators and seed dispersers are planted in large numbers, the likelihood that they will invade undisturbed forests increases dramatically.

Phase-Shifts in Ecosystems

For those concerned with the long-term fate of tropical forests, one of the most vexing possibilities is that seemingly subtle environmental changes may provoke fundamental phase-shifts in ecosystems (Shugart et al., chap. 3 in this volume). Regional desiccation from large-scale deforestation, severe El Niño droughts, proliferating surface wildfires, shifts in regional rainfall patterns caused by global warming, the loss of ecologically pivotal species from defaunation, exotic-species introductions, and other environmental changes might drive tropical forests across a threshold from which they are unable to recover (Avissar et al., chap. 4 in this volume, Baider and Florens, chap. 11 in this volume). The conversion of large expanses of seasonal rainforest into fire-prone scrub and savanna seems especially likely (W. F. Laurance, chap. 5 in this volume, Barlow and Peres, chap. 12 in this volume). An alarming aspect of sudden environmental phase-shifts is that, by the time their effects become apparent, it may already be too late to reverse their course.

This book focuses on tropical forests, but it is important to remember that the fate of forests is also strongly linked to that of marine and aquatic ecosystems. Here we describe three key interactions as examples of these linkages.

Disappearing Mangrove Forests

Mangrove forests are comprised of salt-tolerant trees and shrubs that grow in sheltered, muddy intertidal areas at tropical and subtropical latitudes. Mangrove plants are highly specialized in their physiology, morphology, and reproduction for living in the stressful, dynamic conditions of the intertidal zone (Ellison and Farnsworth 2001). The global extent of mangroves is only about 180,000 km², with slightly more than half of all forests and the greatest diversity of species occurring in Asia and Australasia (Lacerda et al. 1993; Spalding et al. 1997).

Long denigrated as muddy, mosquito-infested swamps, mangroves are today known to play vital roles in coastal ecology, fisheries, and human welfare. Mangroves provide critical habitat for a great diversity of terrestrial and marine wildlife. They also function as nurseries for larvae and juveniles of many commercially important fish and prawn species, and thus are vital for tropical fisheries (Lacerda et al. 1993). Litterfall from mangroves produces an enormous quantity of detritus that contributes to the high productivity of coastal zones (Odum and Heald 1975; D'Croz et al. 1989). Mangrove forests are natural filters for terrestrial runoff, retaining fine sediments and uptaking excess dissolved inorganic nutrients. This function ensures good water quality, which is vital for neighboring coral-reef and sea-grass ecosystems that are easily smothered by sediments or outcompeted by nutrient-loving algae (Ellison and Farnsworth 2001). Mangroves also stabilize coastlines and prevent erosion from waves and coastal currents. Finally, they act as natural barriers that protect coastal human settlements during storms and tsunamis. The devastating loss of lives and property during the 1970 typhoon in Bangladesh, and the 2004 tsunami in Southeast Asia, were very likely exacerbated by extensive destruction of coastal mangroves (Sharma 2005).

Despite their unquestioned importance for terrestrial and marine ecosystems and human welfare, mangrove forests are being rapidly destroyed (fig. 23.5). At least 35% of all mangroves have been lost worldwide (Valiela et al. 2001), and many remaining forests have been fragmented and degraded. Mangroves are being logged for timber, fuelwood,

Figure 23.5. Mangrove forests, like these in central Panama, are being rapidly destroyed in tropical coastal zones worldwide (Photo by William F. Laurance)

and charcoal production; drained and reclaimed to provide land for housing and industry; converted into rice fields, cattle pastures, and aquaculture ponds; and degraded by oil, pesticide, metal, and solid waste pollution (Lacerda et al. 1993; Valiela et al. 2001). The conservation of mangroves is among the greatest challenges for developing countries, which are struggling to manage enormous growth in their coastal populations.

Declining Fisheries and Hunting

The commercial bushmeat trade is a major threat to the survival of tropical wildlife (Wilkie et al. 1992; Fa et al. 2002; Elkan et al., chap. 21 in this volume). A recent study by Brashares et al. (2004) in Ghana, West Africa suggests a strong relationship between the intensity of illegal bushmeat hunting in nature reserves and the local availability of fish from the nearby Gulf of Guinea. First, the authors showed that incursions of hunters into six nature reserves was negatively related to per capita fish supply; in years in which fish were abundant (and thus less expensive) in local markets, hunting in nature reserves declined. Second, increased

William F. Laurance, Carlos A. Peres, Patrick A. Jansen, and Luis D'Croz hunting was strongly associated with a decline of wildlife in the reserves. As hunting pressure rose from 1970 to 1998, the estimated biomass of 41 wildlife species in reserves declined by an astounding 76%, with 16% to 45% of these species becoming locally extinct in the six reserves. Finally, the authors found that the increase in hunting pressure in years with poor fish availability was greater in nature reserves near the coast than in those further inland. This reflects the fact that, in coastal areas where fishing is an important source of employment, individuals often turn to hunting for income and food as fish stocks decline (Brashares et al. 2004).

This study suggests that in tropical nations, hunting pressure may be strongly influenced by the availability of fish as an alternative protein source. Alarmingly, commercial fish stocks are plummeting in many areas of the world, largely because of chronic overharvesting (Pauly et al. 1998; Jackson et al. 2001; Roberts 2002). In the Gulf of Guinea, fish biomass has fallen by more than 50% since 1977. This decline reflects both an increase in local fishing pressure—the population of Ghana has tripled since 1970, reducing per capita fish consumption despite intensified fish harvests—and a massive increase in commercial fishing in west-African waters by European nations (Brashares et al. 2004). Stresses on the oceans are rapidly growing because of a skyrocketing human populace in coastal areas and increasingly aggressive commercial fishing. In large areas of the tropics, this trend could exacerbate pressures on terrestrial wildlife.

Tropical Logging and Sea Turtles

Tropical logging is known to have important impacts on forests, but a recent study (Fay et al., submitted) reveals a previously unanticipated environmental risk of logging, the rapid accumulation of lost timber along beaches and waterways. In much of the humid tropics, harvested logs from timber operations are transported via rivers to coastal areas for export or processing. Logs are loaded onto barges or cabled together into large floating "rafts" that involve up to several thousand stems (Johns 1997). These logs are sometimes lost during transport and yarding near timber mills, and defective logs may also be dumped into waterways by unscrupulous timber operators. Floating logs can damage coral reefs and also pose an important hazard for boat traffic and for people on beaches. In tropical nations, many small-boat and ferry accidents, and a number of deaths, have been attributed to collisions with floating logs (Fay et al., submitted).

The central-African nation of Gabon is experiencing a major timber boom and exhibits a striking accumulation of lost timber along its beaches and major waterways. An aerial survey of Gabon's coastline in 2003 revealed an average of 17 beached logs per km of beach, with some stretches having up to 247 logs per km. Overall, over 48,000 m³ of lost logs, with an estimated commercial value of nearly US$11.1 million, were detected (Fay et al., submitted).

In significant numbers, beached logs can have serious impacts on the ecology of coastal ecosystems. Sea turtles, which must nest above the high-tide line on beaches to be successful and are extremely clumsy on land, are especially vulnerable. Gabon is one of the world's two most important nesting areas for the critically endangered leatherback turtle (*Dermochelys coriacea*), the world's largest turtle species (Sounguet 2002). Over the past three decades, a large proportion of the world's leatherback turtles have been killed by entanglement in fishing gear, ingestion of plastic waste, nest robbing, and other causes (Spotila et al. 1996, 2000; Eckert 1997).

In Gabon, the most important nesting area for sea turtles is the 5 km long Pongara Beach, typically used by more than 1100 female leatherbacks each year (Sounguet 2002). In 2005, more than 30% of this beach was blocked by logs, which had at least three deleterious effects on sea turtles (Fay et al., submitted). First, even a single log can create an impassable barrier for adult females attempting to nest, causing the nesting attempt to be aborted or forcing the female to nest too close to the waterline, where the eggs are killed by seawater. Second, adult turtles can become wedged or trapped by logs, causing death by heat exhaustion or predation (fig. 23.6). Finally, newly hatched turtles, which are highly vulnerable to terrestrial predators as they scuttle to the sea, can be impeded by logs and thereby suffer elevated mortality (Fay et al., submitted).

Threats to nesting turtles are not confined to Gabon. In Costa Rica and Panama, for example, many turtle-nesting beaches are heavily occluded by driftwood—the result of extensive local deforestation—and by plastic waste and other jetsam (Chacon 1999; Ruiz et al. 2005). These studies demonstrate that logging and deforestation not only damage tropical forests, but can cause coastal and marine ecosystems to suffer as well.

TROPICAL PROTECTED AREAS ARE IN TROUBLE

Most of the insidious threats described above can degrade tropical protected areas. It is certainly true that protected areas, including national parks as well as multiple-use areas such as indigenous lands and sustainable-development reserves, generally suffer less deforestation than do lands

Figure 23.6. A female leatherback sea turtle that died while attempting to nest in Gabon. The turtle became entangled in beached logs that were lost during tropical timber operations (Photo © www.dav eliggett.com)

outside reserves (Brandon et al. 1998; Bruner et al. 2001; Nepstad et al. 2006). Nonetheless, given the burgeoning impacts of humankind on the natural world (Sanderson et al. 2002), ecosystems in many reserves already face daunting threats that will only increase in the future.

If the mission of tropical nature reserves is to preserve examples of functionally intact ecosystems, the majority are failing—often dismally (Fagan et al., chap. 22 in this volume). Many reserves are being carved apart and isolated from formerly adjoining forests, and thereby suffer from the many deleterious consequences of habitat fragmentation and edge effects (W. F. Laurance and Bierregaard 1997; W. F. Laurance et al. 2002b). Such effects are especially severe in small and irregularly shaped reserves. Over time, isolated reserves are likely to lose large-bodied animals, top carnivores, forest-interior specialists, elevational migrants, and other species that require large expanses of intact habitat (Terborgh 1975; Woodroffe and Ginsberg 1998; Peres 2001b; W. F. Laurance et al. 2002b).

Current investments in tropical protected areas are grossly inadequate to maintain reserve integrity (Whitten and Balmford, chap. 17 in this volume; Fagan et al., chap. 22 in this volume). Many reserves are being degraded by illegal agricultural encroachment (Oluput and Chapman, chap. 7 in this volume; fig. 23.7), surface fires (W. F. Laurance, chap. 5 in

Figure 23.7. Many tropical nature reserves suffer from human encroachment. Slash-and-burn farming, such as practiced by these Bantu farmers in Gabon, is a common threat to reserves (Photo by William F. Laurance)

this volume; Barlow and Peres, chap. 12 in this volume), predatory logging (Curran et al. 1999; MacKinnon, chap. 16 in this volume), unsustainable harvests of fuelwood and nontimber products (Peres et al. 2003a; Oluput and Chapman, chap. 7 in this volume), illegal gold mining, and chronic overhunting (Peres and Terborgh 1995; Peres and Michalski, chap. 6 in this volume). Reserves are often internally fragmented by roads and other linear clearings that promote species invasions and have surprisingly potent effects on disturbance-sensitive wildlife (S. G. W. Laurance, chap. 14). The great diversity of threats described here also have secondary effects: by causing local extinctions and large alterations in species abundances, such threats can provoke ecological distortions that ramify throughout forest food webs (Terborgh et al. 2001; Terborgh and Nuñez-Iturri, chap. 13 in this volume).

As a general strategy for minimizing potential threats to nature reserves, the maxim "bigger is better" is paramount (table 23.2). Large reserves are less vulnerable to fragmentation and edge effects, are less prone to illegal encroachment by farmers (fig. 23.7), hunters, miners, and loggers, and are more likely to sustain a full complement of extinction-prone species, thereby minimizing ecological distortions (Peres 2005). Not all emerging threats, however, will be mitigated by big reserves. Large reserves are still likely to suffer the effects of large-scale climatic and atmospheric change, although they might be better able to withstand such effects than smaller reserves (W. F. Laurance 2005a). Big reserves can also be invaded by some exotic pathogens, such as the chytrid fungus that has devastated stream-

Table 23.2 Threats to tropical forests and their biota that can be reduced or mitigated by large nature reserves, versus those that cannot

Threats mitigated by large reserves	Threats not mitigated by large reserves
Fragmentation effects	Ecosystem- and community-level
Edge effects	changes from large-scale climatic or
Illegal encroachment by loggers,	atmospheric changes
hunters, farmers, and miners	Some exotic pathogens
Surface fires (except during extreme	Surface fires during extreme droughts
droughts)	
Local-scale climatic and microclimatic	
alterations	
Population sinks for hunted and	
persecuted species in surrounding	
modified lands	
Invasions of most exotic species	

dwelling frogs in pristine upland forests. On balance, however, the best opportunity to maintain the natural composition and dynamics of tropical forests is in effectively protected mega-reserves.

But how big is big enough? This is a difficult question that depends on a variety of factors, such as local environmental conditions, the density of human populations surrounding the reserve, and the effectiveness of reserve management. Our personal bias is that reserves simply cannot be too large, and that many are drastically too small. Even the largest tropical forest reserves in existence today, such as the Tucumumaque Mountains Reserve and Jaú National Park, both of which are in Brazilian Amazonia and easily exceed 2 million hectares in area, are likely to suffer from encroachment and isolation effects if their surrounding landscapes become human-dominated.

Protected-area systems in the tropics suffer from diverse problems. In many nations, reserves are poorly located to preserve important centers of biological diversity and endemism (a problem compounded by our wholly inadequate understanding of the spatial distribution of tropical biodiversity, and of the ecological and historical factors that have shaped it; e.g., Fjeldså and Rahbek 1997; Patton et al. 1997). Moreover, as discussed earlier, most reserves are too small and too irregularly shaped, and hence highly vulnerable to fragmentation, edge effects, and chronic encroachment. The majority of reserves are also seriously underfunded, understaffed, and underprotected (Peres and Terborgh 1995). Finally, despite recent progress, the total land area of tropical forests within strictly protected reserves is still far too small to guarantee against large-scale species extinctions (Fagan et al., chap. 22 in this volume; but see Wright and Muller-Landau [2006] for an iconoclastic perspective). Collectively, these points militate against an attitude of complacence. Even within nature reserves, much of the tropical world, as we know it, could be altered profoundly in the coming decades.

FOCUSING ON FRONTIERS

For forest-dependent species, the world is rapidly shrinking. With surprisingly few exceptions—mainly limited areas of the Amazon, Central Africa, northern Borneo, and eastern New Guinea—tropical forests have already been degraded by roads, infrastructure projects, widespread land transformation, and sizable human populations (Bryant et al. 1997; Sanderson et al. 2002).

In this final section we argue an obvious point: Efforts to manage and protect nature reserves in regions that have already been substantially degraded will be far more difficult and expensive than in areas lacking such pressures (cf. Peres and Terborgh 1995; Peres 2002). This perspective leads naturally to understanding the critical need to maintain large, intact expanses of forest in a roadless condition. Roads are usually the first critical step in the frontier-colonization process (S. G. W. Laurance, chap. 14 in this volume). Moreover, once created, roads frequently open a Pandora's Box of spontaneous activities—unplanned colonization, predatory logging, incursions of hunters and miners, land speculation—that are nearly impossible for governments of developing nations to control.

Nowhere is this dynamic more apparent than in the Brazilian Amazon, which sustains about 40% of the world's remaining tropical rainforest. Since 2000, the Brazilian government has been planning and undertaking a truly unprecedented expansion of its road and highway networks in the Brazilian Amazon, initially under an aggressive national-development strategy called Avança Brasil (Advance Brazil). Under this program some 7500 km of highways will be paved and many unpaved roads will be constructed, along with massive new power lines, gas lines, hydroelectric projects, railroads, river-channelization projects, and other infrastructure. If completed as planned, the total investment for these projects is projected to exceed US$40 billion (W. F. Laurance et al. 2001b). Many of these projects will penetrate deep into the heart of the basin, bisecting large, intact tracts of forest. GIS analyses of relevant infrastructure and biophysical variables suggest that these many new projects will lead to a dramatic increase in rates of forest loss and fragmentation during the next few decades (G. Carvalho et al. 2001; W. F. Laurance et al. 2001b; Soares-Filho et al. 2004, 2006). Indeed, partly as a result of the current highway and road expansion, deforestation has risen sharply in Brazilian Amazonia in recent years (W. F. Laurance et al. 2001a, 2004a, 2005a, 2005b; Fearnside et al. 2005; W. F. Laurance 2005b).

Astonishingly, representatives of the Brazilian government have often maintained that it can control development activities in the remote Amazonian frontier (Amaral 2001; Goidanich 2001; Silveira 2001; Weber 2001). Although some fledgling efforts at frontier governance have taken hold, and should be fostered (Nepstad et al. 2002; W. F. Laurance 2005b), this argument is entirely at variance with current reality and past experience (W. F. Laurance et al. 2002a). In reality, roads and infrastructure projects predispose forests to a diverse panoply of threats, the effects of which are usually irreversible (W. F. Laurance 1998; S. G. W. Laurance, chap. 14 in this volume).

Our final point, then, is this: Every conceivable effort should be made to minimize new roads and infrastructure developments in tropical forests, especially in remote frontier areas. Defending these final wilderness frontiers should be among the highest priorities for tropical conservationists. Given the enormous difficulties and expense involved in managing tropical protected areas after they have been severed away from other forests and engulfed by hostile, human-dominated landscapes, maintaining remote frontier areas is the cheapest and certainly the most effective way to conserve viable samples of tropical ecosystems. To adopt such a strategy, developing nations will suffer the lost opportunity costs of forgone development, and wealthy nations must help to bear these financial burdens (Whitten and Balmford, chap. 17 in this volume), for which many taxpayers are more than willing to contribute (e.g., Horton et al. 2003). In the long run, however, such short-term sacrifices will be far more cost-effective than the enduring, remedial efforts needed to maintain the viability of isolated tropical-forest reserves.

ACKNOWLEDGMENTS

We thank Stephanie Bohlman, Susan Laurance, Robert Ewers, and two anonymous referees for commenting on drafts of this chapter.

LITERATURE CITED

Ab'Sáber, A., J. Goldemberg, L. Rodés, and W. Zulauf. 1990. Identificação de áreas para o florestamento no espaço total do Brasil. *Estudos Avançados* **4**:63–119.

Achard, F., H. D. Eva, H. J. Stibig, P. Mayaux, J. Gallego, T. Richards, and J.-P. Malingreau. 2002. Determination of deforestation rates of the world's humid tropical forests. *Science* **297**:999–1002.

African Forest Law Enforcement and Governance. 2003. Declaration on African forest law enforcement and governance. http://www.worldbank.org/forestry/afleg.

Agrios, G. N. 1997. *Plant Pathology*. Academic Press, New York, USA.

Albacete, C. 2003. Unpublished data from ParksWatch and Trópico Verde, Durham, North Carolina, USA.

Aleixo, A., and J. M. E. Vielliard. 1995. Composição e dinâmica da comunidade de aves da Mata de Santa Genebra, Campinas, SP. *Revista Brasileira de Zoologia* **12**:493–511.

Alencar, A. A., L. A. Solórzano, and D. C. Nepstad. 2004. Modeling forest understory fires in an eastern Amazonian landscape. *Ecological Applications* **14**:S139–S149.

Alexander, H. M. 1992. Fungal pathogens and the structure of plant populations and communities. Pages 481–497 in G. C. Carroll, and D. T. Wicklow, eds., *The fungal community: Its organization and role in the ecosystem*, 2nd ed. Marcel Dekker, New York, USA.

Alleaume, S., C. Hély, J. Le Roux, S. Korontzi, R. J. Swap, H. H. Shugart, and C. O. Justice. 2006. Using MODIS to evaluate heterogeneity of biomass burning and emissions in southern African savannas: Etosha National Park Case Study. *International Journal of Remote Sensing* **26**:4219–4237.

Alpin, K., and P. Kilpatrick. 2000. Chytridiomycosis in southwest Australia: Historical sampling documents the date of introduction, rates of spread and seasonal epidemiology, and sheds new light on chytrid ecology. Proceedings of a conference titled "Getting the Jump! On Amphibian Diseases," Cairns, Queensland, Australia, 26–29 August. http://www.jcu.edu.au/school/phtm/PHTM/frogs/gjoad.htm.

Alvarado, E., and D. V. Sandberg. 2001. *Logging in tropical forest: Literature review on ecological impacts.* Fire and Environmental Research Applications, Pacific Northwest Research Station, USDA Forest Service, Corvallis, Oregon, USA.

Amazonas em Tempo. 1999. Floresta amazônica pode vivar papéis nas bolsas. Manaus, Brazil, 5 October, p. A-4.

Anagnostakis, S. 1987. Chestnut blight: The classical problem of an introduced pathogen. *Mycologia* **79**:23–27.

Andama, E. 2000. The status and distribution of carnivores in Bwindi Impenetrable National Park, south-western Uganda. M.Sc. thesis, Makerere University, Kampala, Uganda.

Anderson, P., A. A. Cunningham, N. G. Patel, F. J. Morales, P. R. Epstein, and P. Daszak. 2004. Emerging infectious diseases of plants: Pathogen pollution, climate change and agricultural drivers. *Trends in Ecology and Evolution* **19**:535–544.

Anderson, R. M., and R. M. May. 1982. *Population biology of infectious diseases.* Springer-Verlag, New York.

Anderson, R. M., and R. M. May. 1986. The invasion, persistence and spread of infectious diseases within animal and plant communities. *Philosophical Transactions of the Royal Society of London B* **314**:533–570.

Andrén, H. 1994. Effects of habitat fragmentation on birds and mammals in landscapes with different proportions of suitable habitat: A review. *Oikos* **71**:355–366.

Andresen, E. 1999. Seed dispersal by monkeys and the fate of dispersed seeds in a Peruvian rain forest. *Biotropica* **31**:145–158.

Angelsen, A., and D. Kaimowitz, eds. 2001. *Agricultural technologies and tropical deforestation.* CAB International, Wallingford, U.K.

Angelsen, A., and D. Kaimowitz. 2004. Is agroforestry likely to reduce deforestation? Pages 87–106 in G. Schroth, G. da Fonseca, C. Harvey, C. Gascon, H. Vasconcelos, and A. Izac, eds., *Agroforestry and biodiversity conservation in tropical landscapes.* Island Press, Washington, D.C., USA.

Anjos, L. 2001. Comunidade de aves florestais: implicações na conservação. Pages 17–37 in J. L. B. Albuquerque, J. F. Cândido Jr., F. C. Straube, and A. L. Roos, eds., *Ornitologia e conservação da ciência às estratégias.* Editora Unisul, Tubarão, Santa Catarina, Brazil.

Anon. 1997. A rain forest imperiled. Editorial, *New York Times*, 15 October.

Anon. 1998a. The burning of Central America. *The Economist*, 30 May.

Anon. 1998b. Fernando Henrique lança programa contra incêndios florestais na Amazônia Legal. Brazilian Ministry of Foreign Relations (MRE), Brasilía, Brazil.

Anon. 1999a. *Programa fogo: Emergência crônica.* Friends of the Earth: Brazilian Amazonia, São Paulo, Brazil.

Anon. 1999b. *World population prospects: The 1998 revision.* United Nations Department of Economic and Social Affairs, New York, USA.

Ape Alliance. 1998. *The African bushmeat trade: A recipe for extinction.* Fauna and Flora International, Cambridge, U.K.

Araújo, A. C., A. D. Nobre, B. Kruijt, J. A. Elbers, R. Dallarosa, P. Stefani, C. von Randow, A. O. Manzi, A. D. Culf, J. H. C. Gash, R. Valentini, and P. Kabat. 2002. Comparative measurements of carbon dioxide fluxes from two nearby towers in a central Amazonian rainforest: The Manaus LBA site. *Journal of Geophysical Research* **107** (D20), no. 8090.

Asner, G. P., M. Keller, R. Pereira, J. C. Zweede, and J. N. M. Silva. 2004a. Canopy damage and recovery after selective logging in Amazonia: Field and satellite studies. *Ecological Applications* **14**:S280–S298.

Asner, G. P., M. Keller, and J. N. M. Silva. 2004b. Spatial and temporal dynamics of canopy gaps following selective logging in the eastern Amazon. *Global Change Biology* **10**:765–783.

Asner, G. P., D. Knapp, E. Broadbent, P. Oliveira, M. Keller, and J. Silva. 2005. Selective logging in the Brazilian Amazon. *Science* **310**:480–482.

Asner, G. P., and P. M. Vitousek. 2005. Remote analysis of biological invasion and biogeochemical change. *Proceedings of the National Academy of Sciences USA* **102**:4383–4386.

Asquith, N. M., M. T. V. Ríos, and J. Smith. 2002. Can forest protection carbon projects improve rural livelihoods? Analysis of the Noel Kempff Mercado Climate Action Project, Bolivia. *Mitigation and Adaptation Strategies for Global Change* **7**:323–337.

Asquith, N. M., S. J. Wright, and M. Clauss. 1997. Does mammal community composition control recruitment in neotropical forests? Evidence from Panama. *Ecology* **78**:941–946.

Atjay, G. L., P. Ketner, and P. Duvigneaud. 1987. Terrestrial primary production and phytomass. Pages 129–181 in B. Bolin, ed., *The Global Carbon Cycle (Scope 13)*. John Wiley, New York, USA.

Auchoybur, G. 2003. The impact of alien weeds on fruit productivity of *Canarium paniculatum*. B.Sc. thesis. University of Mauritius.

Augspurger, C. K. 1983a. Offspring recruitment around tropical trees: Changes in cohort distance with time. *Oikos* **20**:189–196.

Augspurger, C. K. 1983b. Seed dispersal distance of the tropical tree, *Platypodium elegans*, and the escape of its seedlings from fungal pathogens. *Journal of Ecology* **71**:759–771.

Augspurger, C. K. 1984. Seedling survival among tropical tree species: Interactions of dispersal distance, light gaps and pathogens. *Ecology* **65**:1705–1712.

Augspurger, C. K. 1988. Impact of pathogens on natural plant populations. Pages 413–433 in A. Davy, M. Hutchings, and A. Watkinson, eds., *Plant population ecology*. Blackwell Scientific, Oxford, U.K.

Augspurger, C. K., and C. K. Kelly. 1984. Pathogen mortality of tropical seedlings: Experimental studies of the effect of dispersal distance, seedling density and light conditions. *Oecologia* **61**:211–217.

Auzel, P., and D. S. Wilkie. 2000. Wildlife use in northern Congo: Hunting in a commercial logging concession. Pages 413–426 in J. Robinson and E. Bennett, eds., *Hunting for sustainability in tropical forests*. Columbia University Press, New York, USA.

Avissar, R., and Y. Liu. 1996. A three-dimensional numerical study of shallow convective clouds and precipitation induced by land-surface forcing. *Journal of Geophysical Research* **101**:7499–7518.

Avissar, R., and C. A. Nobre. 2002. The Large-Scale Biosphere–Atmosphere Experiment in Amazônia (LBA). *Journal of Geophysical Research* **107**:8034, doi:10.1029/2002 JD002507.

Avissar, R., and T. Schmidt. 1998. An evaluation of the scale at which ground-surface heat flux patchiness affects the convective boundary layer using a large-eddy simulation model. *Journal of Atmospheric Science* **55**:2666–2689.

Avissar, R., P. L. Silva Dias, M. A. F. Silva Dias, and C. A. Nobre. 2002. The Large-Scale Biosphere–Atmosphere Experiment in Amazônia (LBA): Insights and future research needs. *Journal of Geophysical Research* **107**:8086, doi:10.1029/2002JD002704.

Avissar, R., and D. Werth. 2005. Global hydroclimatological teleconnections resulting from tropical deforestation. *Journal of Hydrometeorology* **6**:134–145.

Babaasa, D., R. Bitariho, and A. Kasangaki. 1999. Fire incidences in Bwindi Impenetrable National Park, SW Uganda, June–August 1999. Unpublished report, Institute of Tropical Forest Conservation—Ecological Monitoring Program, Uganda.

Babaasa, D., R. Bitariho, and A. Kasangaki. 2001. Gap characteristics in Bwindi Impenetrable National Park, Uganda. Unpublished report, Institute of Tropical Forest Conservation—Ecological Monitoring Program, Uganda.

Baidya Roy, S., and R. Avissar. 2002. Impact of land use/land cover change on regional hydrometeorology in the Amazon. *Journal of Geophysical Research* **107**:8037, doi:10.1029/2000JD00266.

Baize, S., E. M. Leroy, A. J. Georges, M. C. Georges-Courbot, M. Capron, I. Bedjabaga, J. Lansoud-Soukate, and E. Mavoungou. 2002. Inflammatory responses in Ebola virus–infected patients. *Clinical and Experimental Immunology* **128**:163–168.

Baker, T. R., O. L. Phillips, Y. Malhi, S. Almeida, L. Arroyo, T. Di Fiore, N. Higuchi, T. Killeen, S. G. Laurance, W. F. Laurance, S. L. Lewis, A. Monteagudo, D. A. Neill, N. C. A. Pitman, J. N. M. Silva, and R. V. Martínez. 2004a. Increasing biomass in Amazonian forest plots. *Philosophical Transactions of the Royal Society of London B* **359**:353–365.

Baker, T. R., O. L. Phillips, Y. Malhi, S. Almeida, L. Arroyo, T. Di Fiore, T. Killeen, S. G. Laurance, W. L. Laurance, S. L. Lewis, J. Lloyd, A. Monteagudo, D. A. Neill, S. Patiño, N. C. A. Pitman, J. N. M. Silva, and R. V. Martínez. 2004b. Variation in wood density determines spatial patterns in Amazonian forest biomass. *Global Change Biology* **10**:545–562.

Baldocchi, D. D. 2003. Assessing the eddy covariance technique for evaluating carbon dioxide exchange rates of ecosystems: Past, present and future. *Global Change Biology* **9**:479–492.

Balmford, A., A. Bruner, P. Cooper, R. Costanza, S. Farber, R. E. Green, M. Jenkins, P. Jefferiss, V. Jessamy, J. Madden, K. Munro, N. Myers, S. Naeem, J. Paavola, M. Rayment, S. Rosendo, J. Roughgarden, K. Trumper, and R. K. Turner. 2002. Economic reasons for conserving wild nature. *Science* **397**:950–953.

Balmford, A., K. J. Gaston, S. Blyth, A. James, and V. Kapos. 2003. Global variation in conservation costs, conservation benefits, and unmet conservation needs. *Proceedings of the National Academy of Sciences USA* **100**:1046–1050.

Balmford, A., J. L. Moore, T. Brooks, N. Burgess, L. A. Hanzen, P. Williams, C. Rahbek. 2001. Conservation conflicts across Africa. *Science* **291**:2616–2619.

Balmford, A., and T. Whitten. 2003. Who should pay for tropical conservation, and how could the costs be met? *Oryx* **37**:238–250.

Barber, C. 1998. Forest resource scarcity and social conflict in Indonesia. *Environment* **40**:4–37.

Barber, C. V., and J. Schweithelm. 2000. *Trial by fire*. World Resources Institute, Washington, D.C., USA.

Barbosa, P. 1991. Plant pathogens and nonvector herbivores. Pages 341–382 in P. Barbosa, V. A. Krischik, and C. G. Jones, eds., *Microbial mediation of plant–herbivore interactions*. John Wiley and Sons, New York, USA.

Barbosa, R. I., and P. M. Fearnside. 1999. Incêndios na Amazônia brasileira: Estimativa da emissão de gases do efeito estufa pela queima de diferentes ecossistemas de Roraima na passagem do evento "El Niño" (1997/98). *Acta Amazonica* **29**:513–534.

Barlow, J. 2003. Ecological effects of wildfires in a central Amazonian forest. Ph.D. thesis, University of East Anglia, Norwich, U.K.

Barlow, J., T. Haugaasen, and C. A. Peres. 2002. Effects of ground fires on understorey bird assemblages in Amazonian forests. *Biological Conservation* **105**:157–169.

Barlow, J., B. O. Lagan, and C. A. Peres. 2003a. Morphological correlates of fire-induced tree mortality in a central Amazonian forest. *Journal of Tropical Ecology* **19**:219–299.

Barlow, J., and C. A. Peres. 2004a. Avifaunal responses to single and recurrent wildfires in Amazonian forests. *Ecological Applications* **14**:1358–1373.

Barlow, J., and C. A. Peres. 2004b. Ecological response to El Niño–induced surface fires in central Amazonia: Management implications for flammable tropical forests. *Philosophical Transactions of the Royal Society of London B* **359**:367–380.

Barlow, J., and C. A. Peres. 2006. Effects of single and recurrent wildfires on fruit production and large vertebrate abundance in a central Amazonian forest. *Biodiversity and Conservation* **15**:985–1012.

Barlow, J., C. A. Peres, B. Lagan, and T. Haugaasen. 2003b. Forest biomass collapse following Amazonian wildfires. *Ecology Letters* **6**:6–8.

Barnes, R. F. W., K. L. Barnes, M. P. T. Alers, and A. Blom. 1991. Man determines the distribution of elephants in the rain-forests of northeastern Gabon. *African Journal of Ecology* **29**:54–63.

Barnston, A. G., and P. T. Schikedanz. 1984. The effect of irrigation on warm season precipitation in the southern Great Plains. *Journal of Applied Meteorology* **23**:865–888.

Barret, S. C. 1998. The reproductive biology and genetics of island plants. Pages 18–34 in P. R. Grant, ed., *Evolution on islands*. Oxford University Press, Oxford, U.K.

Barrett, C. B., and P. Arcese. 1995. Are integrated conservation–development projects (ICDPs) sustainable? On the conservation of large mammals in sub-Saharan Africa. *World Development* **23**:1073–1084.

Bataamba, A. M. 1990. The ecology of raptors in and around the Impenetrable Forest, south-western Uganda. M.Sc. thesis, Makerere University, Uganda.

Batjes, N. H. 1998. Mitigation of atmospheric CO_2 concentrations by increased carbon sequestration in the soil. *Biology and Fertility of Soils* **27**:230–235.

Batjes, N. H., and W. G. Sombroek. 1997. Possibilities for carbon sequestration in tropical and subtropical soils. *Global Change Biology* **3**:161–173.

Bawa, K., and R. Seidler. 1998. Natural forest management and the conservation of biological diversity in tropical forests. *Conservation Biology* **12**:46–55.

Beasley, D. W. C., C. T. Davis, H. Guzman, D. L. Vanlandingham, A. P. A. Travassos da Rosa, R. E. Parsons, S. Higgs, R. B. Tesh, and A. D. T. Barrett. 2003. Limited evolution of West Nile virus has occurred during its southwesterly spread in the United States. *Virology* **309**:190–195.

Becker, C. D. 1999. Protecting a Garœa forest in Ecuador: The role of institutions and ecosystem valuation. *Ambio* 28:156–161.

Beier, P., M. Van Drielen, and B. O. Kankam. 2002. Avifaunal collapse in West African forest fragments. *Conservation Biology* 16:1097–1111.

Bell, R. H. V. 1987. Conservation with a human face: Conflict and reconciliation in African land use planning. Pages 79–101 in D. Andersen and R. Grove, eds., *Conservation in Africa: People, policies and practice.* Cambridge University Press, Cambridge, U.K.

Benítez-Malvido, J. 1998. Impact of forest fragmentation on seedling abundance in a tropical rain forest. *Conservation Biology* 12:380–389.

Benítez-Malvido, J., G. García-Guzmán, and I. D. Kossmann-Ferraz. 1999. Foliar pathogens incidence and herbivory in tropical rainforest fragments: An experimental study. *Biological Conservation* 91:143–150.

Benítez-Malvido, J., and I. D. Kossmann-Ferraz. 1999. Litter cover variability affects seedling performance and herbivory. *Biotropica* 31:598–606.

Benítez-Malvido, J., and A. Lemus-Albor. 2005. The seedling community of tropical rain forest edges and interactions with herbivores and leaf-pathogens. *Biotropica* 37:301–313.

Benítez-Malvido, J., M. Martínez-Ramos, and E. Ceccon. 2001. Seed rain vs. seed bank, and the effect of vegetation cover on the recruitment of tree seedlings in tropical successional vegetation. *Dissertationes Botanicae* 346:185–203.

Bennett, A. F. 1991. Roads, roadsides and wildlife conservation: A review. Pages 99–118 in D. A. Saunders and R. J. Hobbs, eds., *Nature conservation 2: The role of corridors.* Surrey Beatty and Sons, Chipping Norton, New South Wales, Australia.

Bennett, E. L. 2002. Is there a link between wild meat and food security? *Conservation Biology* 16:590–592.

Bennett, E. L., and Z. Dahaban. 1995. Wildlife response to disturbances in Sarawak and their implications for forest management. Pages 66–86 in R. B. Primack and T. E. Lovejoy, eds., *Ecology, conservation and management of Southeast Asian rain forests.* Yale University Press, New Haven, Connecticut, USA.

Bennett, E. L., and M. T. Gumal. 2001. The interrelationships of commercial logging, hunting, and wildlife in Sarawak: Recommendations for forest management. Pages 359–374 in R. Fimbel, J. G. Robinson, and A. Grajal, eds., *The cutting edge: Conserving wildlife in logged tropical forests.* Columbia University Press, New York, USA.

Bennett, E. L., and M. Rao. 2002. Wild meat consumption in Asian tropical forest countries: Is this a glimpse of the future for Africa? Pages 39–44 in S. Mainka and M. Trivedi, eds., *Links between biodiversity, conservation, livelihoods and food security: The sustainable use of wild species for meat.* IUCN, Gland, Switzerland.

Bentrupperbäumer, J., and J. Reser. 2000. *Impacts of visitation and use in the Wet Tropics of Queensland World Heritage Area, Stage 2.* Wet Tropics Management Authority and Rainforest CRC, Cairns, Australia.

Bentrupperbäumer, J., and J. Reser. 2002. *Measuring and monitoring impacts of visitation and use in the Wet Tropics World Heritage Area 2001/2002: A site based regional perspective.* Wet Tropics Management Authority and Rainforest CRC, Cairns, Australia.

Berenstain, L. 1986. Responses of long-tailed macaques to drought and fire in eastern Borneo—a preliminary report. *Biotropica* 18:257–262.

Berger, L., R. Speare, P. Daszak, D. E. Greene, A. A. Cunningham, C. L. Goggin, R. Slocombe, M. A. Ragan, A. D. Hyatt, K. R. McDonald, H. B. Hines, K. R. Lips, G. Marantelli, and H. Parkes. 1998. Chytridiomycosis causes amphibian mortality associated with population declines in the rainforests of Australia and Central America. *Proceedings of the National Academy of Sciences USA* 95:9031–9036.

Bermejo, M. 1999. Status and conservation of primates in Odzala National Park, Republic of the Congo. *Oryx* 33:323–331.

Bermingham, E., C. Dick, and C. Moritz, eds. 2005. *Tropical rainforests: Past, present, and future.* University of Chicago Press, Chicago, USA.

Betts, C., D. Jones, S. A. Spall, and I. J. Totterdell. 2000. Acceleration of global warming due to carbon-cycle feedbacks in a coupled climate model. *Nature* 408: 184–187.

Bierregaard, R. O., Jr., and T. E. Lovejoy. 1989. Effects of forest fragmentation on Amazonian understory bird communities. *Acta Amazonica* 19:215–241.

Binder, S., A. M. Levitt, J. J. Sacks, and J. M. Hughes. 1999. Emerging infectious diseases: Public health issues for the 21st century. *Science* 284:1311–1313.

Biodiversidad en América Latina. 2000. Bajo la excusa de proteger el clima, proyecto forestal quiere destruir bosque patagónico. *Noticias,* 2 November. http://www.biodiversidadla.org/noticias83.htm.

Biodiversity Project. 2002. *Ethics for a small planet: A communications handbook on the ethical and theological reasons for protecting biodiversity.* Biodiversity Project, Madison, Wisconsin, USA.

BirdLife International. 2000. *Threatened birds of the world.* Lynx Editions, Barcelona, Spain.

Blake, S. 2001. Forest buffalo prefer clearings to closed-canopy forest in the primary forest of northern Congo. *Oryx* 36:81–86.

Blake, S. 2002. Ecology of forest elephant distribution and its implications for conservation. Ph.D. thesis, University of Edinburgh, Edinburgh, Scotland.

Blanco, J. T., and C. Forner. 2000. Expiring CERs: A proposal to addressing the permanence issue for LUCF projects in the CDM. Unpublished manuscript, Economic and Financial Analysis Group, Ministry of the Environment, Bogotá, Colombia. www.unfccc.de.

Blaustein, A. R., P. D. Hoffman, D. G. Hokit, J. M. Kiesecker, S. C. Walls, and J. B. Hayes. 1994. UV repair and resistance to solar UV-B in amphibian eggs: A link to population declines? *Proceedings of the National Academy of Sciences USA* 91: 1791–1795.

Bodmer, R. E., J. F. Eisenberg, and K. H. Redford. 1997. Hunting and the likelihood of extinction of Amazonian mammals. *Conservation Biology* 11:460–466.

Boer, C. 1989. *Investigations of the steps needed to rehabilitate the areas of East Kalimantan seriously affected by fire: Effects of the forest fires of 1982/83 in East Kalimantan towards wildlife.* Report No. 7, German Forest Service/ITTO/GTZ, Samarinda, Indonesia.

Bonaccorso, E., J. M. Guayasamin, D. Méndez, and R. Speare. In press. Chytridiomycosis in a Venezuelan anuran (Bufonidae: *Atelopus cruciger*). *Herpetological Review.*

Bonnie, R., M. Carey, and A. Petsonk. 2002. Protecting terrestrial ecosystems and the climate through a global carbon market. *Philosophical Transactions of the Royal Society of London A* 360:1853–1873.

Boo, E. 1992. *The Ecotourism boom: Planning for development and management.* World Wildlife Fund, Washington, D.C., USA.

Bookbinder, M. P., E. Dinerstein, A. Rijal, H. Cauley, and A. Rajouria. 1998. Ecotourism's support of biological conservation. *Conservation Biology* 12: 1399–1404.

Borges, S. H., and P. C. Stouffer. 1999. Bird communities in two types of anthropogenic successional vegetation in central Amazonia. *Condor* 101:529–536.

Boscolo, M., J. Buongiorno, and T. Panayotou. 1997. Simulating options for carbon sequestration through improved management of lowland tropical rainforest. *Environment and Development Economics* 2:241–263.

Boucher, D. H., J. Aviles, R. Chepote, O. E. Dominguez Gil, and B. Vilchez. 1991. Recovery of trailside vegetation from trampling in a tropical rain forest. *Environmental Management* 15:257–262.

Bourlière, F., ed. 1983. *Tropical savannas.* Elsevier, Amsterdam, The Netherlands.

Bowles, I., R. Rice, R. Mittermeier, and G. da Fonseca. 1998. Logging and tropical forest conservation. *Science* 280:1899–1900.

Bradley, D. J., G. S. Gilbert, and I. M. Parker. 2003. Susceptibility of clover species to fungal infection: The interaction of leaf surface traits and environment. *American Journal of Botany* 90:857–864.

Bradshaw, R., and F. J. G. Mitchell. 1999. The paleoecological approach to reconstructing former grazing–vegetation interactions. *Forest Ecology and Management* 120:3–12.

Brand, J., T. Healy, A. Keck, B. Minten, and C. Randrianarisoa. 2002. *Truths and myths in watershed management: The effects of deforestation upland on rice productivity in the lowlands.* FOFIFA, Antananarivo, Madagascar and ILO, Cornell University, New York, USA.

Brandon, K. 1996. *Ecotourism and conservation: A review of key issues.* World Bank, Washington, D.C., USA.

Brandon, K. 1997. Policy and practical considerations in land use strategies for biodiversity conservation. Pages 90–114 in R. Kramer, C. van Schaik, and J. Johnson, eds., *Last stand: Protected areas and the defense of tropical biodiversity.* Oxford University Press, New York, USA.

Brandon, K. 1998. Perils to parks: The social context of threats. Pages 415–440 in K. Brandon, K. H. Redford, and S. E. Sanderson, eds., *Parks in peril: People, politics, and protected areas.* Island Press, Washington, D.C., USA.

Brandon, K., K. H. Redford, and S. E. Sanderson, eds. 1998. *Parks in peril: People, politics and protected areas.* Island Press, Washington, D.C., USA.

Brashares, J. S. 2003. Ecological, behavioral and life-history correlates of mammal extinctions in West Africa. *Conservation Biology* 17:733–743.

Brashares, J. S., P. Arcese, and M. K. Sam. 2001. Human demography and reserve size predict wildlife extinction in West Africa. *Proceedings of the Royal Society of London B* 268:2473–2478.

Brashares, J. S., P. Arcese, M. K. Sam, P. B. Coppolillo, A. R. E. Sinclair, and A. Balmford. 2004. Bushmeat hunting, wildlife declines, and fish supply in West Africa. *Science* 306:1180–1183.

Brazil MMA (Ministério do Meio Ambiente). 2003. *Programa piloto para proteção das florestas tropicais do Brasil: PPG-7.* Ministry of Environment, Brasília, Brazil. http://www.mma.gov.br/port/sca/fazemos/ppg7/apresent.html.

Brewer, C. A., and W. K. Smith. 1997. Patterns of leaf surface wetness for montane and subalpine plants. *Plant, Cell and Environment* 20:1–11.

Brittingham, M., and S. Temple. 1983. Have cowbirds caused songbirds to decline? *Bioscience* 33:31–35.

Brochier, B., F. Costy, and P. P. Pastoret. 1995. Elimination of fox rabies from belgium using a recombinant vaccinia-rabies vaccine: An update. *Veterinary Microbiology* 46:269–279.

Brochier, B., M. P Kieny, F. Costy, P. Coppens, B. Bauduin, J. P. Lecocq, B. Languet, G. Chappuis, P. Desmettre, K. Afiademanyo, R. Libois, and P. P. Pastoret. 1991. Large-scale eradication of rabies using recombinant vaccinia rabies vaccine. *Nature* 354:520–522.

Brockie, R. E., L. L. Loope, M. B. Usher, and O. Hamann. 1988. Biological invasions on island nature reserves. *Biological Conservation* 44:9–36.

Broembsen, S. V., and F. J. Kruger. 1985. *Phytophthora cinnamomi* associated with mortality of native vegetation in South Africa. *Plant Disease* 69:715–717.

Brothers, T. S., and A. Spingarn. 1992. Forest fragmentation and alien plant invasion of central Indiana old-growth forests. *Conservation Biology* 6:91–100.

Brown, C., and C. A. Bolin, eds. 2000. *Emerging diseases of animals.* ASM Press, Washington, D.C., USA.

Brown, D. G. 1999. *Addicted to rent. Corporate and spatial distribution of forest resources in Indonesia: Implications for forest sustainability and government policy.* DFID/ITFMP. Jakarta, Indonesia.

Brown, D. G., D. P. Lusch, and K. A. Duda. 1998. Supervised classification of types of glaciated landscapes using digital elevation data. *Geomorphology* 21:233–250.

Brown, J. H., and M. V. Lomolino. 1998. *Biogeography, 2nd edition.* Sinauer Associates, Sunderland, Massachusetts, USA.

Brown, J. R., and J. Carter. 1998. Spatial and temporal patterns of exotic shrub invasion in an Australian tropical grassland. *Landscape Ecology* 13:93–102.

Brown, K. S., Jr., and R. W. Hutchings. 1997. Disturbance, fragmentation, and the dynamics of diversity in Amazonian forest butterflies. Pages 91–110 in W. F. Laurance and R. O. Bierregaard Jr., eds., *Tropical forest remnants: Ecology, management, and conservation of fragmented communities.* University of Chicago Press, Chicago, USA.

Brown, S., and 17 others. 2000a. Project-based activities. Pages 283–338 in R. T. Watson, I. R. Noble, B. Bolin, N. H. Ravindranath, D. J. Verardo, and D. J. Dokken, eds., *Land use, land-use change, and forestry.* Cambridge University Press, Cambridge, U.K.

Brown, S., M. Burnham, M. Delany, R. Vaca, M. Powell, and A. Moreno. 2000b. Issues and challenges for forest-based carbon-offset projects: A case study of the Noel Kempf climate action project in Bolivia. *Mitigation and Adaptation Strategies for Global Change* 5:99–121.

Bruner, A. G., R. E. Gullison, R. E. Rice, and G. A. B. da Fonseca. 2001. Effectiveness of parks in protecting tropical biodiversity. *Science* 291:125–128.

Bryant, D., D. Nielsen, and L. Tangley. 1997. *The last frontier forests: Ecosystems and economies on the edge.* World Resources Institute, Washington, D.C., USA.

Bryant, J., H. Wang, C. Cabezas, G. Ramirez, D. Watts, K. Russell, and A. Barrett. 2003. Enzootic transmission of yellow fever virus in Peru. *Emerging Infectious Diseases* 9: 926–933.

Burdon, J. J. 1987. *Disease and plant population biology*. Cambridge University Press, Cambridge, U.K.

Burdon, J. J. 1991. Fungal pathogens as selective forces in plant populations and communities. *Australian Journal of Ecology* 16:423–432.

Burdon, J. J. 1993a. The role of parasites in plant populations and communities. Pages 165–179 in E. D. Schulze and H. A. Mooney, eds., *Biodiversity and ecosystem function*. Springer-Verlag, Berlin, Germany.

Burdon, J. J. 1993b. The structure of pathogen populations in natural plant communities. *Annual Review of Phytopathology* 31:305–323.

Burgess, N., N. Doggart, and J. C. Lovett. 2002. The Uluguru Mountains of eastern Tanzania: The effect of forest loss on biodiversity. *Oryx* 36:140–152.

Burgman, M. A., S. Ferson, and H. R Akcakaya. 1993. *Risk assessment in conservation biology*. Chapman and Hall, New York, USA.

Burnett, S. E. 1992. Effects of a rainforest road on movements of small mammals: Mechanisms and implications. *Wildlife Research* 19:95–104.

Burrowes, P. A., R. L. Joglar, and D. E. Green. 2004. Potential causes for amphibian declines in Puerto Rico. *Herpetologica* 60:141–154.

Burt, F. J., A. A. Grobbelaar, P. A. Leman, F. S. Anthony, G. V. F Gibson, and R. Swanepoel. 2002. Phylogenetic relationships of southern African West Nile virus isolates. *Emerging Infectious Diseases* 8:820–826.

Butler, B., A. Birtles, R. Pearson, and K. Jones. 1997. *Ecotourism, water quality and Wet Tropics streams*. Report to the Commonwealth Department of Tourism, Canberra, Australia.

Butynski, T. M. 1984. *Ecological survey of the Impenetrable (Bwindi) Forest, Uganda, and recommendations for its conservation and management*. New York Zoological Society, Bronx, New York, USA.

Butynski, T. M., and J. Kalina. 1998. Gorilla tourism: A critical look. Pages 294–313 in E. J. Milner-Gulland and R. Mace, eds., *Conservation of biological resources*. Blackwell Science, Oxford, U.K.

Cabin, R. J., S. G. Weller, D. H. Lorence, S. Cordell, and L. J. Hadway. 2002. Effects of microsite, water, weeding and direct seeding on the regeneration of native and alien species within a Hawaiian dry forest preserve. *Biological Conservation* 104: 181–190.

Cabin, R. J., S. G. Weller, D. H. Lorence, T. W. Flynn, A. K. Sakai, D. Sandquist, and L. J. Hadway. 2000. Effects of long-term ungulate exclusion and recent alien species control on the preservation and restoration of a Hawaiian tropical dry forest. *Conservation Biology* 14:439–453.

Calvet, J. C., R. Santos-Alvala, G. Jaubert, C. Delire, C. Nobre, I. Wright, and J. Noilhan. 1997. Mapping surface parameters for mesoscale modeling in forested and deforested south-western Amazônia. *Bulletin of the American Meteorological Society* 78:413–423.

Cam, E., J. D. Nichols, J. E. Hines, J. R. Sauer, R. Alpizar-Jara, and C. H. Flather. 2002. Disentangling sampling and ecological explanations underlying species–area relationships. *Ecology* 83:1118–1130.

Camargo, J. L. C., and V. Kapos. 1995. Complex edge effects on soil moisture and microclimate in Central Amazonian forest. *Journal of Tropical Ecology* 11: 205–221.

Campbell, K., and H. Hofer. 1995. People and wildlife: Spatial dynamics and zones of interaction. Pages 534–570 in A. R. E. Sinclair and P. Arcese, eds., *Serengeti II:*

Dynamics, management and conservation of an ecosystem. University of Chicago Press, Chicago, USA.

Canaday, C. 1996. Loss of insectivorous birds along a gradient of human impact in Amazonia. *Biological Conservation* **77**:63–77.

Caraballo, A. J. 1996. Outbreak of vampire bat biting in a Venezuelan village. *Revista de Saude Publica* (Brazil) **30**:483–484.

Cardille, J. A, J. A. Foley, and M. H. Costa. 2002. Characterizing patterns of agricultural land use in Amazonia by merging satellite classifications and census data. *Global Biogeochemical Cycles* **16**:1045.

Carey, C. 1993. Hypothesis concerning the causes of the disappearance of boreal toads from the mountains of Colorado. *Conservation Biology* **7**:355–362.

Carpenter, S. R., N. Caraco, D. Correll, R. Howarth, A. Sharpley, and V. H. Smith. 1998. Nonpoint pollution of surface waters with phosphorus and nitrogen. *Ecological Applications* **8**:559–568.

Carret, J.-C., and D. Loyer. 2003. *Madagascar protected area network sustainable financing: Economic analysis perspective.* World Bank, Washington, D.C., USA. www.banque-mondiale.org.mg.

Carrillo, E., G. Wong, and A. D. Cuaron. 2000. Monitoring mammal populations in Costa Rican protected areas under different hunting restrictions. *Conservation Biology* **14**:1580–1591.

Carroll, R. W. 1988. Relative density, range extension, and conservation potential of the lowland gorilla (*Gorilla gorilla gorilla*) in the Dzanga-Sangha region of south-western Central African Republic. *Mammalia* **52**:309–323.

Carvalho, G., A. C. Barros, P. Moutinho, and D. C. Nepstad. 2001. Sensitive development could protect the Amazon instead of destroying it. *Nature* **409**:131.

Carvalho, K. S., and H. L. Vasconcelos. 1999. Forest fragmentation in central Amazonia and its effects on litter-dwelling ants. *Biological Conservation* **91**:151–158.

Carvalho, O., Jr., A. Alencar, D. C. Nepstad, and N. G. Hayashi. 2002. *Forest biomass reduction by logging and fire in the Eastern Amazon.* 12th Scientific Steering Committee Meeting of the Long-term Biosphere–Atmosphere Experiment in the Amazon (LBA) Program, Manaus, Brazil.

Castello, J. D., D. J. Leopold, and P. J. Smallidge. 1995. Pathogens, patterns, and processes in forest ecosystems. *BioScience* **45**:16–24.

Castro, G., I. Locker, V. Russell, L. Cornwell, and E. Fajer. 2000. *Mapping conservation investments: An assessment of biodiversity funding in Latin America and the Caribbean.* Biodiversity Support Program, World Wildlife Fund, Washington, D.C., USA.

Castro, R., F. Tattenbach, L. Gámez, and N. Olson. 1998. *The Costa Rican experience with market instruments to mitigate climate change and conserve biodiversity.* Fundecor and MINAE, San José, Costa Rica.

Central Africa Regional Program for the Environment. 2001. *Taking action to manage and conserve forest resources in the Congo Basin: Results and lessons learned from the first phase (1996–2000).* CARPE Information Series. Biodiversity Support Program. Washington, D.C., USA.

Chacon, D. 1999. Anidación de la tortuga *Dermochelys coriaceae* (Testudines: Dermochelyidae) en playa Gandoca, Costa Rica (1990 a 1997). *Revista de Biologia Tropical* **47**:225–236.

Chamberlin, C. 1978. The migration of the Fang into central Gabon during the nineteenth century: A new interpretation. *International Journal of African Historical Studies* 11:429–456.

Chambers, J. Q., N. Higuchi, J. P. Schimel, L. V. Ferreira, and J. M. Melack. 2000. Decomposition and carbon cycling of dead trees in tropical forests of the central Amazon. *Oecologia* 122:380–388.

Chambers, J. Q., N. Higuchi, E. S. Tribuzy, and S. E. Trumbore. 2001. Carbon sink for a century. *Nature* 410:429.

Chambers, J. Q., and W. L. Silver. 2004. Some aspects of ecophysiological and biogeochemical responses of tropical forests to atmospheric change. *Philosophical Transactions of the Royal Society of London B* 359:463–476.

Chape, S., S. Blyth, L. Fish, P. Fox, and M. Spalding, compilers. 2003. *2003 United Nations list of protected areas.* IUCN, Gland, Switzerland and Cambridge, UK and UNEP-WCMC, Cambridge, U.K.

Chape, S., and D. Sheppard, compilers. 2003. *Draft 2003 State of the World's Protected Areas.* IUCN, Gland, Switzerland and Cambridge, UK and UNEP-WCMC, Cambridge, U.K.

Chapman, C. A., and L. J. Chapman. 1996. Frugivory and the fate of dispersed and non-dispersed seeds of six African tree species. *Journal of Tropical Ecology* 12: 491–504.

Chapman, C. A., M. J. Lawes, L. Naughton-Treves, and T. R. Gillespie. 2003. Primate survival in community-owned forest fragments: Are metapopulation models useful amidst intensive use? Pages 63–78 in L. K. Marsh, ed., *Primates in fragments: Ecology and conservation.* Kluwer Academic/Plenum Publishers, New York, USA.

Chapman, C. A., and C. A. Peres. 2001. Primate conservation in the new millennium: The role of scientists. *Evolutionary Anthropology* 10:16–33.

Charrel, R. N., A. C. Brault, P. Gallian, J. J. Lemasson, B. Murgue, S. Murri, B. Pastorino, H. Zeller, R. de Chesse, P. de Micco, and X. de Lamballerie. 2003. Evolutionary relationship between Old World West Nile virus strains: Evidence for viral gene flow between Africa, the Middle East, and Europe. *Virology* 315:381–388.

Chase, T. N., J. A. Knaff, R. A. Pielke, and E. Kalnay. 2003. Changes in global monsoon circulations since 1950. *Natural Hazards* 29:229–254.

Chave, J., R. Condit, S. Lao, J. P. Caspersen, R. B. Foster, and S. P. Hubbell. 2003. Spatial and temporal variation of biomass in a tropical forest: Results from a large census plot in Panama. *Journal of Ecology* 91:240–252.

Chazdon, R. 1998. Tropical forests: Log "em or leave" em? *Science* 281:1295–1296.

Chazdon, R. L. 1988. Sunflecks and their importance to forest understory plants. *Advances in Ecological Research* 18:1–63.

Cheke, A. S. 1987. An ecological history of the Mascarene Islands. Pages 5–89 in J. W. Diamond, ed., *Studies of the Mascarene Islands birds.* Cambridge University Press, Cambridge, U.K.

Cheke, A. S., T. Gardner, C. G. Jones, A. W. Owadally, and F. Staub. 1984. Did the dodo do it? *Animal Kingdom* 87:4–6.

Chen, J., J. F. Franklin, and T. A. Spies. 1992. Vegetation responses to edge environments in old-growth Douglas-fir forests. *Ecological Applications* 2:387–396.

Chiarello, A. G. 1999. Effects of fragmentation of the Atlantic forest mammal communities in south-eastern Brazil. *Biological Conservation* 89:71–82.

Chichilnisky, G., and G. Heal. 1998. Economic returns from the biosphere. *Nature* **391**:629–630.

Chomitz, K. M., E. Brenes, and L. Constantino. 1999. Financing environmental services: The Costa Rican experience and its implications. *Science of the Total Environment* **240**:157–169.

Chomitz, K. M., and D. A. Gray. 1996. Roads, land use, and deforestation: A spatial model applied to Belize. *World Bank Economics Review* **10**:487–512.

Chomitz, K. M., and K. Kumari. 1998. The domestic benefits of tropical forests: A critical review. *World Bank Research Observer* **13**:13–35.

Christ, C., O. Hittel, S. Matus, and J. Sweeting. 2003. *Tourism and biodiversity: Mapping tourism's global footprint.* United Nations Environment Programme and Conservation International, Washington, D.C., USA.

Chua, K. B., W. J. Bellini, P. A. Rota, B. H. Harcourt, A. Tamin, S. K. Lam, T. G. Ksiazek, P. E. Rollin, S. R. Zaki, W. J. Shieh, C. S. Goldsmith, D. Gubler, J. T. Roehrig, B. T. Eaton, A. R. Gould, J. Olson, H. Field, P. Daniels, A. E. Ling, C. J. Peters, L. J. Anderson, and B. J. Mahy. 2000. Nipah virus: A recently emergent deadly Paramyxovirus. *Science* **288**:1432–1435.

Chun-Yen-Chang. 2003. Landscape structure and bird's diversity in the rural areas of Taiwan. *Journal of Environmental Sciences* (China) **15**:241–248.

Cincotta, R. P., J. Wisnewski, and R. Engelman. 2000. Human population in the biodiversity hotspots. *Nature* **404**:990–992.

Cintra, R. 1997. A test of the Janzen-Connell model with two common tree species in Amazonian forest. *Journal of Tropical Ecology* **13**:641–658.

Clark, D. A. 2002. Are tropical forests an important carbon sink? Reanalysis of the long-term plot data. *Ecological Applications* **12**:3–7.

Clark, D. A., S. C. Piper, C. D. Keeling, and D. H. Clark. 2003. Tropical rain forest tree growth and atmospheric carbon dynamics linked to interannual temperature variation during 1984–2000. *Proceedings of the National Academy of Sciences USA* **100**: 5852–5857.

Clark, D. B., and D. A. Clark. 1985. Seedling dynamics of a tropical tree: Impacts of herbivory and meristem damage. *Ecology* **66**:1884–1892.

Clark, D. B., and D. A. Clark. 1996. Abundance, growth and mortality of very large trees in neotropical lowland rain forest. *Forest Ecology and Management* **80**: 235–244.

Clark, J. S., M. Silman, R. Kern, E. Macklin, and J. HilleRistLamers. 1999. Seed dispersal near and far: Patterns across temperate and tropical forests. *Ecology* **80**: 1475–1494.

Clay, K. 1988. Fungal endophytes of grasses: A defensive mutualism between plants and fungi. *Ecology* **69**:10–16.

Cleaveland, S., M. K. Laurenson, and L. H. Taylor. 2001. Diseases of humans and their domestic mammals: Pathogen characteristics, host range and the risk of emergence. *Philosophical Transactions of the Royal Society of London B* **356**:991–999.

Cleaveland, S., S. Thirgood, and K. Laurenson. 1999. Pathogens as allies in island conservation? *Trends in Ecology and Evolution* **14**:83–84.

Cochrane, M. A. 2001a. In the line of fire: Understanding the impacts of tropical forest fires. *Environment* **43**:28–38.

Cochrane, M. A. 2001b. Synergistic interactions between habitat fragmentation and fire in evergreen tropical forests. *Conservation Biology* **15**:1515–1521.

Cochrane, M. A. 2003. Fire science for rainforests. *Nature* **421**:913–919.

Cochrane, M. A., A. Alencar, M. D. Schulze, C. M. Souza, D. C. Nepstad, P. Lefebvre, and E. Davidson. 1999. Positive feedbacks in the fire dynamics of closed canopy tropical forests. *Science* **284**:1832–1835.

Cochrane, M. A., and W. F. Laurance. 2002. Fire as a large-scale edge effect in Amazonian forests. *Journal of Tropical Ecology* **18**:311–325.

Cochrane, M. A., and M. D. Schulze. 1999. Fire as a recurrent event in tropical forests of the eastern Amazon: Effects on forest structure, biomass, and species composition. *Biotropica* **31**:2–16.

Cochrane, M. A., D. L. Skole, E. A. T. Matricardi, C. Barber, and W. Chomentowski. 2002. *Interaction and synergy between selective logging, forest fragmentation and fire disturbance in tropical forests: Case study Mato Grosso, Brazil.* CGCEO Research Advances No. RA03–02/w. East Lansing, Michigan, USA.

Cochrane, M. A., and C. M. Souza. 1998. Linear mixture model classification of burned forests in the eastern Amazon. *International Journal of Remote Sensing* **19**:3433–3440.

Cohn-Haft, M. A. 1995. Dietary specialization by lowland tropical rainforest birds: Forest interior versus canopy and edge habitats. M.Sc. thesis, Tulane University, New Orleans, USA.

Colebunders, R. and Borchert, N. 2000. Ebola haemorrhagic fever: A review. *Journal of Infection* **40**:16–20.

Colhoun, J. 1973. Effect of environmental factors on plant disease. *Annual Review of Phytopathology* **11**:343–364.

Commonwealth of Australia. 1986. *Tropical rainforests of north Queensland: Their conservation significance.* Special Australian Heritage Publication Series number 3, Australian Government Publishing Service, Canberra, Australia.

Condit, R. 1997. Forest turnover, density, and CO_2. *Trends in Ecology and Evolution* **12**:249–250.

Condit, R., S. P. Hubbell, and R. B. Foster. 1995. Mortality rates of 205 neotropical tree and shrub species and the impact of a severe drought. *Ecological Monographs* **65**:419–439.

Connell, J. H. 1971. On the role of natural enemies in preventing competitive exclusion in some marine animals and in rain forest trees. Pages 298–312 in P. J. den Boer and G. R. Gradwell, eds., *Dynamics of populations.* Centre for Agricultural Publications and Documentation, Wageningen, The Netherlands.

Coode, M. J. E. 1979. Burséraccées. Page 3 in J. Bosser, T. H. Cadet, J. Guého, and W. Marais, eds., *Flore des Mascareignes—La Réunion, Maurice, Rodrigues.* MSIRI/ORSTOM/Kew. L. Carl Achille, Mauritius.

Coode, M. J. E. 1987. Éléocarpacées. Page 4 in J. Bosser, T. H. Cadet, J. Guého, and W. Marais, eds., *Flore des Mascareignes—La Réunion, Maurice, Rodrigues.* MSIRI/ORSTOM/Kew. L. Carl Achille, Mauritius.

Coomes, D. A., and P. J. Grubb. 2000. Impacts of competition in forests and woodlands: A theoretical framework and review of experiments. *Ecological Monographs* **200**:171–207.

Cordeiro, N. J., and H. F. Howe. 2001. Low recruitment of trees dispersed by animals in African forest fragments. *Conservation Biology* **15**:1733–1741.

Cordeiro, N. J., and H. F. Howe. 2003. Forest fragmentation severs mutualism between seed dispersers and an endemic African tree. *Proceedings of the National Academy of Sciences USA* **100**:14052–14056.

Corlett, R. T., and I. M. Turner. 1997. Long-term survival in tropical forest remnants in Singapore and Hong Kong. Pages 333–346 in W. F. Laurance and R. O. Bierregaard Jr., eds., *Tropical forest remnants: Ecology, management, and conservation of fragmented communities*. University of Chicago Press, Chicago, USA.

Costanza, R., R. d'Arge, R. D. Groot, S. Farber, M. Grasso, B. Hannon, K. Limburg, S. Naeem, R. V. O'Neill, J. Paruelo, R. G. Raskin, P. Sutton, and M. van den Belt. 1997. The value of the world's ecosystem services and natural capital. *Nature* 387:253–260.

Council on Foreign Relations Independent Task Force. 2001. *A letter to the President and a memorandum on U.S. policy toward Brazil*. Council on Foreign Relations, New York, USA. www.cfr.org.

Cowie, I. D., and P. A. Werner. 1993. Alien plant species invasive in Kakadu National Park, tropical northern Australia. *Biological Conservation* 63:127–135.

Cox, G. W. 1999. *Alien species in North America and Hawaii: Impacts on natural ecosystems*. Island Press, Washington, D.C., USA.

Cox, M. P., C. R. Dickman, and J. Hunter. 2003. Effects of rainforest fragmentation on non-flying mammals of the Eastern Dorrigo Plateau, Australia. *Biological Conservation* 115:175–189.

Cox, P. A., T. Elmqvist, E. D. Pierson, and W. E. Rainey. 1991. Flying foxes as strong interactors in South Pacific island ecosystems: A conservation hypothesis. *Conservation Biology* 5:448–454.

Cox, P. M., R. A. Betts, M. Collins, P. Harris, C. Huntingford, and C. D. Jones. 2003. *Amazonian dieback under climate-carbon cycle projections for the 21st century*. Hadley Center Technical Note No. 42, Hadley Centre, Wallingford, U.K. http://www. meto.gov.uk/research/ hadleycentre/pubs/HCTN/HCTN_42.pdf.

Cox, P. M., R. A. Betts, M. Collins, P. Harris, C. Huntingford, and C. D. Jones. 2004. Amazonian dieback under climate-carbon cycle projections for the 21st century. *Theoretical and Applied Climatology* 78:137–156.

Cox, P. M., R. A. Betts, C. D. Jones, S. A. Spall, and I. J. Totterdell. 2000. Acceleration of global warming due to carbon-cycle feedbacks in a coupled climate model. *Nature* 408:184–187.

Cramer, W., A. Bondeau, F. I. Woodward, I. C. Prentice, R. A. Betts, V. Brovkin, P. M. Cox, V. Fisher, J. A. Foley, A. D. Friend, C. Kucharik, M. R. Lomas, N. Ramankutty, S. Sitch, B. Smith, A. White, and C. Young-Molling. 2001. Global response of terrestrial ecosystem structure and function to CO_2 and climate change: Results from six dynamic global vegetation models. *Global Change Biology* 7:357–373.

Crooks, K. R., and M. E. Soulé. 1999. Mesopredator release and avifaunal extinctions in a fragmented system. *Nature* 400:563–566.

Crutzen, P. J. 2002. Geology of mankind. *Nature* 415:23.

Cullen, L., Jr., R. E. Bodmer, and C. V. Padua. 2000. Effects of hunting in habitat fragments of the Atlantic forests, Brazil. *Biological Conservation* 95:49–56.

Cunningham, A. A., P. Daszak, and J. P. Rodríguez. 2003. Pathogen pollution: Defining a parasitological threat to biodiversity conservation. *Journal of Parasitology* 89:S78–S83.

Curran, L. M., I. Caniago, G. D. Paoli, D. Astianti, M. Kusneti, M. Leighton, C. E. Nirarita, and H. Haeruman. 1999. Impact of El Niño and logging on canopy tree recruitment in Borneo. *Science* 286:2184–2188.

Curtis, I. A. 2002. Environmentally sustainable tourism and a case for carbon trading in north Queensland hotels and resorts. *Australian Journal of Environmental Management* 9:27–36.

Cutrim, E., D. W. Martin, and R. M. Rabin. 1995. Enhancement of cumulus clouds over deforested lands in Amazônia. *Bulletin of the American Meteorogical Society* 76:1801–1805.

Daily, G. C., ed. 1997. *Nature's services*. Island Press, Washington, D.C., USA.

Daily, G. C., and K. Ellison. 2002. *The new economy of nature: The quest to make conservation profitable*. Shearwater, Washington, USA.

Daily, G. C., and B. H. Walker. 2000. Seeking the great transition. *Nature* 403:243–245.

Dale, V. H., and S. M. Pearson. 1997. Quantifying habitat fragmentation due to land use change in Amazonia. Pages 400–409 in W. F. Laurance and R. O. Bierregaard Jr., eds., *Tropical forest remnants: Ecology, management and conservation of fragmented communities*. University of Chicago Press, Chicago, USA.

Dale, V. H., S. M. Pearson, H. L. Offerman, and R. V. O'Neill. 1994. Relating patterns of land-use change to faunal biodiversity in the central Amazon. *Conservation Biology* 8:1027–1036.

Dalling, J. W., M. D. Swaine, and N. C. Garwood. 1998. Dispersal patterns and seed bank dynamics of pioneer trees in moist tropical forest. *Ecology* 79:564–578.

D'Angelo, S., A. Andrade, S. G. Laurance, W. F. Laurance, and R. Mesquita. 2004. Inferred causes of tree mortality in fragmented and intact Amazonian forests. *Journal of Tropical Ecology* 20:243–246.

Danielsen, F., and M. Heegaard. 1994. The impact of logging and forest conversion on lowland forest birds and other wildlife in Seberida, Riau province, Sumatra. Pages 59–60 in O. Sandbukt and H. Wiriadinata, eds., *Rain forest and resource management*. Indonesian Institute of Sciences, Jakarta, Indonesia.

D'Antonio, C. M., L. A. Meyerson, and J. S. Denslow. 2001. Exotic species and conservation: Research needs. Pages 59–80 in M. E. Soulé and G. H. Orians, eds., *Conservation biology: Research priorities for the next decade*. Island Press, Washington, D.C., USA.

D'Antonio, C. M., and P. M. Vitousek. 1992. Biological invasions by exotic grasses, the grass/fire cycle and global change. *Annual Review of Ecology and Systematics* 23:63–87.

Daszak, P., L. Berger, A. A. Cunningham, A. D. Hyatt, D. E. Green, and R. Speare. 1999. Emerging infectious diseases and amphibian population declines. *Emerging Infectious Diseases* 5:735–748.

Daszak, P., and A. A. Cunningham. 1999. Extinction by infection. *Trends in Ecology and Evolution* 14:279.

Daszak, P., and A. A. Cunningham. 2003. Anthropogenic change, biodiversity loss and a new agenda for emerging diseases. *Journal of Parasitology* 89:S37–S41.

Daszak, P., A. A. Cunningham, and A. D. Hyatt. 2000. Emerging infectious diseases of wildlife: Threats to biodiversity and human health. *Science* 287:443–449.

Daszak, P., A. A. Cunningham, and A. D. Hyatt. 2001. Anthropogenic environmental change and the emergence of infectious diseases in wildlife. *Acta Tropica* 78: 103–116.

Daszak, P., A. A. Cunningham, and A. D. Hyatt. 2003. Infectious disease and amphibian population declines. *Diversity and Distributions* 9:141–150.

Daszak, P., A. Strieby, A. A. Cunningham, J. E. Longcore, C. C. Brown, and D. Porter. 2004. Experimental evidence that the bullfrog (*Rana catesbeiana*) is a potential carrier of chytridiomycosis: An emerging fungal disease of amphibians. *Journal of Herpetology* **14**:201–207.

Davenport, L., W. Y. Brockelman, P. C. Wright, K. Ruf, and F. B. Rubio del Valle. 2002. Ecotourism tools for parks. Pages 279–306 in J. Terborgh, C. van Schaik, L. Davenport, and M. Rao, eds., *Making parks work: Strategies for preserving tropical nature.* Island Press, Washington, D.C., USA.

Davies, A. G., J. F. Oates, and G. L. Dasilva. 1999. Patterns of frugivory in three West African colobine monkeys. *International Journal of Primatology* **20**:327–357.

Davies, G. 2002. Bushmeat and international development. *Conservation Biology* **16**: 587–589.

Day, T. J., and S. M. Turton. 2000. Ecological impacts of recreation in protected areas: A review. Pages 120–122 in J. Bentrupperbäumer and J. Reser, eds., *Impacts of visitation and use in the Wet Tropics of Queensland World Heritage Area, Stage 2.* Wet Tropics Management Authority and Rainforest CRC. Cairns, Australia.

D'Croz, L., J. Del Rosario, and R. Holness. 1989. Degradation of red mangrove (*Rhizophora mangle* L.) leaves in the Bay of Panamá. *Revista de Biologia Tropical* **37**:101–104.

Dean, W. 1997. *A ferro e fogo: A história e a devastação da Mata Atlântica brasileira.* Companhia das Letras, São Paulo, Brazil.

Deem, S. L., W. B. Karesh, and W. Weisman. 2001. Putting theory into practice: Wildlife health in conservation. *Conservation Biology* **15**:1224–1233.

De Lacerda, L. D. 2003. Updating global Hg emissions from small-scale gold mining and assessing its environmental impacts. *Environmental Geology* **43**:308–314.

de Steven, D., and F. E. Putz. 1984. Impact of mammals on early recruitment of a tropical canopy tree, *Dipteryx panamensis*, in Panama. *Oikos* **43**:207–216.

Deutscher Bundestag. 1990. *Protecting the tropical forests: A high-priority international task.* Referat Öffentlichkeitsarbeit, Deutscher Bundestag, Bonn, Germany.

Develey, P. F. 1997. Ecologia de bandos mistos de aves de Mata Atlântica na Estação Ecológica Juréia-Itatins. São Paulo, Brasil. M.Sc. thesis, University of São Paulo, São Paulo, Brazil.

Develey, P. F., and P. C. Stouffer. 2001. Effects of roads on movements by understory birds in mixed-species flocks in central Amazonian Brazil. *Conservation Biology* **15**:1416–1422.

Diamond, J. M. 1972. Biogeographic kinetics: Estimation of relaxation times for avifaunas of southwest Pacific islands. *Proceedings of the National Academy of Sciences USA* **69**:3199–3203.

Diamond, J. M. 1975. The island dilemma: Lessons of modern biogeographic studies for the design of natural reserves. *Biological Conservation* **7**:129–146.

Diaz, M. C. V., D. Nepstad, M. Mendonça, R. Seroa da Motta, A. Alencar, J. Gomes, and R. Ortiz. 2002. O prejuízo oculto do fogo: Custos econômicos das queimadas e incêndios florestais na Amazônia. Unpublished manuscript.

Dickinson, R., and P. Kennedy. 1992. Impacts on regional climate of Amazon deforestation. *Geophysical Research Letters* **19**:1947–1950.

Dickinson, R. E., and A. Henderson-Sellers. 1988. Modeling tropical deforestation: A study of GCM land-surface parameterizations. *Quarterly Journal of the Royal Meteorological Society* **114**:439–462.

Dickman, A. 1992. Plant pathogens and long-term ecosystem changes. Pages 499–520 in G. C. Carroll and D. T. Wicklow, eds., *The fungal community: Its organization and role in the ecosystem*, 2nd ed. Marcel Dekker, New York, USA.

Didham, R. K., J. Ghazoul, N. Stork, and A. J. Davis. 1996. Insects in fragmented forests: A functional approach. *Trends in Ecology and Evolution* 11:255–260.

Didham, R. K., and J. H. Lawton. 1999. Edge structure determines the magnitude of changes in microclimate and vegetation structure in tropical forest fragments. *Biotropica* 31:17–30.

Dinerstein, E., and E. D. Wikramanayake. 1993. Beyond hotspots: How to prioritize investments to conserve biodiversity in the Indo-Pacific region. *Conservation Biology* 7:53–65.

Diprose, G., B. Lottermoser, S. Marks, and T. Day. 2000. Geochemical impacts on road-side soils in the Wet Tropics of Queensland World Heritage Area as a result of transport activities. Pages 66–82 in M. W. Goosem and S. M. Turton, eds., *Impacts of roads and powerlines on the Wet Tropics World Heritage Area*. Rainforest CRC, Cairns, Australia.

Dirzo, R., and A. Miranda. 1991. Altered patterns of herbivory and diversity in the for-est understory: A case study of the possible consequences of contemporary defau-nation. Pages 273–287 in P. W. Price, P. W. Lewinsohn, G. W. Fernandes, and W. W. Benson, eds., *Plant–animal interactions: Evolutionary ecology in tropical and temperate regions*. John Wiley and Sons, New York, USA.

Ditchfield, A. D. 2000. The comparative phylogeography of neotropical mammals: Patterns of intraspecific mitochondrial DNA variation among bats contrasted to nonvolant small mammals. *Molecular Ecology* 9:1307–1318.

do Amaral, S. S. 2001. Threat to the Amazon. *The Independent Newspaper*, London, 26 January.

Dobson, A., and M. Crawley. 1994. Pathogens and the structure of plant communities. *Trends in Ecology and Evolution* 9:393–398.

Dobson, A. P., and R. M. May. 1986. Disease and Conservation. Pages. 345–365 in M. Soule, ed., *Conservation biology: The science of scarcity and diversity*. Sinauer Associates, Sundersland Massachusetts, USA.

Dobson, A. P., J. P. Rodriguez, W. M. Roberts, D. S. Wilcove. 1997. Geographic distribu-tion of endangered species in the United States. *Science* 275:550–553.

Doi, T. 1988. Present status of large mammals in Kutai National Park after a large-scale fire in East Kalimantan. Pages 122–132 in H. Tagawa and N. Wirawan, eds., *Research on the process of earlier recovery of tropical rain forest after a large scale fire in Kalimantan Timur, Indonesia*. Kagoshima University, Japan.

Doran, D. M., and A. McNeilage. 1998. Gorilla ecology and behavior. *Evolutionary Anthropology* 6:120–131.

Doran, J. C., and S. Zhong. 2002. Comments on "Atmospheric disturbance caused by human modification of the landscape." *Bulletin of the American Meteorological Society* 83:277–279.

Dourojeanni, M. J. 2002. Political will for establishing and managing parks. Pages 320–334 in J. Terborgh, C. van Schaik, L. Davenport, and M. Rao, eds., *Making parks work: Strategies for preserving tropical nature*. Island Press, Washington, D.C., USA.

Dowty, P. R. 1999. Modeling biophysical processes in the savannas of southern Africa. Ph.D. thesis, University of Virginia, Charlottesville, USA.

Driml, S. 1997. *Towards sustainable tourism in the Wet Tropics World Heritage Area.* Report to the Wet Tropics Management Authority, Cairns, Australia.

Dublin, H. T., A. R. E. Sinclair, and J. McGlade. 1990. Elephants and fire as causes of multiple stable states in the Serengeti Mara woodlands. *Journal of Animal Ecology* 59:1147–1164.

Dudley, N., and S. Stolton. 2003. *Running pure: The importance of forest protected areas to drinking water.* World Bank and WWF Alliance for Forest Conservation and Sustainable Use, Washington, D.C., USA.

Dunbar, R. B. 2000. El Niño—clues from corals. *Nature* 407:956–959.

Duncan, J. 1999. Phytophthora: An abiding threat to our crops. *Microbiology Today* 26:114–116.

Dunstan, C. E., and B. J. Fox. 1996. The effects of fragmentation and disturbance of rainforest on ground-dwelling small mammals on the Robertson Plateau, New South Wales, Australia. *Journal of Biogeography* 23:187–201.

Durant, J. M., D. Hjermann, T. Anker-Nilssen, G. Beaugrand, A. Mysterud, N. Pettorelli, and N. Stenseth. 2005. Timing and abundance as key mechanisms affecting trophic interactions in variable environments. *Ecology Letters* 8:952–958.

Dutschke, M. 2002. Fractions of permanence: Squaring the cycle of sink carbon accounting. *Mitigation and Adaptation Strategies for Global Change* 7:381–402.

Eagleson, P. S., and R. I. Segarra. 1985. Water-limited equilibrium of savanna vegetation systems. *Water Resources Research* 21:1483–1493.

Easterling, D. R., G. A. Meehl, C. Parmesan, S. A. Changnon, T. R. Karl, and L. O Mearns. 2000. Climate extremes: Observations, modelling, and impacts. *Science* 289:2068–2074.

Eaton, M. J. 2002. Subsistence wildlife hunting in a multi-use forest of the Republic of Congo: Monitoring and management for sustainable harvest. M.S. thesis, University of Minnesota, St. Paul, Minnesota, USA.

Eckert, S. A. 1997. Distant fisheries implicated in the loss of the world's largest leatherback nesting population. *Marine Turtle Newsletter* 78:2–7.

Ehrlich, P. R., A. H. Ehrlich, and G. C. Daily. 1995. *The stork and the plow.* Yale University Press, New Haven, Connecticut, USA.

Elkan, E., S. Elkan, and H. Crowley. 2001. Short term solutions to the bushmeat crisis: Effective widlife management in logging concessions in Central Africa through linkages between international ngos, the private sector, local communities and governments. Wildlife Conservation Society, Republic of Congo. http://www.bush meat.org/may2001.htm.

Elkan, P. W. 2003. Ecology and conservation of bongo antelope (*Tragelaphus eurycerus*) in lowland forest, northern Republic of Congo. Ph.D. thesis, University of Minnesota, Minneapolis, USA.

Ellison, A. M., and E. J. Farnsworth. 2001. Mangrove communities. Pages 423–443 in M. D. Bertness, S. Gaines, and M. Hay, eds., *Marine community ecology.* Sinauer Associates, Sunderland, Massachusetts, USA.

Ellison, K., and G. C. Daily. 2003. Making conservation profitable. *Conservation in Practice* 4:12–19.

Elton, C. 1958. *The ecology of invasions by plants and animals.* Methuen, London, U.K.

Enserik, M. 1999. Biological invaders sweep in. *Science* 285:1834–1836.

Environmental Investigation Agency. 1998. *The politics of extinction.* Environmental Investigation Agency, London, U.K.

Environmental Investigation Agency. 1999. *The final cut*. Environmental Investigation Agency, London, U.K.

Environmental Investigation Agency and Telapak. 2000. *Illegal logging in Tanjung Puting National Park: An update on the final cut report*. Environmental Investigation Agency, London and Telapak, Bogor, Indonesia.

Esquivel, R. E., and J. Carranza. 1996. Pathogenicity of *Phylloporia chrysita* (Aphyllophorales: Hymenochaetaceae) on *Erythrochiton gymnanthus* (Rutaceae). *Revista de Biología Tropical* **44**:137–145.

Emanuel, K. 2005. Increasing destructiveness of tropical cyclones over the past 30 years. *Nature* **436**:686–688.

Eva, H., and E. F. Lambin. 2000. Fire and land-cover change in the tropics: A remote sensing analysis at the landscape scale. *Journal of Biogeography* **27**:765–776.

Eva, H. D., F. Achard, H.-J. Stibig, and P. Mayaux. 2003. Response to comment on "Determination of deforestation rates of the world's humid tropical forests." *Science* **299**:1015b.

Eve, E., F. A. Arguelles, and P. M. Fearnside. 2000. How well does Brazil's environmental law work in practice? Environmental impact assessment and the case of the Itapiranga private sustainable logging plan. *Environmental Management* **26**: 251–267.

Ewel, J. J., D. O'Dowd, J. Bergelson, C. Daehler, C. D'Antonio, L. Gómez, D. Gordon, R. Hobbs, A. Holt, K. Hopper, C. Hughes, M. LaHart, R. Leakey, W. Lee, L. Loope, D. Lorence, S. Louda, A. Lugo, P. McEvoy, D. Richardson, and P. Vitousek. 1999. Deliberate introductions of species: Research needs. *BioScience* **49**:619–630.

Fa, J. E., and C. A. Peres. 2001. Game vertebrate extraction in African and neotropical forests: An intercontinental comparison. Pages 203–241 in J. D. Reynolds, G. M. Mace, K. H. Redford, and J. G. Robinson, eds., *Conservation of exploited species*. Cambridge University Press, Cambridge, U.K.

Fa, J. E., C. A. Peres, and J. Meeuwig. 2002. Bushmeat exploitation in tropical forests: An intercontinental comparison. *Conservation Biology* **16**:232–237.

Facelli, J. M., and S. T. A. Pickett. 1991. Plant litter: Its dynamics and effects on plant community structure. *Botanical Review* **57**:1–32.

Fahrig, L. 1997. Relative effects of habitat loss and fragmentation on population extinction. *Journal of Wildlife Management* **61**:603–610.

Fahrig, L. 2001. How much habitat is enough? *Biological Conservation* **100**:65–74.

Fahrig, L. 2002. Effect of habitat fragmentation on the extinction threshold: A synthesis. *Ecological Applications* **12**:346–353.

Fahrig, L. 2003. Effects of habitat fragmentation on biodiversity. *Annual Review of Ecology, Evolution and Systematic* **34**:487–515.

Fahrig, L., J. H. Pedlar, S. E. Pope, P. D. Taylor, and J. F. Wegner. 1995. Effect of road traffic on amphibian density. *Biological Conservation* **74**:177–182.

FAO (Food and Agriculture Organization of the United Nations). 1993. *Forest resources assessment 1990: Tropical countries*. FAO Forestry Paper 112, FAO, Rome, Italy.

FAO. 1999. *State of the world's forests*. FAO, Rome, Italy.

FAO. 2001a. *Global forest resources assessment 2000*. FAO Forestry Paper 140, FAO, Rome, Italy.

FAO. 2001b. *State of the world's forests*. FAO, Rome, Italy.

Farrell, T. A., and J. L. Marion. 2001. Identifying and assessing ecotourism visitor impacts at eight protected areas in Costa Rica and Belize. *Environmental Conservation* 28:215–225.

Fauchereau, N., S. Trzaska, M. Rouault, and Y. Richard. 2003. Rainfall variability and changes in southern Africa during the 20th century in the global warming context. *Natural Hazards* 29:139–154.

Fay, J. M., and M. Agnagna. 1992. Census of gorillas in northern Republic-of-Congo. *American Journal of Primatology* 27:275–284.

Fay, M., M. Agnagna, and J. M. Moutsambote. 1990. *Surveys of the Nouabale-Ndoki region, northern Congo.* Wildlife Conservation Society, Bronx, New York, USA.

Fay, M., W. F. Laurance, R. J. Parnell, G. P. Sounguet, A. Formia, and M. Lee. Submitted. Industrial logging imperils endangered sea turtles. *Proceedings of the National Academy of Sciences USA.*

Fearnside, P. M. 1986. *Human carrying capacity of the Brazilian rainforest.* Columbia University Press, New York, USA.

Fearnside, P. M. 1989a. The charcoal of Carajás: Pig-iron smelting threatens the forests of Brazil's Eastern Amazon Region. *Ambio* 18:141–143.

Fearnside, P. M. 1989b. Forest management in Amazonia: The need for new criteria in evaluating development options. *Forest Ecology and Management* 27:61–79.

Fearnside, P. M. 1990a. Environmental destruction in the Amazon. Pages 179–225 in D. Goodman and A. Hall, eds., *The future of Amazonia: Destruction or sustainable development?* Macmillan, London, U.K.

Fearnside, P. M. 1990b. Fire in the tropical rain forest of the Amazon basin. Pages 106–116 in J. G. Goldammer, ed., *Fire in the tropical biota: Ecosystem processes and global challenges.* Springer-Verlag, Berlin, Germany.

Fearnside, P. M. 1990c. Predominant land uses in the Brazilian Amazon. Pages 235–251 in A. B. Anderson, ed., *Alternatives to deforestation: Towards sustainable use of the Amazon rain forest.* Columbia University Press, New York, USA.

Fearnside, P. M. 1993. Deforestation in Brazilian Amazonia: The effect of population an land tenure. *Ambio* 22:537–545.

Fearnside, P. M. 1995a. Agroforestry in Brazil's Amazonian development policy: The role and limits of a potential use for degraded lands. Pages 125–148 in M. Clüsener-Godt and I. Sachs, eds., *Brazilian perspectives on sustainable development of the Amazon region.* UNESCO, Paris, France and Parthenon Publishing Group, Carnforth, U.K.

Fearnside, P. M. 1995b. Global warming response options in Brazil's forest sector: Comparison of project-level costs and benefits. *Biomass and Bioenergy* 8:309–322.

Fearnside, P. M. 1995c. Hydroelectric dams in the Brazilian Amazonia as sources of "greenhouse" gases. *Environmental Conservation* 22:7–19.

Fearnside, P. M. 1996a. Amazonian deforestation and global warming: Carbon stocks in vegetation replacing Brazil's Amazon forest. *Forest Ecology and Management* 80:21–34.

Fearnside, P. M. 1996c. Socio-economic factors in the management of tropical forests for carbon. Pages 349–361 in M. J. Apps and D. T. Price, eds., *Forest ecosystems, forest management and the global carbon cycle, NATO ASI series, subseries I "Global environmental change,"* vol. 40. Springer-Verlag, Heidelberg, Germany.

Fearnside, P. M. 1996b. Hydroelectric dams in Brazilian Amazonia: Response to Rosa, Schaeffer and dos Santos. *Environmental Conservation* 23:105–108.

Fearnside, P. M. 1997a. Environmental services as a strategy for sustainable development in rural Amazonia. *Ecological Economics* 20:53–70.

Fearnside, P. M. 1997b. Greenhouse-gas emissions from Amazonian hydroelectric reservoirs: The example of Brazil's Tucuruí Dam as compared to fossil fuel alternatives. *Environmental Conservation* 24:64–75.

Fearnside, P. M. 1997c. Greenhouse gases from deforestation in Brazilian Amazonia: Net committed emissions. *Climatic Change* 35:321–360.

Fearnside, P. M. 1998. Plantation forestry in Brazil: Projections to 2050. *Biomass and Bioenergy* 15:437–450.

Fearnside, P. M. 1999a. Biodiversity as an environmental service in Brazil's Amazonian forests: Risks, value and conservation. *Environmental Conservation* 26:305–321.

Fearnside, P. M. 1999b. Como o efeito estufa pode render dinheiro para o Brasil. *Ciência Hoje* 26 (155): 41–43.

Fearnside, P. M. 1999c. Environmental and social impacts of wood charcoal in Brazil. Pages 177–182 in M. Prado, ed., *Os carvoeiros: The charcoal people of Brazil.* Wild Images Ltd., Rio de Janeiro, Brazil.

Fearnside, P. M. 1999d. Forests and global warming mitigation in Brazil: Opportunities in the Brazilian forest sector for responses to global warming under the "Clean Development Mechanism." *Biomass and Bioenergy* 16:171–189.

Fearnside, P. M. 1999e. Plantation forestry in Brazil: The potential impacts of climatic change. *Biomass and Bioenergy* 16:91–102.

Fearnside, P. M. 1999f. Social impacts of Brazil's Tucuruí Dam. *Environmental Management* 24:483–495.

Fearnside, P. M. 2000a. Effects of land use and forest management on the carbon cycle in the Brazilian Amazon. *Journal of Sustainable Forestry* 12:79–97.

Fearnside, P. M. 2000b. Global warming and tropical land-use change: Greenhouse gas emissions from biomass burning, decomposition and soils in forest conversion, shifting cultivation and secondary vegetation. *Climatic Change* 46:115–158.

Fearnside, P. M. 2000c. Greenhouse gas emissions from land use change in Brazil's Amazon region. Pages 231–249 in R. Lal, J. M. Kimble, and B. A. Stewart, eds., *Global climate change and tropical ecosystems: Advances in soil science.* CRC Press, Boca Raton, Florida, USA.

Fearnside, P. M. 2000d. Uncertainty in land-use change and forestry sector mitigation options for global warming: Plantation silviculture versus avoided deforestation. *Biomass and Bioenergy* 18:457–468.

Fearnside, P. M. 2001a. Environmental impacts of Brazil's Tucuruí Dam: Unlearned lessons for hydroelectric development in Amazonia. *Environmental Management* 27:377–396.

Fearnside, P. M. 2001b. Environmentalists split over Kyoto and Amazonian deforestation. *Environmental Conservation* 28:295–299.

Fearnside, P. M. 2001c. The potential of Brazil's forest sector for mitigating global warming under the Kyoto Protocol. *Mitigation and Adaptation Strategies for Global Change* 6:355–372.

Fearnside, P. M. 2001d. Saving tropical forests as a global warming countermeasure: An issue that divides the environmental movement. *Ecological Economics* 39:167–184.

Fearnside, P. M. 2001e. Soybean cultivation as a threat to the environment in Brazil. *Environmental Conservation* 28:23–38.

Fearnside, P. M. 2002a. Avança Brasil: Environmental and social consequences of Brazil's planned infrastructure in Amazonia. *Environmental Management* **30**: 735–747.

Fearnside, P. M. 2002b. Can pasture intensification discourage deforestation in the Amazon and Pantanal regions of Brazil? Pages 283–364 in C. H. Wood and R. Porro, eds., *Deforestation and land use in the Amazon*. University Press of Florida, Gainesville, Florida, USA.

Fearnside, P. M. 2002c. Controle de desmatamento em Mato Grosso: Um novo modelo para reduzir a velocidade de perda de floresta amazônica. Pages 29–40 in B. Millikan, L. Teixeira, L. Salvo, M. Sacramento, and P. Curvo, eds., *Workshop: Aplicações do sensoriamento remoto e sistemas de informação geográfica no monitoramento e controle do desmatamento na Amazônia Brasileira*. Subprograma dos Recursos Naturais–SPRN and Programa de Apoio a monitoramento e Análise–AMA, Secretaria da Amazônia, Ministério do Meio Ambiente, Brasília, Brazil.

Fearnside, P. M. 2002d. Greenhouse gas emissions from a hydroelectric reservoir (Brazil's Tucuruí Dam) and the energy policy implications. *Water, Air, and Soil Pollution* **133**:69–96.

Fearnside, P. M. 2002e. Time preference in global warming calculations: A proposal for a unified index. *Ecological Economics* **41**:21–31.

Fearnside, P. M. 2002f. Why a 100-year time horizon should be used for global warming mitigation calculations. *Mitigation and Adaptation Strategies for Global Change* **7**:19–30.

Fearnside, P. M. 2003a. Conservation policy in Brazilian Amazonia: Understanding the dilemmas. *World Development* **31**:757–779.

Fearnside, P. M. 2003b. Deforestation control in Mato Grosso: A new model for slowing the loss of Brazil's Amazon forest. *Ambio* **32**:343–345.

Fearnside, P. M. 2004a. A água de São Paulo e a floresta amazônica. *Ciência Hoje* **34** (203): 63–65.

Fearnside, P. M. 2004b. Greenhouse gas emissions from hydroelectric dams: Controversies provide a springboard for rethinking a supposedly "clean" energy source. *Climatic Change* **66**:1–8.

Fearnside, P. M. 2005a. Hidrelétricas planejadas no Rio Xingu como fontes de gases do efeito estufa: Belo Monte (Kararaô) e Altamira (Babaquara). Pages 204–241 in A. O. Sevá Filho, ed. *Tenotã-mõ: Alertas sobre as conseqüências dos projetos hidrelétricos no rio Xingu, Pará, Brasil*. International Rivers Network, São Paulo, Brazil.

Fearnside, P. M. 2005b. Indigenous peoples as providers of environmental services in Amazonia: Warning signs from Mato Grosso. Pages 187–198 in A. Hall, ed. *Global impact, local action: New environmental policy in Latin America*. University of London, School of Advanced Studies, London, U.K.

Fearnside, P. M. 2006a. Dams in the Amazon: Belo Monte and Brazil's hydroelectric development of the Xingu River Basin. *Environmental Management*, DOI: 10.1007/s00267-005-00113-6.

Fearnside, P. M. 2006b. Greenhouse gas emissions from hydroelectric dams: Reply to Rosa et al. *Climatic Change*, DOI: 10.1007/s10584-005-9016-z.

Fearnside, P. M., and R. I. Barbosa. 1998. Soil carbon changes from conversion of forest to pasture in Brazilian Amazonia. *Forest Ecology and Management* **108**:147–166.

Fearnside, P. M., and R. I. Barbosa. 2003. Avoided deforestation in Amazonia as a global warming mitigation measure: The case of Mato Grosso. *World Resource Review* 15:352–361.

Fearnside, P. M., and R. I. Barbosa. 2004. Accelerating deforestation in Brazilian Amazonia: Towards answering open questions. *Environmental Conservation* 31:7–10.

Fearnside, P. M., and J. Ferraz. 1995. A conservation gap analysis of Brazil's Amazonian vegetation. *Conservation Biology* 9:1134–1147.

Fearnside, P. M., N. L. Filho, and F. M. Fernandes. 1993. Rainforest burning and the global carbon budget: Biomass, combustion efficiency and charcoal formation in the Brazilian Amazon. *Journal of Geophysical Research* 98 (D9): 16733–16743.

Fearnside, P. M., and W. M. Guimarães. 1996. Carbon uptake by secondary forests in Brazilian Amazonia. *Forest Ecology and Management* 80:35–46.

Fearnside, P. M., D. A. Lashof, and P. Moura-Costa. 2000. Accounting for time in mitigating global warming through land-use change and forestry. *Mitigation and Adaptation Strategies for Global Change* 5:239–270.

Fearnside, P. M., and W. F. Laurance. 2003. Comment on "Determination of deforestation rates of the world's humid tropical forests." *Science* 299:1015.

Fearnside, P. M., and W. F. Laurance. 2004. Tropical deforestation and greenhouse gas emissions. *Ecological Applications* 14:982–986.

Fearnside, P. M., W. F. Laurance, A. K. Albernaz, H. L. Vasconcelos, and L. V. Ferreira. 2005. A delicate balance in Amazonia. *Science* 307:1045.

Feer, F. 1989. The use of space by 2 sympatric duikers (*Cephalophuscallipygus* and *Cephalophus dorsalis*) in an African rain-forest: The role of activity rhythms. *Revue d'Ecologie* 44:225–248.

Ferraro, P. J. 2001. The local costs of establishing protected areas in low-income nations: Ranomafana National Park, Madagascar. Ph.D. thesis, Georgia State University, Atlanta, Georgia, USA.

Ferraro, P. J., and A. Kiss. 2002. Direct payments to conserve biodiversity. *Science* 298:1718–1719.

Ferraro, P. J., and R. A. Kramer. 1997. Compensation and economic incentives: Reducing pressure on protected areas. Pages 187–211 in R. A. Kramer, C. van Schaik, and J. Johnson, eds., *Last stand: Protected areas and the defence of tropical biodiversity*. Oxford University Press, New York, USA.

Ferraro, P. J., and R. D. Simpson. 2002. The cost-effectiveness of conservation payments. *Land Economics* 78:339–353.

Ferraz, G., G. J. Russell, P. C. Stouffer, R. O. Bierregaard Jr., S. L. Pimm, and T. E. Lovejoy. 2003. Rates of species loss from Amazonian forest fragments. *Proceedings of the National Academy of Sciences USA* 100:14069–14073.

Field, H., P. Young, J. M. Yob, J. Mills, L. Hall, and J. Mackenzie. 2001. The natural history of Hendra and Nipah viruses. *Microbes and Infection* 3:307–314.

Fimbel, R. A., A. Grajal, and J. G. Robinson, eds. 2001. *The cutting edge: Conserving wildlife in logged tropical forests*. Columbia University Press, New York, USA.

Fine, P. V. A. 2002. The invasibility of tropical forests by exotic plants. *Journal of Tropical Ecology* 18.687–705.

Fjeldså, J., and C. Rahbek. 1997. Species richness and endemism in South American birds: Implications for the design of networks of nature reserves. Pages 466–484 in W. F. Laurance and R. O. Bierregaard Jr., eds., *Tropical forest remnants: Ecology, man-*

agement, and conservation of fragmented communities. University of Chicago Press, Chicago, USA.

Flather, C. H., and M. Bevers. 2002. Patchy reaction-diffusion and population abundance: The relative importance of habitat amount and arrangement. *American Naturalist* 159:40–56.

Foody, G. M., M. E. Cutler, J. McMorrow, D. Pelz, H. Tangki, D. S. Boyd, and I. Douglas. 2001. Mapping the biomass of Bornean tropical rain forest from remotely sensed data. *Global Ecology and Biogeography* 10:379–387.

Forest Watch Indonesia/Global Forest Watch. 2002. *The state of the forest: Indonesia.* Forest Watch Indonesia and Global Forest Watch, Washington, D.C., USA.

Forget, P.-M. 1996. Removal of seeds of *Carapa procera* (Meliaceae) by rodents and their fate in rainforest in French Guiana. *Journal of Tropical Ecology* 12:751–761.

Forget, P.-M., F. Mercier, and F. Collinet. 1999. Spatial patterns of two rodent-dispersed rain forest trees *Carapa procera* (Meliaceae) and *Vouacapoua Americana* (Caesalpiniaceae) at Paracou, French Guiana. *Journal of Tropical Ecology* 15:301–313.

Forget, P.-M., and T. Milleron. 1992. Evidence for secondary seed dispersal by rodents in Panama. *Oecologia* 87:596–599.

Forman, R. T. T. 1995. *Land mosaics: The ecology of landscape and regions.* Cambridge University Press, New York, USA.

Forman, R. T. T. 2000. Estimate of the area affected ecologically by the road system in the United States. *Conservation Biology* 14:31–35.

Forman, R. T. T., and R. D. Deblinger. 2000. The ecological road-effect zone of a Massachusetts suburban highway. *Conservation Biology* 14:36–46.

Forman, R. T. T., D. Sealing, J. A. Bissonette, A. P. Clevenger, C. D. Cutshall, V. H. Dale, L. Fahrig, R. France, C. R. Goldman, K. Heanue, J. A. Jones, F. J. Swanson, T. Turrentine, and T. C. Winter. 2002. *Road ecology: Science and solutions.* Island Press, Washington, D.C., USA.

Formenty, P., C. Boesch, M. Wyers, C. Steiner, F. Donati, F. Dind, F. Walker, and B. Le Guenno. 1999. Ebola virus outbreak among wild chimpanzees living in a rain forest of Cote d'Ivoire. *Journal of Infectious Diseases* 179:S120–S126.

Forrester, N. L., B. Boag, S. R. Moss, S. L. Turner, R. C. Trout, P. J. White, P. J. Hudson, and E. A Gould. 2003. Long-term survival of New Zealand rabbit haemorrhagic disease virus RNA in wild rabbits, revealed by RT-PCR and phylogenetic analysis. *Journal of General Virology* 84:3079–3086.

Fossey, D. 1984. Infanticide in mountain gorillas (*Gorilla gorilla berengei*) with comparative notes on chimpanzees. Pages 217–236 in G. Hausfater and S. B. Hrdy, eds., *Infanticide: Comparative and evolutionary perspectives.* Aldine Publishing Company, New York, USA.

Fowler, H. G., A. Silva-Carlos, and E. Venticinque. 1993. Size, taxonomic and biomass distributions of flying insects in central Amazonia: Forest edges vs. understory. *Revista de Biologia Tropical* 41:755–760.

Fox, B. J., and M. D. Fox. 2000. Factors determining mammal species richness on habitat islands and isolates: Habitat diversity, disturbance, species interactions and guild assembly rules. *Global Ecology and Biogeography* 9:19–37.

Fragoso, J. M. V. 1997. Tapir-generated seed shadows: Scale-dependent patchiness in the Amazon fain forest. *Journal of Ecology* 85:519–529.

Freitas, S. R., M. A. F. Silva Dias, and P. L. Silva Dias. 2000. Modeling the convective transport of trace gases by deep and moist convection. *Hybrid Methods in Engineering* **3**:317–330.

Friedmann, F. 1981. Sapotacées. Page 10 in J. Bosser, T. H. Cadet, J. Guého, and W. Marais, eds., *Flore des Mascareignes—La Réunion, Maurice, Rodrigues.* MSIRI/ORSTOM/Kew., L. Carl Achille, Mauritius.

Frith, C. B., and D. W. Frith. 1985. Seasonality of insect abundance in an Australian upland tropical rainforest. *Australian Journal of Ecology* **10**:237–248.

Frumhoff, P. 1995. Conserving wildlife in tropical forests managed for timber. *BioScience* **45**:456–464.

Galloway, J. N., and E. B. Cowling. 2002. Reactive nitrogen and the world: 200 years of change. *Ambio* **31**:64–71.

Gamboa, M. A., and P. Bayman. 2001. Communities of endophytic fungi in leaves of a tropical timber tree (*Guarea guidonia*: Meliaceae). *Biotropica* **33**:352–360.

Gao, F., E. Bailes, D. L. Robertson, Y. L., Chen, C. M. Rodenburg, S. F. Michael, L. B. Cummins, L. O. Arthur, M. Peeters, G. M. Shaw, P. M. Sharp, and B. H. Hahn. 1999. Origin of HIV-1 in the chimpanzee *Pan troglodytes troglodytes*. *Nature* **397**:436–441.

García, J. D. D. 2002. Interaction between introduced rats and a bird–plant system in a relict island forest. *Journal of Natural History* **36**:1247–1258.

García-Guzmán, G. 1990. Estudios sobre ecología de patógenos en el follaje de plantas en la selva de Los Tuxtlas. M.Sc. thesis, Universidad Nacional Autónoma de México, México.

García-Guzmán, G., and J. Benítez-Malvido. 2003. Effect of litter on the incidence of foliar pathogens and herbivory in seedlings of the tropical tree *Nectandra ambigens*. *Journal of Tropical Ecology* **19**:171–177.

García-Guzmán, G., J. Burdon, J. Ash, and R. Cunninghan. 1996. Regional and local patterns of the spatial distribution of the flower infecting smut fungus *Sporisorium amphilophis* (Syd.) Langdon and Fullerton in natural populations of its host *Bothrichloa macra* (Steud) S. T. Blake. *New Phytologist* **132**:459–469.

García-Guzmán, G., and R. Dirzo. 2001. Patterns of leaf-pathogens infection in the understory of a Mexican rain forest: Incidence, spatiotemporal variation, and mechanisms of infection. *American Journal of Botany* **88**:634–645.

Garmendia, A. E., H. J. Van Kruiningen, R. A. French, J. F. Anderson, T. G. Andreadis, A. Kumar, and A. B. West. 2000. Recovery and identification of West Nile virus from a hawk in winter. *Journal of Clinical Microbiology* **38**:3110–3111.

Garret, S. D. 1970. *Pathogenic root-infecting fungi*. Cambridge University Press, Cambridge, U.K.

Gascon, C., T. E. Lovejoy, R. O. Bierregaard Jr., J. R. Malcolm, P. C. Stouffer, H. L. Vasconcelos, W. F. Laurance, B. Zimmerman, M. Tocher, and S. Borges. 1999. Matrix habitat and species richness in tropical forest remnants. *Biological Conservation* **91**:223–229.

Gascon, C., G. B. Williamson, and G. A. B. Fonseca. 2000. Receding edges and vanishing reserves. *Science* **288**:1356–1358.

Gash, J. H. C., and C. A. Nobre. 1997. Climatic effects of Amazonian deforestation: Some results from ABRACOS. *Bulletin of the American Meteorological Society* **78**:823–830.

Gautier-Hion, A., J.-M. Duplantier, R. Quris, F. Feer, C. Sourd, J.-P. DeCoux, G. Dubost, L. Emmons, C. Erard, P. Hecketsweiler, A. Moungazi, C. Roussilhon, and J.-M.

Thiollay. 1985. Fruit characters as a basis of fruit choiceand seed dispersal in a tropical forest vertebrate community. *Oecologia* **65**:324–337.

GEF (Global Environment Facility). 2002. What is the Global Environment Facility? http://gefweb.org/What_is_the_GEF/what_is_the_gef.html.

Gelbard, J. L., and J. Belnap. 2003. Roads as conduits for exotic plant invasions in a semiarid landscape. *Conservation Biology* **17**:420–432.

Gentry, A. H. 1991. The distribution and evolution of climbing plants. Pages 3–49 in F. E. Putz and H. A. Mooney, eds., *The biology of vines*. Cambridge University Press, Cambridge, U.K.

Gentry, A. H., and L. H. Emmons. 1987. Geographical variation in fertility, phenology, and composition of the understory neotropical forests. *Biotropica* **19**:216–27.

Georges, A. J., E. M. Leroy, A. A. Renaut, C. T. Benissan, R. J. Nabias, M. T. Ngoc, P. I. Obiang, J. P. M. Lepage, E. J. Bertherat, D. D. Benoni, E. J. Wickings, J. P. Amblard, J. M. Lansoud-Soukate, J. M. Milleliri, S. Baize, and M. C. Georges-Courbot. 1999. Ebola hemorrhagic fever outbreaks in Gabon, 1994–1997: Epidemiologic and health control issues. *Journal of Infectious Diseases* **179**:S65–S75.

Georges-Courbot, M. C., A. Sanchez, C. Y. Lu, S. Baize, E. Leroy, J. LansoutSoukate, C. TeviBenissan, A. J. Georges, S. G. Trappier, S. R. Zaki, R. Swanepoel, P. A. Leman, P. E. Rollin, C. J. Peters, S. T. Nichol, and T. G. Ksiazek. 1997. Isolation and phylogenetic characterization of Ebola viruses causing different outbreaks in Gabon. *Emerging Infectious Diseases* **3**:59–62.

Georgescu, M., C. P. Weaver, R. Avissar, R. L. Walko, and G. Miguez-Macho. 2003. Sensitivity of model-simulated summertime precipitation over the Mississippi River Basin to the spatial distribution of initial soil moisture. *Journal of Geophysical Research* **108**:8855, doi:10.1029/2002JD003107.

Gerwing, J., J. Johns, and E. Vidal. 1996. Reducing waste during logging and log processing: Forest conservation in eastern Amazonia. *Unasylva* **187**:17–25.

Gerwing, J. J. 2002. Degradation of forests through logging and fire in the eastern Brazilian Amazon. *Forest Ecology and Management* **157**:131–141.

Ghazoul, J., and J. Hill. 2001. The impacts of selective logging on tropical forest invertebrates. Pages 261–288 in R. A. Fimbel, J. G. Robinson, and A. Grajal, eds., *The cutting edge: Conserving wildlife in logged tropical forests*. Columbia University Press, New York, USA.

Gibbs, W. 2001. On the termination of species. *Science* **285**:40–49. Gibson, I. A. S. 1972. *Dothistroma* blight of *Pinus radiata*. *Annual Review of Phytopathology* **10**: 51–72.

Gilbert, G. S. 1995. Rain forest plant diseases: The canopy–understory connection. *Selbyana* **16**:75–77.

Gilbert, G. S. 2002. Evolutionary ecology of plant diseases in natural ecosystems. *Annual Review of Phytopathology* **40**:13–43.

Gilbert, G. S., and S. P. Hubbell. 1996. Plant diseases and the conservation of tropical forests. *BioScience* **46**:98–106.

Gilbert, G. S., S. P. Hubbell, and R. B. Foster. 1994. Density and distance-to-adult effects of a canker disease of trees in a moist tropical forest. *Oecologia* **98**:100–108.

Gittinger, J. 1982. *Economic analysis of agricultural projects*, 2nd ed. The World Bank, Washington, D.C., USA.

Givnish, T. J. 1999. On the causes of gradients in tropical tree diversity. *Journal of Ecology* **87**:193–210.

Glantz, M. H., R. W. Katz, and N. Nicholls, eds. 1991. *Teleconnections linking worldwide climate anomalies*. Cambridge University Press, Cambridge, U.K.

Glanz, W. E. 1991. Mammalian densities at protected versus hunted sites in central Panama. Pages 163–173 in J. G. Robinson and K. H. Redford, eds., *Neotropical wildlife use and conservation*. University of Chicago Press, Chicago, USA.

Glaser, B., J. Lehmann, and W. Zech. 2002. Ameliorating physical and chemical properties of highly weathered soils in the tropics with charcoal: A review. *Biology and Fertility of Soils* 35:219–230.

Goidanich, R. 2001. The future of the Brazilian Amazon. Science Magazine Email Debates, 26 January.

Goldammer, J. G. 1999. Forests on fire. *Science* 284:1782–1783.

Goldsmith, E., ed. 2000. Religion in society, nature and the cosmos. *Ecologist* 30:1–56.

Gonzalez, J. P., E. Nakoune, W. Slenczka, P. Vidal, and J. M. Morvan. 2000. Ebola and Marburg virus antibody prevalence in selected populations of the Central African Republic. *Microbes and Infection* 2:39–44.

Goodwin, H. 1996. In pursuit of ecotourism. *Biodiversity and Conservation* 5: 277–291.

Goosem, M. 1997. Internal fragmentation: The effects of roads, highways and powerline clearings on movement and mortality of rainforest vertebrates. Pages 241–255 in W. F. Laurance and R. O. Bierregaard Jr., eds., *Tropical forest remnants: Ecology, management and conservation of fragmented communities*. University of Chicago Press, Chicago, USA.

Goosem, M. 2000. Effects of tropical rainforest roads on small mammals: Edge changes in community composition. *Wildlife Research* 27:151–163.

Goosem, M. 2001. Effects of tropical rainforest roads on small mammals: Inhibition of crossing movements. *Wildlife Research* 28:351–364.

Goosem, M. 2002. Effects of tropical rainforest roads on small mammals: Fragmentation, edge effects and traffic disturbance. *Wildlife Research* 29:1–13.

Goosem, M. 2004. Linear infrastructure in tropical rainforests: Mitigating impacts on fauna of roads and powerline clearings. Pages 418–432 in D. Lunney, ed., *Conservation of Australia's forest fauna*, 2nd ed. Royal Society of New South Wales, Mosman, Australia.

Goosem, M. W., and S. M. Turton. 1998. *Impacts of roads and powerline corridors on the Wet Tropics of Queensland World Heritage Area, Stage 1*. Wet Tropics Management Authority and Rainforest CRC, Cairns, Australia.

Goosem, M. W., and S. M. Turton. 2000. *Impacts of roads and powerline corridors on the Wet Tropics of Queensland World Heritage Area, Stage 2*. Wet Tropics Management Authority and Rainforest CRC, Cairns, Australia.

Grace, J., J. Lloyd, J. McIntyre, A. C. Miranda, P. Meir, H. Miranda, C. Nobre, J. M. Moncrieff, J. Massheder, Y. Malhi, I. R. Wright, and J. Gash. 1995. Carbon dioxide uptake by an undisturbed tropical rain forest in south-west Amazonia, 1992–1993. *Science* 270:778–780.

Graham, C., C. Moritz, and S. E. Williams. 2006. Habitat history improves prediction of biodiversity in a rainforest fauna. *Proceedings of the National Academy of Sciences USA* 103:632–636.

Grainger, A. 1988. Future supplies of high-grade tropical hardwoods from intensive plantations. *Journal of World Resources Management* 3:15–29.

Gray, C., and F. Ngolet. 1999. Lambaréné, okoumé and the transformation of labor along the middle Ogooué (Gabon), 1870–1945. *Journal of African History* **40**:87–107.

Greiser-Wilke, I., and L. Haas. 1999. Emergence of "new" viral zoonoses. *Deutsche Tierarztliche Wochenschrift* **106**:332–338.

Griffiths, M., and C. P. Van Schaik. 1993. The impact of human traffic on the abundance and activity periods of Sumatran rain forest wildlife. *Conservation Biology* **7**:623–626.

Groff, J. M., A. Mughannam, T S. McDowell, A. Wong, M. J. Dykstra, F. L. Frye, and R. P. Hedrick. 1991. An epizootic of cutaneous zygomycosis in cultured dwarf African clawed frogs (*Hymenochirus curtipes*) due to *Basidiobolus ranarum*. *Journal of Medical and Veterinary Mycology* **29**:215–23.

Groombridge, B., and M. D. Jenkins. 2003. *World atlas of biodiversity*. University of California Press, Berkeley, California, USA.

Grubb, P. J. 1971. Interpretation of the "Massenerhebung" effect on tropical mountains. *Nature* **229**:44–45.

Guariguata, M. R., H. Arias-Le Claire, and G. Jones. 2002. Tree seed fate in a logged and fragmented forest landscape, northeastern Costa Rica. *Biotropica* **34**:405–415.

Gudhardja, E., M. Fatawi, M. Sutisna, T. Mori, and S. Ohta. 2000. *Rainforest ecosystems of East Kalimantan: El Niño, drought, fire and human impacts*. Springer Life Sciences, New York, USA.

Gullison, R., R. Rice, and A. Blundell. 2000. "Marketing" species conservation. *Nature* **404**:923–924.

Guyana Chronicle. 2002. Guyana sells forest for conservation purposes. *Guyana Chronicle*, Georgetown, Guyana, 18 July.

Hackel, J. D. 1999. Community conservation and the future of Africa's wildlife. *Conservation Biology* **13**:726–734.

Hahn, B. H., G. M. Shaw, K. M. de Cock, and P. M. Sharp. 2000. Aids as a zoonosis: Scientific and public health implications. *Science* **287**:607–614.

Halford, T., and P. Auzel. 2003. *Statut des populations de gorilles (*Gorilla gorilla gorilla *Savage & Wyman 1847) et de chimpanzes (*Pan troglodytes troglogdytes *Blumenbach 1779) dans le Sanctuaire a Gorilles de Mengame, Province du Sud, Cameroun: densite, distribution, pressions et conservation*. Report to the Jane Goodall Institute, Washington, D.C., USA.

Hamilton, J. G., E. H. DeLucia, K. George, S. L. Naidu, A. C. Finzi, and W. H. Schlesinger. 2002. Forest carbon balance under elevated CO_2. *Oecologia* **131**: 250–260.

Hammitt, W. E., and D. N. Cole. 1998. *Wildland recreation: Ecology and management*, 2nd ed. John Wiley and Sons, New York, USA.

Hanselmann, R., A. Rodriguez, M. Lampo, L. Fajardo-Ramos, A. A. Aguirre, J. P. Rodriguez, and P. Daszak. 2004. Presence of an emerging pathogen of amphibians in introduced bullfrogs (*Rana catesbeiana*) in Venezuela. *Biological Conservation* **120**:115–119.

Hansen, M. C., R. S. Defries, J. R. G. Townshend, and R. Sohlberg. 2000. Global land cover classification at 1 km spatial resolution using a classification tree approach. *International Journal of Remote Sensing* **21**:1331–1364.

Harcourt, A. H., and S. A. Parks. 2003. Threatened primates experience high human densities: Adding an index of threat to the IUCN Red List criteria. *Biological Conservation* **109**:137–149.

Harcourt, A. H., S. A. Parks, and R. Woodroffe. 2001. Human density as an influence on species/area relationships: Double jeopardy for small African reserves? *Biodiversity and Conservation* **10**:1011–1026.

Hardner, J., P. Frumhoff, and D. Goetze. 2000. Prospects for mitigating carbon, conserving biodiversity, and promoting socioeconomic development through the Clean Development Mechanism. *Mitigation and Adaptation Strategies for Global Change* **5**:61–80.

Hardner, J., and R. Rice. 2002. Rethinking green consumerism. *Scientific American* **286**:88–95.

Hare, B., and M. Meinshausen. 2000. *Cheating the Kyoto Protocol: Loopholes undermine environmental effectiveness*. Greenpeace International, Amsterdam, the Netherlands.

Harrington, T. C., and M. J. Wingfield. 2000. Diseases and the ecology of indigenous and exotic pines. Pages 381–404 in D. M. Richardson, ed., *Ecology and biogeography of Pinus*. Cambridge University Press, Cambridge, U.K.

Hart, T. B., J. A. Hart, and J. S. Hall. 1996. Conservation in the declining nation state: A view from eastern Zaire. *Conservation Biology* **10**:685–686.

Harvell, C. D., C. E. Mitchell, J. R. Ward, S. Altizer, A. P. Dobson, R. S. Ostfeld, and M. D. Samuel. 2002. Climate warming and disease risks for terrestrial and marine biota. *Science* **296**:2158–2162.

Haugaasen, T., J. Barlow, and C. A. Peres. 2003. Surface wildfires in central Amazonia: Short-term impact on forest structure and carbon loss. *Forest Ecology and Management* **179**:321–331.

Haung, W. Y., and R. P. Lippman. 1988. Comparisons between neural net and conventional classifiers. *Proceedings of the International Joint Conference on Neural Networks, San Diego* **4**:485–493.

Hawksworth, D. L. 1991. The fungal dimension of biodiversity: Magnitude, significance and conservation. *Mycological Research* **95**:641–655.

Hayden, B. P. 1998. Ecosystem feedbacks on climate at the landscape scale. *Philosophical Transactions of the Royal Society of London B* **353**:5–18.

Healey, J. R., C. Price, and J. Tay. 2000. The cost of carbon retention by reduced impact logging. *Forest Ecology and Management* **139**:237–255.

Helle, P., and J. Muona. 1985. Invertebrate numbers in edges between clear-fellings and mature forests in northern Finland. *Silva Fennica* **19**:281–294.

Hély, C., P. R. Dowty, S. Alleaume, K. K Caylor, S. Korontzi, R. J. Swap, H. H. Shugart, and C. O. Justice. 2003. Regional fuel load for two climatically contrasting years in southern Africa. *Journal of Geophysical Research—Atmospheres* **108**:8475–8491.

Henderson-Sellers, A., and V. Gornitz. 1984. Possible climatic impacts of land cover transformation, with particular emphasis on tropical deforestation. *Climatic Change* **6**:231–258.

Herms, D. A., and W. J. Mattson. 1992. The dilemma of plants: To grow or defend. *Quarterly Review of Biology* **67**:283–335.

Herzog, H., K. Caldeira, and J. Reilly. 2003. An issue of permanence: Assessing the effectiveness of temporary carbon storage. *Climatic Change* **59**:293–310.

Hilbert, D. W., M. Bradford, T. Parker, and D. A. Westcott. 2004. Golden bowerbird (*Prionodura newtoniana*) habitat in past, present and future climates: Predicted extinction of a vertebrate in tropical highlands due to global warming. *Biological Conservation* **116**:367–377.

Hilbert, D. W., and B. Ostendorf. 2001. The utility of artificial neural network approaches for modelling the distribution of vegetation in past, present and future climates. *Ecological Modelling* **146**:311–327.

Hilbert, D. W., B. Ostendorf, and M. S. Hopkins. 2001. Sensitivity of tropical forests to climate change in the humid tropics of north Queensland. *Austral Ecology* **26**: 590–603.

Hilbert, D. W., and J. van den Muyzenberg. 1999. Using an artificial neural network to characterise the relative suitability of environments for forest types in a complex tropical vegetation mosaic. *Diversity and Distributions* **5**:263–274.

Hill, A. W. 1941. The genus *Calvaria* with an account of the stony endocarp and germination of the seed, and description of a new species. *Annals of Botany* **5**: 587–606.

Hill, J. K., C. D. Thomas, R. Fox, M. G. Telfer, S. G. Willis, J. Asher, and B. Huntley. 2002. Responses of butterflies to twentieth century climate warming: Implications for future ranges. *Proceedings of the Royal Society of London B* **269**:2163–2171.

Hilton-Taylor, C. 2000. *2000 IUCN red list of threatened species*. IUCN, Gland, Switzerland.

Hocking, M., S. Stolton, N. Dudley, and A. Phillips. 2000. *Evaluating effectiveness—A framework for assessing the management of protected areas*. IUCN, Gland, Switzerland.

Hoffmann, A. A., R. J. Hallas, J. A. Dean, and M. Schiffer. 2003. Low potential for climatic stress adaptation in a rainforest *Drosophila* species. *Science* **301**:100–102.

Holdgate, M. W. 1986. Summary and conclusions: Characteristics and consequences of biological invasions. *Philosophical Transactions of the Royal Society of London B* **314**:733–742.

Holdsworth, A. R., and C. Uhl. 1997. Fire in Amazonian selectively logged rain forest and the potential for fire reduction. *Ecological Applications* **7**:713–725.

Holmes, D. 2002. *Where have all the forests gone?* EASES Discussion Paper, World Bank, Washington, D.C., USA.

Holmes, T., G. Blate, J. Zweede, R. Pereira, P. Barreto, F. Boltz, and R. Bauch. 2000. *Financial costs and benefits of reduced impact logging relative to conventional logging in the Eastern Amazon*. Tropical Forest Foundation, Washington, D.C., USA.

Horne, R., and J. Hickey. 1991. Ecological sensitivity of Australian rainforests to selective logging. *Australian Journal of Ecology* **16**:119–129.

Horta, K., R. Round, and Z. Young. 2002. *The Global Environment Facility. The first ten years: Growing pains or inherent flaws?* Environmental Defense, New York, USA.

Horton, B., G. Colarullo, I. Bateman, and C. A. Peres. 2003. Evaluating non-user willingness to pay for a large-scale conservation programme in Amazonia: A UK/Italian contingent valuation study. *Environmental Conservation* **30**:139–146.

Houghton, J. T., Y. Ding, D. J. Griggs, M. Noguer, P. J. van der Linden, X. Dai, K. Maskell, and C. A. Johnson, eds. 2001. *IPCC third assessment report: Climate change 2001*. Cambridge University Press, Cambridge, U.K.

Houghton, R. A. 1999. The annual net flux of carbon to the atmosphere from changes in land use 1850–1990. *Tellus* **51B**:298–313.

Houghton, R. A. 2003. Why are estimates of the terrestrial carbon balance so different? *Global Change Biology* 9:500–509.

Houlahan, J. E., C. S. Findlay, B. R. Schmidt, A. H. Meyer, and S. L. Kuzmin. 2000. Quantitative evidence for global amphibian population declines. *Nature* **404**: 752–755.

Howard, A., R. Rice, and R. Gullison. 1996. Simulated economic returns and selected environmental impacts from four alternative silvicultural prescriptions applied in the neotropics: A case study of the Chimanes Forest, Bolivia. *Forest Ecology and Management* **89**:43–57.

Howard, P. C. 1991. *Nature conservation in Uganda's tropical forest reserves*. IUCN Conservation Library, Gland, Switzerland.

Howden, M., L. Hughes, M. Dunlop, I. Zethoven, D. Hilbert, and C. Chilcott, eds. 2003. *Impacts of global climate change on the rainforest vertebrates of the Australian Wet Tropics*. Climate change impacts on Biodiversity in Australia, Commonwealth of Australia, Canberra.

Howe, H. F. 1977. Bird activity and seed dispersal of a tropical wet forest tree. *Ecology* 58:539–550.

Howe, H. F. 1984. Implications of seed dispersal by animals for tropical reserve management. *Biological Conservation* **30**:261–281.

Howe, H. F. 1986. Seed dispersal by fruit eating birds and mammals. Pages 123–190 in J. Murray, ed., *Seed dispersal*. Academic Press, Sydney, Australia.

Howe, H. F., and M. N. Miriti. 2000. No question: Seed dispersal matters. *Trends in Ecology and Evolution* 15:434–436.

Howe, H. F., E. W. Schupp, and L. C. Westley. 1985. Early consequences of seed dispersal for a neotropical tree (*Virola surinamensis*). *Ecology* **66**:781–791.

Howe, H. F., and J. Smallwood. 1982. Ecology of seed dispersal. *Annual Review of Ecology and Systematics* **13**:201–228.

Hubbell, S. P., R. B. Foster, S. T. O'Brien, K. E. Harms, R. Condit, B. Wechsler, S. J. Wright, and S. Loo de Lao. 1999. Light-gap disturbances, recruitment limitation, and tree diversity in a neotropical forest. *Science* **283**:554–557.

Huenneke, L. F., and P. M. Vitousek. 1990. Seedling and clonal recruitment of the invasive tree *Psidium cattleianum*: Implications for management of native Hawaiian forests. *Biological Conservation* **53**:199–211.

Hughes, C. E., and B. T. Styles. 1989. The benefits and risks of woody legume introductions. Pages 505–531 in C. H. Stirton and J. L. Zarucchi, eds., *Advances in legume biology*. Missouri Botanical Garden, St. Louis, Missouri, USA.

Hughes, L. 2000. Biological consequences of global warming: Is the signal already apparent? *Trends in Ecology and Evolution* 15:56–61.

Hughes, T. P., A. H. Baird, D. R. Bellwood, M. Card, S. R. Connolly, C. Folke, R. Grosberg, O. Hoegh-Guldberg, J. B. C. Jackson, J. Kleypas, J. M. Lough, P. Marshall, M. Nystrom, S. R. Palumbi, J. M. Pandolfi, B. Rosen, and J. Roughgarden. 2003. Climate change, human impacts, and the resilience of coral reefs. *Science* **301**:929–933.

Huijbregts, B., P. De Wachter, L. S. N. Obiang, and M. E. Akou. 2003. Ebola and the decline of gorilla *Gorilla gorilla* and chimpanzee *Pan troglodytes* populations in Minkebe Forest, north-eastern Gabon. *Oryx* 37:437–443.

Humphries, S. E., R. H. Groves, and D. Mitchell. 1991. Plant invasions of Australian ecosystems: A status review and management directions. *Kowari* 2:1–134.

Hunter, C., and H. Green. 1995. *Tourism and the environment: A sustainability relationship?* Routledge, London, USA.

Huntley, B. J., and B. H Walker, eds. 1982. *Ecology of tropical savannas.* Springer-Verlag, Berlin, Germany.

Hurt, G. C., and S. Pacala. 1995. The consequences of recruitment limitation: Reconciling chance, history and competitive differences between plants. *Journal of Theoretical Biology* 176:1–12.

IBAMA. 1998. Estimativa da area de cobertura florestal afetada pelo incendio em Roraima, utilizando dados de multisensores. Unpublished manuscript, Instituto Nacional de Pesquisas Espaciais (INPE)/Divisao de Sensoriamento Remoto (Outobro de 1998).

Ickes, K., and G. B. Williamson. 2000. Edge effects and ecological processes: Are they on the same scale? *Trends in Ecology and Evolution* 15:373.

Infante, L. A. 1999. Complex interactions: Exploring the role of soil borne plant pathogens in tropical seedling communities. M.Sc. thesis, University of California, Berkeley.

INPE (Instituto Nacional de Pesquisas Espaciais). 2003. *Incremento no desflorestamento por estado entre 2000 e 2001.* www.orb.inpe.br.

IPAM (Instituto de Pesquisa Ambiental da Amazônia). 2000. *Manifestação da sociedade civil brasileira sobre as relações entre florestas e mudanças climáticas e as expectativas para a COP-6.* IPAM, Belém, Pará, Brazil. http://www.ipam.org.br/polamb/man belem.htm.

ISSG. 2003. *Global invasive species database.* http://www.issg.org/database/species/ ecol ogy.asp?si=139andfr=1andsts=sss.

IUCN. 1993. *Parks for life: Report of the IVth world congress on national parks and protected areas.* IUCN, Gland, Switzerland.

IUCN. 1998. *1997 United Nations list of protected areas.* IUCN, Gland, Switzerland.

IUCN. 2002. *2002 red list of threatened species.* IUCN, Gland, Switzerland. http://www.redlist.org/search/details.php?species=19310.

IUCN. 2003. *2003 red list of threatened species.* IUCN, Gland, Switzerland. www.redlist.org.

IUCN/UNEP/WWF. 1980. *World conservation strategy: Living resource conservation for sustainable development.* IUCN/UNEP/WWF, Gland, Switzerland.

IUCN/UNEP/WWF. 1991. *Caring for the Earth: A strategy for sustainable living.* IUCN/UNEP/WWF, Gland, Switzerland.

Iverson, J. B. 1987. Tortoises, not dodos, and the Tambalacoque tree. *Journal of Herpetology* 21:229–230.

Jachmann, H. 1998. *Monitoring illegal wildlife use and law enforcement in African savanna rangelands.* The Wildlife Resource Monitoring Unit, Lusaka, Zambia.

Jackson, J. B. C., M. Kirby, W. Berger, K. Bjorndal, L. Botsford, B. Bourque, R. Bradbury, R. Cooke, J. Erlandson, J. Estes, T. Hughes, S. Kidwell, C. Lange, H. Lenihan, J. Pandolfi, C. Peterson, R. Steneck, M. Tegner, and R. Warner. 2001. Historical overfishing and the recent collapse of coastal ecosystems. *Science* 293:629–637.

James, A., K. Gaston, and A. Balmford. 1999a. Balancing the earth's accounts. *Nature* 401:323–324.

James, A., K. Gaston, and A. Balmford. 2000. Why private institutions alone will not do enough to protect biodiversity. *Nature* **404**:120.

James, A., K. Gaston, and A. Balmford. 2001. Can we afford to conserve biodiversity? *BioScience* **51**:43–52.

James, A., M. J. B. Green, and J. R. Paine. 1999b. *A global review of protected areas budgets and staff.* World Conservation Monitoring Centre, Cambridge, U.K.

Jansen, P., M. Bartholomeus, F. Bongers, J. A. Elzinga, J. den Ouden, and S. E. van Wieren. 2002. The role of seed size in dispersal by a scatter-hoarding rodent. Pages 209–225 in D. J. Levey, W. R. Silva, and M. Galetti, eds., *Seed dispersal and frugivory: Ecology, evolution and conservation.* CAB International, Wallingford, U.K.

Jansen, P. A., and P. A. Zuidema. 2001. Logging, seed dispersal by vertebrates, and the natural regeneration of tropical timber trees. Pages 35–59 in R. A. Fimbel, J. G. Robinson, and A. Grajal, eds., *The cutting edge: Conserving wildlife in logged tropical forests.* Columbia University Press, New York, USA.

Janzen, D. H. 1970. Herbivores and the number of tree species in tropical forests. *American Naturalist* **104**:501–528.

Janzen, D. H. 1983. No park is an island: Increase in interference from outside as park size decreases. *Oikos* **41**:402–410.

Janzen, D. H. 1986a. The eternal external threat. Pages 286–303 in M. E. Soulé, ed., *Conservation biology: The science of scarcity and diversity.* Sinauer, Sunderland, Massachussets, USA.

Janzen, D. H. 1986b. The future of tropical ecology. *Annual Review of Ecology and Systematics* **17**:305–324.

Janzen, D. H. 1988. Tropical dry forests: The most endangered major tropical ecosystem. Pages 130–137 in E. O. Wilson, ed., *Biodiversity.* National Academy Press, Washington, D.C., USA.

Jarozs, A. M., and A. L. Davelos. 1995. Effects of disease in wild plant populations and the evolution of pathogen aggressiveness. *New Phytologist* **129**:371–382.

Jepson, P., J. Jarvie, K. MacKinnon, and K. A. Monk. 2001. The end for Indonesia's lowland forests? *Science* **292**:859–861.

Jerozolimski, A., and C. A. Peres. 2003. Bringing home the biggest bacon: A cross-site analysis of the structure of hunter–kill profiles in neotropical forests. *Biological Conservation* **111**:415–425.

Jipp, P., D. C. Nepstad, K. Cassel, and C. R. de Carvalho. 1998. Deep soil moisture storage and transpiration in forests and pastures of seasonally dry Amazônia. *Climatic Change* **39**:395–412.

Johns, A. G. 1988. Effects of selective timber extraction on rainforest structure and composition and some consequences for frugivores and folivores. *Biotropica* **20**:31–37.

Johns, A. G. 1997. *Timber production and biodiversity conservation in tropical rain forests.* Cambridge University Press, Cambridge, U.K.

Johns, J. S., P. Barreto, and C. Uhl. 1996. Logging management in planned and unplanned logging operations and its implications for sustainable timber production in the eastern Amazon. *Forest Ecology and Management* **89**:59–77.

Johnson, M. L., and R. Speare. 2003. Survival of *Batrachochytrium dendrobatidis* in water: Quarantine and disease control implications. *Emerging Infectious Diseases* **9**:922–925.

Johnson, N., and B. Cabarle. 1993. *Surviving the cut: Natural forest management in the humid tropics.* World Resources Institute, Washington, D.C., USA.

Jones, J. A., F. J. Swanson, B. C. Wemple, and K. U. Snyder. 2000. Effects of roads on hydrology, geomorphology, and disturbance patches in stream networks. *Conservation Biology* **14**:76–85.

Jones, M. E. 2000. Road upgrade, road mortality and remedial measures: Impacts on a population of eastern quolls and Tasmanian devils. *Wildlife Research* **27**:289–296.

Jordano, P. 1992. Fruits and frugivory. Pages 105–151 in M. Fenner, ed., *Seeds: The ecology of regeneration in natural plant communities.* CAB International, Wallingford, U.K.

Julliot, C. 1997. Impact of seed dispersal by red howler monkeys *Alouatta seniculus* on the seedling population in the understorey of tropical rain forest. *Journal of Ecology* **85**:431–440.

Jusoff, K. 1989. Physical soil properties associated with recreational use of a forested reserve area in Malaysia. *Environmental Conservation* **16**:339–342.

Justice, C., D. Wilkie, Q. Zhang, J. Brunner, and C. Donoghue. 2001. Central African forests, carbon and climate change. *Climate Research* **17**:229–246.

Kamdem-Toham, A., A. W. Adeleke, N. D. Burgess, R. Carroll, J. D'Amico, E. Dinnerstein, D. M. Olson, and L. Some. 2003. Forest conservation in the Congo Basin. *Science* **299**:246.

Kanowski, J. 2001. Effects of elevated CO_2 on the foliar chemistry of seedlings of two rainforest trees from north-east Australia: Implications for folivorous marsupials. *Austral Ecology* **26**:165–172.

Kanowski, J., M. S. Hopkins, H. Marsh, and J. W. Winter. 2001. Ecological correlates of folivore abundance in north Queensland rainforests. *Wildlife Research* **28**:1–8.

Kanowski, J., A. K. Irvine, and J. W. Winter. 2003. The relationship between the floristic composition of rainforests and the abundance of folivorous marsupials in north-east Queensland. *Journal of Animal Ecology* **72**:627–632.

Kaplan, H., K. Hill, J. Lancaster, and M. A. Hurtado. 2000. A theory of human life history evolution: Diet, intelligence, and longevity. *Evolutionary Anthropology* **9**:156–183.

Kapos, V. 1989. Effects of isolation on the water status of forest patches in the Brazilian Amazon. *Journal of Tropical Ecology* **5**:173–185.

Kapos, V., E. Wandelli, J. Camargo, and G. Ganade. 1997. Edge-related changes in environment and plant responses due to forest fragmentation in central Amazonia. Pages 33–44 in W. F. Laurance and R. O. Bierregaard Jr., eds., *Tropical forest remnants: Ecology, management, and conservation of fragmented communities.* University of Chicago Press, Chicago, USA.

Karanth, K. U., and M. D. Madhusudan. 2002. Mitigating human–wildlife conflicts in southern Asia. Pages 250–264 in J. Terborgh, C. van Schaik, L. Davenport, and M. Rao, eds., *Making parks work: Strategies for preserving tropical nature.* Island Press, Washington, D.C., USA.

Kasangaki, A., D. Babaasa, R. Bitariho, and G. Mugiri. 2001. A survey of burnt areas in Bwindi Impenetrable and Mgahinga Gorilla National Parks, S. W. Uganda: The fires of 2000. Unpublished report, Institute of Tropical Forest Conservation— Ecological Monitoring Program, Uganda.

Kauffman, J. B. 1991. Survival by sprouting following fire in tropical forests of the eastern Amazon. *Biotropica* **23**:219–224.

Kauffman, J. B., and C. Uhl. 1990. Interactions of anthropogenic activities, fire, and rain forests in the Amazon basin. Pages 117–134 in J. Goldammer, ed., *Fire in the tropical biota*. Springer-Verlag, New York, USA.

Kellas, J. D., G. A. Kile, R. G. Jarrett, and J. T. Morgan. 1987. The occurrence and effects of *Armillaria luteobubalina* following partial cutting in mixed eucalypt stands in the Wombat forest, Victoria. *Australian Forest Research* 17:263–276.

Kennedy, J., and G. Weste. 1986. Vegetation changes associated with invasion by *Phytophthora cinnamomi* on monitored sites in the Grampians, Western Victoria. *Australian Journal of Botany* 34:251–279.

Kerr, S., and C. Leining. 2000. *Permanence of LULUCF CERs in the Clean Development Mechanism*. Center for Clean Air Policy, Washington, D.C., USA.

Kiltie, R. 1981. Distribution of palm fruits on a rain forest floor: Why white-lipped peccaries forage near objects. *Biotropica* 13:141–145.

Kinnaird, M. F., and T. G. O'Brien. 1996. Ecotourism in the Tangkoko Duasudara Nature Reserve: Opening Pandora's Box? *Oryx* 30:65–73.

Kinnaird, M. F., and T. G. O'Brien. 1998. Ecological effects of wildfire on lowland rainforest in Sumatra. *Conservation Biology* 12:954–956.

Kinnaird, M. F., E. W. Sanderson, T. G. O'Brien, H. T. Wibisono, and G. Woolmer. 2003. Deforestation trends in a tropical landscape and implications for endangered large mammals. *Conservation Biology* 17:245–257.

Kiss, A., G. Castro, and K. Newcombe. 2002. The role of multilateral institutions. *Philosophical Transactions of the Royal Society of London A* 360:1641–1652.

Klein, B. C. 1989. Effects of forest fragmentation on dung and carrion beetle communities in central Amazonian. *Ecology* 70:1715–1725.

Koenig, P. 1914. Economic flora. Pages 102–109 in A. Macmillan, ed., *Mauritius illustrated*. Les Editions du Pacifique (Reprinted 1991), Singapore.

Körner, C. 2003. Slow in, rapid out: Carbon flux studies and Kyoto targets. *Science* 300:1242–1243.

Körner, C. 2004. Through enhanced tree dynamics carbon dioxide enrichment may cause tropical forests to lose carbon. *Philosophical Transactions of the Royal Society of London B* 359:493–498.

Kramer, R., C. van Schaik, and J. Johnson, eds. 1997. *Last stand: Protected areas and the defense of tropical biodiversity*. Oxford University Press, Oxford, U.K.

Kramer, R. A., and N. Sharma, N. 1997. Tropical forest biodiversity protection: Who pays and why. Pages 162–186 in J. Terborgh, C. van Schaik, L. Davenport, and M. Rao, eds., *Making parks work: Strategies for preserving tropical nature*. Island Press, Washington, D.C., USA.

Krause, R. M. 1992. The origins of plagues: Old and new. *Science* 257:1073–1078.

Krause, R. M. 1994. Dynamics of emergence. *Journal of Infectious Diseases* 170:265–271.

Kremen, C., J. O. Niles, M. G. Dalton, G. C. Daily, P. R. Ehrlich, J. P. Fay, D. Grewal, and R. P. Guillery. 2000. Economic incentives for rain forest conservation across scales. *Science* 288:1828–1832.

Kriebitzsch, W. U., G. van Oleimb, G. Ellenberg, B. Engelschall and J. Heuveldop. 2000. Development of woody plant species in fenced and unfenced plots in deciduous forests on soils of the last glaciation in northernmost Germany. *Allegemaine Forst und Jagdzeitung* 171:1–10.

Kroodsma, R. L. 1982. Edge effect on breeding forest birds along a power-line corridor. *Journal of Applied Ecology* **19**:361–370.

Kumari, K. 1994. Sustainable forest management in Peninsular Malaysia: Towards a total economic valuation approach. Ph.D. thesis, University of East Anglia, Norwich, U.K.

Kuno, G. 2001a. Persistence of arboviruses and antiviral antibodies in vertebrate hosts: Its occurrence and impacts. *Reviews in Medical Virology* **11**:165–190. ·

Kuno, G. 2001b. Transmission of arboviruses without involvement of arthropod vectors. *Acta Virologica* **45**:139–150.

Kuss, F. R. 1986: A review of major factors influencing plant responses to recreation impacts. *Environmental Management* **19**:637–650.

Lacerda, L. D., J. Conde, P. Bacon, C. Alarcón, L. D.'Croz, B. Kjerfve, J. Polanía, and M. Vanucci. 1993. Mangrove ecosystems of Latin America and the Caribbean: A summary. Pages 1–42 in L. E. Lacerda, ed., *Conservation and sustainable utilization of mangrove forests in Latin America and Africa regions*. International Society for Mangrove Ecosystems and International Tropical Timber Organization, ITTO/ISME Project PD114/90(F).

Lahm, S. A. 1993. Ecology and economics of human/wildlife interaction in northeastern Gabon. Ph.D. thesis, New York University, New York, USA.

Laidlaw, R. K. 2000. Effects of habitat disturbance and protected areas on mammals of peninsular Malaysia. *Conservation Biology* **14**:1639–1648.

Laidlaw, W. S., and B. A. Wilson. 2003. Floristic and structural characteristics of a coastal heathland exhibiting symptoms of *Phytophtora cinamomi* infestation in the eastern Otway Ranges, Victoria. *Australian Journal of Botany* **51**:283–293.

Lambert, F. R. 1992. The consequences of selective logging for Bornean lowland forest birds. *Philosophical Transactions of the Royal Society of London B* **335**:443–457.

Lambert, F. R., and N. J. Collar. 2002. The future for Sundaic lowland forest birds: Long-term effects of commercial logging and fragmentation. *Forktail* **18**: 127–146.

Laurance, S. G. W. 2001. The effects of roads and their edges on the movement patterns and community composition of understorey rainforest birds in central Amazonia, Brazil. Ph.D. thesis, University of New England, Armidale, New South Wales, Australia.

Laurance, S. G. W. 2004. Responses of understory rain forest birds to road edges in central Amazonia. *Ecological Applications* **14**:1344–1357.

Laurance, S. G. W., and W. F. Laurance. 1999. Tropical wildlife corridors: Use of linear rainforest remnants by arboreal mammals. *Biological Conservation* **91**: 231–239.

Laurance, S. G. W., and M. Santamaria Gomez. 2005. Clearing width and movements of understory rain forest birds. *Biotropica* **37**:149–152.

Laurance, S. G. W., P. C. Stouffer, and W. F. Laurance. 2004. Effects of road clearings on movement patterns of understory rainforest birds in central Amazonia. *Conservation Biology* **18**:1099–1109.

Laurance, W. F. 1991a. Ecological correlates of extinction proneness in Australian tropical rainforest mammals. *Conservation Biology* **5**:79–89.

Laurance, W. F. 1991b. Edge effects in tropical forest fragments: Application of a model for the design of nature reserves. *Biological Conservation* **57**:205–219.

Laurance, W. F. 1994. Rainforest fragmentation and the structure of small mammal communities in tropical Queensland. *Biological Conservation* **69**:23–32.

Laurance, W. F. 1995. Exotic pathogens and aquatic wildlife: A conservation dilemma. *Search* **26**:300–303.

Laurance, W. F. 1997. Hyper-disturbed parks: Edge effects and the ecology of isolated rainforest reserves in tropical Australia. Pages 71–83 in W. F. Laurance and R. O. Bierregaard Jr., eds., *Tropical forest remnants: Ecology, management, and conservation of fragmented communities*. University of Chicago Press, Chicago, Illinois, USA.

Laurance, W. F. 1998. A crisis in the making: Responses of Amazonian forests to land use and climate change. *Trends in Ecology and Evolution* **13**:411–415.

Laurance, W. F. 1999. Reflections on the tropical deforestation crisis. *Biological Conservation* **91**:109–117.

Laurance, W. F. 2000. Do edge effects occur over large spatial scales? *Trends in Ecology and Evolution* **15**:134–135.

Laurance, W. F. 2001. Future shock: Forecasting a grim fate for the Earth. *Trends in Ecology and Evolution* **16**:531–533.

Laurance, W. F. 2003a. Amazonian forests falling faster. BioMedNet News Feature, 28 July. www.biomednet.com.

Laurance, W. F. 2003b. Slow burn: The insidious effects of surface fires on tropical forests. *Trends in Ecology and Evolution* **18**:209–212.

Laurance, W. F. 2004a. Forest-climate interactions in fragmented tropical landscapes. *Philosophical Transactions of the Royal Society of London B* **359**:345–352.

Laurance, W. F. 2004b. The perils of payoff: Corruption as a threat to global biodiversity. *Trends in Ecology and Evolution* **19**:399–401.

Laurance, W. F. 2005a. When bigger is better: The need for Amazonian megareserves. *Trends in Ecology and Evolution* **20**:645–648.

Laurance, W. F. 2005b. Razing Amazonia. *New Scientist,* October 15, pp. 34–39.

Laurance, W. F. 2006a. The value of trees. *New Scientist*, April 15, p. 24.

Laurance, W. F. 2006b. A change in climate. *Tropinet Newsletter* **17**(2):1–3.

Laurance, W. F., A. Albernaz, and C. Da Costa. 2001a. Is deforestation accelerating in the Brazilian Amazon? *Environmental Conservation* **28**:305–311.

Laurance, W. F., A. K. M. Albernaz, P. M. Fearnside, H. L. Vasconcelos, and L. V. Ferreira. 2005a. Amazonian deforestation models. *Science* **307**:1044.

Laurance, W. F., A. K. M. Albernaz, P. M. Fearnside, H. L. Vasconcelos, and L. V. Ferreira. 2005b. Response to "Underlying causes of deforestation." *Science* **307**:1046–1047.

Laurance, W. F., A. K. M. Albernaz, P. M. Fearnside, H. L. Vasconcelos, and L. V. Ferreira. 2004a. Deforestation in Amazonia. *Science* **304**:1109.

Laurance, W. F., A. K. M. Albernaz, G. Schroth, P. M. Fearnside, E. Ventincinque, and C. Da Costa. 2002a. Predictors of deforestation in the Brazilian Amazon. *Journal of Biogeography* **29**:737–748.

Laurance, W. F., A. Alonso, M. Lee, and P. Campbell. 2006a. Challenges for forest conservation in Gabon, central Africa. *Futures* **38**:454–470.

Laurance, W. F., and R. O. Bierregaard Jr., eds. 1997. *Tropical forest remnants: Ecology, management, and conservation of fragmented communities*. University of Chicago Press, Chicago. Illinois, USA.

Laurance, W. F., and M. A. Cochrane, eds. 2001. Synergistic effects in fragmented land-scapes. Special section in *Conservation Biology* 15:1488–1535.

Laurance, W. F., M. A. Cochrane, S. Bergen, P. M. Fearnside, P. Delamonica, C. Barber, S. D'Angelo, and T. Fernandes. 2001b. The future of the Brazilian Amazon. *Science* 291:438–439.

Laurance, W. F., P. Delamonica, S. G. Laurance, H. L. Vasconcelos, and T. E. Lovejoy. 2000a. Rainforest fragmentation kills big trees. *Nature* 404:836.

Laurance, W. F., and P. M. Fearnside. 1999. Amazon burning. *Trends in Ecology and Evolution* 14:457.

Laurance, W. F., and P. M. Fearnside. 2002. Issues in Amazonian development. *Science* 295:1643–1644.

Laurance, W. F., L. V. Ferreira, J. M. Rankin-de Merona, and S. G. Laurance. 1998. Rain forest fragmentation and the dynamics of Amazonian tree communities. *Ecology* 79:2032–2040.

Laurance, W. F., and C. Gascon. 1997. How to creatively fragment a landscape. *Conservation Biology* 11:577–579.

Laurance, W. F., and S. G. W. Laurance. 1996. Responses of five arboreal mammals to recent selective logging in tropical Australia. *Biotropica* 28:310–322.

Laurance, W. F., S. G. Laurance, L. V. Ferreira, J. M. Rankin-de Merona, C. Gascon, and T. E. Lovejoy. 1997a. Biomass collapse in Amazonian forest fragments. *Science* 278:1117–1118.

Laurance, W. F., T. E. Lovejoy, H. L. Vasconcelos, E. M. Bruna, R. K. Didham, P. C. Stouffer, C. Gascon, R. O. Bierregaard, S. G. Laurance, and E. Sampaio. 2002b. Ecosystem decay of Amazonian forest fragments: A 22-year investigation. *Conservation Biology* 16:605–618.

Laurance, W. F., H. Nascimento, S. G. Laurance, A. Andrade, P. M. Fearnside, and J. Ribeiro. 2006b. Rain forest fragmentation and the proliferation of successional trees. *Ecology* 87:469–482.

Laurance, W. F., K. R. McDonald, and R. Speare. 1996. Epidemic disease and the cata-strophic decline of Australian rain forest frogs. *Conservation Biology* 10:406–413.

Laurance, W. F., K. R. McDonald, and R. Speare. 1997b. In defense of the epidemic disease hypothesis. *Conservation Biology* 11:1030–1034.

Laurance, W. F., A. A. Oliveira, S. G. Laurance, R. Condit, H. E. M. Nascimento, A. C. Sanchez-Thorin, T. E. Lovejoy, A. Andrade, S. D'Angelo, J. E. Ribeiro, and C. W. Dick. 2004b. Pervasive alteration of tree communities in undisturbed Amazonian forests. *Nature* 428:171–174.

Laurance, W. F., D. Perez-Salicrup, P. Delamonica, P. M. Fearnside, S. D'Angelo, A. Jerozolinski, L. Pohl, and T. E. Lovejoy. 2001c. Rain forest fragmentation and the structure of Amazonian liana communities. *Ecology* 82:105–116.

Laurance, W. F., J. M. Rankin-de Merona, A. Andrade, S. G. Laurance, S. D'Angelo, T. E. Lovejoy, and H. L. Vasconcelos. 2003. Rain-forest fragmentation and the phenol-ogy of Amazonian tree communities. *Journal of Tropical Ecology* 19:343–347.

Laurance, W. F., H. L. Vasconcelos, and T. E. Lovejoy. 2000b. Forest loss and fragmen-tation in the Amazon: Implications for wildlife conservation. *Oryx* 34:39–45.

Laurance, W. F., and G. B. Williamson. 2001. Positive feedbacks among forest frag-mentation, drought, and climate change in the Amazon. *Conservation Biology* 15:1529–1535.

Laurance, W. F., G. B. Williamson, P. Delamonica, A. Olivera, C. Gascon, T. E. Lovejoy, and L. Pohl. 2001d. Effects of a strong drought on Amazonian forest fragments and edges. *Journal of Tropical Ecology* 17:771–785.

Laurance, W. F., and E. Yensen. 1991. Predicting the impacts of edge effects in fragmented habitats. *Biological Conservation* 55:77–92.

Lawler, I. R., W. J. Foley, I. E. Woodrow, and S. J. Cork. 1997. The effects of elevated CO_2 atmospheres on the nutritional quality of *Eucalyptus* foliage and its interaction with soil nutrient and light availability. *Oecologia* 109:59–68.

Lawton, J. H. 1995. Population dynamic principles. Pages 147–163 in J. H Lawton and R. M. May, eds., *Extinction rates*. Oxford University Press, Oxford, U.K.

Leader-Williams, N., and S. Albon. 1988. Allocation of resources for conservation. *Nature* 336:533–535.

Lean, J., and P. Rowntree. 1993. A GCM simulation of the impact of Amazonian deforestation on climate using an improved canopy representation. *Quarterly Journal of the Royal Meteorological Society* 119:509–530.

Lean, J., and D. A. Warrilow. 1989. Simulation of the regional climatic impact of Amazon deforestation. *Nature* 342:411–413.

Lederberg, J., R. E. Shope, and S. C. J. Oakes. 1992. *Emerging infections: Microbial threats to health in the United States*. Institute of Medicine, National Academy Press, Washington, D.C., USA.

Lee, J. C., X. S. Yang, M. Schwartz, G. Strobel, and J. Clardy. 1995. The relationship between an endangered North American tree and an endophytic fungus. *Chemistry and Biology* 2:721–727.

Leggett, J., ed. 1990. *Global warming: The Greenpeace report*. Oxford University Press, Oxford, U.K.

LeGuenno, B. 1997. Haemorrhagic fevers and ecological perturbations. *Archives of Virology* 13:S191–S199.

Leighton, M. 1983. The El Niño–Southern Oscillation event in Southeast Asia: Effects of drought and fire in tropical forest in eastern Borneo. Unpublished report, Department of Anthropology, Harvard University, Cambridge, Massachusetts, USA.

Leighton, M., and D. R. Leighton. 1982. The relationship of size and feeding aggregate to size of food patch: Howler monkey *Alouatta palliata* feeding in *Trichilia cipo* trees on Barro Colorado Island. *Biotropica* 14:81–90.

Leighton, M., and N. Wirawan. 1986. Catastrophic drought and fire in Borneo tropical rain forest associated with the 1982–1983 El Niño–Southern Oscillation event. Pages 75–102 in G. T. Prance, ed., *Tropical rain forests and the world atmosphere*. AAAS Symposium 10, Boulder, Colorado, USA.

Lele, U., ed. 2002. *Managing a global resource: Challenges of forest conservation and development*. World Bank Series on Evaluation and Development, 5. World Bank, Washington, D.C., USA.

Lelieveld, J., P. J. Crutzen, V. Ramanathan, M. O. Andreae, C. A. M. Brenninkmeijer, T. Campos, G. R. Cass, R. R. Dickerson, H. Fischer, J. A. de Gouw, A. Hansel, A. Jefferson, D. Kley, A. T. J. de Laat, S. Lal, M. G. Lawrence, J. M. Lobert, O. L. Mayol-Bracero, A. P. Mitra, T. Novakov, S. J. Oltmans, K. A. Prather, T. Reiner, H. Rodhe, H. A. Scheeren, D. Sikka, and J. Williams. 2001. The Indian Ocean experiment: Widespread air pollution from South and Southeast Asia. *Science* 291:1031–1036.

Lemus-Albor, A. 2000. Efecto de la densidad y la distancia al árbol adulto en la incidencia de patógenos foliares en plántulas de *Nectandra ambigens*, en una selva tropical. B.Sc. thesis, Universidad Michoacana de San Nicolás de Hidalgo, Mexico.

Lens, L., S. Van Dongen, K. Norris, M. Githiru, and E. Matthysen. 2002. Avian persistence in fragmented rainforest. *Science* 298:1236–1238.

Leroy, E. M., P. Rouquet, P. Formenty, S. Souquiere, A. Kilbourne, J. M. Froment, M. Bermejo, S. Smit, W. Karesh, R. Swanepoel, S. R. Zaki, and P. E. Rollin. 2004. Multiple Ebola virus transmission events and rapid decline of central African wildlife. *Science* 303:387–390.

Leroy, E. M., S. Souquiere, P. Rouquet, and D. Drevet, 2002. Re-emergence of ebola haemorrhagic fever in Gabon. *Lancet* 359:712–712.

Levey, D. J., and M. M. Byrne. 1993. Complex ant–plant interactions: Rain forest ants as secondary dispersers and post-dispersal seed predators. *Ecology* 74:1802–1812.

Lewis, S. 1998. Treefall gaps and regeneration: A comparison of continuous forest and fragmented forest in central Amazonia. Ph.D. thesis, University of Cambridge, Cambridge, U.K.

Lewis, S. L. 2005. Tropical forests and the changing Earth system. *Philosophical Transactions of the Royal Society of London B.*

Lewis, S. L., Y. Malhi, and O. L. Phillips. 2004a. Fingerprinting the impacts of global change on tropical forests. *Philosophical Transactions of the Royal Society of London B* 359:437–462.

Lewis, S. L., O. L. Phillips, T. R. Baker, J. Lloyd, Y. Malhi, S. Almeida, N. Higuchi, W. F. Laurance, D. A. Neill, J. N. M. Silva, J. Terborgh, A. Torres Lezama, R. Vásquez Martinez, S. Brown, J. Chave, C. Kuebler, P. Núñez Vargas, and B. Vinceti. 2004b. Concerted changes in tropical forest structure and dynamics: Evidence from 50 South American long-term plots. *Philosophical Transactions of the Royal Society of London B* 359:421–436.

Limbong, D., J. Kumampung, J. Rimper, T. Arai, T., and N. Miyazaki. 2003. Emissions and environmental implications of mercury from artisanal gold mining in north Sulawesi, Indonesia. *Science of the Total Environment* 302:227–236.

Lindenmayer, D. B., and G. Luck. 2005. Synthesis: Thresholds in conservation and management. *Biological Conservation* 124:351–354.

Lips, K. R. 1998. Decline of a tropical montane amphibian fauna. *Conservation Biology* 12:106–117.

Lips, K. R. 1999. Mass mortality and population declines of anurans at an upland site in Western Panama. *Conservation Biology* 13:117–25.

Lips, K. R., D. E. Green, and R. Papendick. 2003a. Chytridiomycosis in wild frogs from southern Costa Rica. *Journal of Herpetology* 37:215–218.

Lips, K. R., J. D. Reeve, and L. R. Witters. 2003b. Ecological traits predicting amphibian population declines in Central America. *Conservation Biology* 17:1078–1088.

Liu, J., M. Linderman, Z. Ouyang, L. An, J. Yang, and H. Zhang. 2001. Ecological degradation in protected areas: The case of Wolong Nature Reserve for giant pandas. *Science* 292:98–101.

Liu, J., Z. Ouyang, W. W. Taylor, R. Groop, Y. Tan, and H. Zhang. 1999. A framework for evaluating the effects of human factors on wildlife habitat: The case of giant pandas. *Conservation Biology* 13:1360–1370.

Lively, C. M., and V. Apanius. 1995. Genetic diversity in host-parasite interactions. Pages 421–449 in B. T. Grenfell and A. P. Dobson, eds., *Ecology of infectious diseases in natural populations*. Cambridge University Press, Cambridge, U.K.

Llactayo, W., and N. Pitman. 2003. Unpublished data collected in conjunction with Centro de Conservacion, Investigacion, y Manejo de Áreas Naturales.

Lloyd, B. D. 2003. Intraspecific phylogeny of the New Zealand short-tailed bat *Mystacina tuberculata* inferred from multiple mitochondrial gene sequences. *Systematic Biology* 52:460–476.

Lloyd, J., and G. D. Farquhar. 1996. The CO_2 dependence of photosynthesis, plant growth responses to elevated atmospheric CO_2 concentrations and their inter-action with plant nutrient status. *Functional Ecology* 10:4–32.

Lodge, D. J. 1996. Microorganisms. Pages 53–108 in D. P. Regan and R. B. Waide, eds., *The food web of a tropical forest*. University of Chicago Press, Chicago, USA.

Lodge, D. J., and S. Cantrell. 1995. Fungal communities in wet tropical forests: Variation in time and space. *Canadian Journal of Botany* 73 (Suppl. 1):1391–1398.

Lodge, D. M. 1993. Biological invasions: Lessons for ecology. *Trends in Ecology and Evolution* 8:133–137.

Loiselle, B. A., and W. G. Hoppes. 1983. Nest predation in insular and mainland lowland rain forest in Panama. *Condor* 85:93–95.

Lomborg, B. 2001. *The skeptical environmentalist*. Cambridge University Press, Cambridge, U.K.

Longcore, J. E., A. P. Pessier, and D. Nichols. 1999. *Batrachochytrium dendrobatidis* gen. et sp. nov., a chytrid pathogenic to amphibians. *Mycologia* 91:219–27.

Loope, L. L., and D. Mueller-Dombois. 1989. Characteristics of invaded islands. Pages 257–280 in J. A. Drake, H. A. Mooney, F. DiCastri, R. H. Groves, F. J. Kruger, M. Rejmánek, and M. Williamson, eds., *Biological invasions: A global perspective*. John Wiley and Sons, Chichester, U.K.

Los, S. O., C. O. Justice, and C. J. Tucker. 1994. A global 1 degree by 1 degree NDVI data set for climate studies derived from the GIMMS continental NDVI data. *International Journal of Remote Sensing* 15:3493–3518.

Lovejoy, T. E., R. O. Bierregaard Jr., A. B. Rylands, J. R. Malcolm, C. E. Quintela, L. H., Harper, K. S. Brown Jr., A. H. Powell, G. V. N. Powell, H. O. Schubart, and M. B. Hays. 1986. Edge and other effects of isolation on Amazon forest fragments. Pages 257–285 in M. E. Soulé, ed., *Conservation biology: The science of scarcity and diversity*. Sinauer, Sunderland, Massachusetts, USA.

Lovejoy, T. E., J. M. Rankin, R. O. Bierregaard Jr., K. S. Brown Jr., L. H. Emmons, and M. E. Van der Voort. 1984. Ecosystem decay of Amazon forest fragments. Pages 295–325 in M. H. Nitecki, ed., *Extinctions*. University of Chicago Press, Chicago, USA.

Lubchenco, J., A. M. Olson, L. B. Brubaker, S. R. Carpenter, M. M. Holland, S. P. Hubbell, S. A. Levin, A. MacMahon, P. A. Matson, J. M. Melillo, H. A. Mooney, C. H. Peterson, H. R. Pulliam, L. A. Real, P. J. Regal, and P. G. Risser. 1991. The sustainable biosphere initiative: An ecological research agenda. *Ecology* 72:371–412.

Lugo, A. E., M. Applefield, D. Pool, and R. McDonald. 1983. The impact of Hurricane David on the forests of Dominica. *Canadian Journal of Forest Research* 132:201–211.

Luizão, F. J., and H. O. R. Schubart. 1987. Litter production and decomposition in a terra firme forest of central Amazonia. *Experiencia* **43**:259–265.

Lundberg, J., and F. Moberg. 2003. Mobile link organisms and ecosystem functioning: Implications for ecosystem resilience and management. *Ecosystems* **6**:87–98.

Lynam, A. J. 1997. Rapid decline of small mammal diversity in monsoon evergreen forest fragments in Thailand. Pages 222–240 in W. F. Laurance and R. O. Bierregaard Jr., eds., *Tropical forest remnants: Ecology, management, and conservation of fragmented communities*. University of Chicago Press, Chicago, USA.

Mabaza, G. 2003. *Suivi Ebola dans le Departement de la Zadie*. Report to World Wildlife Fund—Gabon, Libreville, Gabon.

Mabberley, D. J. 1997. *The plant book*. Cambridge University Press, Cambridge, U.K.

MacArthur, R., and E. O. Wilson. 1967. *The theory of island biogeography*. Princeton University Press, New Jersey, USA.

MacDonald, L. H., D. M. Anderson, and W. E. Dietrich. 1997. Paradise threatened: Land use and erosion on St. John, U.S. Virgin Islands. *Environmental Management* **21**:851–863.

Machado, R. B., and G. A. B. Da Fonseca. 2000. The avifauna of Rio Doce valley, southeastern Brazil, a highly fragmented area. *Biotropica* **32**:914–924.

Mackie, C. 1984. The lessons behind East Kalimantan's forest fires. *Borneo Research Bulletin* **16**:63–74.

MacKinnon, J. 1997. *Protected area systems review of the Indomalayan realm*. World Bank, Asian Bureau for Conservation (ABC) and World Conservation Monitoring Centre (WCMC), Cambridge, U.K.

MacKinnon, J., and K. MacKinnon. 1986. *Review of the protected areas system in the Indo-Malayan Realm*. IUCN, Gland, Switzerland.

MacKinnon, J., K. MacKinnon, G. Child, and J. Thorsell. 1986. *Managing protected areas in the tropics*. IUCN, Gland, Switzerland.

MacKinnon, J., and K. Phillipps. 1993. *A field guide to the birds of Borneo, Sumatra, Java and Bali*. Oxford University Press, Oxford, U.K.

MacKinnon, K. 1997. The ecological foundations of biodiversity protection. Pages 36–63 in R. Kramer, C. van Schaik, and J. Johnson, eds., *Last stand: Protected areas and the defense of tropical biodiversity*. Oxford University Press, Oxford, U.K.

MacKinnon, K. 2001. Integrated conservation and development projects: Can they work? *Parks* **11**:1–5.

MacKinnon, K. 2005. Parks, peoples and policies: Conflicting agendas for forests in Southeast Asia. Pages 558–582 in E. Bermingham, C. W. Pick, and C. Moritz, eds. *Tropical rainforests: Past, present, and future*. University of Chicago Press, Chicago, USA.

MacKinnon, K., G. Hatta, H. Halim, and A. Mangalik. 1996. *The ecology of Kalimantan*. Periplus, Singapore.

MacPhee, R. D. E., and C. Flemming. 1999. Requiem æternam: The last five hundred years of mammalian species extinction. Pages 333–371 in R. D. E. MacPhee, ed., *Extinctions in near time*. Kluwer Academic/Plenum publishers, New York, USA.

Maddison, A. 1995. *Monitoring the world economy, 1820–1992*. Organization for Economic Cooperation and Development, Paris, France.

Maeher, D. S., E. D. Land, and M. E. Roelke. 1991. Mortality patterns of panthers in southwest Florida. *Proceedings of the Annual Conference of Southeastern Association of Fish and Wildlife Agencies* **45**:201–207.

Magin, C. M. 2003. Dominica's frogs are croaking. *Oryx* 37:406.

Mahlman, J. D. 1997. Uncertainties in projections of human-caused climate warming. *Science* 278:1416–1417.

Mahony, M. 1996. The decline of the green and golden bell frog *Litoria aurea* viewed in the context of declines and disappearances of other Australian frogs. *Australian Zoologist* 30:237–47.

Mahy, B. W. J., and C. C. Brown. 2000. Emerging zoonoses: Crossing the species barrier. *Scientific and Technical Review of the Office Internationale des Epizooties* 19:33–40.

Malcolm, J. R. 1994. Edge effects in central Amazonian forest fragments. *Ecology* 75:2438–2445.

Malcolm, J. R. 1997. Biomass and diversity of small mammals in Amazonian forest fragments. Pages 207–221 in W. F. Laurance and R. O. Bierregaard Jr., eds., *Tropical forest remnants: Ecology, management, and conservation of fragmented communities.* University of Chicago Press, Chicago, USA.

Malcolm, J. R. 1998. A model of conductive heat flow in forest edges and fragmented landscapes. *Climatic Change* 39:487–502.

Malcolm, J. R. 2001. Extending models of edge effects to diverse landscape configurations. Pages 346–357 in R. O. Bierregaard Jr., C. Gascon, T. E. Lovejoy, and R. Mesquita, eds., *Lessons from Amazonia: The ecology and management of a fragmented forest.* Yale University Press, New Haven, Connecticut, USA.

Malhi, Y., T. R. Baker, O. L. Phillips, S. Almeida, E. Alvarez, L. Arroyo, J. Chave, C. Czimczik, A. Di Fiore, N. Higuchi, T. Killeen, S. G. Laurance, W. F. Laurance, S. L. Lewis, L. M. Mercado, A. Monteagudo, D. A. Neill, P. Núñez Vargas, S. Patiño, N. C. A. Pitman, A. Quesada, N. Silva, A. Torres Lezama, J. Terborgh, R. Vásquez M., B. Vinceti, and J. Lloyd. 2004. The above-ground coarse woody productivity of 104 neotropical forest plots. *Global Change Biology* 10:563–591.

Malhi, Y., and J. Grace. 2000. Tropical forests and atmospheric carbon dioxide. *Trends in Ecology and Evolution* 15:332–337.

Malhi, Y., P. Meir, and S. Brown. 2002a. Forests, carbon and global climate. *Philosophical Transactions of the Royal Society of London A* 360:1567–1591.

Malhi, Y., E. Pegoraro, A. D. Nobre, M. G. P. Pereira, J. Grace, A. D. Culf, and R. Clement. 2002b. Energy and water dynamics of a central Amazonian rain forest. *Journal of Geophysical Research—Atmospheres* 107 (D20), no. 8061.

Malhi, Y., O. L. Phillips, T. R. Baker, S. Almeida, T. Frederiksen, J. Grace, N. Higuchi, T. Killeen, W. F. Laurance, C. Leaño, S. Lewis, J. Lloyd, P. Meir, A. Monteagudo, D. Neill, P. Núñez V., S. N. Panfil, N. Pitman, A. Rudas-Ll., R. Salomão, S. Saleska, N. Silva, M. Silveira, W. G. Sombroek, R. Valencia, R. Vásquez M., I. Vieira, and B. Vinceti. 2002c. An international network to understand the biomass and dynamics of Amazonian forests (RAINFOR). *Journal of Vegetation Science* 13: 439–450.

Malhi, Y., and J. Wright. 2004. Spatial patterns and recent trends in the climate of tropical rainforest regions. *Philosophical Transactions of the Royal Society of London B* 359:311–329.

Marcot, B. G. 1992. Conservation of Indian forests. *Conservation Biology* 6:12–16.

Marengo, J. A., W. R. Soares, C. Saulo, and M. Nicolini. 2004. Climatology of the low-level jet East of the Andes derived from NCEP-NCAR reanalyses: Characteristics and temporal variability. *Journal of Climate* 17:2261–2280.

Marland, G., K. Fruit, and R. A. Sedjo. 2001. Accounting for sequestered carbon: The question of permanence. *Environmental Science and Policy* **4**:259–268.

Marron, C.-H. 1999. The impact of ecotourism. *Flora and Fauna News* **11**:14.

Marsden, S. J., M. Whiffin, and M. Galetti. 2001. Bird diversity and abundance in forest fragments and *Eucalyptus* plantations around an Atlantic forest reserve, Brazil. *Biodiversity and Conservation* **10**:737–751.

Mason, D. 1996. Responses of Venezuelan understory birds to selective logging, enrichment strips, and vine cutting. *Biotropica* **28**:296–309.

Mason, S. J. 1997. Recent changes in El Niño Southern Oscillation events and their implications for southern African climate. *Transactions of the Royal Society of South Africa* **52**:377–403.

Mason, S. J. 2001. El Niño, climate change, and southern African climate. *Environmetrics* **12**:327–345.

Matlack, G. R. 1994. Vegetation dynamics of the forest edge: Trends in space and successional time. *Ecology* **82**:113–123.

Mattos, L., A. Faleiro, and C. Pereira. 2001. *Uma proposta alternativa para o desenvolvimento da agricultura familiar rural na Amazônia: O caso do PROAMBIENTE.* IV Encontro Nacional da Sociedade Internacional de Economia Ecológica-ECO-ECO, NEPAM, Universidade Estadual de Campinas (UNICAMP), Campinas, São Paulo, Brazil. http://www.nepam.unicamp.br/ecoeco/artigos/ encontros/encontro4_ple naria.html.

Maunder, M., A. Culham, and C. Hankamer. 1997. Picking up the pieces: Biological conservation on degraded oceanic island. Pages 317–344 in P. L. Fiedler and P. M. Kareiva, eds., *Conservation biology for the coming decade.* Thompson International Publishing, New York, USA.

May, R. M., J. H. Lawton, and N. E. Stork. 1995. Assessing extinction rates. Pages 1–24 in J. H. Lawton and R. M. May, eds., *Extinction rates.* Oxford University Press, Oxford, U.K.

Mayen, F. 2003. Haematophagous bats in Brazil, their role in rabies transmission, impact on public health, livestock industry and alternatives to an indiscriminate reduction of bat population. *Journal of Veterinary Medicine Series B—Infectious Diseases and Veterinary Public Health* **50**:469–472.

Mayer, J. H. 1989. *Socioeconomic aspects of the forest fire of 1982/83 and the relation of local communities towards forestry and forest management.* FR Report No. 8, Samarinda, Indonesia.

Mazzoni, R., A. A. Cunningham, P. Daszak, A. Apolo, E. Perdomo, and G. Speranza. 2003. Emerging pathogen of wild amphibians in frogs (*Rana catesbeiana*) farmed for international trade. *Emerging Infectious Diseases* **9**:995–998.

McCallum, H., and A. Dobson. 1995. Detecting disease and parasite threats to endangered species and ecosystems. *Trends in Ecology and Evolution* **10**:190–194.

McCarthy, J. F. 1999. "Wild logging": The rise and fall of logging networks and biodiversity conservation projects on Sumatra's rainforest frontier. Occasional Paper 31, CIFOR, Bogor, Indonesia. www.cifor.cgiar.org.

McCarthy, M. 2000. Ebola outbreak continues in Uganda. *Lancet* **356**:1499.

McClanahan, T. R. 1999. Is there a future for coral reef parks in poor tropical countries? *Coral Reefs* **18**:321–325.

McCullough, D. R. 1996. Spatially structured populations and harvest theory. *Journal of Wildlife Management* **60**:1–9.

McGarigal, K., and S. A. Cushman. 2002. Comparative evaluation of experimental approaches to the study of habitat fragmentation effects. *Ecological Applications* **12**:335–345.

McGarigal, K., and B. J. Marks. 1995. *FRAGSTATS: Spatial pattern analysis program for quantifying landscape structure*. USDA Forest Service Gen. Tech. Rep. PNW-351, Washington, D.C., USA.

McGarigal, K., and W. C. McComb. 1995. Relationships between landscape structure and breeding birds in the Oregon coast range. *Ecological Monographs* **65**:235–260.

McKey, D. 1975. The ecology of coevolved seed dispersal systems. Pages 159–209 in L. E. Gilbert and P. H. Raven, eds., *Coevolution of animals and plants*. University of Texas Press, Austin, USA.

McLaughlin, J. F., J. J. Hellmann, C. L. Boggs, and P. R. Ehrlich. 2002. Climate change hastens population extinctions. *Proceedings of the National Academy of Sciences USA* **99**:6070–6074.

McNeely, J. A. 1988. *Economics and biological diversity: Developing and using economic incentives to conserve biological resources*. IUCN, Gland, Switzerland.

McNeely, J. A. 1989. Protected areas and human ecology: How national parks can contribute to sustaining societies of the twenty-first century? Pages 150–157 in D. Western and M. C. Pearl, eds., *Conservation for the twenty-first century*. Oxford University Press, Oxford, U.K.

McNeely, J., and S. Scherr. 2001. *Common ground, common future: How ecoagriculture can help feed the world and save wild biodiversity*. IUCN, Gland, Switzerland, and Future Harvest, Washington, D.C., USA.

McNeilage, A., A. Plumptre, A. Brock-Doyle, and A. Vedder. 1998. *Bwindi Impenetrable National Park, Uganda: Gorilla and large mammal census, 1997*. Wildlife Conservation Society. Working Paper 14, Uganda.

McShane, T., and M. Wells, eds. 2004. *Getting biodiversity projects to work: Towards more effective conservation and development*. Columbia University Press, New York, USA.

Meehan, H. J., K. R. McConkey, and D. Drake. 2002. Potential disruptions to seed dispersal mutualisms in Tonga, Western Polynesia. *Journal of Biogeography* **29**:695–712.

Meggers, B. J. 1994. Archeological evidence for the impact of mega-Niño events on Amazonian forests during the past two millennia. *Climatic Change* **28**:321–338.

Meinshausen, M., and B. Hare. 2000. *Temporary sinks do not cause permanent climate benefits*. Greenpeace International, Amsterdam, The Netherlands. www.carbonsinks.de.

Merlen, G. 1995. Use and misuse of the seas around the Galápagos archipelago. *Oryx* **29**:99–106.

Mesquita, R. C. G., K. Ickes, G. Ganade, and G. B. Williamson. 2001. Alternative successional pathways in the Amazon Basin. *Journal of Ecology* **89**:528–573.

Metzger, J. P. 2000. Tree functional group richness and landscape structure in a Brazilian tropical fragmented landscape. *Ecological Applications* **10**:1147–1161.

Meyer, J. Y. 2000. Preliminary review of the invasive plants in the Pacific islands (SPREP Member Countries). Pages 86–114 in G. Sherley, ed., *Invasive species in the Pacific: A technical review and draft regional strategy*. SPREP, Samoa.

Meyer, J. Y., and J. Florence. 1996. Tahiti's native flora endangered by the invasion of *Miconia calvescens* DC. (Melastomataceae). *Journal of Biogeography* 23:775–781.

Meyers, S., J. Sathaye, B. Lehman, K. Schumacher, O. van Vliet, and J. R. Moreira. 2000. *Preliminary assessment of potential CDM early start projects in Brazil.* LBNL-46120. Lawrence Berkeley National Laboratory, Berkeley, California, USA.

Michalski, F., R. Boulhosa, A. Faria, and C. A. Peres. 2006. Human-wildlife conflicts in a fragmented Amazonian forest landscape: Determinants of large felid depredation on livestock. *Animal Conservation* 9:179–188.

Michalski, F., and C. A. Peres. 2005. Anthropogenic determinants of primate and carnivore local extinctions in a fragmented forest landscape of southern Amazonia. *Biological Conservation* 124:383–396.

Michelmore, F., K. Beardsley, R. F. W. Barnes, and I. Douglas-Hamilton. 1994. A model illustrating the changes in forest elephant numbers caused by poaching. *African Journal of Ecology* 32:89–99.

Midgley, J. J. 2003. Is bigger better in plants? The hydraulic costs of increasing size in trees. *Trends in Ecology and Evolution* 18:5–6.

Milgroom, M. G., K. R. Wang, Y. Zhou, S. E. Lipari, and S. Kaneko. 1996. Intercontinental population structure of the chestnut blight fungus, *Cryphonectria parasitica*. *Mycologia* 88:179–190.

Mills, C., K. Harms, R. Condit, D. King, J. Thompson, F. He, H. Muller-Landau, P. Ashton, E. Losos, L. Comita, S. Hubbell, J. LaFrankie, S. Bunyavejchewin, H. Dattaraja, S. Davies, S. Esufali, R. Foster, N. Gunatilleke, S. Gunatilleke, P. Hall, A. Itoh, R. John, S. Kinatiprayoon, S. Loo de Lao, M. Massa, C. Nath, M. Supardi, A. Rahman, R. Sukumar, H. Suresh, I. Sun, S. Tan, T. Yamakura, and J. Zimmerman. 2006. Nonrandom processes maintain diversity in tropical forests. *Science* 311:527–531.

Milly, P. C. D., R. T. Wetherald, K. A. Dunne, and T. L. Delworth. 2002. Increasing risk of great floods in a changing climate. *Nature* 415:514–517.

Milner-Gulland, E. J., E. L. Bennett, and the SCB 2002 Annual Meeting Wild Meat Group. 2003. Wild meat: The bigger picture. *Trends in Ecology and Evolution* 18:351–357.

Mittermeier, R. A., C. G. Mittermeier, P. R. Gil, and J. Pilgrim. 2003. *Wilderness: Earth's last wild places.* University of Chicago Press, Chicago, USA.

Mol, J. H., and P. Ouboter. 2004. Downstream effects of erosion from small-scale gold mining on the instream habitat and fish community of a small neotropical rainforest stream. *Conservation Biology* 18:201–214.

Mol, J. H., J. Ramlal, C. Lietar, and M. Verloo. 2001. Mercury contamination in freshwater, estuarine, and marine fishes in relation to small-scale gold mining in Suriname, South America. *Environmental Research* 86:183–197.

Molofsky, J., and C. Augspurger. 1992. The effect of leaf litter on early seedling establishment in a tropical forest. *Ecology* 73:68–77.

Molyneux, D. H. 2003. Common themes in changing vector-borne disease scenarios. *Transactions of the Royal Society of Tropical Medicine and Hygiene* 97:129–132.

Monath, T. P. 1999. Ecology of Marburg and Ebola viruses: Speculations and directions for future research. *Journal of Infectious Diseases* 179:S127–S138.

Mondet, B. 2001. Yellow fever epidemiology in Brazil: New considerations. *Bulletin de la Societe de Pathologie Exotique* 94:260–267.

Monzoni, M., A. Muggiatti, and R. Smeraldi. 2000. *Mudança climática: Tomando posições*. Friends of the Earth/Amigos da Terra, Programa Amazônia, São Paulo, Brazil. http://www.amazonia.org.br/ef/Mudanca%20Climatica.pdf.

Moore, N., and S. Rojstaczer. 2001. Irrigation induced rainfall and the Great Plains. *Journal of Applied Meteorology* 40:1297–1309.

Morehouse, E. A., T. Y. James, A. R. D. Ganley, R. Vilgalys, L. Berger, P. J. Murphy, and J. E. Longcore. 2003. Multilocus sequence typing suggests that the chytrid pathogen of amphibians is a recently emerged clone. *Molecular Ecology* 12:395–403.

Morellato, P. C., and H. F. Leitao. 1996. Reproductive phenology of climbers in a southeastern Brazilian forest. *Biotropica* 28:180–191.

Mori, S. A., and P. Becker. 1991. Flooding affects survival of Lecythidaceae in terra firme forest near Manaus, Brazil. *Biotropica* 23:87–90.

Moritz, C. 2002. Strategies to protect biological diversity and the evolutionary processes that sustain it. *Systematic Biology* 51:238–254.

Morse, S. S. 1993a. *Emerging viruses*. Oxford University Press, Oxford, U.K.

Morse, S. S. 1993b. Examining the origins of emerging viruses. Pages 10–28 in S. S. Morse, ed., *Emerging viruses*. Oxford University Press, Oxford, U.K.

Morvan, J. A, E. Nakoune, V. Deubel, and M. Colyn. 2000. Ebola virus and forest ecosystem. *Bulletin de la Societe de Pathologie Exotique* 93:172–175.

Motala, M. S. 1999. A preliminary survey on the degradation of the Macchabé forest after sixty years. B.Sc. thesis, University of Mauritius.

Moukassa, A. 2001. *Etude demographique et socio-economique dans la zone peripherique du Parc National Nouabale-Ndoki (Kabo, Pokola, Loundougou, Mokabi)*. Wildlife Conservation Society, Bronx, New York.

Moulton, M. P., and S. L. Pimm. 1986. Species introductions to Hawaii. Pages 231–249 in H. A. Mooney and J. A. Drake, eds., *Ecology of biological invasions of North America and Hawaii*. Springer-Verlag, New York, USA.

Moura-Costa, P., and C. Wilson. 2000. An equivalence factor between CO_2 avoided emissions and sequestration—description and applications in forestry. *Mitigation and Adaptation Strategies for Global Change* 5:51–60.

Moutinho, P., and S. Schwartzman, eds. 2005. *Tropical deforestation and climate change*. Instituto de Pesquisa Ambiental da Amazônia, Belém, Brazil.

Muchaal, P. K., and G. Ngandjui. 1999. Impact of village hunting on wildlife populations in the western Dia Reserve, Cameroon. *Conservation Biology* 13:385–396.

Mueller-Landau, H. C., S. J. Wright, O. Calderon, S. P. Hubbell, and R. B. Foster. 2002. Assessing recruitment limitation: Concepts, methods and case-studies from a tropical forest. Pages 36–53 in D. J. Levey, W. R. Silva, and M. Galetti, eds., *Seed dispersal and frugivory: Ecology, evolution and conservation*. CAB International, Wallingford, UK.

Mugisha, A. R. 2002. Evaluation of community-based conservation approaches: The management of protected areas in Uganda. Ph.D. thesis, University of Florida, Gainesville, Florida, USA.

Murcia, C. 1995. Edge effects in fragmented forests: Implications for conservation. *Trends in Ecology and Evolution* 10:58–62.

Murphy, F. A. 1998. Emerging zoonoses. *Emerging Infectious Diseases* 4:429–435.

Musil, C. F. 1993. Effect of invasive Australian acacias on the regeneration, growth, and nutrient chemistry of South African lowland fynbos. *Journal of Applied Ecology* 30:361–372.

Mutschmann, F., L. Berger, P. Zwart, and C. Gaedicke. 2000. Chytridiomycosis of amphibians: First report from Europe. *Beliner und Munchener Tierarztliche wochenschrift* **113**:380–383.

Myers, N. 1987. The extinction spasm impending: Synergisms at work. *Conservation Biology* **1**:14–21.

Myers, N. 1988a. Synergistic interactions and environment. *BioScience* **39**:506.

Myers, N. 1988b. Threatened biotas: "Hotspots" in tropical forests. *Environmentalist* **8**:187–208.

Myers, N. 1989. *Deforestation rates in tropical forests and their climatic implications.* Friends of the Earth, London, U.K.

Myers, N. 1996. Environmental services of biodiversity. *Proceedings of the National Academy of Sciences USA* **93**:2764–2769.

Myers, N. 1998. Lifting the veil on perverse subsidies. *Nature* **392**:327–328.

Myers, N., and J. Kent. 2001. *Perverse subsidies: How tax dollars can undercut the environment and the economy.* Island Press, Washington, D.C., USA.

Myers, N., and J. Kent. 2003. New consumers: The influence of affluence on the environment. *Proceedings of the National Academy of Sciences USA* **100**:4963–4968.

Myers, N., R. A. Mittermeier, C. G. Mittermeier, G. A. B. da Fonseca, and J. Kent. 2000. Biodiversity hotspots for conservation priorities. *Nature* **403**:853–858.

Nadin-Davis, S. A., M. I. Sampath, G. A. Casey, R. R. Tinline, and A. I. Wandeler. 1999. Phylogeographic patterns exhibited by Ontario rabies virus variants. *Epidemiology and Infection* **123**:325–336.

Nakounne, E., B. Selekon, and J. Morvan. 2001. Microbiological surveillance: Viral haemorrhagic fevers in the Central African Republic; updated serological data for human beings. *Bulletin de la Societe de Pathologie Exotique* **93**:340–347.

Namias, J. 1978. Multiple causes of the North American abnormal winter 1976–77. *Monthly Weather Review* **106**:279–295.

Nascimento, H. E. M., and W. F. Laurance. 2004. Biomass dynamics in Amazonian forest fragments. *Ecological Applications* **14**:S127–S138.

Nasi, R., S. Wunder, and J. Campos. 2002. *Forest ecosystem services: Can they pay our way out of deforestation.* Global Environmental Facility Discussion Paper, Washington, D.C., USA.

National Research Council. 1999. *Our common journey: A transition toward sustainability.* National Academy Press, Washington, D.C., USA.

Naughton-Treves, L., J. L. Mena, A. Treves, N. Alvarez, and V. C. Radeloff. 2003. Wildlife survival beyond park boundaries: The impact of slash-and-burn agriculture and hunting on mammals in Tambopata, Peru. *Conservation Biology* **17**:1106–1117.

Nelson, K. C., and B. H. J. de Jong. 2003. Making global initiatives local realities: Carbon mitigation projects in Chiapas, Mexico. *Global Environmental Change* **13**:19–30.

Nepstad, D., G. Carvalho, A. C. Barros, A. Alencar, J. P. Capobianco, J. Bishop, P. Moutinho, P. Lefebvre, U. L. Silva Jr., and E. Prins. 2001. Road paving, fire regime feedbacks, and the future of Amazon forests. *Forest Ecology and Management* **154**:395–407.

Nepstad, D. C. 1998. *Origin, incidence, and implications of Amazon fires.* U.S. Global Change Research Program Seminar Series, Washington, D.C., USA, 30 March.

Nepstad, D. C., C. Carvalho, E. Davidson, P. Jipp, P. Lefebre, G. Negreiros, E. Silva, T. Stone, S. Trumbore, and S. Vieira. 1994. The role of deep roots in the hydrological cycles of Amazonian forests and pastures. *Nature* **372**:666–669.

Nepstad, D. C., A. McGrath, A. Alencar, A. Barros, G. Carvalho, M. Santilli, and M. Vera Diaz. 2002. Frontier governance in Amazonia. *Science* **295**:629–631.

Nepstad, D. C., A. G. Moreira, and A. A. Alencar. 1999a. *Flames in the rain forest: Origins, impacts and alternatives to Amazonian fires.* World Bank, Brasilia, Brazil.

Nepstad, D. C., S. Schwartzman, B. Bamberger, M. Santilli, D. Ray, P. Schlesinger, P. LeFebvre, A. Alencar, E. Prinz, G. Fiske, and A. Rolla. 2006. Inhibition of Amazon deforestation and fire by parks and indigenous lands. *Conservation Biology* **20**:65–76.

Nepstad, D. C., A. Verissimo, A. Alencar, C. Nobre, E. Lima, P. Lefebre, P. Schlesinger, C. Potter, P. Moutinho, E. Medoza, M. Cochrane, and V. Brooks. 1999b. Large-scale impoverishment of Amazonian forests by logging and fire. *Nature* **398**:505–508.

Newmark, W. D. 1995. Extinction of mammal populations in Western North American national parks. *Conservation Biology* **9**:512–526.

Newmark, W. D. 1996. Insularization of Tanzanian parks and the local extinction of large mammals. *Conservation Biology* **10**:1549–1556.

Newmark, W. D., and J. L. Hough. 2000. Conserving wildlife in Africa: Integrated conservation and development projects and beyond. *BioScience* **50**:585–592.

Newsome, D., S. A. Moore, and R. K. Dowling. 2002. *Natural area tourism: Ecology, impacts and management.* Channel View Publications, Clevedon, U.K.

Nicholson, S. E. 2001. Climatic and environmental change in Africa during the last two centuries. *Climate Research* **17**:123–144.

Nichols-Orians, C. M. 1991. Environmentally-induced differences in plant traits: Consequences to a leaf-cutter ant. *Ecology* **72**:1609–1623.

Niesten, E., P. C. Frumhoff, M. Manion, and J. J. Hardner. 2002. Designing a carbon market that protects forests in developing countries. *Philosophical Transactions of the Royal Society of London A* **360**:1875–1888.

Niesten, E., S. Ratay, and R. Rice. 2004a. Achieving biodiversity conservation using conservation concessions to substitute or complement agroforestry. Pages 135–150 in G. Schroth, G. Fonseca, C. Harvey, C. Gascon, H. Vasconcelos, and A. Izac, eds., *Agroforestry and biodiversity conservation in tropical landscapes.* Island Press, Washington, D.C., USA.

Niesten, E., and R. Rice. 2004. Sustainable forest management and direct incentives for conservation. *Revue Tiers Monde* **45**:129–152.

Niesten, E., R. Rice, and J. Hardner. 2004b. Globalization and direct incentives for conservation. Pages 602–617 in C. Reinhard Meier-Walser and P. Stein, eds., *Globalisierung und Perspektiven internationaler Verantwortung.* Saur-Verlag, Munich, Germany.

Nieuwstadt, M. G. L., and D. Sheil. 2002. Separating the effects of severe drought and subsequent fire on tree survival in a lowland dipterocarp rainforest in East Kalimantan, Indonesia. Pages 27–45 in M. G. L. Nieuwstadt, ed., *Trial by fire: Postfire development of a tropical dipterocarp forest.* Printpartners Ipskam B.V., Enschede, Germany.

Niles, J. O., S. Brown, J. Pretty, A. S. Ball, and J. Fay. 2002. Potential carbon mitigation and income in developing countries from changes in use and management of agricultural and forest lands. *Philosophical Transactions of the Royal Society of London A* **360**:1621–1639.

Nix, H. A. 1991. Biogeography: Pattern and process. In H. A. Nix and M. A. Switzer, eds., *Rainforest animals: Atlas of vertebrates endemic to the Wet Tropics*. Australian National Parks and Wildlife Service, Canberra, Australia.

Nobre, C. A., P. Sellers, and J. Shukla. 1991. Amazonian deforestation and regional climate change. *Journal of Climate* **4**:411–413.

Norby, R. J., P. J. Hanson, E. G. O'Neill, T. J. Tschaplinski, J. F. Weltzin, R. A. Hansen, W. X. Cheng, S. D. Wullschleger, C. A. Gunderson, N. T. Edwards, and D. W. Johnson. 2002. Net primary productivity of a CO_2-enriched deciduous forest and the implications for carbon storage. *Ecological Applications* **12**:1261–1266.

Norby, R. J., S. D. Wullschleger, C. A. Gunderson, D. W. Johnson, and R. Ceulemans. 1999. Tree responses to rising CO_2 in field experiments: Implications for the future forest. *Plant Cell and Environment* **22**:683–714.

Norris, D. 2002. Spatial influences on terrestrial mammal communities in forest fragments around Alta Floresta, Brazil. M.Sc. thesis, University of East Anglia, Norwich, U.K.

Norton-Griffiths, M., and C. Southey. 1995. The opportunity costs of biodiversity conservation in Kenya. *Ecological Economics* **12**:125–139.

Noss, A. J. 1997. Challenges to nature conservation with community development in central African forests. *Oryx* **31**:180–188.

Noss, A. J. 1998a. The impacts of BaAka net hunting on rainforest wildlife. *Biological Conservation* **86**:161–167.

Noss, A. J. 1998b. The impacts of cable snare hunting on wildlife populations in the forests of the Central African Republic. *Conservation Biology* **12**:390–398.

Noss, A. J. 1999. Censusing rainforest game species with communal net hunts. *African Journal of Ecology* **37**:1–11.

Novaro, A. J., K. H. Redford, and R. E. Bodmer. 2000. Effect of hunting in source–sink systems in the neotropics. *Conservation Biology* **14**:713–721.

Nriagu, J. O., W. Pfeiffer, O. Malm, C. De Souza, and G. Mierle. 1992. Mercury pollution in Brazil. *Nature* **356**:389.

Nunn, C. L., S. Altizer, K. Sechrest, K. E. Jones, R. A. Barton, and J. L. Gittleman. 2004. Parasites and the evolutionary diversification of primate clades. *American Naturalist* **164**:S90–S103.

Oates, J. F. 1996. Habitat alteration, hunting and the conservation of folivorous primates in African forests. *Australian Journal of Ecology* **21**:1–9.

Oates, J. F. 1999. *Myth and reality in the rain forest: How conservation strategies are failing in West Africa*. University of California Press, Berkeley, USA.

Oates, J. F. 2002. West Africa: Tropical forest parks on the brink. Pages 57–75 in J. Terborgh, C. van Schaik, M. Rao, and L. Davenport, eds., *Making parks work: Strategies for preserving tropical nature*. Island Press, Washington, D.C., USA.

Oates, J. F., M. Abedi-Lartey, W. S. McGraw, T. T. Struhsaker, and G. H. Whitesides. 2000. Extinction of a West African red colobus monkey. *Conservation Biology* **14**:1526–1532.

Oba, G., E. Post, N. C. Stenseth, and W. J. Luisigi. 2000a. The role of small ruminants in arid zone environments. *Annals of the Arid Zone* **39**:305–332.

Oba, G., E. Post, P. O. Syvertsen, and N. C. Stenseth. 2000b. Bush cover and range conditions in relation to landscape and grazing in southern Ethiopia. *Landscape Ecology* 15:525–546.

O'Brien, T. G., and M. F. Kinnaird. 2000. Differential vulnerability of large birds and mammals to hunting in north Sulawesi, Indonesia, and the outlook for the future. Pages 199–230 in J. G. Robinson and E. L. Bennett, eds., *Hunting for sustainability in tropical forests*. Columbia University Press, New York, USA.

O'Brien, T. G., M. Kinnaird, A. Nurcahyo, M. Prasetyaningrum, and M. Iqbal. 2003. Fire, demography and the persistance of siamang (*Symphalangus syndactylus*: Hylobatidae) in a Sumatran rainforest. *Animal Conservation* 6:115–121.

Odum, W. E., and E. J. Heald. 1975. The detritus based food web of an estuarine mangrove community. *Estuarine Research* 1:265–286.

OECD. 1995. *The economic appraisal of environmental projects and policies*. Organization for Economic Cooperation and Development, Paris, France.

OECD. 2002. *Handbook of biodiversity valuation: A guide for policy makers*. Organization for Economic Cooperation and Development, Paris, France.

Oliveira Filho, F. J. B. 2001. Padrão de desmatamento e evolução da estrutura da paisagem em Alta Floresta (MT). M.Sc. thesis, University of Sao Paulo, São Paulo, Brazil.

Oliver, W. L. R., and I. B. Santos. 1991. Threatened endemic mammals of the Atlantic forest region of south-east Brazil. Jersey Wildlife Preservation Trust, *Special Science Report* 4:1–126.

Osterhaus, A. 2000. Circulation of viruses and inter-species contaminations in wild animals. *Bulletin de la Societe de Pathologie Exotique* 93:156–156.

Otterman, J., A. Manes, S. Rabin, P. Alpert, and D. O. C. Starr. 1990. An increase of early rains in southern Israel following land-use change. *Boundary-Layer Meteorology* 53:333–351.

Owadally, A. W. 1979. The dodo and the Tambalacoque tree. *Science* 203:1363–1364.

Page, W., and G. A. D.'Argent. 1997. *A vegetation survey of Mauritius (Indian Ocean) to identify priority rainforest areas for conservation management*. IUCN/MWF report, Mauritius.

Pain, D. J., and P. F. Donald. 2002. Outside the reserve: Pandemic threats to bird biodiversity. Pages 157–179 in K. Norris and D. J. Pain, eds., *Conserving bird biodiversity: General principles and their application*. Cambridge University Press, Cambridge, U.K.

Palmer, M., and V. Finlay. 2003. *Faith in conservation: New approaches to religions and the environment*. World Bank, Washington, D.C., USA.

Palmer, T. N., and J. Raianen. 2002. Quantifying the risk of extreme seasonal precipitation events in a changing climate. *Nature* 415:512–514.

Palombit, R. A. 1997. Inter- and intraspecific variation in the diets of sympatric siamang (*Hylobates syndactylus*) and lar gibbons (*Hylobates lar*). *Folia Primatologica* 68:321–337.

Parmesan, C. 1996. Climate change and species' ranges. Nature 382:765–766.

Parmesan, C., and G. Yohe. 2003. A globally coherent fingerprint of climate change impacts across natural systems. *Nature* 421:37–42.

Parris, M., and D. R. Baud. 2004. Interactive effects of a heavy metal and chytrid fungal pathogen on gray tree frog larvae (*Hyla chrysoscelis*). *Copeia* 2004:344–350.

Paruelo, J. M., M. Garbulsky, J. Guerschman, and E. Jobbágy. 2004. Two decades of Normalized Difference Vegetation Index changes in South America: Identifying the imprint of global change. *International Journal of Remote Sensing* 25:2793–2806.

Pattanavibool, A., and P. Dearden. 2002. Fragmentation and wildlife in montane evergreen forests, northern Thailand. *Biological Conservation* 107:155–164.

Patton, J. L., M. N. da Silva, M. C. Lara, and M. M. Mustrangi. 1997. Diversity, differentiation, and the historical biogeography of small mammals of the neotropical forests. Pages 455–465 in W. F. Laurance and R. O. Bierregaard Jr., eds., *Tropical forest remnants: Ecology, management, and conservation of fragmented communities.* University of Chicago Press, Chicago, USA.

Pauly, D., V. Christensen, J. Dalsgaard, R. Froese, and F. Torres. 1998. Fishing down marine food webs. *Science* 279:860–863.

Pearce, D., E. Putz, and J. Vanclay. 2000. *A sustainable forest future.* CSERGE Working Paper GEC99-15. University College London, London, U.K.

Pearl, R. 1925. *The biology of population growth.* Knopf, New York, USA.

Pearman, P. B. 2002. The scale of community structure: Habitat variation and avian guilds in tropical forest understory. *Ecological Monographs* 72:19–39.

Pearson, H. 2003. West Nile Virus may have felled Alexander the Great: Conservation and medicine collide in the jungle. News@Nature.com, 28 November. http://www.nature.com/news/2003/031124/full/031124-12.html.

Peluso, N. L. 1992. The ironwood problem: (Mis)management and development of an extractive rainforest product. *Conservation Biology* 6:210–219.

Peres, C. A. 1990. Effects of hunting on western Amazonian primate communities. *Biological Conservation* 54:47–59.

Peres, C. A. 1999a. Effects of hunting and habitat quality on Amazonian primate communities. Pages 268–283 in J. G. Fleagle, C. Janson, and K. E. Reed, eds., *Primate communities.* Cambridge University Press, Cambridge, U.K.

Peres, C. A. 1999b. Ground fires as agents of mortality in a central Amazonian forest. *Journal of Tropical Ecology* 15:535–541.

Peres, C. A. 2000a. Effects of subsistence hunting on vertebrate community structure in Amazonian forests. *Conservation Biology* 14:240–253.

Peres, C. A. 2000b. Identifying keystone plant resources in tropical forests: The case of gums from *Parkia* pods. *Journal of Tropical Ecology* 16:287–317.

Peres, C. A. 2001a. Paving the way to the future of Amazonia. *Trends in Ecology and Evolution* 16:217–219.

Peres, C. A. 2001b. Synergistic effects of subsistence hunting and habitat fragmentation on Amazonian forest vertebrates. *Conservation Biology* 15:1490–1505.

Peres, C. A. 2002. Expanding networks of conservation areas in our last tropical forest frontiers: The case of Brazilian Amazonia. Pages 137–148 in J. Terborgh, C. van Schaik, M. Rao, and L. Davenport, eds., *Making parks work: Strategies for preserving tropical nature.* Island Press, Washington, D.C., USA.

Peres, C. A. 2005. Why we need megareserves in Amazonia. *Conservation Biology* 19:728–733.

Peres, C. A., C. Baider, P. Zuidema, L. Wadt, K. Kainer, D. Gomes-Silva, R. Salamão, L. Simões, E. Francioso, F. C. Valverde, R. Gribel, G. Shepard, M. Kanashiro, P. Coventry, D. Yu, A. Watkinson, and R. Freckleton. 2003a. Demographic threats to the sustainability of Brazil nut exploitation. *Science* 302:2112–2114.

Peres, C. A., J. Barlow, and T. Haugaasen. 2003b. Vertebrate responses to surface fires in a central Amazonian forest. *Oryx* 37:97–109.

Peres, C. A., J. Barlow, and W. F. Laurance. 2006. Detecting anthropogenic disturbance in tropical forests. *Trends in Ecology and Evolution* 21:227–229.

Peres, C. A., and I. R. Lake. 2003. Extent of nontimber resource extraction in tropical forests: Accessibility to game vertebrates by hunters in the Amazon basin. *Conservation Biology* 17:521–535.

Peres, C. A., and M. G. van Roosmalen. 2002. Patterns of primate frugivory in Amazonia and the Guianan shield: Implications to the demography of large-seeded plants in overhunted tropical forests. Pages 407–423 in D. J. Levey, W. R. Silva, and M. Galetti, eds., *Seed dispersal and frugivory: Ecology, evolution and conservation.* CAB International, Wallingford, UK.

Peres, C. A., and J. Terborgh. 1995. Amazonian nature reserves: An analysis of the defensibility status of existing conservation units and design criteria for the future. *Conservation Biology* 9:34–46.

Pessier, A. P., D. K. Nichols, J. E. Longcore, and M. S. Fuller. 1999. Cutaneous chytridiomycosis in poison dart frogs (*Dendrobates spp.*) and White's tree frogs (*Litoria caerulea*). *Journal of Veterinary Diagnostics and Investigation* 11:194–199.

Peters, H. A. 2001. *Clidemia hirta* invasion at the Pasoh Forest Reserve: An unexpected plant invasion in an undisturbed tropical forest. *Biotropica* 33:60–68.

Peterson, A. T., M. A. Ortega-Huerta, J. Bartley, V. Sanchez-Cordero, J. Sorberon, R. H. Buddermeler, and D. R. B. Stockwell. 2002. Future projections for Mexican faunas under global climate change scenarios. *Nature* 416:626–629.

Peterson, G. D., and M. Heemskerk. 2001. Deforestation and forest regeneration following small-scale gold mining in the Amazon: The case of Suriname. *Environmental Conservation* 28:117–126.

Petithuguenin, P. 1995. Regeneration of cocoa cropping systems: The Ivorian and Togolese experience. Pages 89–106 in F. Ruf and P. S. Siswoputranto, eds., *Cocoa cycles: The economics of cocoa supply.* Woodhead Publishing, Cambridge, U.K.

Philipp, E., and K. Fabricius. 2003. Photophysiological stress in scleractinian corals in response to short-term sedimentation. *Journal of Experimental Marine Biology and Ecology* 287:57–78.

Phillips, O. L. 1995. Evaluating turnover in tropical forests. *Science* 268:894–895.

Phillips, O. L. 1996. Long-term environmental change in tropical forests: Increasing tree turnover. *Environmental Conservation* 23:235–248.

Phillips, O. L., T. R. Baker, L. Arroyo, N. Higuchi, T. Killeen, W. F. Laurance, S. L. Lewis, J. Lloyd, Y. Malhi, A. Monteagudo, D. Neill, P. Núñez Vargas, N. Silva, J. Terborgh, R. Vásquez Martínez, M. Alexiades, S. Almeida, S. Brown, J. Chave, J. Comiskey, C. I. Czimczik, A. Di Fiore, T. Erwin, C. Kuebler, S. G. Laurance, H. E. M. Nascimento, M. Palacios, S. Patiño, N. Pitman, J. Olivier, C. A. Quesada, M. Saldias, A. Torres Lezama, and B. Vinceti, B. 2004. Pattern and process in Amazon tree turnover, 1976–2001. *Philosophical Transactions of the Royal Society of London B* 359:381–407.

Phillips, O. L., and A. H. Gentry. 1994. Increasing turnover through time in tropical forests. *Science* 263:954–958.

Phillips, O. L., Y. Malhi, N. Higuchi, W. F. Laurance, P. V. Nuñez, R. Vásquez M., S. G. Laurance, L. V. Ferriera, M. Stern, S. Brown, and J. Grace. 1998. Changes in the carbon balance of tropical forest: Evidence from long-term plots. *Science* **282**:439–442.

Phillips, O. L., Y. Malhi, B. Vinceti, T. Baker, S. L. Lewis, N. Higuchi, W. F. Laurance, P. N. Vargas, R. V. Martínez, S. G. Laurance, L. V. Ferreira, M. Stern, S. Brown, and J. Grace. 2002a. Changes in the biomass of tropical forests: Evaluating potential biases. *Ecological Applications* **12**:576–587.

Phillips, O. L., R. V. Martínez, L. Arroyo, T. R. Baker, T. Killeen, S. L. Lewis, Y. Malhi, A. M. Mendoza, D. Neill, P. N. Vargas, M. Alexiades, C. Cerón, A. Di Fiore, T. Erwin, A. Jardim, W. Palacios, M. Saldias, and B. Vinceti. 2002b. Increasing dominance of large lianas in Amazonian forests. *Nature* **418**:770–774.

Phillips, O. L., and D. Sheil. 1997. Forest turnover, diversity and CO_2. *Trends in Ecology and Evolution* **12**:404.

Pickett, S. T. A., and J. N. Thompson. 1978. Patch dynamics and the design of nature reserves. *Biological Conservation* **13**:27–37.

Pimentel, D., C. Wilson, C. McCullum, R. Huang, P. Dwen, J. Flack, Q. Tran, T. Saltman, and B. Cliff. 1997. Economic and environmental benefits of biodiversity. *BioScience* **47**:747–757.

Pimm, S. L., H. L. Jones, and J. M. Diamond. 1988. On the risk of extinction. *American Naturalist* **132**:757–785.

Pimm, S. L., G. J. Russell, J. Gittleman, and T. M. Brooks. 1995. The future of biodiversity. *Science* **269**:347–350.

Pinard, M. A., and F. E. Putz. 1996. Retaining forest biomass by reducing logging damage. *Biotropica* **28**:278–295.

Pinard, M. A., and F. E. Putz. 1997. Monitoring carbon sequestration benefits associated with a reduced impact logging project in Malaysia. *Mitigation and Adaptation Strategies for Global Change* **2**:203–215.

Pinard, M. A., F. Putz, and J. Licona. 1999. Tree mortality and vine proliferation following a wildfire in a subhumid tropical forest in eastern Bolivia. *Forest Ecology and Management* **116**:247–252.

Pinto, L. P. S., and A. B. Rylands. 1997. Geographic distribution of the golden-headed lion tamarin, *Leontopithecus chysomelas*: Implications for its management and conservation. *Folia Primatologica* **68**:161–180.

Piperno, D. R., and P. Becker. 1996. Vegetational history of a site in the central Amazon Basin derived from phytolith and charcoal records from natural soils. *Quaternary Research* **45**:202–209.

Pitman, N., and P. Jorgensen. 2002. Estimating the size of the world's threatened flora. *Science* **298**:989.

PLANTAR. 2003. *Projetos de crédito de carbono.* www.plantar.com.br.

Plowright, W. 1982. The effects of rinderpest and rinderpest control on wildlife in Africa. *Symposium of the Zoological Society of London* **50**:1–28.

Posey, D. 1999. *Cultural and spiritual values of biodiversity: A complementary contribu-tion to the Global Biodiversity Assessment.* UNEP/Intermediate Technology, London, U.K.

Post, E., R. O. Peterson, N. C. Stenseth, and B. E. McClaren. 1999. Ecosystem consequences of wolf behavioural response to climate. *Nature* **451**:905–901.

Potter, C. S., J. T. Randerson, C. B. Field, P. A. Matson, P. M. Vitousek, H. A. Mooney, and S. A. Klooster. 1993. Terrestrial ecosystem production: A process model based on global satellite and surface data. *Global Biogeochemical Cycles* **7**:811–841.

Poulsen, J. R., C. J. Clark, E. F. Connor, and T. B. Smith. 2002. Differential resource use by primates and hornbills: Implications for seed dispersal. *Ecology* **83**:228–240.

Pounds, J. A., M. Bustamente, L. Coloma, J. Consuegra, M. Fogden, P. Foster, E. La Marca, K. Masters, A. Merino-Viteri, R. Puschendorf, et al. 2006. Widespread amphibian extinctions from epidemic disease driven by global warming. *Nature* **439**:161–167.

Pounds, J. A., and M. L. Crump. 1994. Amphibian declines and climate disturbance: The case of the golden toad and the harlequin frog. *Conservation Biology* **8**: 72–85.

Pounds, J. A., M. P. Fogden, and J. H. Campbell. 1999. Biological response to climate change on a tropical mountain. *Nature* **398**:611–615.

Pourtier, R. 1989. *Le Gabon*. Harmattan, Paris, France.

Prance, G. 1996. *The Earth under threat: A Christian perspective*. Wild Goose Press, Glasgow, Scotland.

Prentice, I. C., and 60 others. 2001. The carbon cycle and atmospheric carbon dioxide. Pages 183–237 in *Climate Change 2001: The Scientific Basis*. Cambridge University Press, Cambridge, U.K.

Pressey, R. L., R. M. Cowling, and M. Rouget. 2003. Formulating conservation targets for biodiversity pattern and process in the Cape Floristic Region, South Africa. *Biological Conservation* **112**:99–127.

Prince, S. D., S. J. Goetz, and S. N. Goward. 1995. Monitoring primary production from earth observing satellites. *Water, Air, and Soil Pollution* **82**:509–522.

Privette, J. L., J. Nickeson, D. Landis, and J. Morisette, eds. 2001. *SAFARI 2000*. CD-ROM Series, Volume 1. NASA, Washington, D.C., USA.

Purvis A, J. L. Gittleman, G. Cowlishaw, and G. M. Mace. 2000. Predicting extinction risk in declining species. *Proceedings of the Royal Society of London B* **267**: 1947–1952.

Putz, F. E. 1998. Halt the Homogeocene: A frightening future filled with too few species. *The Palmetto* **18**:7–10.

Putz, F. E., G. Blate, K. Redford, R. Fimbel, and J. Robinson. 2001a. Tropical forest management and conservation of biodiversity: An overview. *Conservation Biology* **15**:7–20.

Putz, F. E., and M. A. Pinard. 1993. Reduced-impact logging as a carbon-offset method. *Conservation Biology* **7**:755–759.

Putz, F. E., L. K. Sirot, and M. A. Pinard. 2001b. Tropical forest management and wildlife: Silviculture effects on forest structure, fruit production, and locomotion of arboreal animals. Pages 11–34 in R. A. Fimbrel, A. Grajal, and J. G. Robinson, eds., *The cutting edge: Conserving wildlife in logged tropical forests*. Columbia University Press, New York, USA.

Putz, F. E., and D. Windsor. 1987. Liana phenology on Barro Colorado Island, Panama. *Biotropica* **19**:334–341.

Quammen, D. 1996. *The song of the dodo*. Pimlico, London, U.K.

Quiatt, D., V. Reynolds, and E. J. Stokes. 2002. Snare injuries to chimpanzees (*Pan troglodytes*) at 10 study sites in East and West Africa. *African Journal of Ecology* **40**:303–305.

Rabin, R. M., S. Stadler, P. J. Wetzel, and D. J. Stensrud. 1990. Observed effects of landscape variability on convective clouds. *Bulletin of the American Meteorological Society* **71**:272–280.

Radford, J. Q., A. F. Bennett, and G. Cheers. 2005. Landscape-level thresholds of habitat cover for woodland-dependent birds. *Biological Conservation* 124:317–337.

Rajvanshi, A., V. B. Mathur, G. C. Teleki, and S. K. Mukherjee. 2001. *Roads, sensitive habitats, and wildlife: Environmental guidelines for India and South Asia.* Wildlife Institute of India, Dehradun, India.

Ramakrishnan, P., K. G. Saxena, and U. M. Chandrashekaran. 1998. *Conserving the sacred for biodiversity management.* Oxford University Press, Delhi, India.

Ramanathan, V., P. J. Crutzen, J. T. Kiehl, and D. Rosenfeld. 2001. Aerosols, climate, and the hydrological cycle. *Science* 294:2119–2124.

Raman Shankar, T. R., G. Rawat, and A. Johnsingh. 1998. Recovery of tropical rainforest avifauna in relation to vegetation succession following shifting cultivation in Mizoram, north-east India. *Journal of Applied Ecology* 35:214–231.

Ramlugun, G. D. 2003. The influence of degree of invasion on native tree diversity in Mauritian upland forest. B.Sc. thesis, University of Mauritius.

Ramos da Silva, R., and R. Avissar. 2006. The hydrometeorology of a deforested region of the Amazon basin. *Journal of Hydrometeorology.*

Ramos da Silva, R., D. Werth, and R. Avissar. In press. The future of the Amazon Basin hydrometeorology. *Journal of Climate.*

Ranvestel, A. W., K. R. Lips, C. M. Pringle, M. R. Whiles, and R. J. Bixby. 2004. Neotropical tadpoles influence stream benthos: Evidence for the ecological consequences of decline in amphibian populations. *Freshwater Biology* 49:274–285.

Raven, P. H. 2002. Science, sustainability, and the human prospect. *Science* 297:1477.

Rayner, P. J., and R. M. Law. 1999. The interannual variability of the global carbon cycle. *Tellus Series B Chemical and Physical Meteorology* 51:210–212.

Read, J. M., D. B. Clark, E. M. Venticinque, and M. P. Moreira. 2003. Application of 1-m and 4-m resolution satellite data to research and management in tropical forests. *Journal of Applied Ecology* 42:592–600.

Redford, K. H. 1992. The empty forest. *BioScience* 42:412–422.

Redford, K. H., and P. Feinsinger. 2001. The half-empty forest: Sustainable use and the ecology of interactions. Pages 370–399 in J. D. Reynolds, G. M. Mace, K. H. Redford, and J. G. Robinson, eds., *Conservation of exploited species.* Cambridge University Press, Cambridge, U.K.

Reed, K. D., G. R. Ruth, J. A. Meyer, and S. K. Shukla. 2000. *Chlamydia pneumoniae* infection in a breeding colony of African clawed frogs (*Xenopus tropicalis*). *Emerging Infectious Diseases* 6:196–199.

Reid, J., and R. Rice. 1997. Assessing natural forest management as a tool for tropical forest conservation. *Ambio* 26:382–386.

Reiter, P., M. Turell, R. Coleman, B. Miller, G. Maupin, J. Liz, A. Kuehne, J. Barth, J. Geisbert, D. Dohm, J. Glick, J. Pecor, R. Robbins, P. Jahrling, C. Peters, and T. Ksiazek. 1999. Field investigations of an outbreak of Ebola hemorrhagic fever, Kikwit, Democratic Republic of the Congo, 1995: Arthropod studies. *Journal of Infectious Diseases* 179:S148–S154.

Rejmánek, M. 1989. Invasibility of plant communities. Pages 369–388 in J. A. Drake, H. Mooney, F. di Castri, R. Groves, R. Kruger, M. Rejmánek, and M. Williams, eds., *Biological invasions: A global perspective.* John Wiley and Sons, Chichester, U.K.

Rejmánek, M. 1996. Species richness and resistance to invasions. Pages 153–172 in G. Orians, R. Dirzo, and H. Cushman, eds., *Biodiversity and ecosystem processes in tropical forests*. Springer-Verlag, Berlin, Germany.

Rejmánek, M. 1999. Invasive plant species and invasible ecosystems. Pages 79–102 in O. T. Sandlund, P. Schei, and A. Viken, eds., *Invasive species and biodiversity management*. Kluwer Academic Press, Dordrecht, The Netherlands.

Renjifo, L. M. 2001. Effect of natural and anthropogenic landscape matrices on the abundance of subandean bird species. *Ecological Applications* 11:14–31.

Repetto, R. 1988. *The forest for the trees: Government policies and misuse of forest resources*. Cambridge University Press, Cambridge, U.K.

Retallack, G. J. 2001. A 300-million-year record of atmospheric carbon dioxide from fossil plant cuticles. *Nature* 411:287–290.

Reynolds, J. D., G. M. Mace, K. H. Redford, and J. G. Robinson, eds. 2001. *Conservation of exploited species*. Cambridge University Press, Cambridge, U.K.

Reynolds, R. T., J. M. Scott, and R. A. Nussbaum. 1980. A variable circular-plot method for estimating bird numbers. *Condor* 82:309–313.

Ribon, R., J. E. Simon, and G. T. Mattos. 2003. Bird extinctions in Atlantic forest fragments on the Viçosa region, southeastern Brazil. *Conservation Biology* 17: 1827–1839.

Rice, R., R. Gullison, and J. Reid. 1997. Can sustainable management save tropical forests? *Scientific American* 276:34–39.

Rice, R., C. Sugal, S. Ratay, and G. da Fonseca. 2001. *Sustainable forest management: A review of conventional wisdom*. CABS/Conservation International, Washington, D.C., USA.

Ricklefs, R. E., and G. L. Miller. 1999. *Ecology*, 4th ed. W. H. Freeman, San Francisco, USA.

Rizzo, D. M., M. Garbelotto, J. M. Davidson, G. W. Salaughter, S. T. Koike. 2002. *Phytophtora ramorum* as the cause of extensive mortality of *Querqus ssp.* and *Lothocarpus densiflorus* in California. *Plant Disease* 86:205–214.

Roberts, C. M. 2002. Deep impact: The rising toll of fishing in the deep sea. *Trends in Ecology and Evolution* 17:242–245.

Roberts, L. 1998. *World resources*. Oxford University Press, New York, USA.

Robinson, J. G. 1993. The limits to caring: Sustainable living and the loss of biodiversity. *Conservation Biology* 7:20–28.

Robinson, J. G. 1996. Hunting wildlife in forest patches: An ephemeral resource. Pages 111–130 in J. Schellas and R. Greenberg, eds., *Forest patches in tropical landscapes*. Island Press, Washington, D.C., USA.

Robinson, J. G., and E. L. Bennett, eds. 2000. *Hunting for sustainability in tropical forests*. Columbia University Press, New York, USA.

Robinson, J. G., and R. E. Bodmer. 1999. Towards wildlife management in tropical forests. *Journal of Wildlife Management* 63:1–13.

Robinson, J. G., K. H. Redford, and E. L. Bennett. 1999. Wildlife harvest in logged tropical forests. *Science* 284:595–596.

Robinson, S. K., F. R. Thompson III, T. M. Donovan, D. R. Whitehead, and J. Faaborg. 1995. Regional forest fragmentation and the nesting success of migratory birds. *Science* 267:1987–1990.

Rodenbeck, C., S. Houweling, M. Gloor, and M. Heimann. 2003. CO_2 flux history 1982–2001 inferred from atmospheric data using a global inversion of atmospheric transport. *Atmospheric Chemistry and Physics* 3:1919–1964.

Rodrigues, A. S. L., and K. J. Gaston. 2001. How large do reserve networks need to be? *Ecology Letters* 4:602–609.

Rodríguez, K. F. 1994. The fungal endophytes of the Amazonian palm *Euterpe oleraceae*. *Mycologia* 86:376–385.

Rodríguez, K. F., and G. Samuels. 1999. Fungal endophytes of *Spondias mombin* leaves in Brazil. *Journal of Basic Microbiology* 39:131–135.

Roldan, A. I., and J. A. Simonetti. 2001. Plant–mammal interactions in tropical Bolivian forests experiencing different hunting pressures. *Conservation Biology* 15:617–623.

Romo, M. 1997. Seasonal variation in fruit consumption and seed dispersal by canopy bats (*Artibeus* spp.) in a lowland forest in Perú. *Vida Silvestre Neotropical* 5:110–119.

Ron, S. R., and A. Merino. 2000. Amphibian declines in Ecuador: Overview and first report of chytridiomycosius from South America. *Froglog* 42:2–3.

Root, T. L., D. P. MacMynowski, M. Mastrandrea, and S. Schneider. 2005. Human-modified temperatures induce species changes. *Proceedings of the National Academy of Sciences USA* 102:7465–7469.

Root, T. L., J. T. Price, K. R. Hall, S. H. Schneider, C. Rosenzweig, and J. A. Pounds. 2003. Fingerprints of global warming on wild animals and plants. *Nature* 421: 57–60.

Rosa, L. P., B. M. Sikar, M. A. dos Santos, and E. M. Sikar. 2002. *Emissões de dióxido de carbono e de metano pelos reservatórios hidrelétricos brasileiros. Relatório de referência—Inventário brasileiro de emissões antrópicas de gases de efeito estufa*. Ministério da Ciência e Tecnologia, Brasília, Brazil. http://www.mct.gov.br/clima.

Rosa, L. P., M. A. dos Santos, B. Matvienko, E. O. dos Santos, and E. Sikar. 2004. Greenhouse gas emissions by hydroelectric reservoirs in tropical regions. *Climatic Change* 66:9–21.

Rosa, L. P., M. A. dos Santos, B. Matvienko, E. O. dos Santos, and E. Sikar. 2006. Scientific errors in the Fearnside comments on greenhouse gas emissions (GHG) from hydroelectric dams and response to his political claims. *Climatic Change*, DOI:10.1007/s10584-005-9046-6.

Rosenfeld, D. 1999. TRMM observed first direct evidence of smoke from forest fires inhibiting rainfall. *Geophysical Research Letters* 26:3105–3108.

Rosenzweig, M. L. 1995. *Species diversity in space and time*. Cambridge University Press, Cambridge, U.K.

Ross, K. A., B. J. Fox, and M. D. Fox. 2002. Changes to plant species richness in forest fragments: Fragment age, disturbance and fire history may be as important as area. *Journal of Biogeography* 29:749–765.

Royer, D. L., S. L. Wing, D. J. Beerling, D. W. Jolley, P. L. Koch, L. J. Hickey, and R. A. Berner. 2001. Paleobotanical evidence for near present-day levels of atmospheric CO_2 during part of the Tertiary. *Science* 292:2310–2313.

Rudel, T. K. 2005. *Tropical forests: Regional paths of destruction and regeneration in the late twentieth century*. Columbia University Press, Chichester, U.K.

Ruf, F. 1995. From forest rent to tree-capital: Basic "laws" of cocoa supply. Pages 1–53 in F. Ruf and P. S. Siswoputranto, eds., *Cocoa cycles: The economics of cocoa supply*. Woodhead Publishing, Cambridge, U.K.

Ruf, F., and G. Schroth. 2004. Chocolate forests and monocultures: An historical review of cocoa growing and its conflicting role in tropical deforestation and forest conservation. Pages 107–134 in G. Schroth, G. Fonseca, C. Harvey, C. Gascon, H. Vasconcelos, and A. Izac, eds., *Agroforestry and biodiversity conservation in tropical landscapes*. Island Press, Washington, D.C., USA.

Ruimy, A., G. Dedieu, and B. Saugier. 1996. TURC: A diagnostic model of continental gross primary productivity and net primary productivity. *Global Biogeochemical Cycles* 10:269–285.

Ruiz, A., R. A. Merel, and M. Diaz. 2005. *Plan de acción para recuperación de las tortugas marinas de Panamá*. Widecast, Latin America Program, Panama City, Republic of Panama.

Rumelhart, D. E., and J. L. McClelland. 1986. *Parallel Distributed Processing: Explorations in the Microstructures of Cognition*. MIT Press, Cambridge, Massachusetts, USA.

Rumiz, D. I., D. Guinart S., and L. Solar O. 2001. Logging and hunting in community forests and corporate concessions: Two contrasting case studies in Bolivia. Pages 333–358 in R. A. Fimbel, A. Grajal, and J. G. Robinson, eds., *The cutting edge: Conserving wildlife in logged tropical forests*. Columbia University Press, New York, USA.

Rylands, A. B., and A. Keuroghlian. 1988. Primate populations in continuous forest and forest fragments in central Amazonia. *Acta Amazonica* 18:291–307.

Sá, R. M. L., L. V. Ferreira, R. Buschbacher, and G. Batmanian. 2000. *Áreas protegidas ou espaços ameaçados? O grau de implementação e a vulnerabilidade das unidades de conservação federais brasileiras de uso indireto*. Série Técnica III. World Wildlife Fund—Brasil, Brasília, Brazil.

Saatchi, S., D. Agosti, K. Alger, J. Delabie, and J. Musinsky. 2001. Examining fragmentation and loss of primary forest in the southern Bahian Atlantic forest of Brazil with radar imagery. *Conservation Biology* 15:867–875.

Safford, R. J. 1997. A survey of the occurrence of native vegetation remnants on Mauritius in 1993. *Biological Conservation* 80:181–188.

Sala, O. E., R. A. Golluscio, W. K. Lauenroth, and A. Soriano. 1989. Resource partitioning between shrubs and grasses in the Patagonian steppe. *Oecologia* 81:501–505.

Salafsky, N., R. Margoluis, K. H. Redford, and J. Robinson. 2002. Improving the practice of conservation: A conceptual framework and research agenda for conservation science. *Conservation Biology* 16:1469–1479.

Salati, E., J. Marques, and L. C. Molion. 1978. Origin and distribution of rain in the Amazon Basin. *Interciencia* 3:200–206.

Salati, E., and P. B. Vose. 1984. Amazon Basin: A system in equilibrium. *Science* 225:129–138.

Saldarriaga, J., and D. C. West. 1986. Holocene fires in the northern Amazon basin. *Quaternary Research* 26:358–366.

Saleh, C. 1997. Wildlife survey report from burned and unburned forest areas in central Kalimantan. Unpublished report, WWF Indonesia Program, Bogor, Indonesia.

Sanchez, M. 2002. Colombia's coffee research center braces for hard times. Reuters World Environment News, 28 January. www.planetark.org.

Sanderson, E. W., M. Jaiteh, M. A. Levy, K. H. Redford, A. Wannebo, and G. Woolmer. 2002. The human footprint and the last of the wild. *BioScience* 52:891–904.

Sandor, R. L., E. C. Bettelheim, and I. R. Swingland. 2002. An overview of a free-market approach to climate change and conservation. *Philosophical Transactions of the Royal Society of London A* **360**:1607–1620.

Sanford, R. L., J. Saldarriaga, K. Clark, C. Uhl, and R. Herrera. 1985. Amazon rainforest fires. *Science* **227**:53–55.

Santilli, M., P. Moutinho, S. Schwartzman, D. Nepstad, L. Curran, and C. Nobre. 2005. Tropical deforestation and the Kyoto Protocol: an editorial essay. *Climatic Change* **71**:267–276.

Santos, J. R., M. Pardi Lacruz, L. Araujo, and H. Xaud. 1998. El processo de quiema de biomassa de bosque tropical y de sabanas en la Amazonia Brasileira: experiencas de monitoreo com dados ópticos y de microndas. *Revista Serie Geografica* **7**:97–108.

Santos, L. A. O., and L. M. M. de Andrade, eds. 1990. *Hydroelectric dams on Brazil's Xingu River and indigenous peoples*. Report 30, Cultural Survival, Cambridge, Massachusetts, USA.

Sathirathai, S. 1998. *Economic valuation of mangroves and the roles of local communities in the conservation of natural resources: Case study of Serat Thani, south of Thailand*. Economy and Environment Program for South-East Asia, Singapore.

Saville, N. M. 2000. *Battling with bee brood disease in Apis cerana in W. Nepal: Some findings*. Proceedings of the 7th International Bee Research Association conference on tropical bees: Management and diversity and 5th Asian Apicultural Association, 19–25 March, Chang Mai, Thailand.

Sawyer, D. 1993. Economic and social consequences of malaria in new colonization projects in Brazil. *Social Science and Medicine* **37**:1131–1136.

Schaffner Cappello, R. 2004. *Prospection de la forêt de Souanké-Garabinzam-Ivindo Trinational Dja-Odzala-Minkébé (TRIDOM)*. Report to World Wildlife Fund—Gabon, Libreville, Gabon.

Schemske, D. W., and N. Brokaw. 1981. Treefalls and the distribution of understory birds in a tropical forest. *Ecology* **62**:938–945.

Schlaepfer, M. A., and T. A. Gavin. 2001. Edge effects on lizards and frogs in tropical forest fragments. *Conservation Biology* **15**:1079–1090.

Schlamadinger, B., M. Obersteiner, A. Michaelowa, M. Grubb, C. Azar, Y. Yamagata, D. Goldberg, P. Read, M. U. F. Kirschbaum, P. M. Fearnside, T. Sugiyama, E. Rametsteiner, and K. Böswald. 2001. Capping the cost of compliance with the Kyoto Protocol and recycling revenues into land-use projects. *Scientific World* **1**:271–280.

Schlesinger, W. H., J. F. Reynolds, G. L. Cunningham, L. F. Huenneke, W. M. Jarrell, R. A. Virginia and W. G. Whitford. 1990. Biological feedbacks in global desertification. *Science* **247**:1043–1048.

Schmitz, D. C., D. Simberloff, R. Hofstetter, R. Haller, and D. Sutton. 1997. Theecological impact of nonindigenous plants. Pages 39–62 in D. Simberloff, D. Schmitz, and T. C. Brown, eds., *Strangers in paradise: Impact and management of nonindigenous species in Florida*. Island Press, Washington, D.C., USA.

Schneider, C., and S. E. Williams. 2005. Combining spatial analyses of species diversity with molecular genetics to examine historical determinants of species diversity. Pages 401–424 In E. Bermingham, C. Dick, and C. Moritz, eds., *Tropical rainforests: Past, present, and future*. University of Chicago Press, Chicago, USA.

Schneider, C. J., M. Cunningham, and C. Moritz. 1998. Comparitive phylogeography and the history of endemic vertebrates in the wet tropics rainforests of Australia. *Molecular Ecology* 7:487–498.

Schnitzer, S. A., and F. Bongers. 2002. The ecology of lianas and their role in forests. *Trends in Ecology and Evolution* 17:223–230.

Scholes, R. J. 1998. *The South African 1:250 000 maps of areas of homogeneous grazing potential*. Division of Water, Environment and Forest Technology, CSIR Technical Report, Pretoria, South Africa.

Scholes, R. J., J. Kendall, and C. O. Justice. 1996. The quantity of biomass burned in southern Africa. *Journal of Geophysical Research* 101:23667–23676.

Scholes, R. J., and B. H. Walker. 1993. *An African savanna: Synthesis of the Nylsvley study*. Cambridge University Press, Cambridge, U.K.

Schroth, G., G. A. B. da Fonseca, C. A. Harvey, C. Gascon, H. L. Vasconcelos, and A. M. N. Izac, eds. 2004. *Agroforestry and biodiversity conservation in tropical landscapes*. Island Press, Washington, D.C., USA.

Schubert, S. D., R. B. Rood, and J. Pfaendtner. 1993. An assimilated data set for Earth science applications. *Bulletin of the American Meteorological Society* 74:2331–2342.

Schupp, E. W., T. Milleron, and S. Russo. 2002. Dissemination limitation and the origin and maintenance of species-rich tropical forests. Pages 19–33 in D. J. Levey, W. R. Silva, and M. Galetti, eds., *Seed dispersal and frugivory: Ecology, evolution and conservation*. CAB International, Wallingford, U.K.

Schwartz, M. W., S. M. Hermann, and C. Vogel. 1995. The catastrophic loss of *Torreya taxifolia*: Assessing environmental induction of disease hypotheses. *Ecological Applications* 5:501–516.

Segura, O., and K. Kindegard. 2001. Joint implementation in Costa Rica: A case study at the community level. *Journal of Sustainable Forestry* 12:61–78.

Sellers, P. J. 1985. Canopy reflectance, photosynthesis and transpiration. *International Journal of Remote Sensing* 6:135–1372.

Sellers, P. J. 1987. Canopy reflectance, photosynthesis and transpiration II. *Remote Sensing of the Environment* 21:143–183.

Sellers, P. J., Y. Mintz, Y. C. Sud, and A. Dalcher. 1986. The design of a simple biosphere model (SiB) for use within general circulation models. *Journal of Atmospheric Science* 43:505–531.

Semazzi, F. H. M., and Y. Song. 2001. A GCM study of climate change induced by deforestation in Africa. *Climate Research* 17:169–182.

Seroa da Motta, R., and C. Ferraz. 2000. *Brazil: CDM opportunities and benefits*. World Resources Institute, Washington, D.C., USA.

Serrão, E. A. S., and J. M. Toledo. 1990. The search for sustainability in Amazonian pastures. Pages 195–214 in A. B. Anderson, ed., *Alternatives to deforestation: Towards sustainable use of the Amazon rain forest*. Columbia University Press, New York, USA.

Shabbar, A., B. Bonsal, and M. Khandekar. 1997. Canadian precipitation patterns associated with the Southern Oscillation. *Journal of Climate* 10:3016–3027.

Sharma, D. 2005. Tsunami, mangroves and market economy. *India Together*. http://www.trrcindia.org/static/dsh-tsunami.pdf.

Sharp, P. M., E. Bailes, F. Gao, B. E. Beer, V. M. Hirsch, B. H. Hahn. 2000. Origins and evolution of AIDS viruses: Estimating the time-scale. *Biochemical Society Transactions* **28**:275–282.

Shea, R. W., B. W. Shea, J. B. Kauffman, D. E. Ward, C. I. Haskins, and M. C. Scholes. 1996. Fuel biomass and combustion factors associated with fires in savanna ecosystems of South Africa and Zambia. *Journal of Geophysical Research* **101** (D19): 23551–23568.

Shearer, B. L., and J. T. Tippett. 1989. Jarrah dieback: The dynamics and management of *Phytophthora cinnamomi* in the jarrah (*Eucalyptus marginata*) forest of south-western Australia. Research Bulletin 3, Department of Conservation and Land Management, Western Australia.

Sheil, D. 1995. Evaluating turnover in tropical forests. *Science* **268**:894.

Sheil, D. 1997. *Monitoring for the Kibale and Semuliki National Parks.* A report for the Kibale and Semuliki Conservation and Development Project.

Shoo L. P., S. E. Williams, and J.-M. Hero. 2005a. Decoupling of trends in distribution area and population size of species with climate change. *Global Change Biology* **11**:1469–1476.

Shoo L. P., S. E. Williams, and J.-M. Hero. 2005b. Climate change and the rainforest birds of the Australian Wet Tropics: Using total population size as a sensitive predictor of the impacts of increasing temperature. *Biological Conservation* **125**: 335–343.

Shugart, H. H., W. R. Emanuel, D. C. West, and D. L. DeAngelis. 1980. Environmental gradients in a beech–yellow poplar stand simulation model. *Mathematical Biosciences* **50**:163–170.

Shukla, J., C. Nobre, and P. J. Sellers. 1990. Amazon deforestation and climate change. *Science* **247**:1322–1325.

Sick, H. 1997. *Ornitologia Brasileira.* Editora Nova Fronteira, Rio de Janeiro, Brazil.

Siegert, F., G. Ruecker, A. Hinrichs, and A. Hoffmann. 2001. Increased damage from fires in logged forests during droughts caused by El Niño. *Nature* **414**:437–440.

Sieving, K. E. 1992. Nest predation and differential insular extinction among selected forest birds of central Panama. *Ecology* **73**:2310–2328.

Silbergeld, E. K., J. B. Sacci, and A. F. Azad. 2000. Mercury exposure and murine response to *Plasmodium yoelii* infection and immunization. *Immunopharmacology and Immunotoxicology* **22**:685–695.

Sinclair, A. R. E., and A. E. Byrom. 2006. Understanding ecosystem dynamics for conservation of biota. *Journal of Animal Ecology* **75**:64–79.

Silva, J. M. C., and M. Tabarelli. 2000. Tree species impoverishment and the future flora of the Atlantic forest of northeast Brazil. *Nature* **404**:72–74.

Silva Dias, M. A. F., E. Williams, L. Pereira, A. Pereira Filho, and P. Matsuo. 2002. Shallow convection response to land use and topography in SW Amazon in the wet season. *Journal of Geophysical Research.*

Silva Dias, P. L., and P. Regnier. 1996. Simulation of mesoscale circulations in a deforested area of Rondônia in the dry season. Pages 531–547 in J. Gash, C. Nobre, J. Roberts, and R. Victoria, eds., *Amazonian deforestation and climate.* John Wiley and Sons, San Francisco, USA.

Silveira, J. P. 2001. Development in the Brazilian Amazon. *Science* **292**:1651–1652.

Simberloff, D. 1995. Why do introduced species appear to devastate islands more than mainland areas? *Pacific Science* **49**:87–97.

Simberloff, D. J., D. Doak, M. Groom, S. Trombulak, A. Dobson, S. Gatewood, M. E. Soulé, M. Gilpin, C. Martinez del Rio, and L. Mills. 1999. Regional and continental restoration. Pages 65–98 in M. E. Soulé and J. Terborgh, eds., *Continental conservation: Scientific foundations of regional reserve networks*. Island Press, Washington, D.C., USA.

Simioni, G., X. Le Roux, J. Gignoux, and H. Sinoquet. 2000. Treegrass: A 3D, process-based model for simulating plant interactions in tree–grass ecosystems. *Ecological Modelling* **131**:47–63.

Sinclair, A. R. E., D. Ludwig, and C. W. Clark. 2000. Conservation in the real world. *Science* **289**:1875.

Sizer, N., and E. V. J. Tanner. 1999. Responses of woody plant seedlings to edge formation in a lowland tropical rainforest, Amazonia. *Biological Conservation* **91**: 135–142.

Sizer, N. C., E. V. J. Tanner, and I. Kossman Ferraz. 2000. Edge effects on litterfall mass and nutrient concentrations in forest fragments in central Amazonia. *Journal of Tropical Ecology* **16**:853–863.

Skole, D. S., and C. J. Tucker. 1993. Tropical deforestation and habitat fragmentation in the Amazon: Satellite data from 1978 to 1988. *Science* **260**:1905–1910.

Slik, J. W. F., and K. Eichhorn. 2003. Fire survival of lowland tropical rain forest trees in relation to stem diameter and topographic position. *Oecologia* **137**:446–455.

Slik, J. W. F., R. Verburg, and P. Kessler. 2002. Effects of fire and selective logging on the tree species composition of lowland dipterocarp forest in East Kalimantan, Indonesia. *Biodiversity and Conservation* **11**:85–98.

Smeraldi, R., and A. Veríssimo. 1999. *Hitting the target: Timber consumption in the Brazilian market and promotion of forest certification*. Amigos da Terra—Programa Amazônia, São Paulo, Brazil.

Smith, A. P. 1997. Deforestation, fragmentation, and reserve design in western Madagascar. Pages 415–441 in W. F. Laurance and R. O. Bierregaard Jr., eds., *Tropical forest remnants: Ecology, management, and conservation of fragmented communities*. University of Chicago Press, Chicago, USA.

Smith, D. L., B. Lucey, L. A. Waller, J. E. Childs, and L. A Real. 2002. Predicting the spatial dynamics of rabies epidemics on heterogeneous landscapes. *Proceedings of the National Academy of Sciences USA* **99**:3668–3672.

Smith, F. A., H. Browning, and U. Shepherd. 1998. The influence of climate change on the body mass of woodrats *Neotoma* in an arid region of New Mexico, USA. *Ecography* **21**:140–148.

Smith, F. D. M., R. M. May, R. Pellew, T. H. Johnson, and K. S. Walter. 1993. Estimating extinction rates. *Nature* **364**:494–496.

Smith, J., K. Obidzinski, Subarudi, and I. Suramenggala. 2003. Illegal logging, collusive corruption and fragmented governments in Kalimantan, Indonesia. *International Forestry Review* **5**:293–302.

Smith, R. A., and S. M. Turton. 1995. *Environmental impacts on campsites and walking tracks in and alongside the World Heritage Daintree National Park*. Wet Tropics Management Authority, Cairns, Australia.

Smith, R. J., R. D. J. Muir, M. J. Walpole, A. Balmford, and N. Leader-Williams. 2003. Governance and the loss of biodiversity. *Nature* **426**:67–80.

Smith, R. J., and M. J. Walpole. 2005. Should conservationists pay more attention to corruption? *Oryx* **39**:251–256.

Smith, T. M., P. N. Halpin, H. H. Shugart, and C. M. Secrett. 1995. Global Forests. Pages 146–179 in K. M. Strzepek and J. B. Smith, eds., *If climate changes: International impacts of climate change*. Cambridge University Press, Cambridge, U.K.

Smythe, N. 1989. Seed survival in the palm *Astrocaryum standleyanum*: Evidence for dependence upon its seed dispersers. *Biotropica* **21**:50–56.

Soares-Filho, B., A. Alencar, D. Nepstad, G. Cerqueira, M. Diaz, S. Rivero, L. Solórzano, and E. Voll. 2004. Simulating the response of land-cover changes to road paving and governance along a major Amazonian highway: The Santarém–Cuiabá corridor. *Global Change Biology* **10**:745–764.

Soares-Filho, B., D. C. Nepstad, C. M. Curran, G. C. Cerqueira, R. A. Garcia, C. A. Ramos, E. Voll, A. McDonald, P. LeFebvre, and P. Schlesinger. 2006. Modeling conservation in the Amazon basin. *Nature* **440**:520–523.

Sombroek, W. G. 2001. Spatial and temporal patterns of Amazon rainfall: Consequences for planning of agricultural occupation and the protection of primary forests. *Ambio* **30**:388–396.

Sombroek, W. G., D. C. Kern, T. Rodrigues, M da S. Cravo, T. J. Cunha, W. Woods, and B. Glaser. 2002. Terra Preta and Terra Mulata, pre-Colombian kitchen middens and agricultural fields, their sustainability and replication. In R. Dudal, ed., *Symposium 18: Anthropogenic factors of soil formation*. 17th World Congress of Soil Science, Bangkok, Thailand.

Sombroek, W. G., F. O. Nachtergaele, and A. Hebel. 1993. Amounts, dynamics and sequestering of carbon in tropical and subtropical soils. *Ambio* **22**:417–426.

Sombroek, W. G., M. L. Ruivo, P. M. Fearnside, B. Glaser, and J. Lehmann. 2003. Anthropogenic dark earths as carbon stores and sinks. Pages 125–139 in J. Lehmann, ed., *Current advances on terra preta research in Amazonia*. Kluwer, Dordrecht, The Netherlands.

SOS Mata Atlântica/INPE. 1993. *Atlas da evolução dos remanescentes florestais da Mata Atlântica e ecossistemas associados no período de 1985–1990*. São Paulo, Brazil.

Soulé, M. E., D. T. Bolger, A. C. Alberts, J. Wright, M. Sorice, and S. Hill. 1988. Reconstructed dynamics of rapid extinctions of chaparral requiring birds in urban habitat islands. *Conservation Biology* **2**:75–92.

Soulé, M. E., and R. Noss. 1998. Rewilding and biodiversity as complementary tools for continental conservation. *Wild Earth* **8**:18–28.

Soulé, M. E., and M. A. Sanjayan. 1998. Conservation targets: Do they help? *Science* **279**:2060–2061.

Soulé, M. E., and J. Terborgh, eds. 1999. *Continental conservation: Scientific foundations of regional reserve networks*. Island Press, Washington, D.C., USA.

Sounguet, G. P. 2002. *Rapport d'activités programme tortues marines: Site de Pongara, saison 2001–2002*. Aventures Sans Frontières, Libreville, Gabon.

Spalding, M. D., F. Blasco, and C. D. Field, eds. 1997. *World mangrove atlas*. The International Society for Mangrove Ecosystems, Okinawa, Japan.

Spergel, B. 2002. Financing protected areas. Pages 364–382 in J. Terborgh, C. van Schaik, L. Davenport, and M. Rao, eds., *Making parks work: Strategies for preserving tropical nature*. Island Press, Washington, D.C., USA.

Spironello, W. R. 1999. The Sapotaceae community ecology in a Central Amazonian forest: Effects of seed dispersal and seed predation. PhD. Dissertation, University of Cambridge, U.K.

Spotila, J. R., A. Dunham, A. Leslie, A. Steyermark, P. Plotkin, and F. Paladino. 1996. Worldwide population decline of *Dermochelys coriaceae*: Are leatherback turtles going extinct? *Chelonian Conservation Biology* 2:209–222.

Spotila, J. R., R. Reina, A. Steyermark, P. Plotkin, and F. Paladino. 2000. Pacific leatherback turtles face extinction. *Nature* 405:529–530.

Stallknecht, D. E., E. W. Howerth, M. L. Kellogg, C. F. Quist, and T. Pisell. 1997. In vitro replication of epizootic hemorrhagic disease and bluetongue viruses in white-tailed deer peripheral bood mononuclear cells and virus-cell association during in vivo infections. *Journal of Wildlife Diseases* 33:574–583.

Stanford, C. B. 1998. The social behavior of chimpanzees and bonobos: Empirical evidence and shifting assumptions. *Current Anthropology* 39:399–420.

STCP. 2002. *Relatório de impacto ambiental do projeto de florestamento com* Acacia mangium *em uma área de 30.000 ha localizada no Estado de Roraima*. STCP Engenharia de Projetos Ltda./Ouro Verde Agropastoril Ltda. Curitiba, Paraná, Brazil.

Steklis, D., and N. Gerald-Steklis. 2001. Status of Virunga mountain gorilla population. Pages 391–412 in M. M. Robbins, P. Sicotte, and K. J. Stewart, eds., *Mountain gorillas: Three decades of research at Karisoke*. Cambridge University Press, Cambridge, U.K.

Stephenson, P. J. 1993. The impacts of tourism on nature reserves in Madagascar: Perinet, a case study. *Environmental Conservation* 20:262–265.

Stevens, W. E. 1968. *The conservation of wildlife in West Malaysia*. Federal Game Department, Ministry of Lands and Mines, Serembang, Malaysia.

Stevenson, P. R., M. Quinones, and J. Ahumada. 1998. Annual variation in fruiting pattern using two different methods in a lowland tropical forest, Tinigua National Park, Colombia. *Biotropica* 30:129–134.

Still, C. J., N. F. Foster, and S. H. Schneider. 1999. Simulating the effects of climate change on tropical montane cloud forests. *Nature* 398:608–610.

Stohr, K., and F. M. Meslin. 1996. Progress and setbacks in the oral immunisation of foxes against rabies in Europe. *Veterinary Record* 139:32–35.

Stokes, G. M., and S. E. Schwartz. 1994. The Atmospheric Radiation Measurement (ARM) program: Programmatic background and design of the cloud and radiation test bed. *Bulletin of the American Meteorological Society* 75:1201–1220.

Stokstad, E. 2004. Forest loss makes monkeys sick. *Science* 305:1230–1231.

Stolton, S., and N. Dudley, eds. 1999. *Threats to forest protected areas*. IUCN, Gland, Switzerland and UNEP-WCMC, Cambridge, U.K.

Stolton, S., M. Hocking, N. Dudley, K. MacKinnon, and T. Whitten. 2003. *Reporting progress at protected area sites*. World Bank/WWF Forest Alliance, Washington, D.C., USA.

Stork, N. E. 2005. The theory and practice of planning for long-term conservation of biodiversity in the Wet Tropics of Australia: Integrating ecological and economic sustainability. Pages 507–526 in E. Bermingham, C. Dick, and C. Moritz, eds., *Tropical rainforests: Past, present, and future*. University of Chicago Press, Chicago, USA.

Stotz, D. F., J. W. Fitzpatrick, T. A. Parker III, and D. K. Moskovits. 1996. *Neotropical birds: Ecology and conservation*. University of Chicago Press, Chicago, USA.

Stouffer, P. C., and R. O. Bierregaard Jr. 1995a. Effects of forest fragmentation on understory hummingbirds in Amazonian Brazil. *Conservation Biology* 9:1085–1094.

Stouffer, P. C., and R. O. Bierregaard Jr. 1995b. Use of Amazonian forest fragments by understory insectivorous birds. *Ecology* 76:2429–2443.

Strahm, W. A. 1993. The conservation and restoration of Mauritius and Rodrigues. Ph.D. thesis, University of Reading, Reading, U.K.

Stratford, A. J., and P. C. Stouffer. 1999. Local extinctions of terrestrial insectivorous birds in a fragmented landscape near Manaus, Brazil. *Conservation Biology* 13:1416–1423.

Struhsaker, T. T. 1997. *Ecology of an African forest: Logging in Kibale and the conflict between conservation and exploitation*. University of Florida Press, Gainesville, Florida, USA.

Stuart, S., J. S. Chanson, N. A. Cox, B. E. Young, A. S. L. Rodrigues, D. L. Fischman, and R. W. Waller. 2004. Status and trends of amphibian declines and extinctions worldwide. *Science* 306:1783–1786.

Sud, Y., R. Yang, and G. Walker. 1996. Impact of in situ deforestation in Amazonia on the regional climate: General circulation model simulation study. *Journal of Geophysical Research* 101:7095–7109.

Sullivan, N. J., T. W. Geisbert, J. B. Geisbert, L. Xu, Z. Y. Yang, M. Roederer, R. A. Koup, P. B. Jahrling, and G. J. Nabel. 2003. Accelerated vaccination for Ebola virus haemorrhagic fever in non-human primates. *Nature* 424:681–684.

Suplee, C. 1999. El Niño/La Niña: Mature's vicious cycle. *National Geographic Magazine*, March, pages 21–34.

Sussman, R. T., and I. Tattersall. 1981. Behavior and ecology of *Macaca fascicularis* in Mauritius: A preliminary study. *Primates* 22:192–205.

Sutton, A. 1994. *Slavery in Brazil—A link in the chain of modernization*. Anti-Slavery International, London, U.K.

Suzuki, A. 1988. The socioecological study of orangutans and forest conditions after the big forest fire and drought, 1983 in Kutai National Park, Indonesia. Unpublished report.

Swaine, M. D. 1992. Characteristics of dry forest in West Africa and the influence of fire. *Journal of Vegetation Science* 3:365–374.

Swaine, M. D., and T. C. Whitmore. 1988. On the definition of ecological species groups in tropical rain forest. *Vegetatio* 75:81–86.

Swanepoel, R., P. A. Leman, F. J. Burt, N. A. Zachariades, L. E. O. Braack, T. G. Ksiazek, P. E. Rollin, S. R. Zaki, and C. J. Peters. 1996. Experimental inoculation of plants and animals with Ebola virus. *Emerging Infectious Diseases* 2:321–325.

Swap, R. J., H. J. Annegarn, J. T. Suttles, J. Haywood, M. C. Helmlinger, C. Hély, P. V. Hobbs, B. N. Holben, J. Ji, M. King, T. Landmann, W. Maenhaut, L. Otter, B. Pak, S. J. Piketh, S. Platnick, J. Privette, D. Roy, A. M. Thompson, D. E. Ward, and R. Yokelson. 2002. The southern African regional science initiative (SAFARI 2000): Dry-season field campaign: An overview. *South African Journal of Science* 98:125–130.

Symington, M. M. 1987. Ecological and social correlates of party size in the black spider monkey, *Ateles paniscus chamek*. Ph.D. thesis, Princeton University, Princeton, New Jersey, USA.

SYSTAT. 2000. SYSTAT Software Inc., Richmond, California, USA.

Tabarelli, M., W. Mantovani, and C. A. Peres. 1999. Effects of habitat fragmentation on plant guild structure in the montane Atlantic forest of southeastern Brazil. *Biological Conservation* **91**:119–128.

Takamatsu, H., P. S. Mellor, P. P. C. Mertens, P. A. Kirkham, J. N. Burroughs, and R. M. E. Parkhouse. 2003. A possible overwintering mechanism for bluetongue virus in the absence of the insect vector. *Journal of General Virology* **84**:227–235.

Takhtjan, A. L. 1980. Outline of the classification of flowering plants (Magnoliophyta). *Botanical Review* **46**:225–359.

Talbot, L. M., S. M. Turton, and A. W. Graham. 2003. Trampling resistance of tropical rainforest soils and vegetation in the Wet Tropics of north east Queensland, Australia. *Journal of Environmental Management* **69**:63–69.

Taylor, L. H., S. M. Latham, and M. E. J. Woolhouse. 2001. Risk factors for human disease energence. *Philosophical Transactions of the Royal Society of London B* **356**: 983–989.

Temple, S. A. 1977. Plant-animal mutualism: Coevolution with dodo leads to near extinction of plant. *Science* **197**:885–886.

Teo, D. H. L., H. T. W. Tan, and R. T. Corlett. 2003. Continental rain forest fragments in Singapore resist invasion by exotic plants. *Journal of Biogeography* **30**:305–310.

Terborgh, J. 1974. Preservation of natural diversity: The problem of extinction prone species. *BioScience* **24**:715–722.

Terborgh, J. 1975. Faunal equilibria and the design of wildlife preserves. Pages 369–380 in F. B. Golley and E. Medina, eds., *Tropical ecological systems: Trends in terrestrial and aquatic research.* Springer-Verlag, New York, USA.

Terborgh, J. 1983. *Five new world primates: A study in comparative ecology.* Princeton University Press, Princeton, New Jersey, USA.

Terborgh, J. 1999. *Requiem for nature.* Island Press, Washington, D.C., USA.

Terborgh, J., and M. A. Boza. 2002. Internationalization of nature conservation. Pages 383–394 in J. Terborgh, C. van Schaik, L. Davenport, and M. Rao, eds., *Making parks work: Strategies for preserving tropical nature.* Island Press, Washington, D.C., USA.

Terborgh, J., J. A. Estes, P. Paquet, K. Ralls, D. Boyd-Heger, B. J. Miller, and R. F. Noss. 1999. The role of top carnivores in regulating terrestrial ecosystems. Pages 39–64 in M. E. Soulé and J. Terborgh, eds., *Continental conservation: Scientific foundations of regional reserve networks.* Island Press, Washington, D.C., USA.

Terborgh, J., L. Lopez, P. Nuñez V., M. Rao, G. Shahabuddin, G. Orihuela, M. Riveros, R. Ascanio, G. Adler, T. Lambert, and L. Balbas. 2001. Ecological meltdown in predator-free forest fragments. *Science* **294**:1923–1926.

Terborgh, J., N. Pitman, M. Silman, H. Schichter, and P. Nuñez V. 2002a. Maintenance of tree diversity in tropical forests. Pages 1–17 in D. J. Levey, W. R. Silva, and M. Galetti, eds., *Seed dispersal and frugivory: Ecology, evolution and conservation.* CAB International, Wallingford, U.K.

Terborgh, J., and C. P. van Schaik. 1997. Minimizing species loss: The imperative of protection. Pages 15–35 in R. Kramer, C. van Schaik, and J. Johnson, eds., *Last stand: Protected areas and the defense of tropical biodiversity.* Oxford University Press, Oxford, U.K.

Terborgh, J., C. van Schaik, L. Davenport, and M. Rao, eds. 2002b. *Making parks work: Strategies for preserving tropical nature.* Island Press, Washington, D.C., USA.

Thibault, M., and S. Blaney. 2003. The oil industry as an underlying factor in the bushmeat crisis in central Africa. *Conservation Biology* 17:1807–1813.

Thiollay, J. M. 1992. Influence of selective logging on bird species diversity in Guianian rain forest. *Conservation Biology* 6:47–63.

Thiollay, J. M. 1997. Disturbance, selective logging and bird diversity: A neotropical forest study. *Biodiversity and Conservation* 6:1155–1173.

Thiollay, J. M. 1999. Responses of an avian community to rain forest degradation. *Biodiversity and Conservation* 8:513–534.

Thomas, C. D., A. Cameron, R. E. Green, M. Bakkenes, L. J. Beaumont, Y. C. Collingham, B. F. N. Erasmus, M. Ferreira de Siqueira, A. Grainger, L. Hannah, L. Hughes, B. Huntley, A. S. van Jaarsveld, G. F. Midgley, L. Miles, M. A. Ortega-Huerta, A. T. Peterson, O. L. Phillips, and S. E. Williams. 2004. Extinction risk from climate change. *Nature* 427:145–148.

Thompson, R. 1880. *Report on the forest of Mauritius: Their present condition and future management*. Mercantile Record Co., Port Louis, Mauritius.

Thompson, S. 1988. Range expansion by alien weeds in the coastal farmlands of Guyana. *Journal of Biogeography* 15:109–118.

Tilman, D., J. Fargione, B. Wolff, C. D'Antonio, A. Dobson, R. Howarth, D. Schindler, W. H. Schlesinger, D. Simberloff, and D. Swackhamer. 2001. Forecasting agriculturally driven global environmental change. *Science* 292:281–284.

Timmermann, A., J. Oberhuber, A. Bacher, M. Esch, M. Latif, and E. Roeckner. 1999. Increased El Niño frequency in a climate model forced by future greenhouse warming. *Nature* 398:694–697.

Todd, S. W., and M. T. Hoffman. 1999. A fence-line contrast reveals effects of heavy grazing on plant diversity and community relations in Namaqualand, South Africa. *Plant Ecology* 142:169–178.

Toledo, E. 1997. *New species for utilization*. Tropical Forest Update 7, International Tropical Timber Organization, Yokohama, Japan.

Tracey, J. G. 1982. *The vegetation of the humid tropical region of north Queensland*. CSIRO Publications, Melbourne, Australia.

Tracey, J. G., and L. J. Webb. 1975. *Vegetation of the humid tropical region of north Queensland*. Fifteen maps at a scale of 1:100 000 plus key and notes. CSIRO, Indooroopilly, Australia.

Travers, S. E., G. S. Gilbert, and E. T. Perry. 1998. The effect of rust infection on reproduction in a tropical tree (*Faramea occidentalis*). *Biotropica* 30:438–443.

Traveset, A., and M. Verdú. 2002. A meta-analysis of gut treatment on seed germination. Pages 339–350 in D. J. Levey, W. R. Silva, and M. Galetti, eds., *Seed dispersal and frugivory: Ecology, evolution and conservation*. CABI Publishing, Wallingford, U.K.

Trenberth, K., G. Branstator, D. Karoly, A. Kumar, N.-C. Lau, and C. Ropelewski. 1998. Progress during TOGA in understanding and modeling global teleconnections associated with tropical sea surface temperatures. *Journal of Geophysical Research* 103:14291–14324.

Trenberth, K. E., and T. J. Hoar. 1996. The 1990–1995 El Niño–Southern Oscillation event: Longest on record. *Geophysical Research Letters* 23:57–60.

Trenberth, K. E., and T. J. Hoar. 1997. El Niño and climate change. *Geophysical Research Letters* 24:3057–3060.

Trollope, W. S. W., L. A. Trollope, A. L. F. Potgieter, and N. Zambatis. 1996. SAFARI-92 characterization of biomass and fire behavior in the small experimental burns in the Kruger National Park. *Journal of Geophysical Research* **101** (D19): 23531–23539.

Trombulak, S. C., and C. A. Frissell. 2000. Review of ecological effects of roads on terrestrial and aquatic communities. *Conservation Biology* **14**:18–30.

Trzcinski, M. K., L. Fahrig, and G. Merriam. 1999. Independent effects of forest cover and fragmentation on the distribution of forest breeding birds. *Ecological Applications* **9**:586–593.

Tucker, C. J., J. M. Wilson, R. Mahoney, A. Anyamba, K. Linthicum, and M. F. Myers. 2002. *Photogrammetric Engineering and Remote Sensing* **68**:147–152.

Turner, I. M. 2001. *The ecology of trees in the tropical rain forest.* Cambridge University Press, Cambridge, U.K.

Turner, I. M., K. Chua, J. Ong, B. Soong, and H. Tan. 1996. A century of plant species loss from an isolated fragment of lowland tropical rain forest. *Conservation Biology* **10**:1229–1244.

Turner, R. K., J. Paavola, P. Cooper, S. Farber, V. Jessamy, and S. Georgiou. 2002. *Valuing nature: Lessons learned and future directions.* CSERGE, University of East Anglia, Norwich, U.K.

Turner, W., S. Spector, N. Gardiner, M. Fladeland, E. Sterling, and M. Steininger. 2003. Remote sensing for biodiversity science and conservation. *Trends in Ecology and Evolution* **18**:306–314.

Turton, S. M. 2005. Managing environmental impacts of recreation and tourism in rainforests of the Wet Tropics of Queensland World Heritage Area. *Geographical Research* **43**:140–151.

Turton, S. M., and H. J. Freiburger. 1997. Edge and aspect effects on the microclimate of a small tropical forest remnant on the Atherton Tableland, northeastern Australia. Pages 45–54 in W. F. Laurance and R. O. Bierregaard Jr., eds., *Tropical forest remnants: Ecology, management, and conservation of fragmented communities.* University of Chicago Press, Chicago, USA.

Turton, S. M., T. Kluck, and T. J. Day. 2000. Ecological impacts of visitors at day-use and camping areas. Pages 123–134 in J. Bentrupperbäumer and J. Reser, eds., *Impacts of visitation and use in the Wet Tropics of Queensland World Heritage Area, Stage 2.* Wet Tropics Management Authority and Rainforest CRC, Cairns, Australia.

Tutin, C. E. G. 2000. Ecology and social organisation of African rainforest primates: Relevance for understanding the transmission of retroviruses. *Bulletin de la Societe de Pathologie Exotique* **93**:157–161.

Tutin, C. E. G. 2001. Saving the gorillas (*Gorilla gorilla*) and chimpanzees (*Pan troglodytes*) of the Congo Basin. *Reproduction Fertility and Development* **13**:469–476.

Tutin, C. E. G., and M. Fernandez. 1984. Nationwide census of gorilla (*Gorilla gorilla*) and chimpanzee (*Pan troglodytes*) populations in Gabon. *American Journal of Primatology* **6**:313–336.

Tutin, C. E. G., L. J. T. White, and A. Mackanga-Missandzou. 1997. The use of rainforest mammals of natural forest fragments in an equatorial African savanna. *Conservation Biology* **11**:1190–1203.

Uhl, C. 1998. Perspectives on wildfire in the humid tropics. *Conservation Biology* 12:942–943.

Uhl, C., and R. Buschbacher. 1985. A disturbing synergism between cattle ranch burning practices and selective tree harvesting in the eastern Amazon. *Biotropica* 17:265–268.

Uhl, C., and J. B. Kauffman. 1990. Deforestation, fire susceptibility, and potential tree responses to fire in the Eastern Amazon. *Ecology* 71:437–449.

Uhl, C., A. Veríssimo, M. M. Mattos, Z. Brandino, and I. C. G. Vieira. 1991. Social, econonomic, and ecological consequences of selective logging in an Amazon frontier: The case of Tailândia. *Forest Ecology and Management* 46:243–273.

Uhl, C., and I. C. G. Vieira. 1989. Ecological impacts of selective logging in the Brazilian Amazon: A case study from the Paragominas Region of the State of Pará. *Biotropica* 21:98–106.

Umina, P. A., A. R. Weeks, M. Kearney, S. McKechnie, and A. Hoffmann. 2005. A rapid shift in a classic clinal pattern in *Drosophila* reflecting climate change. *Nature* 308:691–693.

UNEP-WCMC. 2000. *Forest protection analysis*. World Conservation Monitoring Center, Cambridge, U.K. http://www.unep-wcmc.org/forest/world.htm.

UN-FCCC (United Nations Framework Convention on Climate Change). 1997. *Kyoto Protocol to the United Nations framework convention on climate change*. Document FCCC/CP/1997;7/Add1. www.unfccc.de (English) and http://www.mct.gov. br/clima (Portuguese).

UN-FCCC (United Nations Framework Convention on Climate Change). 2001. *Review of the implementation of commitments and other provisions of the convention, and preparations for the first session of the conference of the Parties serving as the meeting of the Parties to the Kyoto Protocol* (Decision 8/CP.4), Decision 5/CP 6, Implementation of the Buenos Aires Plan of Action, FCC/CP/2001/L.11. UN-FCCC, Bonn, Germany.

Uryu, Y., O. Malm, I. Thornton, I. Payne, and D. Cleary. 2001. Mercury contamination of fish and its implications for other wildlife of the Tapajós Basin, Brazilian Amazon. *Conservation Biology* 15:438–446.

Usher, M. B. 1991. Biological invasions into tropical nature reserves. Pages 21–35 in P. S. Ramakrishnan, ed., *Ecology of biological invasions in the tropics*. International Scientific Publications, New Delhi, India.

Usher, M. B., F. J. Kruger, I. MacDonald, L. Loope, and R. Brockie. 1988. The ecology of biological invasions into nature reserves: An introduction. *Biological Conservation* 44:1–8.

Valentine, P. S., and D. S. Cassells. 1991. Recreation management issues for tropical rain forests. Pages 9–14 in N. Goudberg, ed., *Proceedings of the Institute for Tropical Rainforest Studies*. James Cook University, Townsville, Australia.

Valiela, I., J. L. Bowen, and J. K. York. 2001. Mangrove forests: One of the world's threatened major tropical environments. *BioScience* 51:807–851.

Van Bael, S. A., J. D. Brawn, and S. K. Robinson. 2003. Birds defend trees from herbivores in a neotropical forest canopy. *Proceedings of the National Academy of Sciences USA* 100:8304–8307.

van Beers, C. P., and A. P. G. de Moor. 1999. *Addicted to subsidies: How governments use your money to destroy the Earth and pamper the rich*. Institute for Research on Public Expenditure, The Hague, The Netherlands.

Van Riper, C., S. G. Van Riper, L. M. Goff, and M. Laird. 1986. The epizootiology and ecological significance of malaria in Hawaiian land birds. *Ecological Monographs* 56:327–344.

van Schaik, C., and H. D. Rijksen. 2002. Integrated conservation and development projects: Problems and potential. Pages 15–29 in J. Terborgh, C. van Schaik, L. Davenport, and M. Rao, eds., *Making parks work: Strategies for preserving tropical nature*. Island Press, Washington, D.C., USA.

van Schaik, C. P., J. Terborgh, and B. Dugelby. 1997. The silent crisis: The state of rain forest nature preserves. Pages 64–89 in R. Kramer, C. van Schaik, and J. Johnson, eds., *Last stand: Protected areas and the defense of tropical biodiversity*. Oxford University Press, Oxford, U.K.

Van Straaten, P. 2000. Mercury contamination associated with small-scale gold mining in Tanzania and Zimbabwe. *Science of the Total Environment* 259:105–113.

Van Vliet, O. P. R., A. P. C. Faaij, and C. Dieperink. 2003. Forestry projects under the clean development mechanism? Modelling of the uncertainties in carbon mitigation and related costs of plantation forestry projects. *Climatic Change* 61:123–156.

van Wilgen, B. W., R. M. Cowling, R. M., and C. Burgers. 1996. Valuation of ecosystem services. *BioScience* 46:184–189.

van Wilgen, B. W., D. C. Le Maitre, and R. M. Cowling. 1998. Ecosystem services, efficiency, sustainability and equity: South Africa's working for water program. *Trends in Ecology and Evolution* 13:378.

van Wilgen, B. W., and R. J. Scholes. 1997. The vegetation and fire regimes of southern hemisphere Africa. Pages 27–46 in B. W. van Wilgen, M. O. Andreae, J. G. Goldammer, and J. A Lindesay, eds., *Fire in southern African savannas: Ecological and atmospheric perspectives*. Witswatersrand University Press, Johannesburg, South Africa.

Vasconcelos, H. L. 1999. Levels of leaf herbivory in Amazonian trees from different stages in forest regeneration. *Acta Amazonica* 19:91–99.

Vasconcelos, H. L., and F. Luizão. 2004. Litterfall dynamics and nutrient concentrations in a fragmented Amazonian forest. *Ecological Applications* 14:884–892.

Vasquez, R., and A. H. Gentry. 1989. Use and misuse of forest-harvested fruits in the Iquitos area. *Conservation Biology* 3:350–361.

Vaughan, R. E., and P. O. Wiehe. 1937. Studies on the vegetation of Mauritius I. A preliminary survey of the plant communities. *Journal of Ecology* 25:289–343.

Vaughan, R. E., and P. O. Wiehe. 1941. Studies on the vegetation of Mauritius III. The structure and development of the upland climax forest. *Journal of Ecology* 29: 127–136.

Vázquez-Yanes, C., A. Orozco-Segovia, E. Rincón, M. E. Sánchez-Coronado, P. Huante, J. R. Toledo, and V. L. Barradas. 1990. Light beneath the litter in a tropical forest: Effect on seed germination. *Ecology* 71:1952–1958.

Veloso, H. P., A. L. R. Rangel Filho, and J. C. A. Lima. 1991. *Classificação da vegetação brasileira adaptada a um sistema universal*. IBGE, Rio de Janeiro, Brazil.

Veríssimo, A., P. Barreto, M. Mattos, R. Tarifa, and C. Uhl. 1992. Logging impacts and prospects for sustainable forest management in an old Amazonian frontier: The case of Paragominas. *Forest Ecology and Management* 55:169–199.

Viana, V. M., A. A. J. Tabanez, and J. L. F. Batista. 1997. Dynamics and restoration of forest fragments in the Brazilian Atlantic Moist Forest. Pages 351–365 in W. F.

Laurance and R. O. Bierregaard Jr., eds., *Tropical forest remnants: Ecology, management, and conservation of fragmented communities*. University of Chicago Press, Chicago, USA.

Viisteensaari, J., S. Johansson, V. Kaarakka, and O. Luukkanen. 2000. Is the alien tree species *Maesopsis eminii* Engl. (Rhamnaceae) a threat to tropical forest conservation in the East Usambaras, Tanzania? *Environmental Conservation* 27:76–81.

Villard, M. A., K. Trzcinski, and G. Merriam. 1999. Fragmentation effects on forest birds: Relative influence of woodland cover and configuration on landscape occupancy. *Conservation Biology* 13:774–783.

Visser, M. E., and C. Both. 2005. Shifts in phenology due to global climate change: The need for a yardstick. *Proceedings of the Royal Society of London B* 272: 2561–2569.

Vitousek, P., P. R. Ehrlich, A. H. Ehrlich, and P. Matson. 1986. Human appropriation of the products of photosynthesis. *BioScience* 36:368–373.

Vitousek, P. M. 1990. Biological invasions and ecosystem processes: Towards an integration of population biology and ecosystem studies. *Oikos* 57:7–13.

Vitousek, P. M., J. Aber, R. Howarth, G. Likens, P. Matson, D. Schindler, W. Schlesinger, and D. Tilman. 1994. Human alteration of the global nitrogen cycle: Sources and consequences. *Ecological Applications* 7:737–750.

Vitousek, P. M., H. Mooney, J. Lubchenco, and J. M. Melillo. 1997. Human domination of the Earth's ecosystems. *Science* 277:494–499.

Vitousek, P. M., L. Walker, L. Whiteaker, D. Mueller-Dumbois, and P. Matson. 1987. Biological invasion by *Myrica faya* alters ecosystem development in Hawaii. *Science* 238:802–804.

Walker, B., S. Carpenter, J. Anderies, N. Abel, G. Cumming, M. Janssen, L. Lebel, J. Norberg, G. D. Peterson, and R. Pritchard. 2002. Resilience management in social-ecological systems: A working hypothesis for a participatory approach. *Conservation Ecology* 6:14. http://www.consecol.org/vol6/iss1/art14.

Walker, B. H. 1983. Is succession a viable concept in African savanna ecosystems? Pages 430–447 in D. C. West, H. H. Shugart, and D. B. Botkin, eds., *Forest succession: Concepts and application*. Springer-Verlag, New York, USA.

Walker, B. H., and I. Noy-Meir. 1982. Aspects of the stability and resilience of savanna ecosystems. Pages 556–590 in B. J. Huntley and B. H. Walker, eds., *Ecology of tropical savannas*. Springer-Verlag, Berlin, Germany.

Walker, G. K., Y. Sud, and R. Atlas. 1995. Impact of ongoing Amazonian deforestation on local precipitation: A GCM simulation study. *Bulletin of the American Meteorological Society* 76:346–361.

Wallace, J. M., and D. S. Gutzler. 1981. Teleconnections in the geopotential height field during the northern hemisphere winter, 1981. *Monthly Weather Review* 109:784–812.

Wallin, T. R., and C. P. Harden. 1996. Estimating trail-related soil erosion in the humid tropics: Jatun Sacha, Ecuador, and La Selva, Costa Rica. *Ambio* 25:517–522.

Walpole, M. J., and H. J. Goodwin. 2000. Local economic impacts of dragon tourism in Indonesia. *Annals of Tourism Research* 27:559–576.

Walpole, M. J., and N. Leader-Williams. 2001. Masai Mara tourism reveals partnership benefits. *Nature* 413:771–771.

Walsh, J. F., D. H. Molyneux, and M. H. Birley. 1993. Deforestation: Effects on vector-borne disease. *Parasitology* 106:S55–S75.

Walsh, K. J. E., and B. F. Ryan. 2000. Tropical cyclone intensity increases near Australia as a result of climate change. *Journal of Climatology* 13:3029–3036.

Walsh, P. D., K. A. Abernethy, M. Bermejo, R. Beyers, P. De Wachter, M. E. Akou, B. Huijbregts, D. I. Mambounga, A. K. Toham, A. M. Kilbourn, S. A. Lahm, S. Latour, F. Maisels, C. Mbina, Y. Mihindou, S. N. Obiang, E. N. Effa, M. P. Starkey, P. Telfer, M. Thibault, C. E. G. Tutin, L. J. T. White, and D. S. Wilkie. 2003. Catastrophic ape decline in western equatorial Africa. *Nature* 422:611–614.

Walter, H. 1971. *Ecology of tropical and subtropical vegetation*. Oliver and Boyd, Edinburgh, Scotland.

Waltert, M., J. Lien, K. Faber, M. Muhlenberg. 2002. Further declines of threatened primates in the Korup Project Area, south-west Cameroon. *Oryx* 36:257–265.

Walther, G., E. Post, P. Convey, A. Menzel, C. Parmesan, T. J. C. Beebee, J.-M. Fromentin, O. Hoegh-Guldberg, and F. Bairlein. 2002. Ecological responses to recent climate change. *Nature* 416:389–395.

Warfield, K. L., C. M. Bosio, B. C. Welcher, E. M. Deal, M. Mohamadzadeh, A. Schmaljohn, M. J Aman, and S. Bavari. 2003. Ebola virus-like particles protect from lethal Ebola virus infection. *Proceedings of the National Academy of Sciences USA* 100:15889–15894.

Wassen, R. J., and M. Claussen. 2002. Earth system models: A test using the mid-Holocene in the southern hemisphere. *Quaternary Science Reviews* 21:819–824.

Waterman, P. G. 1983. Distribution of secondary metabolites in rain forest plants: Toward an understanding of cause and effect. Pages 167–179 in S. L. Sutton, T. C. Whitmore, and A. C. Chadwick, eds., *Tropical rain forest: Ecology and management*. Blackwell Scientific, Oxford, U.K.

WCD (World Commission on Dams). 2000. *Dams and development: A new framework for decision-making*. Earthscan, London, U.K.

WCED. 1987. *Our common future*. Report on the World Commission on Environment and Development. Oxford University Press, Oxford.

Weaver, C. P., and R. Avissar. 2001. Atmospheric disturbances caused by human modification of the landscape. *Bulletin of the American Meteorological Society* 82:269–281.

Weaver, C. P., and R. Avissar. 2002. Reply to "Comments on 'Atmospheric disturbances caused by human modification of the landscape.'" *Bulletin of the American Meteorological Society* 83:280–283.

Webb, C. O., and D. R. Peart. 2001. High seed dispersal rates in faunally intact tropical rain forest: Theoretical and conservation implications. *Ecology Letters* 4:491–499.

Webb, L. J. 1959. A physiognomic classification of Australian rain forests. *Journal of Ecology* 47:551–570.

Webb, L. J., and J. G. Tracey. 1981. The rainforests of northern Australia. Pages 67–101 in R. H. Groves, ed., *Australian vegetation*. Cambridge University Press, Cambridge, U.K.

Webb, N. R. 1989. Studies on the invertebrate fauna of fragmented heathland in Dorset, U. K., and the implications for conservation. *Biological Conservation* 47:153–65.

Weber, D. 2001. Ministério contesta estudo sobre devastação. *O Estado de S. Paulo* Newspaper, São Paulo, Brazil, 21 January.

Webster, P. J., G. J. Holland, J. A. Curry, and H.-R. Chang. 2005. Changes in tropical cyclone number, duration, and intensity in a warming environment. *Science* 309:1844–1846.

Wells, D. R. 1984. The forest avifauna of western Malesia and its conservation. Pages 213–222 in J. W. Diamond and T. E. Lovejoy, eds., *Conservation of tropical birds*. ICBP, Cambridge, U.K.

Wells, M. 1992. Biodiversity conservation, affluence and poverty: Mismatched costs and benefits and efforts to remedy them. *Ambio* 21:237–243.

Wells, M. 1997. *Economic perspectives on nature tourism, conservation and development*. World Bank, Washington, D.C., USA.

Wells, M., and K. Brandon. 1992. *People and parks: Linking protected area management with local communities*. World Bank, WWF-US, and US-AID, Washington, D.C., USA.

Wells, M., S. Guggenheim, A. Khan, W. Wardojo, and P. Jepson. 1999. *Investing in biodiversity: A review of Indonesia's integrated conservation and development projects*. World Bank, Washington, D.C., USA.

Wenny, D. G. 2000. Seed dispersal, seed predation, and seedling recruitment of a neotropical montane tree. *Ecological Monographs* 70:331–351.

Wenny, D. G. 2001. Advantages of seed dispersal: A re-evaluation of directed dispersal. *Evolutionary Ecology* 3:51–74.

Werren, G. L., S. Goosem, J. G. Tracey, and J. P. Stanton. 1995. The Australian Wet Tropics centre of plant diversity. Pages 500–506 in S. D. Davies, V. H. Heywood, and A. C. Hamilton, eds., *World's centre of plant diversity*, vol. 2. Oxford University Press, Oxford, U.K.

Werth, D., and R. Avissar. 2002. The local and global effects of Amazon deforestation. *Journal of Geophysical Research* 107:8087, doi:10.1029/2001JD000717.

West, G. B., J. H. Brown, and B. J. Enquist. 1999. A general model for the structure and allometry of vascular plant systems. *Nature* 400:664–667.

Weste, G. 2003. The dieback cycle in Victorian forests: A 30-year study of changes caused by *Phytophtora cinnamoni* in Victorian open forests, woodlands and heathlands. *Australian Plant Pathology* 32:247–256.

Weste, G., K. Brown, J. Kennedy, and T. Walshe. 2002. *Phytophtora cinnamoni* infestation: A 24-year study of vegetation change in forests and woodlands of the Grampians, Western Victoria. *Australian Journal of Botany* 50:247–274.

White, A., X. Sun, K. Canby, J. Xu, C. Barr, E. Katsigris, G. Bull, C. Cossalter, and S. Nillson. 2006. *China and the global market for forest products: Trends, implications, and steps to transform the trade to benefit forests and livelihoods*. Forest Trends Center for International Forestry Research, and Chinese Center for Agricultural Policy, Washington, D.C., USA.

White, A. T., H. P. Vogt, and T. Arin. 2000. Philippine coral reefs under threat: The economic losses caused by reef destruction. *Marine Pollution Bulletin* 40: 598–605.

White, L., and A. Edwards. 2000. Methods for assessing the status of animal populations. Pages 225–276 in L. White and A. Edwards, eds., *Conservation research in the African rain forests: A technical handbook*. Wildlife Conservation Society, New York, USA.

White, L. J. T. 1992. Vegetation history and logging disturbance: Effects on rain forest mammals in the Lope Reserve, Gabon (with special emphasis on elephants and apes). Ph.D. thesis, University of Edinburgh, Edinburgh, Scotland.

White, L. J. T., and C. E. G. Tutin. 2001. Why chimpanzees and gorillas respond differently to logging: A cautionary tale from Gabon. Pages 449–462 in W. Weber,

L. J. T. White, A. Vedder, and L. Naughton-Treves, eds., *African rain forest ecology and conservation*. Yale University Press, New Haven, Connecticut, USA.

White, P. J., R. A. Norman, and P. J. Hudson. 2002. Epidemiological consequences of a pathogen having both virulent and avirulent modes of transmission: The case of rabbit haemorrhagic disease virus. *Epidemiology and Infection* **129**:665–677.

Whitmore, T. C. 1975. *Tropical rain forests of the Far East*. Clarendon Press, Oxford, U.K.

Whitmore, T. C. 1991. Invasive woody plants in perhumid tropical climates. Pages 35–41 in P. S. Ramakrishnan, ed., *Ecology of biological invasions in the tropics*. International Scientific Publications, New Delhi, India.

Whitmore, T. C. 1997. Tropical forest disturbance, disappearance, and species loss. Pages 3–12 in W. F. Laurance and R. O. Bierregaard Jr., eds., *Tropical forest remnants: Ecology, management, and conservation of fragmented communities*. University of Chicago Press, Chicago, USA.

Wichmann, M. C., J. Groeneveld, F. Jeltsch, and V. Grimm. 2005. Mitigation of climate change impacts on raptors by behavioral adaptation: Ecological buffering mechanisms. *Global and Planetary Change* **47**:273–281.

Wielicki, B. A., T. Wong, R. P. Allan, A. Slingo, J. T. Kiehl, B. J. Soden, C. T. Gordon, A. J. Miller, S. K. Yang, D. A. Randall, F. Robertson, J. Susskind, and H. Jacobowitz. 2002. Evidence for large decadal variability in tropical mean radiative energy budget. *Science* **295**:841–844.

Wikramanayake, E., E. Dinerstein, C. J. Loucks, D. M. Olson, J. Morrison, J. Lamoreux, M. McKnight, and P. Hedao. 2002. *Terrestrial ecoregions of the Indo-Pacific: A conservation assessment*. Island Press, Washington, D.C., USA.

Wilcove, D. S. 1985. Nest predation in forest tracts and the decline of migratory songbirds. *Ecology* **66**:1211–1214.

Wilcove, D. S., D. Rothstein, J. Dubow, A. Phillips, and E. Losos. 1998. Quantifying threats to imperiled species in the United States. *BioScience* **48**:607–615.

Wilkie, D., E. Shaw, F. Rotberg, G. Morelli, and P. Auzel. 2000. Roads, development, and conservation in the Congo Basin. *Conservation Biology* **14**:1614–1622.

Wilkie, D. S., and J. F. Carpenter. 1999. Bushmeat hunting in the Congo Basin: An assessment of impacts and options for mitigation. *Biodiversity and Conservation* **8**:927–955.

Wilkie, D. S., J. G. Sidle, and G. C. Boundzanga. 1992. Mechanized logging, market hunting, and a bank loan in Congo. *Conservation Biology* **6**:570–580.

Wilkie, D. S., J. G. Sidle, G. C. Boundzanga, P. Auzel, and S. Blake. 2001. Defaunation, not deforestation: Commercial logging and market hunting in northern Congo. Pages 375–400 in R. A. Fimbel, A. Grajal, and J. G. Robinson, eds., *The cutting edge: Conserving wildlife in logged tropical forests*. Columbia University Press, New York, USA.

Williams, S. E. 1997. Patterns of mammalian species richness in the Australian tropical rainforests: Are extinctions during historical contractions of the rainforest the primary determinant of current patterns in biodiversity? *Wildlife Research* **24**:513–530.

Williams, S. E. 2003. Impacts of global climate change on the rainforest vertebrates of the Australian Wet Tropics. Pages 50–52 in M. Howden, L. Hughes, M. Dunlop, I. Zethoven, D. Hilbert, and C. Chilcott, eds., *Climate change impacts on biodiversity in Australia*. Commonwealth of Australia, Canberra, Australia.

Williams, S. E., E. E. Bolitho, and S. Fox. 2003. Climate change in Australian tropical rainforests: An impending environmental catastrophe. *Proceedings of the Royal Society of London B* 270:1887–1892.

Williams, S. E., and J.-M. Hero. 1998. Rainforest frogs of the Australian Wet Tropics: Guild classification and the ecological similarity of declining species. *Proceedings of the Royal Society of London B* 265:597–602.

Williams, S. E., and J.-M. Hero. 2001. Multiple determinants of Australian tropical frog biodiversity *Biological Conservation* 68:1–10.

Williams, S. E., and R. G. Pearson. 1997. Historical rainforest contractions, localised extinctions and patterns of vertebrate endemism in the rainforests of Australia's Wet Tropics. *Proceedings of the Royal Society of London B* 264:709–716.

Williams, S. E., R. G. Pearson, and P. J. Walsh. 1996. Distributions and biodiversity of the terrestrial vertebrates of Australia's Wet Tropics: A review of current knowledge. *Pacific Conservation Biology* 4:327–362.

Williams-Linera, G. 1990. Vegetation structure and environmental conditions of forest edges in Panama. *Journal of Ecology* 78:356–373.

Williams-Linera, G., V. Domínguez-Gastelú, and M. E. García-Zurita. 1998. Microenvironment and floristics of different edges in a fragmented tropical rainforest. *Conservation Biology* 12:1091–1102.

Williamson, M. 1996. *Biological invasions.* Chapman and Hall, London, U.K.

Willis, E. O. 1974. Populations and local extinctions of birds on Barro Colorado Island, Panama. *Ecological Monographs* 44:153–169.

Willis, E. O. 1979. The composition of avian communities in remanescent woodlots in southern Brazil. *Papéis Avulsos de Zoologia* 33:1–25.

Wills, C. 1996. *Black goddess, yellow fever: The coevolution of people and plagues.* Addison-Wesley Publishing, Reading, Massachusetts.

Willson, M. F., and A. Traveset. 2000. The ecology of seed dispersion. Pages 85–110 in M. Fenner, ed., *Seeds: The ecology of regeneration in plant communities.* CAB International, Wallingford, U.K.

Wilson, E. O. 1992. *The diversity of life.* Harvard University Press, Cambridge, Massachusetts, USA.

Wilson, E. O., and F. Peter, eds. 1988. *Biodiversity.* National Academy Press, Washington, D.C., USA.

Wilson, M. F. 1989. Vertebrate dispersal syndromes in some Australian and New Zealand plant communities, with geographic comparisons. *Biotropica* 21:133–147.

Wilson, R. 2000. The impact of anthropogenic disturbance on four species of arboreal, folivorous possums in the rainforests of north-east Queensland, Australia. Ph.D. thesis, James Cook University, Cairns, Australia.

Winter, J. W. 1988. Ecological specialization of mammals in Australian tropical and sub-tropical rainforest: Refugial and ecological determinism. Pages 127–138 in R. Kitching, ed., *The ecology of Australia's wet tropics.* Surrey Beatty, Sydney, Australia.

Winter, J. W. 1997. Responses of non-volant mammals to late Quaternary climatic changes in the wet tropics region of north-eastern Australia. *Wildlife Research* 24:493–511.

With, A. K., and T. O. Crist. 1995. Critical thresholds in species' responses to landscape structure. *Ecology* 76:2446–2459.

Witmer, M. C., and A. S. Cheke. 1991. The dodo and the Tambalacoque tree: An obligate mutualism reconsidered. *Oikos* 6:133–137.

Wong, M. 1986. Trophic organization of understory birds in a Malaysian dipterocarp forest. *Auk* **103**:100–116.

Woodroffe, R., and J. R. Ginsberg. 1998. Edge effects and the extinction of populations inside protected areas. *Science* **280**:2126–2128.

Woods, P. 1989. Effects of logging, drought, and fire on structure and composition of tropical forests in Sabah, Malaysia. *Biotropica* **21**:290–298.

Woods, W. I., and B. Glaser, convenors. 2001. *Terra preta symposium*. Benicassim, Spain, 13 June.

Woodward, F. I., T. M. Smith, and W. R. Emanuel. 1995. A global land primary productivity and phytogeography model. *Global Biogeochemical Cycles* 9:471–490.

World Bank. 1992. *Rain Forest Trust Fund resolution, background note: Part I, Introduction and objectives*. World Bank, Washington, D.C., USA. www.worldbank.org.

World Bank. 2000. *Supporting the web of life: The World Bank and biodiversity: A portfolio update (1988–1999)*. World Bank, Washington, D.C., USA.

World Bank. 2001. *Indonesia: Environment and natural resource management in a time of transition*. World Bank, Washington, D.C., USA.

World Bank. 2002a. *Biodiversity conservation in forest ecosystems: World Bank assistance 1992–2002*. World Bank, Washington, D.C., USA.

World Bank. 2002b. *Globalization, growth, and poverty: Building an inclusive world economy*. World Bank, Washington, D.C., USA.

World Bank. 2002c. *World development indicators 2002*. World Bank, Washington, D.C., USA.

World Bank. 2003. *Kerinci-Seblat integrated conservation and development project*. World Bank, Washington, D.C., USA.

World Bank. 2004. *Ensuring the future: The World Bank and biodiversity*. World Bank, Washington, D.C., USA.

World Bank. 2005a. *East Asia regional forestry strategy*. East Asia and Pacific Region, Washington, D.C., USA.

World Bank. 2005b. *Going, going, gone: The illegal trade in wildlife in East and Southeast Asia*. World Bank, Washington, D.C., USA.

World Bank. 2006. *Mountains to coral reefs: The World Bank and biodiversity, 1988–2005*. World Bank, Washington, D.C., USA.

World Conservation Monitoring Centre. 1992. *Global biodiversity: Status of Earth's living resources*. Chapman and Hall, London, U.K.

World Travel and Tourism Council. 2002. *Annual report 2002/2003*. http://www.wttc.org/publications/anRpt2002.htm.

Wright, S. J. 2003. The myriad consequences of hunting for vertebrates and plants in tropical forests. *Perspectives in Plant Ecology, Evolution and Systematics* 6: 73–86.

Wright, S. J. 2005. Tropical forests in a changing environment. *Trends in Ecology and Evolution* **20**:553–560.

Wright, S. J., C. Carrasco, O. Calderón, and S. Paton. 1999. The El Niño Southern Oscillation, variable fruit production, and famine in a tropical forest. *Ecology* **80**:1632–1647.

Wright, S. J., and H. C. Duber. 2001. Poachers and forest fragmentation alter seed dispersal, seed survival, and seedling recruitment in the palm *Attalea butyraceae*, with implications for tropical tree diversity. *Biotropica* 33:583–595.

Wright, S. J., and H. Muller-Landau. 2006. The future of tropical forest species. *Biotropica* 38:287–301.

Wright, S. J., H. Zeballos, I. Dominguez, M. M. Gallardo, M. C. Moreno, and R. Ibanez. 2000. Poachers alter mammal abundance, seed dispersal, and seed predation in a neotropical forest. *Conservation Biology* 14:227–239.

WTMA. 1999. *Wet Tropics Management Authority annual report 1998–99*. Wet Tropics Management Authority, Cairns, Australia.

WTMA. 2002. *Wet Tropics Management Authority annual report 2001–02*. Wet Tropics Management Authority, Cairns, Australia.

Wu, Z.-X., and R. E. Newell. 1998. Influence of sea surface temperature on air temperature in the tropics. *Climate Dynamics* 14:275–290.

Wuethrich, B. 2000. Conservation biology: Combined insults spell trouble for rainforests. *Science* 289:35–37.

WWF Climate Change Campaign. 2000. *Make-or-break the Kyoto Protocol*. World Wildlife Fund–US, Washington, D.C., USA. http://www.panda.org/climate.

WWF-Brazil. 2003. http://www.wwf.org.br/amazonia/.

Wyatt, J. L., and M. R. Silman. 2004. Distance-dependance in two Amazonian palms: Effects of spatial and temporal variation in seed predator communities. *Oecologia* 140:26–35.

Wyse-Jackson, P. S., Q. Cronk, and J. A. N. Parnell. 1988. Notes on the regeneration of two rare Mauritian endemic trees. *Journal of Tropical Ecology* 29:98–106.

Xia, Q., J. Zhao, and K. D. Collerson. 2001. Mid-Holocene climatic variations in Tasmania, Australia: Multi-proxy records in a stalagmite from Lynds cave. *Earth and Planetary Science Letters* 194:177–187.

Xue, Y., and J. Shukla. 1993. The influence of land surface properties on Sahel climate: Part I. Desertification. *Journal of Climate* 6:2232–2245.

Yaron, G. 2001. Forest plantations or small-scale agriculture? An economic analysis of alternative land use options in the Mount Cameroon area. *Journal of Environmental Planning and Management* 44:85–108.

Young, T. P., N. Patridge, and A. Macrae. 1995. Long-term glades in *Acacia* bushland and their edge effects in Laikipia, Kenya. *Ecological Applications* 5:97–108.

Zhang, Q. F., C. O. Justice, and P. V. Desanker. 2002. Impacts of simulated shifting cultivation on deforestation and the carbon stocks of the forests of central Africa. *Agriculture Ecosystems and Environment* 90:203–209.

Zhang, S. 1995. Activity and ranging patterns in relation to fruit availability by brown capuchins (*Cebus apella*) in French Guiana. *International Journal of Primatology* 16: 489–507.

Zhang, S. Y., and L. Wang. 1995. Comparison of three fruit census methods in French Guiana. *Journal of Tropical Ecology* 11:281–294.

Roni Avissar
Department of Civil and Environmental
    Engineering
Duke University
Durham, NC 27708
USA
avissar@duke.edu

Cláudia Baider
The Mauritius Herbarium
Mauritius Sugar Industry Research
    Institute
Réduit
Mauritius
cbaider@msiri.intnet.mu

Timothy R. Baker
Earth and Biosphere Institute
Geography
University of Leeds
Leeds LS2 9JT
UK
timb@geog.leeds.ac.uk

Andrew Balmford
Conservation Biology Group
Department of Zoology
University of Cambridge

Downing Street
Cambridge CB2 3EJ
UK
a.balmford@zoo.cam.ac.uk

Jos Barlow
Centre for Ecology, Evolution, and
    Conservation
School of Environmental Sciences
University of East Anglia
Norwich NR4 7TJ
UK
j.barlow@uea.ac.uk

Julieta Benítez-Malvido
Center for Ecosystem Research
National Autonomous University of
    Mexico (UNAM)
Antigua Carretera a Pátzcuaro No. 8701
Ex-Hacienda de San José de la Huerta
Morelia Michoacán
México C.P. 58190
jbenitez@oikos.unam.mx

K. K. Caylor
Department of Civil and Environmental
    Engineering
Princeton University

Princeton, NJ 08540
USA
kcaylor@princeton.edu

Colin A. Chapman
Department of Zoology
University of Florida
Gainesville, FL 32611
USA
chapman@zoo.ufl.edu

Andrew A. Cunningham
Institute of Zoology
Zoological Society of London
Regent's Park
London NW1 4RY
UK
a.cunningham@ioz.ac.uk

Peter Daszak
Consortium for Conservation
    Medicine
Wildlife Trust
61 Route 9W
Palisades, NY 10952
USA
daszak@conservationmedicine.org

Luis D'Croz
Smithsonian Tropical Research Institute
Apartado 2072
Balboa
Republic of Panamá
dcrozl@si.edu

Pedro Ferreira Develey
Departamento de Ecologia
Instituto de Biociências
Universidade de São Paulo
Rua do Matão, 321, Travessa 14
CEP 05508-900
São Paulo, SP
Brazil
pdeveley@uol.com.br

P. R. Dowty
Department of Environmental
    Sciences
University of Virginia
Charlottesville, VA 22903
USA

Paul W. Elkan
Wildlife Conservation
    Society
B.P. 14537
Brazzaville
Republic of Congo
pelkan@wcs.org

Sarah W. Elkan
Wildlife Conservation
    Society
B.P. 14537
Brazzaville
Republic of Congo
selkan@wcs.org

Chris Fagan
ParksWatch
Center for Tropical Conservation
Nicholas School of the Environment and
    Earth Sciences
Duke University
3705-C Erwin Road
Durham, NC 27705
USA
chrisfagan33@yahoo.com

Philip M. Fearnside
National Institute for Amazonian
    Research (INPA)
C.P. 478
Manaus AM 69011
Brazil
pmfearn@inpa.gov.br

F. B. Vincent Florens
Department of Biosciences
Faculty of Science
University of Mauritius
Réduit
Mauritius
v.florens@uom.ac.mu

C. Hély
CEREGE
Europole de l'Arbois
B.P. 80
13545 Aix en Provence cedex 4
France
hely@cerege.fr

David W. Hilbert
CSIRO Tropical Forest Research Centre
Atherton, Queensland 4883
Australia
david.hilbert@csiro.au

Patrick A. Jansen
Community and Conservation Ecology
    Group
University of Groningen
P.O. Box 14
9750 AA Haren
The Netherlands
p.a.jansen@biol.rug.nl

Susan G. W. Laurance
Smithsonian Tropical Research Institute
Apartado 2072
Balboa
Republic of Panamá
laurances@si.edu

William F. Laurance
Smithsonian Tropical Research Institute
Apartado 2072
Balboa
Republic of Panamá
laurancew@si.edu

Aurora Lemus-Albor
Center for Ecosystem Research
National Autonomous University of
    Mexico (UNAM)
Antigua Carretera a Pátzcuaro No. 8701
Ex-Hacienda de San José de la Huerta
Morelia Michoacán
México C.P. 58190
alemus@oikos.unam.mx

Simon L. Lewis
Earth and Biosphere Institute
Geography
University of Leeds
Leeds LS2 9JT
UK
s.lewis@pobox.com

Thomas E. Lovejoy
H. John Heinz III Center for Science,
    Economics, and the Environment

1001 Pennsylvania Ave. NW
Suite 735 South
Washington, DC 20004
USA
lovejoy@heinzctr.org

Kathy MacKinnon
Environment Department
The World Bank
1818 H Street
Washington, DC 20433
USA
kmackinnon@worldbank.org

Richard Malonga
Wildlife Conservation Society
B.P. 14537
Brazzaville
Republic of Congo
wcscongo@yahoo.fr

Jean Paul Metzger
Departamento de Ecologia
Instituto de Biociências
Universidade de São Paulo
Rua do Matão, 321, Travessa 14
CEP 05508-900
São Paulo, SP
Brazil
jpm@ib.usp.br

Fernanda Michalski
Centre for Ecology, Evolution, and
    Conservation
School of Environmental Sciences
University of East Anglia
Norwich NR4 7TJ
UK
f.michalski@uea.ac.uk

Antoine Moukassa
Wildlife Conservation Society
B.P. 14537
Brazzaville
Republic of Congo
amoukassa@wcs.org

Marcel Ngangoue
Wildlife Conservation Society
B.P. 14537

Brazzaville
Republic of Congo
wcscongo@yahoo.fr

Eduard T. Niesten
Conservation Economics Program
Conservation International
1919 M Street NW
Suite 600
Washington, DC 20036
USA
e.niesten@conservation.org

Gabriela Nunez-Iturri
Department of Biological Sciences
M/C 066
University of Illinois at Chicago
Chicago, IL 60607
USA
gnunez3@uic.edu

William Olupot
Institute of Tropical Forest Conservation
P.O. Box 44
Kabale
Uganda
wolupot@yahoo.com

Nikkita Patel
Consortium for Conservation Medicine
Wildlife Trust
61 Route 9W
Palisades, NY 10952
USA
patel@conservationmedicine.org

Carlos A. Peres
Centre for Ecology, Evolution, and
    Conservation
School of Environmental Sciences
University of East Anglia
Norwich NR4 7TJ
UK
c.peres@uea.ac.uk

Oliver L. Phillips
Earth and Biosphere Institute
Geography
University of Leeds
Leeds LS2 9JT
UK

oliverp@geography.leeds.ac.uk

Renato Ramos da Silva
Department of Civil and Environmental
    Engineering
Duke University
Durham, NC 27708
USA
renato@duke.edu

Richard E. Rice
Conservation Economics Program
Conservation International
1919 M Street NW
Suite 600
Washington, DC 20036
USA
d.rice@conservation.org

H. H. Shugart
Department of Environmental Sciences
University of Virginia
Charlottesville, VA 22903
USA
hhs@virginia.edu

James L. D. Smith
Department of Fisheries and Wildlife and
    Conservation Biology
University of Minnesota
St. Paul, MN 55414
USA
jlds@fw.umn.edu

Nigel E. Stork
Cooperative Research Centre for Tropical
    Rainforest Ecology and Management
James Cook University
Cairns, Queensland 4870
Australia
nigel.stork@jcu.edu.au

R. J. Swap
Department of Environmental Sciences
University of Virginia
Charlottesville, VA 22903
USA
swapper@virginia.edu

John Terborgh
Center for Tropical Conservation

Nicholas School of the Environment and
    Earth Science
Duke University
Box 90381
Durham, NC 27709
USA
manu@acpub.duke.edu

Stephen M. Turton
Cooperative Research Centre for Tropical
    Rainforest Ecology and
    Management
James Cook University
Cairns, Queensland 4870
Australia
steve.turton@jcu.edu.au
Peter D. Walsh
Department of Primatology
Max Planck Institute for Evolutionary
    Anthropology
Deutscher Platz 6
Leipzig 04103, Germany
walsh@eva.mpg.de

David Werth
Department of Civil and Environmental
    Engineering
Duke University
Durham, NC 27708
USA
werth@duke.edu

Tony Whitten
The World Bank
1818 H Street NW
Washington, DC 20043
USA
twhitten@worldbank.org

Stephen E. Williams
Cooperative Research Centre for Tropical
    Rainforest Ecology
School of Tropical Biology
James Cook University
Townsville, Queensland 4811
Australia
stephen.williams@jcu.edu.au

apes: conservation of populations, 190; controlling hunting of, 192; rapid population declines, 194; translocation of, 195
aquatic ecosystems, 452
aquatic fauna, affected by gold mining, 449
Arabian Peninsula, 75
arboreal folivores, 47
arc of deforestation, 231; in the Brazilian Amazon, 231
area-sensitive species, 123
armadillos, 114
Atelopus, 156
Atlantic forest, avian communities in, 220
atmospheric pressure, 67, 94
atmospheric soot, 447
Attalea butyracea, 251
Australia, 8; birds, 43; climate change, 35; endemic fauna, 36; frogs, 43; mammals, 43; Mesophyll vine forests, 39; rainforests, 37; reptiles, 43; warming, 37; Wet Tropics, endemic species, 36
Avança Brasil, infrastructure development, 254, 264, 461
AVHRR, 438
avian poxvirus, 449
avifauna, 238. *See also* birds
avoided deforestation, 355, 368; responses to fire disturbance, 238

Barisan Selatan Park (Sumatra), 297, 298
bark, thick and thin, 97
barrier effects, 254, 259
basal area, of fruiting trees, 235
Batrachochytrium dendrobatidis, 159
bearded saki monkeys (*Chiropotes albinasus*), 237
beehives, 84
Belo Monte Dam (Brazil), 366–67
bioclimatic modeling, 8, 33, 40; BIOCLIM model, 42
biodiversity: conservation, 345–47; failure of indirect approaches, 345–47; hotspots, 418; impacts of global climate change, 34; loss, 291; root causes of loss of, 298; in Southeast Asia, 291; tropical biodiversity and protection of, 418–19
biogeochemical cycles, 1–2
Biological Dynamics of Forest Fragment Project, 91
biological invasion, 451; homogenization of the world's biota, 199. *See also* alien species

biomass, 17; aboveground (AGB), 20, 21; growth and loss, 23. *See also* forest
birds, 43, 111, 125, 270, 278; army antfollowers, 257; Atlantic forests fragmentation, 220; community composition, 279; edge/gap specialists, 257, 259; frugivores, 257, 259; habitat loss, BIOCLIM models, 43–44; insectivores, 257; large, as seed dispersers, 244; understory insectivores, 259
Black River Gorges National Park (Mauritius), 202
Boa Vista (Brazil), 96
body size, 119
bongo, population conservation, 413
Borneo (Indonesia), 221, 291
Bowen ratio, 93
Brazil: Amazonia, 254, 360; Atlantic forest, 107, 271; avifauna of, 271; climate-change mitigation options, 359–74; demographic collapse, 427; "Forest Code," 372; global biodiversity hotspot, 271; National Space Agency (INPE), 88, 112, 373; roads in, 254; settlement schemes, 260
Brazil nut trees (*Bertholletia excelsa*), 427, 444
Brownsberg Nature Reserve (Suriname), 448
bruchid beetles, as seed predators, 251
Brundtland Report, 377
burning: effects on birds, 238–39; severity, 115, 120, 229, 239; and tree mortality, 96, 135; and tree survival, 229
bushmeat harvest: alternatives to, 404–6; crisis in, 312, 394; hunting, 393; sustainable yield of meat, 191; in timber concessions, 393–414; trade, controlling hunting, 192
Bwindi Impenetrable National Park (Uganda), 84, 127, 128

Calvaria major, 200. *See also* Tambalacoque tree (*Sideroxylon grandiflorum*)
Canarium paniculatum (Burseraceae), 211
cane toads, 147
canker diseases, 169
canopy edges, 255
Cape Floristic Region (South Africa), 422; and road clearings, 255
capuchin monkey, 124
carbon, 368; in the Amazon, 310; avoiding deforestation, 310; committed emissions, 368; estimates from deforesta-

tion, 368; fluxes, 30; land-use options, 355–56; -offset, funds, 308; projects, 314; storage, 310

carbon credits, 340, 354; additionality, 370; through avoided deforestation, 358; and the Kyoto Protocol, 340; paying for tropical conservation, 331; permanence of stocks, 368–70; purchasing, 354; schemes, 331

carbon dioxide, 17; atmospheric, 17, 28; concentrations, 16; high levels, 47

carbon sinks, 9, 16, 17, 31, 337

Caribbean sea urchins, 144

carnivores: large, 88, 124, 423; local extinctions of, 115–19

cascading effects, 172, 421; of local extinctions, 421; phytophagous insects, 172

cassiavera (*Cinnamomum burmanni*), 296

catastrophic wildfires, 83, 85

cattle: herding, 59; pastures, 83

Central Africa, 75, 393; commercial hunting, 393

Central Amazonia (Brazil), 255

Cephalophus spp., 406

certification, wildlife management in logging operations, 409

Chajul (Mexico), 170

charcoal, as soil carbon storage, 365

chestnut blight, 151

chimpanzee populations: conservation, 190; Ebola, 190; *Pan troglodytes*, 148, 155, 175, 394; recovery rates, 191; sustainable yield of meat, 191. *See also* great apes

China, 8, 300; demand for forest resources, 300

chlorofluorocarbons, 7

chytrid fungus, 459

Chytridiomycosis, 159. *See also* infectious diseases

Clean Development Mechanism (CDM), 315, 339, 357; of the Kyoto Protocol, 357

Clidemia hirta (Melastomataceae), 451

climate change, 8, 10–11, 33, 34, 97, 187, 353; adaptation strategies, 355–56; in the Amazon, 353; Australia, 35; avoiding deforestation, 368–75; and disease outbreaks, 184; and the emergence of Ebola epizootics, 187; emerging threats from, 7–13; and forest dynamics, 446; and forest productivity, 9, 15–31; frequency of storms, 447; frugivores and

fruit abundance, 447; Hadley Center's model, 371; impact on biodiversity, 34; international context of, 356–59; and liana biomass, 446; mitigation of, 353–75; mitigation options for Amazonia, 359–74; montane species, 446; negotiations, 358; phenology, 49; politics of mitigating, 357–59; potential and predicted distributions, 38; savannas, 53; scenarios, 63; species committed to extinction, 45; stress, 39; unpredictability, 47; vegetation, 63

cloud formation and cover, 94

cloud layer, 48

coastal zones, productivity of, 453

Cocha Cashu Biological Station (Peru), 244

commercial hunting, 175; and habitat loss, 181; law enforcement, 193; in logging operations, 393; logging roads, 179; prey selectivity, 179; protected areas, 192

community-based conservation projects, 141; local communities and role in wildlife management, 412

Congo Basin, 222, 393

connectivity, 105

conservation: as an alternative to exploitation, 337–52; the concept of, 345–47; concessions, 347–48; expenditure, 321–23, 329; funding, 329; future prospects and challenges, 348–51; incentive agreements, 309–10, 337, 345, 350; NGOs, annual budgets of, 351; science, in planning, 421; targets, 312–13; trust funds, 334

Conservation International (CI), 309, 347

continuous forests, species richness in, 278

convective heat loss, 73

coral reefs, 254

core-area model, 100

core forest habitats, 285; remaining under different climate scenarios, 40

core fragment size, 115

core-protection areas, 434; in tropical forests, 417–34

corridors, 196, 384

corruption, 216, 300, 335, 409; in Indonesia, 300; and law-enforcement, 409

Costa Rica, 108, 156

cracids, 114, 125

cryptic degradation, 441; disturbances, 216; surface fires, 217

Culex quinquefasciatus, 155

roads (*continued*)
in temperate vs. tropical areas, 253; terrestrial mammals, 260; traffic, 257; understory rainforest birds, 257, 259; in the United States, 254; wind turbulence, 255
roadside habitat, 386
rodents, 114, 406; as game species in Central Africa, 406
Rondônia (Brazil), 70, 71, 94, 264
roots, shallow, 67
Roraima (Brazil), 226
Royal Chitwan National Park (Nepal), 325
rural people, 105
rust fungus, 166

Sahel, 63
São Paulo (Brazil), 271, 220
saplings: mortality, 84, 92; recruitment of, 244
SARs. *See* species-area relationships (SARs)
savannas: dynamics of, 55; ecosystems, 53, 55; productivity, 60
seasonality, of rainfall, 48
seasonally dry forests, 225
sea-surface temperature, 69
sea turtles, 455–56
secondary (or disturbed) forests, 283
seed dispersal, 218; in the absence of frugivores, 241–52; birds, 283; distances, in common and rare tree species, 235–49; by dodos, 146; gut-dispersed tree species, 251; interactions with alien invaders, 207; by large primates, 443; lost interactions, 146; low germination rate, 206; modes, 245; plant recruitment zone, 243; pulp removal by vertebrate dispersers, 206; reproductive collapse, 146; by secondary dispersers, 241
seed germination, 205–6, 218
seedling density-dependent mortality, 442
seedling herbivory, by deer and feral pigs, 207
seedling recruitment, 218; bottleneck, 203
seed predation, 248–52; by bruchid beetles, 251; interactions with alien invaders, 207; by monkeys, 208
seed scarification, 205
seed shadows, 246
seed treatment, 205
selective logging, 88, 105, 439; conventional vs. sustainable forest management, 341;

detection of canopy damage, 439; Southeast Asia, 292
semi-arid regions, 10–11, 53
semi-arid shrubland, 63
siamangs (*Symphalangus syndactylus*), 237
silverbacks, 191
simultaneous environmental changes, 86
Sinop, Mato Grosso (Brazil), 109
size-class distribution, 204
Skeptical Environmentalist, 2
skinks, 44
slash-and-burn agriculture, 105; farmers, 88; farming, 95
slow loris (*Nycticebus coucang*), 237
smoke plumes, 94
soils, 387; carbon sequestration and climate-change mitigation, 364; compaction, 255; degradation, 298; impacts from tourism, 387–88; mineralization, 28; nutrients, 55
solar radiation, 28
source-sink dynamics, 105, 220
South America: carbon sink, 20; savannas, 54; tropical forest plots, 20–22; tropical grasslands, 54; woodlands, 54
Southern Africa, 8; vegetation changes, 64
Southeast Asia, 221, 232–33, 291; biodiversity in crisis, 291–304; selective logging, 292
Southeastern Australia, 122
soybean farming, industrial, 89
spatial distribution of saplings, 246
spatial requirements of forest vertebrates, 121
species: committed to extinction, 45; in continuous forests, 278; edge, 123; endemic vertebrates, 36, 40–41, 43; exotic, 143, 147, 199; exotic pathogens, 144; in fragmented landscapes, 280; functional groups, 124; introduced, 147, 199; invasive, 143, 199; large-bodied, extinctions in protected areas, 421; life-history sensitivity, 271; light-demanding, 18; matrix-intolerant, 117, 123; matrix-tolerant, 121; non-game, 118; persistence, 106, 280; population viability, 121; regionally endemic, 40; richness, 83, 278; shade-tolerant, 18; species-area curves, 44, 115–20; vulnerability, 43; wide-ranging, 121; wildlife populations, 106
species–area relationships (SARs), 44, 83, 106, 115, 117, 420; human perturbation of, 118; z-value, slopes of, 117